Communication System Principles

Communication System Principles

Peyton Z. Peebles, Jr.
The University of Tennessee

1976

Addison-Wesley Publishing Company
Advanced Book Program
Reading, Massachusetts

London • Amsterdam • Don Mills, Ontario • Sydney • Tokyo

First printing, 1976
Second printing, 1979

Library of Congress Cataloging in Publication Data

Peebles, Peyton Z.
Communication system principles.

Includes bibliographical references and index.
1. Telecommunication. 2. Signal theory (Telecommunication) I. Title.
TK5101.P34 621.38 76–26915
ISBN 0–201–05758–1

Manufactured in the United States of America.

ABCDEFGHIJ-HA-79

To my wife Barbara
and sons Peyton, III
and Edward, for cheerful
endurance of time which
could otherwise have been
spent together.

Contents

Chapter 3

Deterministic Signal Transfer through Networks

Chapter 4

Statistical Concepts and the Description of Random Signals and Noise

Chapter 5

Amplitude Modulation

Chapter 6

Angle Modulation

Chapter 8

Carrier Modulation by Digital Signals

Chapter 9

System Power Transfer and Sensitivity

Preface

This book has been written primarily to introduce the junior or senior electrical engineering student to the principles of communication systems. However, by including the more advanced topics and problems marked with a star (★), the text may also be used for a first-year graduate course. The book is an outgrowth of four courses that I have taught at the University of Tennessee. Two of these, one graduate and one undergraduate, deal with theoretical principles of communication systems. Of the other two, both undergraduate, one is concerned with the practical aspects of communications, while the second covers statistical concepts, random signals, and noise. Organization and development of notes for these courses has determined the content of the book.

Regarding content, the work deals mainly with theoretical principles. The only reader background required to study the book is the elementary knowledge of network theory, electronics, and field theory typical of juniors. More than twenty types of systems, including variations, are considered. Most of the emphasis has been placed on modern pulse and digital modulation systems (Chapters 7 and 8). However, a thorough exposition is also accorded phase and frequency modulation (Chapter 6) and amplitude modulation (Chapter 5). Prior to discussion of these systems, the necessary background is developed for working with signals and noise. Nonrandom signals (Chapter 2) and the responses of networks to these signals (Chapter 3) are considered first, since they are most easily understood. Random signals, noise, and the corresponding network responses are then developed (Chapter 4) in some detail. In fact, Chapter 4 is self-contained and has been used for a separate one-quarter course in random signals and noise at the University of Tennessee. The book closes with a discussion of some of the more practical system principles (Chapter 9) which relate to achievable overall system signal-to-noise ratio.

Regarding method of presentation, I believe that students want and need examples and problems to aid in learning. Thus, I have included many examples as an integral part of the book. There are also over 275 problems covering most of the subjects in the text. A solutions manual is available to instructors on request of the publisher. It is also my belief that students dislike "sketchy" explanations as well as "overdone" discussions. I have tried to strike a reasonable compromise in this book by giving the reader some "meat" to "chew" on while avoiding lengthy developments that do not relate well to a learning process.

Regarding course structure, the book has been written with a good deal of flexibility. For example, in all chapters on communication systems the notation $\overline{f^2(t)}$ is used to represent the power in any signal $f(t)$. For students well versed in statistical concepts this power is interpreted in a statistical sense through expectation. This approach is appealing for a graduate course, a course where Chapter 4 (statistical concepts) is covered, or a course where students already have a strong statistical background. On the other hand, in courses where Chapter 4 is not covered and students do *not* have any statistical background, $\overline{f^2(t)}$ may be interpreted as a simple time average. Thus, the most important portions of the book may profitably be studied for a variety of student backgrounds in random signal theory.

To emphasize the above points, the following table lists some of the ways I feel the book may be used to teach either undergraduates with no statistical background or undergraduates having statistical background derived from a separate course:

Length*	Chapters Covered†	Comments
1 quarter 1 semester	1, 2 (rapidly), 3, 5, 6 1, 2, 3, 5, 6, 9	Analog system emphasis.
1 quarter 1 semester	1, 2 & 3 (rapidly), 7, 8 1, 2, 3, 7, 8, 9	Digital system emphasis.
1 quarter	1, 5, 6, 9	Analog system emphasis. Student background in signals and networks assumed.
1 semester	1, 5, 6, 7, 8	Analog and digital systems. Student background in signals and networks assumed.

*Based on three class hours per week.
†All starred material deleted; some selective topic deletion by instructor to match particular school's requirements.

Chapter 4 can also form the basis of a separate course in statistical concepts:

Length*	Chapters Covered†	Comments
1 quarter	1, 4, 9	Course in random signals and noise.
1 semester	1, 2, 3, 4, 9	Course in signals and noise.

*Based on three class hours per week.
†All starred material deleted; some selective topic deletion by instructor to match particular school's requirements.

At the graduate level, the entire book may be covered in a one-year course.

In a book of this size, errors are nearly unavoidable. Through careful proofreading I hope to have minimized their number. Through the kind efforts of a number of colleagues, who reviewed various portions of the manuscript and gave helpful comments, I hope to have minimized their significance. I am grateful to Dr. N. M. Blachman of the Sylvania Corporation, Dr. A. B. Glenn of the Mitre Corporation, Dr. J. B. O'Neal, Jr., of North Carolina State University, and Drs. R. C. Gonzalez, J. C. Hung, and J. D. Tillman, all of the University of Tennessee, for such efforts. My appreciation and thanks are also extended to Mr. G. C. Guerrant for his excellent work in preparing the figures; to the several typists, coordinated by Mrs. Mary Bearden, who typed several difficult versions of the manuscript, and Dr. J. M. Googe, Head, Electrical Engineering Department, for making their services available.

PEYTON Z. PEEBLES, JR.

Chapter 1

Introduction to Book

This book considers the transfer of information from one point to another by communication systems. Man has been trying to communicate electrically almost since the birth of electrical technology about 1800 when Volta discovered the primary battery. The earlier efforts involved wire communications using the telegraph perfected by Samuel Morse in 1837. Electromagnetic waves had their start in 1864 when James Clerk Maxwell predicted electromagnetic radiation in his classic work "A Dynamical Theory of the Electromagnetic Field." Maxwell's work remained theory until 1887 when Heinrich Hertz verified radiation and the birth of modern communications technology occurred.

Except possibly for the telephone and telegraph, most of the major communication methods have used electromagnetic waves in the atmosphere. The ordinary broadcast amplitude modulation (AM) system was developed prior to 1920, and the frequency modulation (FM) system had its start in 1936 when E. H. Armstrong gave advantages of its use in "A Method of Reducing Disturbances in Radio Signalling by a System of Frequency Modulation" (*Proceedings of the IRE,* May, 1936).

The early 1920's saw the evolution of television (TV). First public broadcasts were in 1927 in England and in 1930 in the United States. The important modern method of pulse code modulation (PCM) was conceived by Alec Reeves in 1937 [1].* Another milestone occurred in 1955 when J. R. Pierce proposed satellite communication systems; the first, Telstar I, was launched in 1962. Today communication by satellite is commonplace.

The above historical sketch points out how young the communications field is and how fast it is expanding. In the space of about ninety years we have progressed from simple knowledge of wave propagation all the way to satellite communication links, spacecraft links, and even interplanetary communications. These are truly exciting times and a fantastic adventure for those in the field.

The purpose of this book is to describe the fundamental principles and methods which may be applied to nearly all modern communication systems. Thus, we seek mainly to develop building blocks which make up a system. In trying to accomplish our purpose, however, the development of these blocks will, in most cases, be couched within descriptions of actual systems so as to not lose sight of overall system concepts.

In this chapter we begin by introducing certain initial assumptions, definitions, and concepts used throughout.

1.1. SOME BASIC DEFINITIONS

Our intuition tells us that *information* is knowledge.† For our purposes this will be an adequate definition of information. Thus, an *information-bearing signal* may be any waveform

ISBN 0-201-05758-1

References start on p. 8.

*Bracketed numbers refer to references listed at the end of each chapter.

†*Knowledge* is used here in the sense of something gained as opposed to something already possessed. Thus, one gains information if he gains something of which he was previously uncertain, so information

that conveys knowledge. We shall be careful to distinguish an information-bearing signal from an *information signal,* the latter being an electrical waveform representing some *information source.* For example, musical sounds from a symphony orchestra may be considered as an information source, while the voltage produced by a nearby microphone would be an infomation signal; finally, any broadcast radio wave carrying the music would be an information-bearing signal.

Baseband or Low-Pass Signals

The definition of an information signal becomes more precise if we observe that a waveform may be described through its frequency components or spectral content. A *baseband signal* is one that has spectral content extending from (or near) dc up to some finite value usually less than a few megahertz. A baseband signal, also sometimes called a *low-pass signal,* will be implied whenever we use the term information signal. Similarly, the term *message* will be synonymous with a baseband information signal.

For information signals of interest there will always be some frequency* W_f (radians per second) above which the frequency content is negligible. It will often be convenient to assume the frequency content is *zero* for frequencies beyond W_f; such waveforms are said to be *bandlimited signals* in the band W_f.

Band-Pass Signals

Many information-bearing waveforms are *band-pass signals*—that is, those that have their spectral content clustered in a band of frequencies near a value called the *carrier frequency.* Similar to low-pass waveforms, there will be frequencies between which all significant spectral content falls. Usually the width of the significant band is small in relation to the lowest frequency; this corresponds to a *narrowband* band-pass waveform. If the spectral content is zero outside the significant band, we have a bandlimited band-pass signal.

Frequency Band Designations

Most communication systems involve a carrier frequency. To describe where in the *frequency spectrum* a particular carrier frequency is located, it is convenient to divide the spectrum into bands. Several methods of band definition are in use. In one method a band is assigned a number N such that [2]

$$0.3(10^N) \text{ hertz} < \text{band } N \leqslant 3(10^N) \text{ hertz}. \tag{1.1-1}$$

Thus, band 3 exists from 300 Hz to and including 3 kHz. Figure 1.1-1 is helpful in visualizing band division of the spectrum.

Also spotted on Fig. 1.1-1 are some allowable operating bands for several systems. A detailed list is given in reference [2]. For example, any ordinary AM broadcast radio station will operate at a carrier frequency [determined and allocated by the Federal Communications Commission (FCC)] somewhere between 535 kHz and 1605 kHz.

---

may be viewed as involving uncertainty. This view forms the basis of an advanced formal theory of communication called *information theory.*

*Throughout this book we shall interchangeably use the word frequency to imply either f, having the unit of hertz, or ω, having the unit of radians per second. Which is implied will be obvious from the context, but the latter will be the most frequently used notation.

ISBN 0-201-05758-1

ISBN 0-201-05758-1

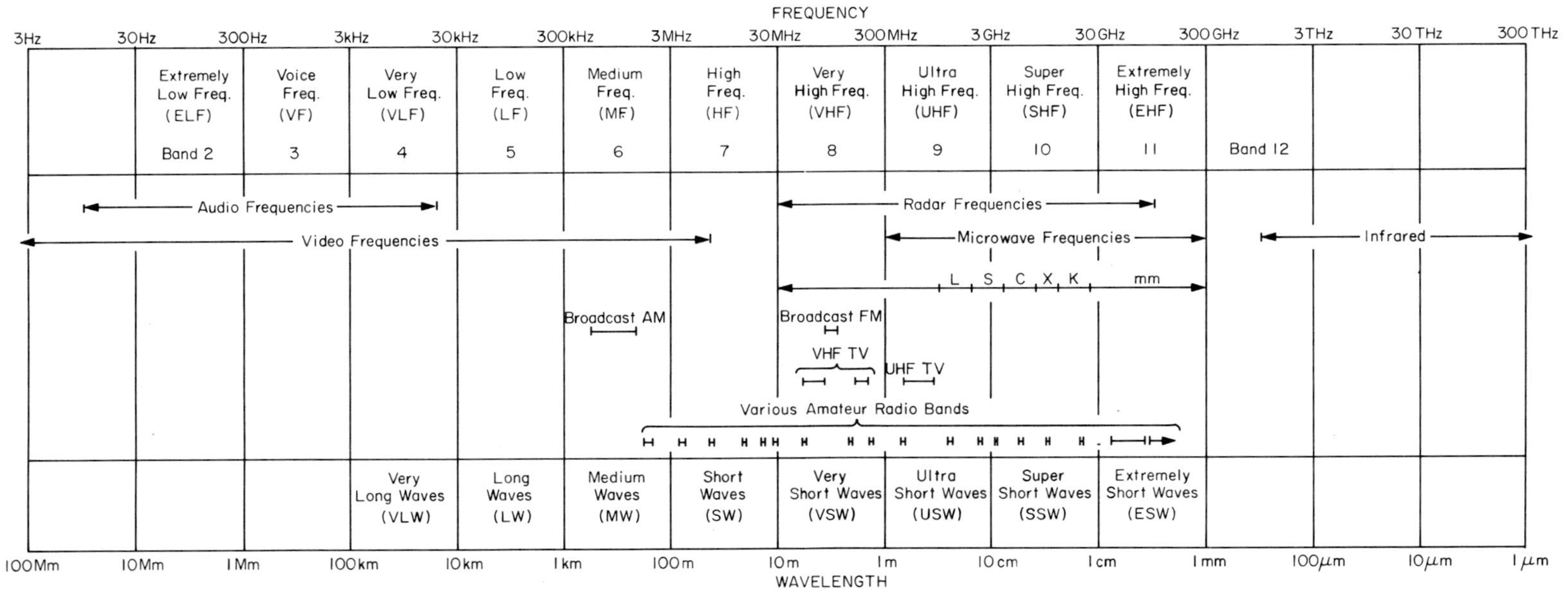

Fig. 1.1-1. Frequency spectrum.

Letter designations based on frequency, such as VLF for very low frequencies and HF for high frequencies, are sometimes used in practice. Other letter-type designations, based on wavelength, are also occasionally used. For example, the shortwave (SW) band is the same as the HF band which, in turn, is the same as band 7, as seen from Fig. 1.1-1. Wavelength λ is related to frequency f by

$$\lambda = c/f, \tag{1.1-2}$$

where c is the speed of light. If c is in meters per second and f is in hertz, then λ is in meters given approximately* by

$$\lambda = 3(10^8)/f. \tag{1.1-3}$$

Much of the practical technology in bands 9 and higher has evolved from research in the radar field. During World War II letter designations were assigned to the operating frequency for security purposes. There is still considerable use of these letters, even though there appears to be no general agreement as to what band a letter actually applies. By combining definitions given in one of the most widely used radar texts [3] for frequencies up to 26.5 GHz and definitions used by one of the leaders in manufacturing millimeter-wave equipment [4] for frequencies above 26.5 GHz, we obtain Table 1.1-1.

TABLE 1.1-1. Letter Designations for Frequency Bands

Name of Band	Frequency Band (GHz)
VHF	0.03- 0.3
UHF	0.3 - 1.0
L	1.0 - 2.0
S	2.0 - 4.0
C	4.0 - 8.0
X	8.0 - 12.5
K_u	12.5 - 18.0
K	18.0 - 26.5
R	26.5 - 40.0
F	40.0 - 60.0
E	60.0 - 90.0
V	90.0 -140.0

1.2. MODULATION

Modulation is a process by which an information signal $f(t)$ may be converted to a more useful form. Since a large portion of the work in this book is related to modulation, it is helpful to categorize modulation broadly according to three basic *methods*. Each method is then subdivided into various *types* of modulation.

Carrier-Wave Modulation Method

One of the most common modulation methods is to alter a *carrier wave* of the form

*The error in using $c = 3(10^8)$ m/s for the speed of light is less than 0.1% of the presently known value.

ISBN 0-201-05758-1

$$A \cos(\omega_0 t + \theta_0), \qquad (1.2\text{-}1)$$

where θ_0 is an arbitrary constant, A is the wave peak amplitude, and ω_0 is the carrier frequency. More-or-less obvious modulation approaches are to make A a function of the information signal $f(t)$ or to add a phase term to the cosine argument which depends on $f(t)$. These methods may be characterized as *continuous-wave* (c-w) modulation. The result of c-w modulation is usually a band-pass signal because the carrier frequency is often much higher than the largest significant frequency component in $f(t)$.

Pulse Modulation Method

Another modulation method, *pulse modulation,* involves changing some parameter of a train of pulses. For example, the amplitude of the pulses could be varied as some function of the information signal $f(t)$. Alternatives would be to vary pulse duration or position.

Pulse modulation does not involve a carrier. The resulting modulated signal is still baseband but is no longer the original information signal.

Both c-w and pulse modulation may be classed as *analog modulation* which involves varying the modulated parameter continuously as a linear function of the information signal.

Coded or Digital Modulation Method

Coded modulation involves changing a characteristic of a pulse train but is a markedly different method from pulse modulation. The method involves first *sampling* the information signal, *quantizing* the sample by rounding off to the closest of a number of discrete levels, and finally generating a prescribed number of pulses according to a *code* related to the nearest discrete level. In the simplest system the code may determine the presence or absence of pulses in the prescribed number, or, alternatively, it may determine pulse sign. Both cases correspond to pulses having only two possible levels. More complicated systems are also feasible with pulses having a larger number of discrete levels.

Quantizing and coding amounts to *digitizing* the signal; thus, coded modulation will also be referred to as *digital modulation.* It is one of the most modern and useful methods of modulation available today.

Types of Modulation

Having briefly classified modulation according to methods, we turn to the definition of some specific types. In the c-w method, if A varies as a linear function of the information signal $f(t)$, the result is *amplitude modulation* (AM). If A varies proportional to $f(t)$, AM with *suppressed carrier* (AM-SC) is generated. There are several variations of suppressed carrier (SC) modulation which depend on how the modulated signal spectral components are dispersed in relation to the carrier frequency ω_0. These are called *double-sideband, single-sideband,* and *vestigial-sideband* suppressed carrier (AM-DSB-SC, AM-SSB-SC, and AM-VSB-SC, respectively) modulation.

Other types of c-w modulation involve adding a phase term to the argument of (1.2-1) which is either proportional to $f(t)$ or proportional to the integral of $f(t)$. The former is called *phase modulation* (PM), and the latter is known as *frequency modulation* (FM). FM is the type of modulation used in the ordinary broadcast FM system. PM and FM may collectively be called *angle modulation.*

In this book three types of pulse modulation are considered. They are *pulse amplitude modulation* (PAM), *pulse duration modulation* (PDM), and *pulse position modulation* (PPM).

ISBN 0-201-05758-1

In PAM or PDM respective pulse amplitudes or durations are varied according to a linear function of the information signal. In PPM pulse position is proportional to the amplitude of the message signal. Other types of pulse modulation exist. However, we consider only these three, since they are the most important.

Many types of digital modulation have been developed. *Pulse code modulation* (PCM) is perhaps the most important. PCM may be either *binary*, where pulses have only two voltage levels, or μ-ary, where pulses may take on μ possible levels. Several important variations of binary PCM are *delta modulation* (DM), *delta–sigma modulation* (D-SM), *adaptive delta modulation* (ADM), and *differential pulse code modulation* (DPCM). These variations involve coding only information about *changes* in the information signal.

Finally, there are types of modulation that correspond to continuous-wave modulation of a carrier by either pulse or digital information-bearing signals, the latter being more representative of modern systems. In the digital case *amplitude shift keying* (ASK), *frequency shift keying* (FSK), and *phase shift keying* (PSK) correspond to modulating amplitude, frequency, and phase of a carrier wave, respectively. *Differential PSK* (DPSK) is a variation of PSK which allows simplifications in receiving equipment.

From the above discussions, it is apparent that a large number of modulation types is available. All of these are discussed in subsequent work.

Reasons for Modulation

Speaking broadly, the primary reason for modulation is to convert the information signal to a more useful form. There are many specific reasons which may be divided into three main categories: for necessity or convenience, for performance increase, and for efficient spectrum utilization.

In the former category it is often necessary to use modulation to translate the useful band of frequencies up to a large carrier frequency so that efficient electromagnetic radiation is possible from an antenna having reasonable size. Efficiency requires antenna physical dimensions of about 1/10 wavelength or larger. To illustrate, suppose we wish to use an antenna of about 1 meter in size; from (1.1-3) we require a carrier frequency of at least 30 MHz, since $\lambda = 10$ meters. On the convenience side, modulation provides a means whereby a number of signals may be added or *multiplexed* together for simultaneous transmission. Through multiplexing we are able to send many messages between two points while using a single communication system.

In the second category it is possible to improve system performance by choice of modulation method. As will be seen, performance is limited by the presence of random *noise* and *interference.* The effect of these unwanted waveforms can be suppressed by using certain forms of modulation.

Finally, modulation serves as a means of efficient spectrum utuilization. By choice of carrier frequency a designer could, in principle, elect to operate in any spectrum band, hoping that no one else would design for the same frequency. If someone else did, the two systems might hopelessly interfere with each other. To avoid such conditions and to assure efficient use of the radio spectrum, the FCC controls frequency allocations in the United States. One needs only to look at a list of current allocations [2] to see how crowded the spectrum is and how important is the job of the FCC.

1.3. COMMUNICATION SYSTEM ELEMENTS AND SURVEY OF BOOK

Fundamentally every communication system requires three elements, a sending or transmitting station, a receiving station, and a transmission medium to connect the two together. Figure 1.3-1 illustrates system elements in block diagram form.

ISBN 0-201-05758-1

In this book we discuss mainly the elements between points A and B and between C and D, since most of the principles are learned from these elements. These functions relate to the methods of modulation and demodulation. In Chapter 9 we shall, however, briefly discuss the transducers, receiver, and medium as part of an overall study of the power transferred to the receiver.

Prior to study of actual communication systems one must first become well based in analysis methods for the waveforms to be processed. Chapter 2 develops some of the fundamental ways of describing *deterministic* (nonrandom) *signals*, while Chapter 3 considers the transmission of these signals through networks. Chapter 4 completes the basic work on waveforms by introducing characterization methods for *random signals* and *noise*.

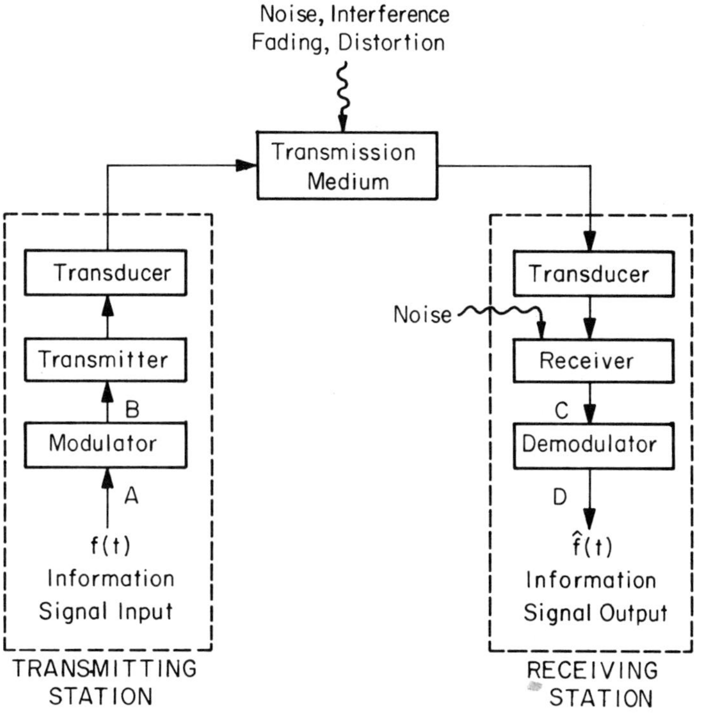

Fig. 1.3-1. Elements of a communication system.

Transmitting and Receiving Stations

From Fig, 1.3-1 we see that each transmitting station element has a corresponding receiving station element which performs an inverse function. The purpose of the modulator is to generate the modulated signal using the input information signal $f(t)$. The inverse operation of the demodulator is to recover a version $\hat{f}(t)$ of the information signal transmitted using the received version of the modulated signal. The various signals in the receiver may be different from their transmitting station counterparts owing to noise and other factors.

The principles of modulation and demodulation applicable to amplitude modulation are developed in Chapter 5. Those for angle (frequency and phase) modulation are covered in Chapter 6, while pulse and digital modulation are considered in Chapters 7 and 8.

The transmitter is usually a high-level stage with enough power output to provide good receiving station demodulated signals in the presence of noise. The transmitter and modulator

ISBN 0-201-05758-1

may sometimes be combined. The corresponding receiving station element is the receiver. It is a low-power device, usually made up of a low-noise amplifier and mixer.*

The transducers are devices for coupling the stations to the medium. The usual transducer is an antenna, where the transition is electrical signal to electromagnetic wave in the atmosphere or the reverse, while the usual medium is the earth's atmosphere. More generally, the transducers are matches to the medium. For example, in sonar they would be voltage-to-pressure devices. Antennas are discussed in Chapter 9 as part of the larger problem of determining the amount of signal and noise powers that exist at the receiver.

Transmission Medium

The transmission medium or *channel* connects the two stations. The most elementary channel is simply a pair of wires such as in telegraph or telephone systems. Other common channels are the earth's atmosphere, vacuum space, or combinations of the two as in a satellite link.

The channel may affect the transmitted signal in many ways. It can add noise or interference to the received waveform. Together these are called additive interference as opposed to *multiplicative* effects such as fading (a time-varying channel attenuation and phase shifting of the desired wave). The channel may also distort the desired signal; a common cause of distortion is limited channel bandwidth.

Although we shall not discuss the channel as a separate entity in the following chapters, we shall touch upon some of its attenuation and noise effects as part of system performance discussions included in Chapter 9.

Finally, we close this book survey and chapter by noting that, at the end of each of the following chapters, there are problems designed to exercise the reader's grasp of the principles discussed. To aid in working the problems, selected mathematical formulas and tables are included in several appendixes located at the end of the book.

REFERENCES

[1] Carlson, A. B., *Communication Systems: An Introduction to Signals and Noise in Electrical Communication,* McGraw-Hill Book Co., New York, 1968.

[2] International Telephone and Telegraph Corp., *Reference Data for Radio Engineers,* Fifth Edition, Howard W. Sams Co., New York, 1968.

[3] Skolnik, M. I., *Radar Handbook,* McGraw-Hill Book Co., New York, 1970.

[4] Hitachi, Ltd., MM Wave Test Equipment (commercial catalog).

*A mixer is a device that shifts the carrier frequency to another (usually lower) value.

ISBN 0-201-05758-1

Chapter 2

Deterministic Signal Representations

2.0. INTRODUCTION

Before a fruitful investigation of communication systems can be undertaken, it is necessary to have a good understanding of the ways in which signals may be described. Signals may be classified broadly as random and nonrandom or deterministic; in this chapter only deterministic signals are discussed. Random or nondeterministic signals, such as noise, are reserved for Chapter 4.

As will be seen, there are mainly two different ways that we may choose to describe a deterministic signal, and the difficulty of solving a given problem involving the signal may depend on which description we choose. The two ways may be categorized as the time-domain and frequency-domain methods. The time-domain method is more or less self-explanatory; that is, the amplitude-time history of a signal is given either graphically or, more usually, by an equation. The frequency-domain method involves use of the Fourier transform, which allows the signal to be described on the basis of its "frequency content." The function giving the signal's frequency content will be known as the *Fourier spectrum* or often just called the *spectrum.*

In the sections to follow we shall discuss both time and frequency descriptions as well as some properties of these characterizations that will assist in the solution of many problems. We shall begin with the time-domain representation of a periodic signal by a Fourier series, since either sinusoids or exponential signals are involved in this representation and these concepts should be familiar to the reader.

2.1. FOURIER SERIES REPRESENTATION OF PERIODIC SIGNALS

The periodic signal is an important class of signal encountered in all areas of engineering and science. Such a signal possesses a characteristic behavior over an interval of time, known as the period, and repeats this behavior during all other contiguous periods from $-\infty$ to $+\infty$ in time. The most obvious and fundamental examples of periodic functions are the sine and cosine waves.

Trigonometric Series

Since sine and cosine waves are so fundamental in nature, it comes to mind that a number of such waves might be used to describe more complex and almost arbitrary* signals. This is

ISBN 0-201-05758-1

References start on p. 52.

*The definition of "almost arbitrary" is subsequently given in the discussion of Dirichlet conditions.

indeed the case, and the series of terms that results is known as the *Fourier series*† when an infinite number of terms is employed to express the desired function exactly.

Thus an almost arbitrary periodic function $f(t)$ of time having a period T may be represented by the series [1, 2]

$$f(t) = \frac{a_0}{2} + a_1 \cos\left(\frac{2\pi}{T}t\right) + a_2 \cos\left(\frac{4\pi}{T}t\right) + \cdots$$

$$+ b_1 \sin\left(\frac{2\pi}{T}t\right) + b_2 \sin\left(\frac{4\pi}{T}t\right) + \cdots$$

$$= \frac{a_0}{2} + \sum_{n=1}^{\infty} a_n \cos\left(\frac{n2\pi}{T}t\right) + \sum_{n=1}^{\infty} b_n \sin\left(\frac{n2\pi}{T}t\right), \tag{2.1-1}$$

where a_n and b_n are constants called the *Euler* [1] or *Fourier coefficients.* Equation (2.1-1) may also be written as

$$f(t) = \frac{a_0}{2} + \sum_{n=1}^{\infty} d_n \cos\left(\frac{n2\pi}{T}t + \phi_n\right), \tag{2.1-2}$$

where

$$d_n = \sqrt{a_n^2 + b_n^2}, \tag{2.1-3}$$

$$\phi_n = -\tan^{-1}(b_n/a_n). \tag{2.1-4}$$

The Fourier series is valid for any choice of time origin. For example, the rectangular pulse train shown in Fig. 2.1-1 can be expressed by eq. (2.1-1) for either choice of origins shown as (*a*) and (*b*); however, the Fourier series coefficients will depend on the origin choice.

Fourier Series Coefficients

The Fourier series coefficients may be found by properly integrating the signal over one complete period. The coefficients are given by

$$a_n = \frac{2}{T}\int_{-T/2}^{T/2} f(t)\cos\left(\frac{n2\pi}{T}t\right) dt, \qquad n = 0, 1, 2, \ldots, \tag{2.1-5a}$$

$$b_n = \frac{2}{T}\int_{-T/2}^{T/2} f(t)\sin\left(\frac{n2\pi}{T}t\right) dt, \qquad n = 1, 2, 3, \ldots. \tag{2.1-5b}$$

In (2.1-5) the interval T of integration is centered on the origin for convenience. It would be equally valid if we chose to integrate from $t_0 - (T/2)$ to $t_0 + (T/2)$ instead, where t_0 is some constant.

†Named for Joseph Fourier (1768–1830), who apparently first studied the properties of such series.

ISBN 0-201-05758-1

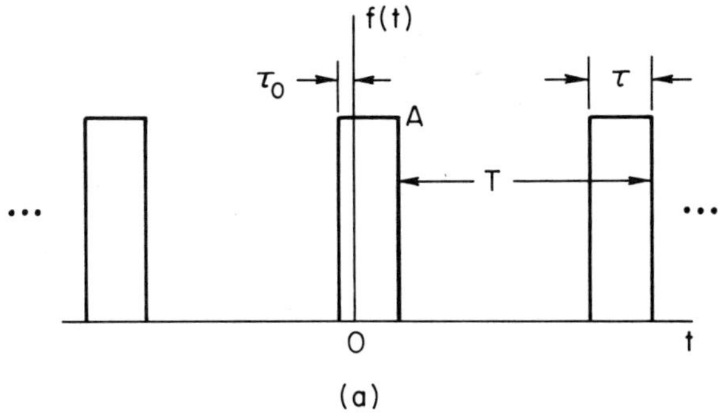

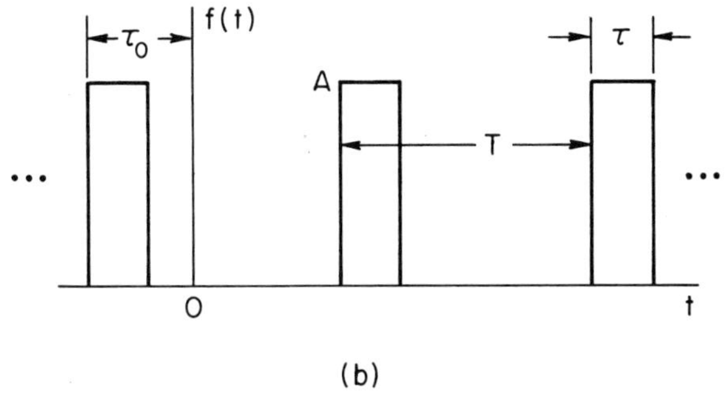

Fig. 2.1-1. Periodic signal consisting of rectangular pulses for two choices of time origin.

The reader may verify quite easily that (2.1-5) is valid. The procedure for (2.1-5a) is to multiply (2.1-1) by $\cos\left(\frac{m2\pi}{T}t\right)$ on both sides, integrate over one period, and, after a variable change, utilize the identities:

$$\int_{-\pi}^{\pi} \sin(nx)\cos(mx)\,dx = 0, \qquad \text{all } m \text{ and } n = 0, 1, \ldots, \tag{2.1-6a}$$

$$\begin{aligned}\int_{-\pi}^{\pi} \cos(nx)\cos(mx)\,dx &= 0, && m \neq n\\ &= \pi, && m = n \text{ but } n \neq 0,\end{aligned} \tag{2.1-6b}$$

$$\begin{aligned}\int_{-\pi}^{\pi} \cos(mx)\,dx &= 0, && m > 0\\ &= 2\pi, && m = 0,\end{aligned} \tag{2.1-6c}$$

$$\begin{aligned}\int_{-\pi}^{\pi} \sin(nx)\sin(mx)\,dx &= 0, && m \neq n\\ &= \pi, && m = n \text{ but } n \neq 0,\end{aligned} \tag{2.1-6d}$$

$$\int_{-\pi}^{\pi} \sin(mx)\,dx = 0, \qquad m = 0, 1, \ldots. \tag{2.1-6e}$$

ISBN 0-201-05758-1

The procedure for verifying (2.1-5b) is identical except multiplication is by sin $\left(\frac{m2\pi}{T}t\right)$.

Coefficients of a piecewise continuous signal such as that shown in Fig. 2.1-1 must be obtained by piecewise integration in (2.1-5). We shall illustrate this fact by some examples.

Example 2.1-1

Let the Fourier coefficients of the pulse train of Fig. 2.1-1 be found. Using (2.1-5), we obtain

$$a_n = \frac{2}{T}\int_{-\tau_0}^{\tau-\tau_0} A\cos\left(\frac{n2\pi}{T}t\right)dt$$

$$= \frac{A}{n\pi}\left\{\sin\left[\frac{n2\pi}{T}(\tau-\tau_0)\right] + \sin\left(\frac{n2\pi}{T}\tau_0\right)\right\}, \qquad n = 0, 1, 2, \ldots,$$

$$b_n = \frac{2}{T}\int_{-\tau_0}^{\tau-\tau_0} A\sin\left(\frac{n2\pi}{T}t\right)dt$$

$$= \frac{A}{n\pi}\left\{-\cos\left[\frac{n2\pi}{T}(\tau-\tau_0)\right] + \cos\left(\frac{n2\pi}{T}\tau_0\right)\right\}, \qquad n = 1, 2, 3, \ldots.$$

Evaluation of these results for various values of n is straightforward except possibly for $n = 0$. L'Hospital's rule may be employed for functions of the form 0/0 to obtain

$$a_0 = 2A\tau/T.$$

Example 2.1-2

If $\tau_0 = \tau/2$ in Fig. 2.1-1(*a*), we obtain a pulse train that is centered on the origin. From example 2.1-1 we have

$$a_0 = 2A\tau/T,$$

$$a_n = a_0\frac{\sin(n\pi\tau/T)}{(n\pi\tau/T)}, \qquad n = 1, 2, \ldots,$$

$$b_n = 0, \qquad n = 1, 2, \ldots.$$

Notice that the Fourier series contains no sine terms, only cosine terms. More will be said about this later when we consider functions with even symmetry.

Example 2.1-3

As another example of a piecewise continuous function, consider Fig. 2.1-2. The signal is defined over the interval $-T/2$ to $+T/2$ by

ISBN 0-201-05758-1

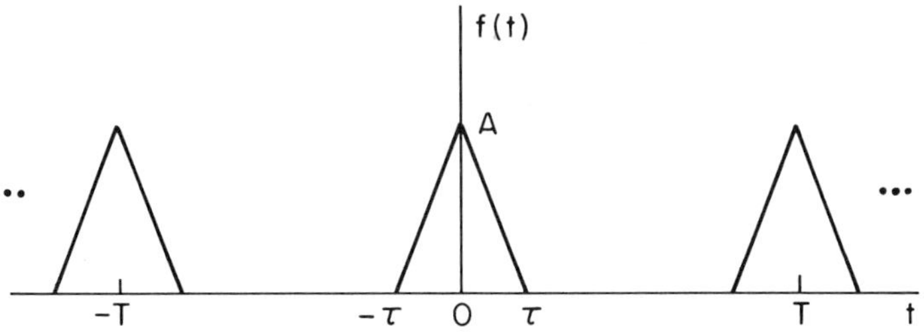

Fig. 2.1-2. Train of triangular pulses used in example 2.1-3.

$$f(t) = A\left(1 - \frac{t}{\tau}\right), \qquad 0 < t \leqslant \tau$$

$$= A\left(1 + \frac{t}{\tau}\right), \qquad -\tau \leqslant t \leqslant 0$$

$$= 0, \qquad \tau < |t| \leqslant T/2.$$

The Fourier coefficients become, using known integral forms (see Appendix B),

$$a_n = \frac{2}{T}\int_{-\tau}^{0} A\left(1 + \frac{t}{\tau}\right)\cos\left(\frac{n2\pi}{T}t\right)dt + \frac{2}{T}\int_{0}^{\tau} A\left(1 - \frac{t}{\tau}\right)\cos\left(\frac{n2\pi}{T}t\right)dt$$

$$= \frac{2A\tau}{T}\frac{\sin^2(n\pi\tau/T)}{(n\pi\tau/T)^2}, \qquad n = 0, 1, \ldots,$$

$$b_n = 0, \qquad n = 1, 2, \ldots.$$

Again we see a case of the Fourier series having only cosine terms.

Example 2.1-4

As an example resulting in a series having only sine terms, let us find the Fourier series coefficients for the signal of Fig. 2.1-3. They are, using tabulated integral forms (Appendix B),

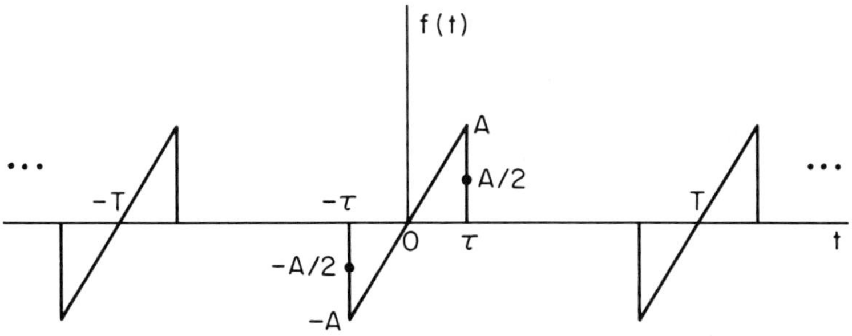

Fig. 2.1-3. Train of sawtooth pulses used in example 2.1-4.

ISBN 0-201-05758-1

$$a_n = \frac{2}{T}\int_{-\tau}^{\tau} \frac{A}{\tau} t \cos\left(\frac{n2\pi}{T} t\right) dt = 0, \qquad n = 0, 1, 2, \ldots,$$

$$b_n = \frac{2}{T}\int_{-\tau}^{\tau} \frac{A}{\tau} t \sin\left(\frac{n2\pi}{T} t\right) dt$$

$$= \frac{2A}{n\pi}\left[\frac{\sin(n2\pi\tau/T)}{(n2\pi\tau/T)} - \cos\left(\frac{n2\pi\tau}{T}\right)\right], \qquad n = 1, 2, \ldots.$$

Example 2.1-5

If $\tau = T/2$ in example 2.1-4, the signal becomes a sawtooth wave. The coefficients b_n reduce to

$$b_n = (2A/n\pi)(-1)^{n+1}, \qquad n = 1, 2, \ldots.$$

Dirichlet Conditions

Not all functions can be described by a Fourier series. However, those, which we shall call "almost" arbitrary, that belong to the class of functions satisfying the so-called *Dirichlet conditions* are known to be describable by a Fourier series. These conditions are summarized in the following theorem paraphrased from Wylie [1]:

> Theorem—If $f(t)$ is a bounded function of period T, and if in any one period it has at most a finite number of maxima and minima and a finite number of discontinuities, then the Fourier series of $f(t)$ converges to $f(t)$ at all points where $f(t)$ is continuous, and converges to the average of the right- and left-hand limits of $f(t)$ at each point where $f(t)$ is discontinuous.

As an illustration of this theorem, the Fourier series of the signal of Fig. 2.1-3 will everywhere yield the exact value of $f(t)$ except at the discontinuous points. At times $nT + \tau$, $n = 0, \pm 1, \pm 2, \ldots$, the result will be $A/2$ as shown. At the other points of discontinuity the result will be $-A/2$.

Nearly all periodic signals of engineering interest satisfy the Dirichlet conditions and have Fourier series representations.

The Complex Fourier Series

In the single-frequency steady-state analysis of networks, it is found that the use of an exponential signal will simplify many calculations. Since an "almost" arbitrary* periodic signal is nothing but a sum of individual single-frequency terms, we may obtain an exponential form of the Fourier series which will simplify linear network calculations when periodic signals are involved.

*We shall drop the "almost" arbitrary notation and henceforth say that any periodic signal has a Fourier series, understanding that by "any" we mean one that satisfies the Dirichlet conditions or is otherwise known to have a Fourier series.

ISBN 0-201-05758-1

If the identities†

$$\cos(x) = \frac{e^{jx} + e^{-jx}}{2}, \tag{2.1-7}$$

$$\sin(x) = \frac{e^{jx} - e^{-jx}}{2j} \tag{2.1-8}$$

are substituted into the series (2.1-1), the following series results:

$$f(t) = \frac{a_0}{2} + \sum_{n=1}^{\infty} \left(\frac{a_n - jb_n}{2}\right) e^{j\frac{n2\pi}{T}t} + \sum_{n=1}^{\infty} \left(\frac{a_n + jb_n}{2}\right) e^{-j\frac{n2\pi}{T}t}. \tag{2.1-9}$$

If complex coefficients are defined as

$$C_0 = a_0/2, \tag{2.1-10a}$$

$$C_n = \frac{1}{2}(a_n - jb_n), \qquad n = 1, 2, \ldots, \tag{2.1-10b}$$

$$C_{-n} = \frac{1}{2}(a_n + jb_n), \qquad n = 1, 2, \ldots, \tag{2.1-10c}$$

then

$$f(t) = C_0 + \sum_{n=1}^{\infty} C_n e^{j\frac{n2\pi}{T}t} + \sum_{n=1}^{\infty} C_{-n} e^{-j\frac{n2\pi}{T}t}. \tag{2.1-11}$$

Now the second sum involving positive values of n is of the same form as the first sum involving negative values of n. Thus we may write (2.1-11) in the compact form

$$f(t) = \sum_{n=-\infty}^{\infty} C_n e^{j\frac{n2\pi}{T}t}. \tag{2.1-12}$$

By using (2.1-5) in (2.1-10) it is easily verified that the coefficients C_n are given by

$$C_n = \frac{1}{T} \int_{-T/2}^{T/2} f(t) e^{-j\frac{n2\pi}{T}t}\, dt, \qquad n = 0, \pm 1, \pm 2, \ldots. \tag{2.1-13}$$

Note that the angles of the complex coefficients C_n are given by (2.1-4).

It should be emphasized that $f(t)$, as we have assumed so far, is a real quantity so that the series of (2.1-12) still gives a real result. The individual terms in the series, however, are complex in general.

†Throughout this book we define $j = \sqrt{-1}$.

ISBN 0-201-05758-1

Even and Odd Functions

A signal is said to have *even symmetry* about the time origin if

$$f(t) = f(-t). \tag{2.1-14}$$

Examples of such a signal are shown in Figs. 2.1-1 (when $\tau_0 = \tau/2$) and 2.1-2.

A signal is said to have *odd symmetry* about the time origin if

$$f(t) = -f(-t). \tag{2.1-15}$$

Figure 2.1-3 shows an example of an odd function.

To investigate the Fourier series of an even function we may substitute (2.1-14) into (2.1-5b). The result is an odd integrand, and the integral evaluates to zero. The effect on (2.1-5a) when (2.1-14) is used is to produce an even integrand. Thus, for an even signal,

$$a_n = \frac{4}{T}\int_0^{T/2} f(t)\cos\left(\frac{n2\pi}{T}t\right)dt, \qquad n = 0, 1, 2, \ldots, \tag{2.1-16a}$$

$$b_n = 0, \qquad n = 1, 2, \ldots, \tag{2.1-16b}$$

$$f(t) = \frac{a_0}{2} + \sum_{n=1}^{\infty} a_n \cos\left(\frac{n2\pi}{T}t\right). \tag{2.1-17}$$

In other words, the Fourier series of an even signal function has only cosine (even) terms.

By similar reasoning, the Fourier series of an odd function has only sine (odd) terms, and

$$a_n = 0, \qquad n = 0, 1, 2, \ldots, \tag{2.1-18a}$$

$$b_n = \frac{4}{T}\int_0^{T/2} f(t)\sin\left(\frac{n2\pi}{T}t\right)dt, \qquad n = 1, 2, \ldots, \tag{2.1-18b}$$

$$f(t) = \sum_{n=1}^{\infty} b_n \sin\left(\frac{n2\pi}{T}t\right). \tag{2.1-19}$$

Arbitrary Real and Complex Signals

An arbitrary real signal $f(t)$ can be thought of as comprising an even part $f_e(t)$ and an odd part $f_o(t)$. We shall show that this is possible by finding the two parts in terms of $f(t)$. We have by assumption

$$f(t) = f_e(t) + f_o(t), \tag{2.1-20}$$

so

$$f(-t) = f_e(-t) + f_o(-t) = f_e(t) - f_o(t). \tag{2.1-21}$$

ISBN 0-201-05758-1

This last form of (2.1-21) results from the definitions of even and odd functions. Addition and subtraction of (2.1-20) and (2.1-21) give

$$f_e(t) = \frac{1}{2}[f(t) + f(-t)], \tag{2.1-22}$$

$$f_o(t) = \frac{1}{2}[f(t) - f(-t)]. \tag{2.1-23}$$

Hence if we have an arbitrary signal we can *construct* even and odd functions to describe the signal. From these, the sine and cosine parts of the signal's Fourier series may be independently found if desired.

An arbitrary complex signal, $f(t)$, may be written as

$$f(t) = r(t) + jx(t), \tag{2.1-24}$$

where $r(t)$ and $x(t)$ are arbitrary real functions each having Fourier series representations. It is seen that the Fourier series of a complex signal is the sum of the Fourier series of the real part added to j times that of the imaginary part. Clearly, the definitions of even and odd functions above apply, without change, to complex signals.

Integration and Differentiation of Fourier Series

In a formal manner one may always differentiate or integrate term by term a Fourier series. Whether or not the resultant series truly represents (converges to) the derivative or integral of the signal, represented by the original Fourier series, depends on the rate of convergence of the original series. Speaking generally, integration will produce more rapid convergence, while differentiation produces less rapid convergence.

As examples, the series terms for the signals of Figs. 2.1-1 and 2.1-3 decrease in magnitude as $1/n$, while those for Fig. 2.1-2 decrease as $1/n^2$. Although it is not within our scope to consider the problem of series convergence, we can make some general statements [1] that give insight into when a Fourier series may be validly differentiated or integrated to represent the signal derivative or integral: (1) The integral of any periodic function that satisfies the Dirichlet conditions can always be found by the termwise integration of the function's Fourier series; (2) if a periodic function is everywhere continuous and its derivative $f'(t)$ satisfies the Dirichlet conditions, then, wherever it exists, $f'(t)$ may be found by termwise differentiation of the Fourier series of $f(t)$. This second point can be extended to higher derivatives.

Example 2.1-6

Let us show that termwise differentiation of the Fourier series representing $f(t)$ of Fig. 2.1-2 does represent the derivative $f'(t)$. The true derivative is plotted in Fig. 2.1-4. This derivative can be considered as the sum of positive and negative pulse trains of the form shown in Fig. 2.1-1. The coefficients of the Fourier series of the two trains are given in example 2.1-1 if A and τ_0 are replaced by A/τ and τ, respectively, for the positive train and $-A/\tau$ and zero for the negative train. Computations will show that the Fourier series for $f'(t)$ is

ISBN 0-201-05758-1

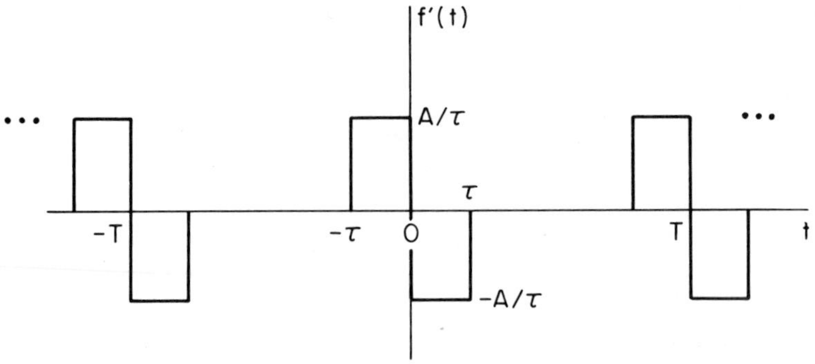

Fig. 2.1-4. Derivative of the signal of Fig. 2.1-2 applicable to example 2.1-6.

$$f'(t) = \sum_{n=1}^{\infty} \frac{-4A}{n\pi\tau} \sin^2\left(\frac{n\pi\tau}{T}\right) \sin\left(\frac{n2\pi}{T}t\right).$$

Since the signal's series has the coefficients of example 2.1-3, we have

$$f(t) = \sum_{n=0}^{\infty} \frac{2A\tau}{T} \frac{\sin^2(n\pi\tau/T)}{(n\pi\tau/T)^2} \cos\left(\frac{n2\pi t}{T}\right).$$

Differentiation of this result term by term will now give $f'(t)$.

Finite Range Expansions

In some problems we may be interested in describing only a portion of a given periodic signal. For example, the whole signal may be given as curve (*a*) of Fig. 2.1-5, but we may be interested only in the intervals τ wide as shown. Since a Fourier series is needed only to describe a finite range of the given signal, we do not care what the series converges to elsewhere. For example, we could assume $f(t)$ is: (1) zero, (2) given by curve (*b*), (3) given by curve (*c*), or (4) an infinite number of other curves. A series found under any of these assumptions will still represent the signal in the desired intervals but will converge to the assumed waveform elsewhere. The assumption made, however, will affect the type of series obtained. Assump-

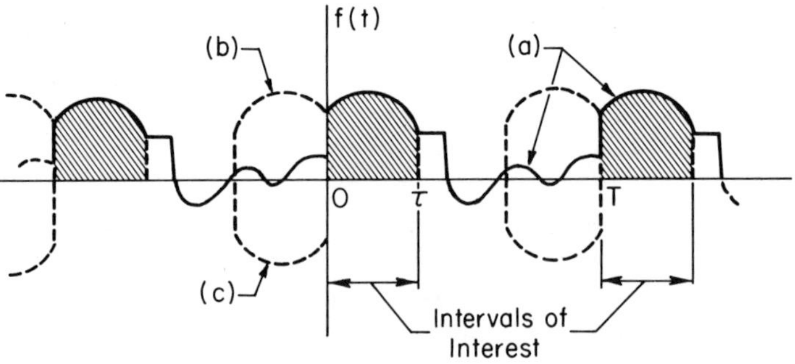

Fig. 2.1-5. Finite ranges of interest in a periodic waveform.

ISBN 0-201-05758-1

tions 2 and 3 will produce series with only cosine and sine terms, respectively, because of symmetry.

The assumption will also have an effect on convergence. While no general conclusions may be stated concerning the best choice for analysis, it does appear that the choice that gives the "smoothest" overall signal tends to produce the most rapid series convergence (see Wylie [1]).

Finite range expansions may also be used to represent a portion of a nonperiodic waveform. In this case one must be interested in only one interval of time and not in repetitive intervals as described above. In the single-interval representation the period may be arbitrarily selected.

Fourier Coefficient Spectrum

Not only does the Fourier series represent the signal of interest, it also gives the frequencies present in the waveform and their magnitudes and phases. Except for dc the lowest frequency present is $\omega_T = 2\pi/T$, where again T is the period. In general all integral multiples $n\omega_T$ of ω_T are present, although for certain symmetries some frequencies are not present.

A plot of the Fourier series coefficient magnitudes, $|a_0|$ and d_n, as a function of frequency will be called a *coefficient magnitude spectrum.* The locus of the tips of the coefficient magnitude *lines* is called the *envelope* of the spectrum. Figure 2.1-6 illustrates two coefficient magnitude spectra for the signals of examples 2.1-2 and 2.1-3. A ratio $T/\tau = 4$ was selected for illustrative purposes.

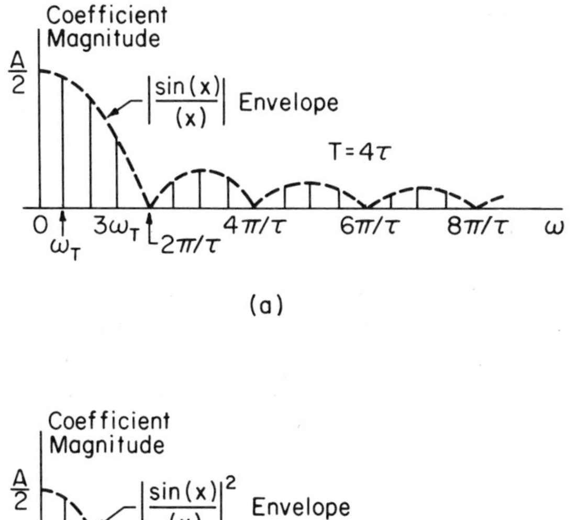

(a)

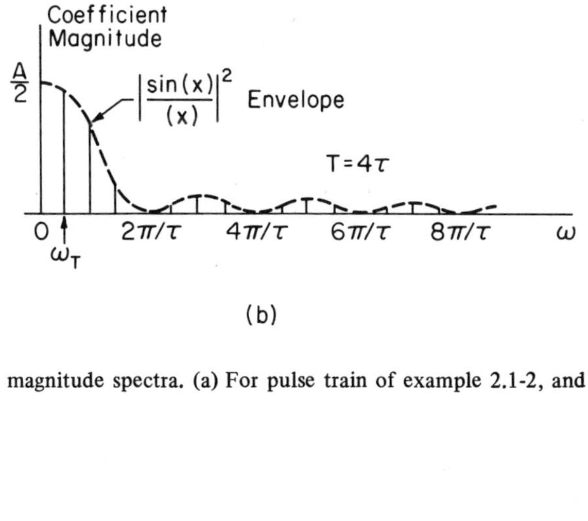

(b)

Fig. 2.1-6. Coefficient magnitude spectra. (a) For pulse train of example 2.1-2, and (b) for pulse train of example 2.1-3.

ISBN 0-201-05758-1

A plot of coefficient phases, given by ϕ_n of (2.1-4), will be called a *coefficient phase spectrum.* Together with the coefficient magnitude spectrum, it forms the *coefficient spectrum* of the Fourier series. Figure 2.1-7 illustrates the coefficient phase spectrum for the signal of example 2.1-2. That of example 2.1-3 gives zero at all frequencies so that Fig. 2.1-6(*b*) is the full coefficient spectrum of the triangular pulse train.

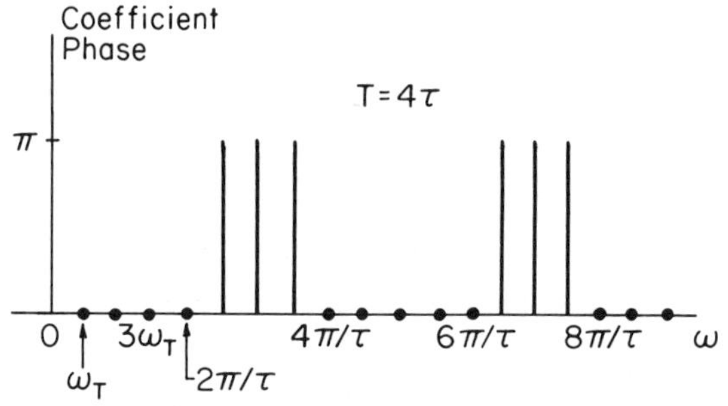

Fig. 2.1-7. Coefficient phase spectrum for the train of rectangular pulses of example 2.1-2.

A more convenient coefficient spectrum, from the standpoint of applications, is that associated with the complex form of the Fourier series. The coefficients are given by (2.1-13), and the series by (2.1-12). When n is negative the sign may be associated with frequency because of the exponents in the series terms. Thus, the coefficient C_{-n} is associated with the negative frequency $-n2\pi/T$. As an example, the plots of coefficient magnitude and phase spectra for a train of rectangular pulses of width τ, amplitude A, and period T are illustrated in Fig. 2.1-8.

Coefficient spectra are called *half-sided* when the plot involves only positive frequencies (as with d_n) and *double-sided* when both negative and positive frequencies are involved (as with $|C_n|$).

2.2. POWER SIGNALS AND AVERAGE POWER

Periodic signals are assumed to exist for all time from $-\infty$ to $+\infty$. Such signals are called *power signals,* since they have finite average power and infinite energy. They are contrasted with *energy signals,* which have finite energy but zero average power (a bounded, finite-time duration signal would be an energy signal). Power and energy signals are considered later in regard to their spectral properties. In this section we only consider some time-average properties of power signals.

Power of Real Signal

Let a real periodic signal $f(t)$ have a period T_0. The instantaneous power of the waveform will be* $f^2(t)$, while the energy expended during the time interval dt is $f^2(t)\,dt$. The total energy over a time interval $2T$, where T is not necessarily the period, is

$$\text{energy in time } 2T = \int_{-T}^{T} f^2(t)\,dt. \tag{2.2-1}$$

*In communication system work it is customary to discuss energy and power on a per-ohm basis. True power in a real waveform $f(t)$ is *proportional* to $f^2(t)$, while power on a per-ohm basis *equals* $f^2(t)$. We may

ISBN 0-201-05758-1

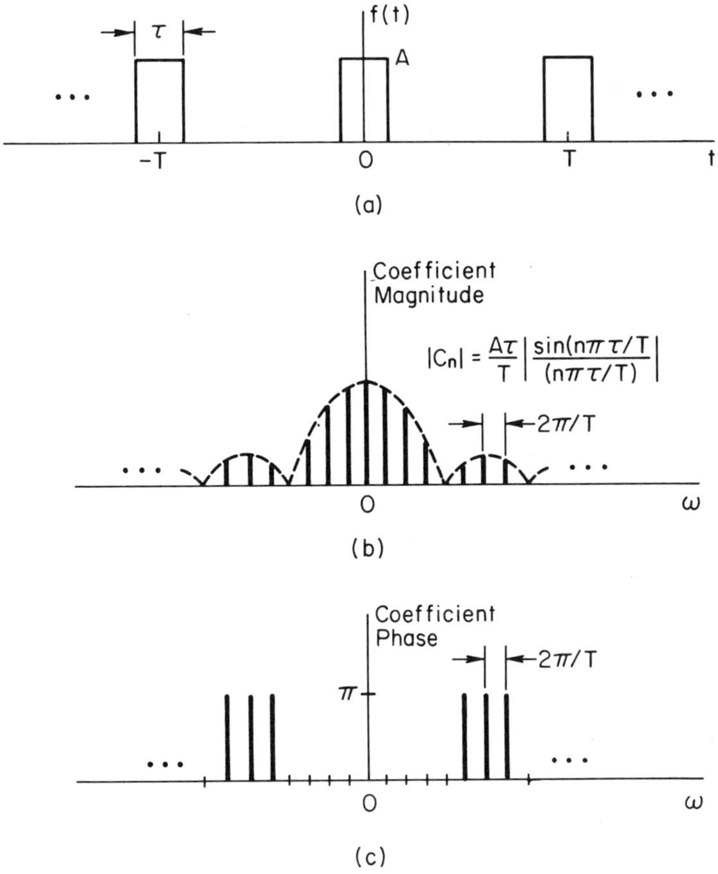

Fig. 2.1-8. Coefficient spectrum for complex Fourier series of a train of rectangular pulses. (a) Pulse train, (b) coefficient magnitude spectrum, and (c) coefficient phase spectrum.

The average power during the interval $2T$ is then

$$\text{average power in time } 2T = \frac{1}{2T}\int_{-T}^{T} f^2(t)\,dt. \tag{2.2-2}$$

Average power P may now be defined as the average when all time is considered. Thus,

$$P = \lim_{T\to\infty}\left\{\frac{1}{2T}\int_{-T}^{T} f^2(t)\,dt\right\}. \tag{2.2-3}$$

This result is valid for any real power signal, although we assumed periodic signals at the start. We may, for example, later apply (2.2-3) to random noise signals in Chapter 4.

For periodic signals the integral in (2.2-3) will be the same over any period T_0. Allowing the limit to be taken in a manner such that $2T$ is an integral number N of periods, the total

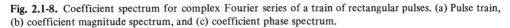

think of the per-ohm power or per-ohm energy as true values if the system under consideration has a one-ohm real impedance.

ISBN 0-201-05758-1

integral of $f^2(t)$ will be N times the integral over one period. Since $2T = NT_0$, it is easy to see that for periodic signals (2.2-3) reduces also to

$$P = \frac{1}{T_0} \int_{-T_0/2}^{T_0/2} f^2(t)\, dt. \tag{2.2-4}$$

More specifically, if the real Fourier series (2.1-1) is substituted into either (2.2-3) or (2.2-4), it may be shown that

$$P = \left(\frac{a_0}{2}\right)^2 + \sum_{n=1}^{\infty} \frac{(a_n^2 + b_n^2)}{2}. \tag{2.2-5}$$

In terms of the series form of (2.1-2) this result becomes

$$P = \left(\frac{a_0}{2}\right)^2 + \sum_{n=1}^{\infty} \frac{d_n^2}{2}. \tag{2.2-6}$$

Finally, in terms of the complex Fourier series (2.1-12), average power may also be expressed as

$$P = C_0^2 + 2 \sum_{n=1}^{\infty} C_n C_{-n}$$

$$= \sum_{n=-\infty}^{\infty} C_n C_{-n} = \sum_{n=-\infty}^{\infty} |C_n|^2. \tag{2.2-7}$$

Power of Complex Signal

In a manner analogous to real signals, a complex signal $f(t)$ has average power defined by

$$P = \lim_{T \to \infty} \left\{ \frac{1}{2T} \int_{-T}^{T} |f(t)|^2\, dt \right\}. \tag{2.2-8}$$

Expressing $f(t)$ in terms of its real part $r(t)$ and imaginary part $x(t)$, where $f(t) = r(t) + jx(t)$, the time average power is

$$P = P_r + P_x, \tag{2.2-9}$$

where

$$P_r = \lim_{T \to \infty} \left\{ \frac{1}{2T} \int_{-T}^{T} r^2(t)\, dt \right\}, \tag{2.2-10}$$

ISBN 0-201-05758-1

$$P_x = \lim_{T\to\infty} \left\{ \frac{1}{2T} \int_{-T}^{T} x^2(t)\, dt \right\}. \tag{2.2-11}$$

Thus, the average power of a complex wave is the sum of the average powers of the real and imaginary parts of the waveform.

2.3. THE FOURIER TRANSFORM

We shall, in this section, consider a very powerful tool for describing both the periodic signal and the nonperiodic waveform for which no Fourier series exists. It is the *Fourier transform.* Instead of describing a signal in the time domain, as in a Fourier series, the Fourier transform provides a frequency-domain description [2].

The Fourier Transform Pair

The Fourier transform, or *Fourier* (frequency) *spectrum,* $F(\omega)$, of a signal $f(t)$ is

$$F(\omega) = \int_{-\infty}^{\infty} f(t) e^{-j\omega t}\, dt. \tag{2.3-1}$$

Given the spectrum of a signal, the signal may be recovered by the *inverse Fourier transform,*

$$f(t) = \frac{1}{2\pi} \int_{-\infty}^{\infty} F(\omega) e^{j\omega t}\, d\omega. \tag{2.3-2}$$

These two results form a *Fourier transform pair.* Extensive tables have been compiled of transforms of various signals [3]. The Fourier transform is valid for real or complex signals, and, as we shall see below, the spectrum is in general complex even if $f(t)$ is real.

We shall often refer to the relations between signal and transform by use of a double-ended arrow:

$$f(t) \leftrightarrow F(\omega). \tag{2.3-3}$$

Expression (2.3-3) can be read several ways, such as: $f(t)$ has the spectrum $F(\omega)$, or the time function corresponding to the spectrum $F(\omega)$ is $f(t)$.

Before proceeding with further details on Fourier transforms, it will be helpful to consider a few examples involving nonperiodic waveforms which will also serve to define some special functions.

Example 2.3-1

Let us find the spectrum of the single pulse of Fig. 2.3-1(a). Using (2.3-1):

$$F(\omega) = \int_{-\tau_0}^{\tau-\tau_0} A e^{-j\omega t}\, dt = \frac{A e^{-j\omega t}}{-j\omega} \Bigg|_{-\tau_0}^{\tau-\tau_0} = A\tau \frac{\sin(\omega\tau/2)}{(\omega\tau/2)} e^{j\omega\left(\tau_0 - \frac{\tau}{2}\right)}.$$

ISBN 0-201-05758-1

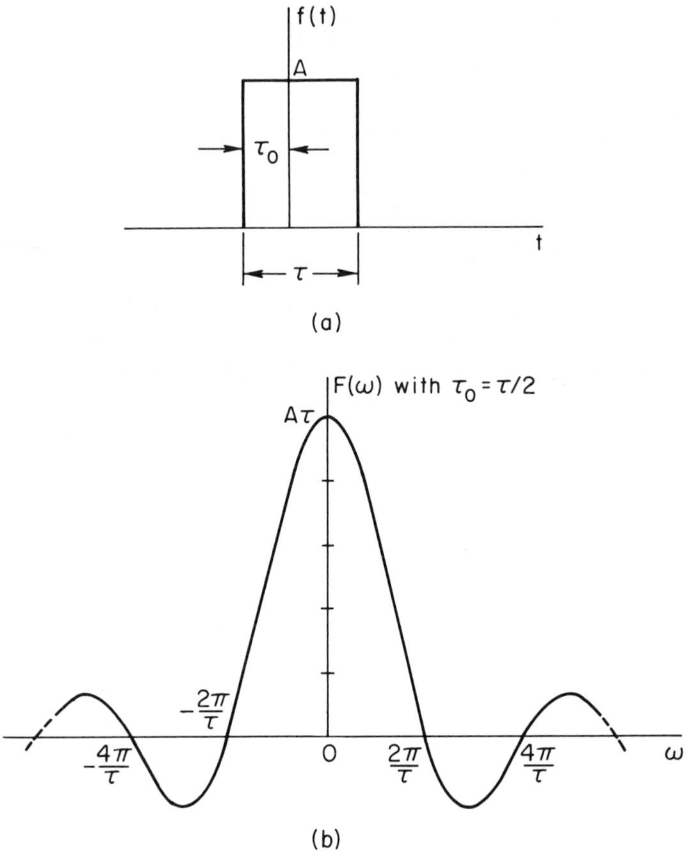

Fig. 2.3-1. A rectangular pulse (a) and its Fourier spectrum (b) when $\tau_0 = \tau/2$.

If the pulse of the above example has even symmetry, $\tau_0 = \tau/2$, and

$$F(\omega) = A\tau \text{ Sa } (\omega\tau/2). \tag{2.3-4}$$

where we define

$$\text{Sa } (x) = \sin (x)/x. \tag{2.3-5}$$

Sa (x) is called the *sampling function* and will be used later in connection with the sampling theorem. The spectrum of (2.3-4) is plotted in Fig. 2-3-1(b).

Let us also define [4] the unit-height unit-width rectangular pulse as

$$\begin{aligned} \text{rect } (t) &= 1, \qquad |t| < \frac{1}{2} \\ &= 0, \qquad |t| > \frac{1}{2}. \end{aligned} \tag{2.3-6}$$

ISBN 0-201-05758-1

Then our example shows that

$$\text{rect}\,(t) \leftrightarrow \text{Sa}\,(\omega/2). \tag{2.3-7}$$

Example 2.3-2

As another example, we find the spectrum of the triangular pulse of Fig. 2.3-2(*a*). From (2.3-1), using known integral forms (see Appendix B),

$$F(\omega) = \int_{-\tau}^{0} A\left(1 + \frac{t}{\tau}\right) e^{-j\omega t}\,dt + \int_{0}^{\tau} A\left(1 - \frac{t}{\tau}\right) e^{-j\omega t}\,dt$$

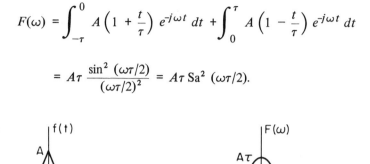

$$= A\tau \frac{\sin^2(\omega\tau/2)}{(\omega\tau/2)^2} = A\tau\,\text{Sa}^2\,(\omega\tau/2).$$

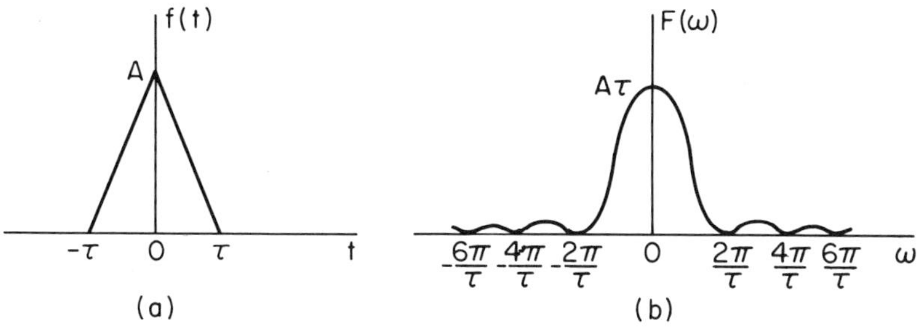

Fig. 2.3-2. Triangular pulse (a) and its Fourier spectrum (b).

This spectrum is plotted in Fig. 2-3-2(*b*).

A function tri (t) is sometimes defined by [4]

$$\text{tri}\,(t) = 1 - |t|, \qquad |t| < 1$$

$$= 0, \qquad |t| > 1. \tag{2.3-8}$$

Then from our example above

$$\text{tri}\,(t) \leftrightarrow \text{Sa}^2\,(\omega/2), \tag{2.3-9}$$

$$A\,\text{tri}\,(t/\tau) \leftrightarrow A\tau\,\text{Sa}^2\,(\omega\tau/2). \tag{2.3-10}$$

Example 2.3-3

As a final example of finding the spectrum of a signal we consider the waveform of Fig. 2.3-3(*a*). The spectrum is again given by eq. (2.3-1):

ISBN 0-201-05758-1

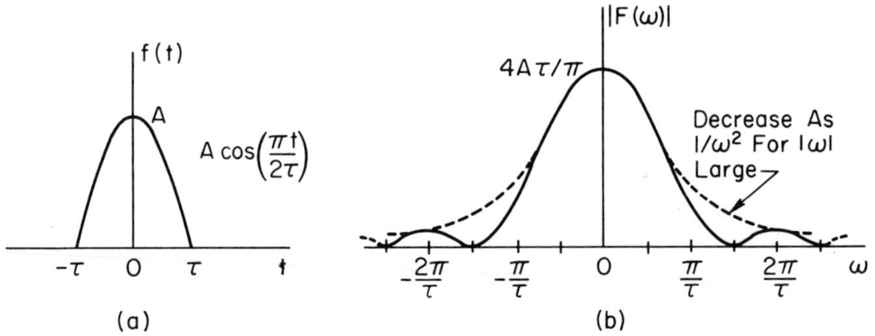

Fig. 2.3-3. Cosine pulse (a) and its Fourier spectrum magnitude (b).

$$F(\omega) = \int_{-\tau}^{\tau} A \cos\left(\frac{\pi t}{2\tau}\right) e^{-j\omega t}\, dt.$$

By expanding cos $(\pi t/2\tau)$ using (2.1-7), this integral is reduced to two simple integrals. The result is

$$F(\omega) = \frac{A\pi}{\tau}\left[\frac{\cos(\omega\tau)}{(\pi/2\tau)^2 - \omega^2}\right].$$

This spectrum is sketched in Fig. 2.3-3(*b*).

Development of the Fourier Transform Pair

The Fourier transform describes the spectral content of a signal $f(t)$. At this point in our discussions it is applied to the description of *aperiodic signals*—that is, those waveforms that are not periodic. Earlier examples have illustrated how such waveforms possess a continuous spectrum. They therefore, in principle, contain all frequencies, and the Fourier transform serves to describe the relative strength of the various frequencies. Later, we shall find that even periodic signals, having only discrete frequencies, have Fourier transforms.

In order to justify how a Fourier transform may be used to describe aperiodic signals, let us begin by considering a familiar waveform, the periodic signal illustrated in Fig. 2.3-4(*a*). Over its central period T_1 it is assumed to have an arbitrary shape corresponding to the aperiodic waveform of interest, $f(t)$. Now if waveform shape is preserved but the period of the periodic signal is made longer, as illustrated in (*b*), it is clear that, in the limit when $T \to \infty$, the periodic waveform $f_2(t)$ becomes the aperiodic signal $f(t)$. We make use of this fact to develop the Fourier transform pair.

The periodic signal $f_2(t)$ may be expressed in terms of its complex Fourier series,

$$f_2(t) = \sum_{n=-\infty}^{\infty} C_n e^{jn\omega_T t}, \qquad (2.3\text{-}11)$$

where $\omega_T = 2\pi/T$. The Fourier coefficients are given by

ISBN 0-201-05758-1

$$C_n = \frac{1}{T} \int_{-T/2}^{T/2} f_2(t) e^{-jn\omega_T t}\, dt. \tag{2.3-12}$$

Next, if we substitute (2.3-12) for C_n into (2.3-11), we may write

$$f_2(t) = \frac{1}{2\pi} \sum_{n=-\infty}^{\infty} \left\{ \int_{-T/2}^{T/2} f_2(t) e^{-jn\omega_T t}\, dt \right\} e^{jn\omega_T t} \frac{2\pi}{T}. \tag{2.3-13}$$

If T is now allowed to become infinite, $2\pi/T$ becomes differential frequency $d\omega$, $n\omega_T$ may be replaced by ω, and the summation becomes an integral while $f_2(t)$ becomes $f(t)$, the aperiodic signal. Thus, we obtain

$$f(t) = \frac{1}{2\pi} \int_{-\infty}^{\infty} \left\{ \int_{-\infty}^{\infty} f(t) e^{-j\omega t}\, dt \right\} e^{j\omega t}\, d\omega, \tag{2.3-14}$$

which is known as the *Fourier integral.*

The braced quantity in (2.3-14) is a function of ω. We *define* this function to be the Fourier spectrum $F(\omega)$ of the signal $f(t)$. Hence, (2.3-14) may be expressed in the equivalent form

$$f(t) = \frac{1}{2\pi} \int_{-\infty}^{\infty} F(\omega) e^{j\omega t}\, d\omega, \tag{2.3-15}$$

where

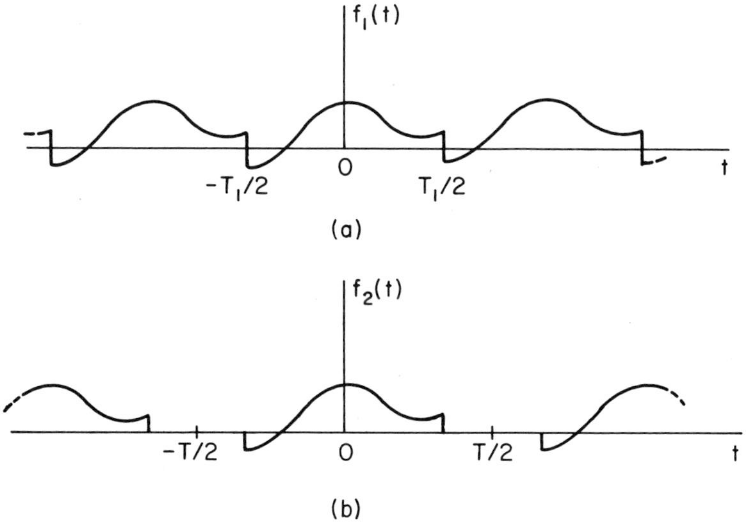

Fig. 2.3-4. A train of signals (a), and (b) the same train with larger period.

ISBN 0-201-05758-1

$$F(\omega) = \int_{-\infty}^{\infty} f(t)e^{-j\omega t}\, dt. \tag{2.3-16}$$

These two expressions form the Fourier transform pair.

Sufficient Conditions for Existence of the Fourier Transform

Not all signals possess a Fourier transform, although all waveforms of any engineering interest do. A condition, which guarantees the existence of the Fourier transform $F(\omega)$ of a signal $f(t)$, states that if $f(t)$ satisfies the Dirichlet conditions in every finite interval* and if

$$\int_{-\infty}^{\infty} |f(t)|\; dt < \infty, \tag{2.3-17}$$

then $F(\omega)$ exists [1, 2]. This result is only sufficient and not necessary, however, for some important signals do exist that have transforms but do not satisfy these conditions. An example is the impulse function $\delta(t)$ to be subsequently discussed.

General Spectrum Forms

In general, $F(\omega)$ is a complex function for either real or complex $f(t)$. Let these quantities have real and imaginary parts defined by

$$f(t) = r(t) + jx(t), \tag{2.3-18}$$

$$F(\omega) = R(\omega) + jX(\omega). \tag{2.3-19}$$

Using these definitions with the Fourier transform pair it is possible to obtain the general forms

$$R(\omega) = \int_{-\infty}^{\infty} [r(t) \cos(\omega t) + x(t) \sin(\omega t)]\; dt, \tag{2.3-20}$$

$$X(\omega) = \int_{-\infty}^{\infty} [x(t) \cos(\omega t) - r(t) \sin(\omega t)]\; dt, \tag{2.3-21}$$

and

$$r(t) = \frac{1}{2\pi}\int_{-\infty}^{\infty} [R(\omega) \cos(\omega t) - X(\omega) \sin(\omega t)]\; d\omega, \tag{2.3-22}$$

$$x(t) = \frac{1}{2\pi}\int_{-\infty}^{\infty} [X(\omega) \cos(\omega t) + R(\omega) \sin(\omega t)]\; d\omega. \tag{2.3-23}$$

*Signals satisfying these conditions are also said to have *bounded variation.*

ISBN 0-201-05758-1

The first two of these forms may be used to determine the symmetry properties of real, imaginary, and more general signals.

Spectral Symmetry of Real Signals

Here $f(t) = r(t)$ and $x(t) = 0$. Examination of (2.3-20) and (2.3-21) will reveal that $R(\omega)$ becomes an even function of ω while $X(\omega)$ is an odd function. Thus,

$$R(-\omega) = R(\omega), \qquad (2.3\text{-}24)$$

$$X(-\omega) = -X(\omega), \qquad (2.3\text{-}25)$$

which means that the spectrum has symmetry defined by

$$F(-\omega) = F^*(\omega) \qquad (2.3\text{-}26)$$

for real signals, where the asterisk denotes the complex conjugate operation.

If $f(t)$ is a real waveform with even symmetry about the time origin, $X(\omega) = 0$, and the spectrum is found to be real with even symmetry about the frequency origin. This case is analogous to a real, even, periodic signal having only cosine (even) terms in its Fourier series.

If $f(t)$ is both real and odd we find that $R(\omega) = 0$ and the spectrum is purely imaginary and odd. An analogy may be drawn here to a real, odd, periodic signal which has only sine (odd) terms in its Fourier series.

Spectral Symmetry of Other Signals

If $f(t)$ is an imaginary signal, $f(t) = jx(t)$ and $r(t) = 0$. From (2.3-20) and (2.3-21) we find that $R(\omega)$ and $X(\omega)$ are odd and even functions of ω, respectively. Thus,

$$R(-\omega) = -R(\omega), \qquad (2.3\text{-}27)$$

$$X(-\omega) = X(\omega), \qquad (2.3\text{-}28)$$

which means that

$$F(-\omega) = -F^*(\omega). \qquad (2.3\text{-}29)$$

If, in addition to being imaginary, $f(t)$ has even symmetry, then $R(\omega) = 0$. For odd symmetry $X(\omega) = 0$.

By combining the above results for spectral symmetry of real and imaginary signals, it can be seen that a general complex signal having even or odd symmetry will possess a complex Fourier transform with even or odd spectral symmetry, respectively.

2.4. PROPERTIES OF FOURIER TRANSFORMS

There is a number of useful properties of Fourier transforms that will allow some problems to be solved almost by inspection. In this section we shall summarize many of these properties, some of which may be more or less obvious to the reader.

ISBN 0-201-05758-1

Linearity

If $f(t)$ is the weighted sum of N signals $\alpha_n f_n(t)$, where the α_n are weighting coefficients (constants), then $f(t)$ has the transform $F(\omega)$ given by

$$f(t) = \sum_{n=1}^{N} \alpha_n f_n(t) \leftrightarrow F(\omega) = \sum_{n=1}^{N} \alpha_n F_n(\omega), \tag{2.4-1}$$

where

$$f_n(t) \leftrightarrow F_n(\omega). \tag{2.4-2}$$

In other words, the transform of a linear sum of signals is the linear sum of the individual transforms.

Time and Frequency Shifting

If $f(t)$ has the transform $F(\omega)$, the signal $f(t - t_0)$, which is $f(t)$ delayed by an amount t_0, has a Fourier transform given by

$$f(t - t_0) \leftrightarrow F(\omega)e^{-j\omega t_0}. \tag{2.4-3}$$

This is easily proved by use of (2.3-2). Hence, delay of a waveform introduces a component of phase to the original spectrum which is linear with frequency. Such a result occurs when a signal is passed through a delay line.

In an analogous manner, if the spectrum of a signal is shifted to a higher radian frequency ω_0, use of (2.3-1) shows that the signal is similarly shifted in frequency:

$$f(t)e^{j\omega_0 t} \leftrightarrow F(\omega - \omega_0). \tag{2.4-4}$$

Frequency shifting, as we shall see later, occurs when a signal is passed through a frequency mixer or is used to modulate a carrier signal.

Scaling

If α is a real constant, we have

$$f(\alpha t) \leftrightarrow \frac{1}{|\alpha|} F\left(\frac{\omega}{\alpha}\right), \tag{2.4-5}$$

a result which is easily proved by use of (2.3-1).

Symmetry

It may often be necessary to use tables of Fourier transforms backwards. That is, we may have a time function which has a form found listed in the *spectrum* column in the tables. The symmetry property of Fourier transforms allows us to find the desired transform. It states that, if $f(t)$ has a transform $F(\omega)$, then

ISBN 0-201-05758-1

$$F(t) \leftrightarrow 2\pi f(-\omega). \tag{2.4-6}$$

We prove (2.4-6) by writing (2.3-2) as

$$2\pi f(-t) = \int_{-\infty}^{\infty} F(\omega)e^{-j\omega t}\, d\omega$$

$$= \int_{-\infty}^{\infty} F(x)e^{-jxt}\, dx, \tag{2.4-7}$$

since ω is just a dummy variable for integration. Now replacing t by ω and x by t gives (2.4-6).

Differentiation

Differentiation of (2.3-2) with respect to time produces

$$\frac{df(t)}{dt} \leftrightarrow (j\omega)F(\omega). \tag{2.4-8}$$

Repeated differentiation gives

$$\frac{d^n f(t)}{dt^n} \leftrightarrow (j\omega)^n F(\omega). \tag{2.4-9}$$

As we noted earlier, differentiation of a Fourier series produced a more slowly convergent series, and the result could be assured to represent the signal derivative only under certain restrictions. Similar problems exist in the application of (2.4-9), which only says that, if the transform of the nth derivative exists, it is given by $(j\omega)^n F(\omega)$. It does not guarantee such existence.

An analogous result derives from repeated differentiation of (2.3-1) with respect to ω:

$$(-jt)^n f(t) \leftrightarrow \frac{d^n F(\omega)}{d\omega^n}. \tag{2.4-10}$$

Integration

Let $f(t)$ be a signal with transform $F(\omega)$, where $F(0) = 0$; i.e.,

$$\int_{-\infty}^{\infty} f(t)\, dt = 0. \tag{2.4-11}$$

For such a class of waveforms

$$\int_{-\infty}^{t} f(\tau)\, d\tau \leftrightarrow \frac{F(\omega)}{j\omega}, \tag{2.4-12}$$

ISBN 0-201-05758-1

which can be seen by formal integration of (2.3-2). Later, after the introduction of impulse functions, we shall be in a position to eliminate the requirement of $F(0) = 0$. The result is (2.5-11).

Let $f(t)$ belong to the class of signals where $f(0) = 0$. The analogous case to time integration now occurs for frequency integration [5]:

$$\frac{f(t)}{-jt} \leftrightarrow \int_{-\infty}^{\omega} F(x)\, d(x). \tag{2.4-13}$$

The restriction $f(0) = 0$ is removed later to obtain (2.5-12).

Conjugation

If $f(t)$ has the transform $F(\omega)$, then the complex conjugate of $f(t)$ and its spectrum are related as

$$f^*(t) \leftrightarrow F^*(-\omega). \tag{2.4-14}$$

Convolution

Assume that two signals $f_1(t)$ and $f_2(t)$ have Fourier transforms $F_1(\omega)$ and $F_2(\omega)$, respectively. The signal $f(t)$ having a transform $F(\omega)$ given by the product $F_1(\omega)F_2(\omega)$ is

$$f(t) = \int_{-\infty}^{\infty} f_1(\tau)f_2(t - \tau)\, d\tau \leftrightarrow F_1(\omega)F_2(\omega) = F(\omega). \tag{2.4-15}$$

The integral

$$f(t) = \int_{-\infty}^{\infty} f_1(\tau)f_2(t - \tau)\, d\tau \tag{2.4-16}$$

is called the *convolution integral.*

The reverse situation involves the product of two time functions:

$$f(t) = f_1(t)f_2(t) \leftrightarrow \frac{1}{2\pi}\int_{-\infty}^{\infty} F_1(u)F_2(\omega - u)\, du = F(\omega). \tag{2.4-17}$$

Note that the frequency convolution integral involves a $1/2\pi$ factor.

In the next chapter we shall find that convolution plays an important role in defining the response of linear systems to various input signals.

Correlation

Again let two signals $f_1(t)$ and $f_2(t)$ be Fourier transformable with transforms $F_1(\omega)$ and $F_2(\omega)$, respectively. The signal $f(t)$ having a transform $F(\omega) = F_1^*(\omega)F_2(\omega)$ is given by

ISBN 0-201-05758-1

$$f(t) = \int_{-\infty}^{\infty} f_1^*(\tau) f_2(\tau + t)\, d\tau \leftrightarrow F_1^*(\omega) F_2(\omega) = F(\omega). \tag{2.4-18}$$

The integral

$$f(t) = \int_{-\infty}^{\infty} f_1^*(\tau) f_2(\tau + t)\, d\tau \tag{2.4-19}$$

is known as the *correlation integral.* If $f_1(t)$ and $f_2(t)$ are different functions, as implied, (2.4-19) is the *cross-correlation integral.* If $f_1(t)$ and $f_2(t)$ are the same we have the *autocorrelation integral:*

$$f(t) = \int_{-\infty}^{\infty} f_1^*(\tau) f_1(\tau + t)\, d\tau \leftrightarrow |F_1(\omega)|^2 = F(\omega). \tag{2.4-20}$$

This last result is useful in analyzing the response of certain optimal systems (the matched filters of Chapter 4).

The result analogous to (2.4-18) for correlation in the frequency domain is

$$f(t) = f_1^*(t) f_2(t) \leftrightarrow \frac{1}{2\pi} \int_{-\infty}^{\infty} F_1^*(u) F_2(u + \omega)\, du = F(\omega). \tag{2.4-21}$$

If the two signals are the same,

$$f(t) = |f_1(t)|^2 \leftrightarrow \frac{1}{2\pi} \int_{-\infty}^{\infty} F_1^*(u) F_1(u + \omega)\, du = F(\omega). \tag{2.4-22}$$

Parseval's Theorem

The correlation integral of (2.4-18) may be written as

$$\int_{-\infty}^{\infty} f_1^*(\tau) f_2(\tau + t)\, d\tau = \frac{1}{2\pi} \int_{-\infty}^{\infty} F_1^*(\omega) F_2(\omega) e^{j\omega t}\, d\omega \tag{2.4-23}$$

by straightforward use of the Fourier transform. Since the result must hold for all t, let $t = 0$. We get

$$\int_{-\infty}^{\infty} f_1^*(\tau) f_2(\tau)\, d\tau = \frac{1}{2\pi} \int_{-\infty}^{\infty} F_1^*(\omega) F_2(\omega)\, d\omega, \tag{2.4-24}$$

which is known as *Parseval's theorem.*

ISBN 0-201-05758-1

An especially useful result corresponds to letting $f_1(\tau) = f_2(\tau)$:

$$\int_{-\infty}^{\infty} |f_1(\tau)|^2 \, d\tau = \frac{1}{2\pi} \int_{-\infty}^{\infty} |F_1(\omega)|^2 \, d\omega. \tag{2.4-25}$$

The left side of this relationship is the energy in the signal $f_1(t)$. Thus, we find that energy can also be found from a frequency-domain integration.

2.5. THE IMPULSE FUNCTION

The *impulse function* is an important concept or tool which allows the easy solution of many problems that would otherwise be quite difficult. One purpose in this text has been to choose the most understandable explanation of a subject, even though more mathematically elegant explanations exist. Attempting to apply this purpose to impulse functions presents a problem.

Many definitions exist for the impulse or *delta function.* The most direct, and actually the most mathematically sound (Papoulis [2]), is a purely mathematical one based on the theory of distributions. However, such an approach leaves the reader void of any physical interpretation of what an impulse function is. On the other hand, a lengthy development of a physical picture may give some readers an unclear idea of the mathematical convenience of delta functions. Thus, we shall here present both sides of the picture.

Mathematical Definition and Properties

A relatively new theory called the theory of distributions is discussed briefly by Papoulis [2] with regard to impulse functions. We only summarize the results important to the present problem. An *impulse* or *delta function,* $\delta(t)$, may be defined formally on the basis of its integral property:

$$\int_{-\infty}^{\infty} \phi(t)\delta(t - t_0) \, dt = \phi(t_0), \tag{2.5-1}$$

where $\phi(t)$ is an arbitrary function continuous at the given point t_0.

Although we shall find only limited use for the derivatives of the impulse function, these may similarly be defined by [2]:

$$\int_{-\infty}^{\infty} \phi(t) \frac{d^n\delta(t - t_0)}{dt^n} \, dt = (-1)^n \frac{d^n\phi(t_0)}{dt^n}. \tag{2.5-2}$$

Some of the properties of the impulse function may be developed from (2.5-1). If $\phi(t) = 1$, a constant independent of t, then

$$\int_{-\infty}^{\infty} \delta(t - t_0) \, dt = 1. \tag{2.5-3}$$

This says that the *area* under the impulse function is unity.

ISBN 0-201-05758-1

Next assume that $\phi(t)$, which can be arbitrary, is an extremely narrow pulse of amplitude A, centered at $t = t_0$, and width 2ϵ, where $\epsilon \to 0$.

$$\int_{-\infty}^{\infty} A \operatorname{rect}\left(\frac{t - t_0}{2\epsilon}\right) \delta(t - t_0)\, dt = A \int_{t_0-\epsilon}^{t_0+\epsilon} \delta(t - t_0)\, dt = A. \qquad (2.5\text{-}4)$$

Since the right side is independent of ϵ, we conclude that the "width" of $\delta(t - t_0)$ must be *less* than 2ϵ. As ϵ may be arbitrarily small, the impulse function must then be characterized by zero width. If the width is zero and the area is unity we must also have an infinite *amplitude* at $t = t_0$ (area $= \infty \cdot 0$, which is an indeterminate form with the limit 1).

Finally we may consider $\delta(t)$ to be an even function of t. This property can be deduced from (2.5-1) using a simple variable change.

Physical Interpretation—Approximation of Signals

To gain a physical picture of why and how impulses are useful in signal descriptions, let us first establish the relationship between an impulse and a unit step function. If $\delta(t)$ is integrated we obtain the unit step due to the above stated properties of $\delta(t)$:

$$u(t) = \int_{-\infty}^{t} \delta(\tau)\, d\tau. \qquad (2.5\text{-}5)$$

Alternatively,

$$\delta(t) = du(t)/dt. \qquad (2.5\text{-}6)$$

Consider now a signal $g(t)$ having a derivative $g'(t)$ as shown in Fig. 2.5-1(b) and (a), respectively. Let $g(t)$ be approximated in intervals $\Delta\tau$ wide by the step function approximation $\hat{g}(t)$ shown. The times of occurrence of the steps are $t_n = n\,\Delta\tau$, and the step amplitudes are $g(n\,\Delta\tau) - g[(n-1)\,\Delta\tau]$ at these times. The derivative $\hat{g}'(t)$ of the approximation, shown in (c), is a sequence of impulses having strengths $g(n\,\Delta\tau) - g[(n-1)\,\Delta\tau]$, which can be seen by applying (2.5-6). The derivative $\hat{g}'(t)$ may then be written as

$$\hat{g}'(t) = \sum_{n=-\infty}^{\infty} \{ g(n\,\Delta\tau) - g[(n-1)\,\Delta\tau] \}\, \delta(t - n\,\Delta\tau). \qquad (2.5\text{-}7)$$

Note next that if $\Delta\tau$ is small enough our step function approximation to $g(t)$ gets better and in the limit zero error occurs. For small $\Delta\tau$ we have

$$g(n\,\Delta\tau) - g[(n-1)\,\Delta\tau] \approx g'(n\,\Delta\tau) \cdot \Delta\tau \qquad (2.5\text{-}8)$$

from the definition of the derivative. We may then write (2.5-7) as

$$\hat{g}'(t) = \sum_{n=-\infty}^{\infty} g'(n\,\Delta\tau)\, \Delta\tau\, \delta(t - n\,\Delta\tau). \qquad (2.5\text{-}9)$$

ISBN 0-201-05758-1

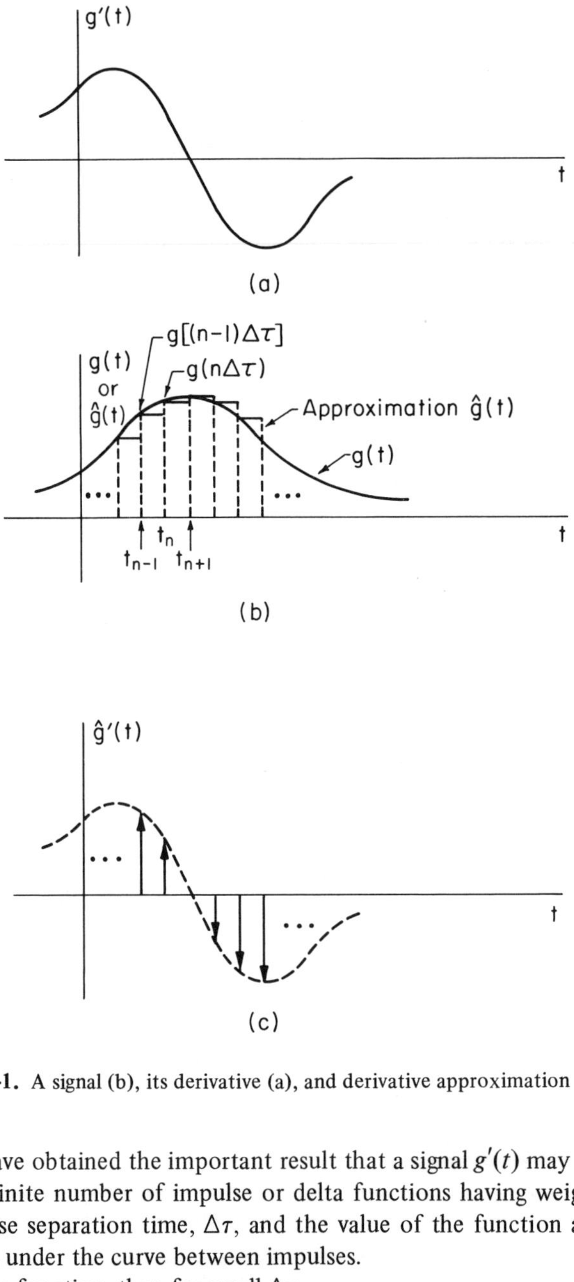

Fig. 2.5-1. A signal (b), its derivative (a), and derivative approximation (c).

In other words, we have obtained the important result that a signal $g'(t)$ may be approximated as the sum of an infinite number of impulse or delta functions having weights given by the product of the impulse separation time, $\Delta\tau$, and the value of the function at the time of the impulse–i.e., the area under the curve between impulses.

If $f(t) = g'(t)$ is the function, then for small $\Delta\tau$,

$$f(t) = \sum_{n=-\infty}^{\infty} f(n\,\Delta\tau)\,\Delta\tau\,\delta(t - n\,\Delta\tau). \tag{2.5-10}$$

The result, (2.5-10), can also be seen directly from the defining equation (2.5-1) for an impulse by letting the integral be approximated by a summation. Alternatively (2.5-10) becomes (2.5-1) in the limit, since $\Delta\tau \to d\tau$, $n\,\Delta\tau \to \tau$, the sum becomes an integral, and the impulse function may be considered even.

ISBN 0-201-05758-1

In the course of this discussion it has been shown how a zero-degree polynomial (constant) approximation over each interval $\Delta\tau$ wide can be used to represent a signal. Higher degree approximations may also be used [2, 6]. An approximation to the spectrum of a signal may be obtained by these methods and use of (2.4-9), as discussed by Papoulis [2].

Having introduced impulse functions, we may revise the integration properties of Fourier transforms. When $F(0) \neq 0$ we may now write (2.4-12) as [2]

$$\int_{-\infty}^{t} f(\tau)\, d\tau \leftrightarrow \pi F(0)\delta(\omega) + \frac{F(\omega)}{j\omega}. \tag{2.5-11}$$

When $f(0) \neq 0$, (2.4-13) may similarly be generalized. The result is [5]

$$\pi f(0)\delta(t) - \frac{f(t)}{jt} \leftrightarrow \int_{-\infty}^{\omega} F(x)\, dx. \tag{2.5-12}$$

2.6. USEFUL TRANSFORMS INVOLVING IMPULSES

We now develop some useful transforms, most of which involve impulses. The derivations of most of the examples are chosen to illustrate the usefulness of the Fourier transform properties given in Section 2.4.

Transform of an Impulse

Let $f(t) = \delta(t)$. The Fourier transform of the impulse is

$$F(\omega) = \int_{-\infty}^{\infty} \delta(t) e^{-j\omega t}\, dt = e^{0} = 1 \tag{2.6-1}$$

from use of (2.5-1). Thus,

$$\delta(t) \leftrightarrow 1, \tag{2.6-2}$$

and the impulse function is characterized by an infinite width, flat spectrum of unit magnitude. The spectrum is illustrated in Fig. 2.6-1(a), where the impulse is shown as a vertical arrow with the amplitude proportional to its strength.

If the impulse is time shifted to occur at $t = t_0$, application of the time shifting property (2.4-3) easily gives

$$\delta(t - t_0) \leftrightarrow e^{-j\omega t_0}. \tag{2.6-3}$$

By application of the symmetry property (2.4-6) to (2.6-2) we have the inverse transform of an impulse at $\omega = 0$.

$$1 \leftrightarrow 2\pi\delta(\omega). \tag{2.6-4}$$

Figure 2.6-1(b) illustrates (2.6-4).

ISBN 0-201-05758-1

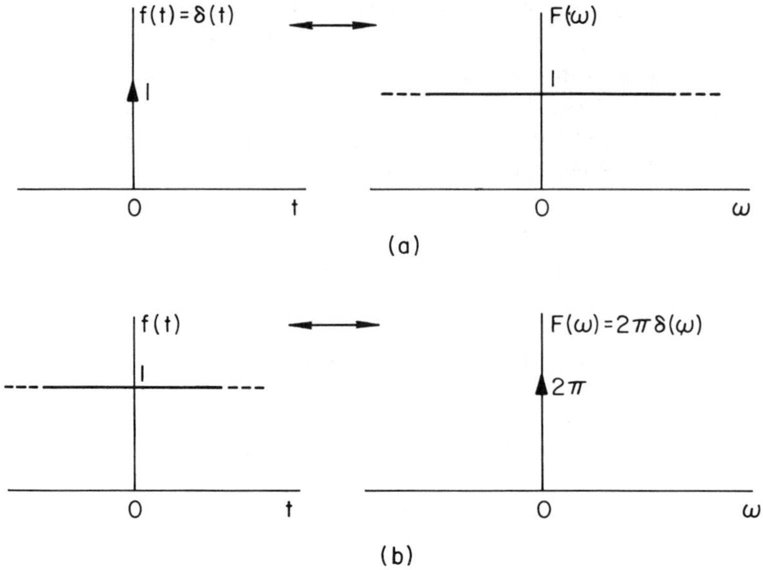

Fig. 2.6-1. Transforms of impulses in time (a) and frequency (b).

Frequency shifting of the impulse may be handled by applying the frequency shifting property of (2.4-4). We obtain

$$e^{j\omega_0 t} \leftrightarrow 2\pi\delta(\omega - \omega_0). \tag{2.6-5}$$

Transform of a Constant

From the linearity property the product of a constant A and a signal $f(t)$ has a transform $AF(\omega)$ so that both sides of all the transform pairs listed above may be multiplied by A. Choosing (2.6-4) in particular results in the transform of a constant A:

$$A \leftrightarrow A2\pi\delta(\omega). \tag{2.6-6}$$

Transform of Sines and Cosines

The Fourier transform of an exponential has already been given in (2.6-5). Since one may write

$$\cos(\omega_0 t) = \frac{1}{2}(e^{j\omega_0 t} + e^{-j\omega_0 t}), \tag{2.6-7}$$

$$\sin(\omega_0 t) = \frac{1}{2j}(e^{j\omega_0 t} - e^{-j\omega_0 t}), \tag{2.6-8}$$

we readily obtain

$$\cos(\omega_0 t) \leftrightarrow \pi[\delta(\omega - \omega_0) + \delta(\omega + \omega_0)], \tag{2.6-9}$$

ISBN 0-201-05758-1

$$\sin(\omega_0 t) \leftrightarrow \frac{\pi}{j}\,[\delta(\omega - \omega_0) - \delta(\omega + \omega_0)]. \tag{2.6-10}$$

Notice that $\cos(\omega_0 t)$ is a real even function of t which produces a real even spectrum, which agrees with our previous conclusion in Section 2.3. Similarly, the sine function is real and odd, and we had also shown that its spectrum must come out imaginary and odd.

Transform of the Signum Function

The *signum function* sgn (t) is defined by

$$\begin{aligned} \operatorname{sgn}(t) &= 1, \qquad 0 < t \\ &= -1, \qquad 0 > t \end{aligned} \tag{2.6-11}$$

as shown in Fig. 2.6-2(a). We shall show that the Fourier transform of this signal is $-j2/\omega$ as plotted in Fig. 2.6-2(b).

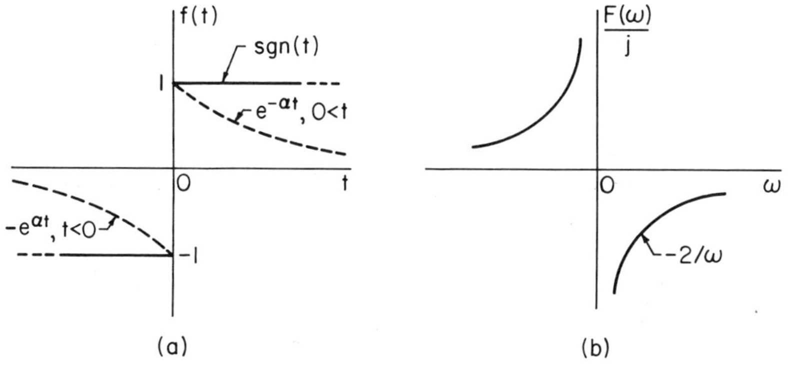

Fig. 2.6-2. The signum function (a) and its Fourier transform (b).

Sgn (t) can be considered the limit of the exponential waveform of Fig. 2.6-2(a) as $\alpha \to 0$. Prior to the limit, the spectrum $F_\alpha(\omega)$ of the odd exponential is

$$F_\alpha(\omega) = -\int_{-\infty}^{0} e^{(\alpha - j\omega)t}\,dt + \int_{0}^{\infty} e^{-(\alpha + j\omega)t}\,dt = \frac{-j2\omega}{\alpha^2 + \omega^2}. \tag{2.6-12}$$

Allowing $\alpha \to 0$, we obtain

$$\operatorname{sgn}(t) \leftrightarrow \lim_{\alpha \to 0}\,[F_\alpha(\omega)] = 2/j\omega. \tag{2.6-13}$$

From the symmetry property the inverse transform of sgn (ω) can be found:

$$j/\pi t \leftrightarrow \operatorname{sgn}(\omega). \tag{2.6-14}$$

ISBN 0-201-05758-1

Transform of a Step

As a last illustration of transforms involving impulse functions we find the spectrum of the unit step function $u(t)$ defined by

$$u(t) = 1, \quad 0 < t$$
$$= 0, \quad 0 > t. \tag{2.6-15}$$

We may write $u(t)$ as

$$u(t) = \frac{1}{2} + \frac{1}{2}\,\text{sgn}\,(t). \tag{2.6-16}$$

Using the linearity property and above results allows us to write the spectrum by inspection:

$$u(t) \leftrightarrow \pi\delta(\omega) + \frac{1}{j\omega}. \tag{2.6-17}$$

The real part $R(\omega)$ of $F(\omega)$ is $\pi\delta(\omega)$. The imaginary part $X(\omega)$ is $-1/\omega$. We could also have arrived at (2.6-17) by using (2.5-11) when (2.6-2) is substituted for $f(\tau)$.

2.7. SPECTRUM OF A PERIODIC SIGNAL

Our discussions of periodic signals have so far centered about time characterization through the Fourier series. With the introduction of the Fourier transform we are now able to develop the spectrum of such signals.

As we have found previously, a periodic signal $f(t)$ having period $T = 2\pi/\omega_T$ may be written as a complex Fourier series:

$$f(t) = \sum_{n=-\infty}^{\infty} C_n e^{jn\omega_T t}, \tag{2.7-1}$$

where the coefficients C_n are found from (2.1-13). Utilizing (2.6-5) we obtain the spectrum of a periodic signal:

$$F(\omega) = 2\pi \sum_{n=-\infty}^{\infty} C_n \delta(\omega - n\omega_T). \tag{2.7-2}$$

The spectrum comprises an infinite number of impulses at $\omega = n\omega_T$ having amplitudes $2\pi C_n$. The main differences between the Fourier spectrum plot and the coefficient spectrum plot are the use of impulses in place of "lines" and a constant 2π.

2.8. THE POWER DENSITY SPECTRUM

A periodic signal is an eternal signal existing from $-\infty$ to $+\infty$ in time. Consequently it has infinite energy but will have some average power as defined by (2.2-8). As indicated earlier, such signals are classified as *power signals* in contrast with signals of finite duration, which

ISBN 0-201-05758-1

have finite energy but zero power on the infinite average and are called *energy signals.*

We now develop the *power density spectrum* of a power signal, which is a function showing the distribution of average power in the signal as a function of frequency. Let $f(t)$ be the eternal signal and $f_T(t)$ represent $f(t)$ in the interval $-T < t < T$. We also assume

$$f_T(t) \leftrightarrow F_T(\omega). \tag{2.8-1}$$

Parseval's theorem states that the energy E† in $f_T(t)$ is

$$E = \int_{-\infty}^{\infty} |f_T(t)|^2\, dt = \frac{1}{2\pi} \int_{-\infty}^{\infty} |F_T(\omega)|^2\, d\omega. \tag{2.8-2}$$

However, since

$$\int_{-\infty}^{\infty} |f_T(t)|^2\, dt = \int_{-T}^{T} |f(t)|^2\, dt, \tag{2.8-3}$$

the average power in the signal $f(t)$ is, using (2.2-8),

$$P = \lim_{T\to\infty} \left\{ \frac{1}{2T} \int_{-T}^{T} |f(t)|^2\, dt \right\} = \frac{1}{2\pi} \int_{-\infty}^{\infty} \lim_{T\to\infty} \left\{ \frac{|F_T(\omega)|^2}{2T} \right\} d\omega. \tag{2.8-4}$$

We shall define a quantity $\mathcal{S}_f(\omega)$ as the power density spectrum of $f(t)$ given by

$$\mathcal{S}_f(\omega) = \lim_{T\to\infty} \left\{ \frac{|F_T(\omega)|^2}{2T} \right\}. \tag{2.8-5}$$

The choice of notation should be clear, for if we write (2.8-4) in the form

$$P = \frac{1}{2\pi} \int_{-\infty}^{\infty} \mathcal{S}_f(\omega)\, d\omega, \tag{2.8-6}$$

we obtain power by frequency summation and $\mathcal{S}_f(\omega)$ must represent the frequency distribution, or density, of power with frequency.

For real signals $F_T(-\omega) = F_T^*(\omega)$ and $|F_T(\omega)|^2 = F_T(\omega)F_T(-\omega)$. Hence, $\mathcal{S}_f(-\omega) = \mathcal{S}_f(\omega)$, so $\mathcal{S}_f(\omega)$ is an even function and we may also write

$$P = \frac{1}{\pi} \int_{0}^{\infty} \mathcal{S}_f(\omega)\, d\omega$$

$$= 2 \int_{0}^{\infty} \mathcal{S}_f(2\pi f)\, df. \tag{2.8-7}$$

†As noted earlier, power and energy in this book are nearly always taken as per-ohm values—that is, we assume a resistive impedance of one ohm. Situations to the contrary are clearly indicated when they occur.

ISBN 0-201-05758-1

The power density spectrum of a signal depends only on the magnitude of the spectrum and not on its phase. There is an infinite number of signals that may have the same power spectrum. However, for a *given* signal there is only one power spectrum.

Power Density and Autocorrelation Function Relationship

It is especially convenient to relate power density spectrum to the waveform $f(t)$. The relationship involves the definition of the *autocorrelation function* of the real signal $f(t)$. Suppose we expand (2.8-5) by using the Fourier transform relationship for $F_T(\omega)$. Since

$$F_T(\omega) = \int_{-T}^{T} f(t)e^{-j\omega t}\,dt, \tag{2.8-8}$$

we may write (2.8-5) as

$$\mathcal{S}_f(\omega) = \lim_{T\to\infty} \left\{ \frac{1}{2T} \int_{-T}^{T} f(x)e^{-j\omega x}\,dx \int_{-T}^{T} f(t)e^{j\omega t}\,dt \right\} \tag{2.8-9}$$

because $|F_T(\omega)|^2 = F_T(\omega)F_T^*(\omega)$ and $f(t)$ is assumed real. Now we change x to the new variable $\tau = x - t$ to obtain

$$\mathcal{S}_f(\omega) = \lim_{T\to\infty} \int_{-T-t}^{T-t} \left\{ \frac{1}{2T} \int_{-T}^{T} f(t)f(t+\tau)\,dt \right\} e^{-j\omega\tau}\,d\tau. \tag{2.8-10}$$

Allowing T in the outer integral to become infinite, we may now express the above result in the form

$$\mathcal{S}_f(\omega) = \int_{-\infty}^{\infty} \left\{ \lim_{T\to\infty} \left[\frac{1}{2T} \int_{-T}^{T} f(t)f(t+\tau)\,dt \right] \right\} e^{-j\omega\tau}\,d\tau. \tag{2.8-11}$$

After performing the integration and taking the limit, the quantity within the braces becomes a function of τ. We call this function the *time autocorrelation function* $\mathcal{R}_f(\tau)$ of $f(t)$:

$$\mathcal{R}_f(\tau) = \lim_{T\to\infty} \left\{ \frac{1}{2T} \int_{-T}^{T} f(t)f(t+\tau)\,d\tau \right\}. \tag{2.8-12}$$

Thus, we recognize from (2.8-11) that the power density spectrum of a real power signal $f(t)$ is the Fourier transform of the signal autocorrelation function:

$$\mathcal{S}_f(\omega) = \int_{-\infty}^{\infty} \mathcal{R}_f(\tau)e^{-j\omega\tau}\,d\tau. \tag{2.8-13}$$

ISBN 0-201-05758-1

The inverse Fourier transform relationship

$$\mathcal{R}_f(\tau) = \frac{1}{2\pi}\int_{-\infty}^{\infty} \mathcal{S}_f(\omega)e^{j\omega\tau}\, d\omega \tag{2.8-14}$$

also holds, although it is not proved here.

The autocorrelation function may be viewed as a quantity expressing the similarity or *correlation* of a signal with a displaced version of itself. For very small displacements τ we would expect a large value of $\mathcal{R}_f(\tau)$, since the waveform is not expected to change much over a small time interval. Indeed, this is the case, and the maximum of $\mathcal{R}_f(\tau)$ occurs for $\tau = 0$.

For some larger value of τ we expect $\mathcal{R}_f(\tau)$ to decrease significantly (to, say, 0.707 times the maximum). If the absolute value of τ at the 0.707 point is large, $\mathcal{S}_f(\omega)$ will tend to have small spectral spread, while a small absolute value of τ corresponds to a broad power density spectrum. These characteristics are similar to a time waveform and its Fourier transform (spectrum). Thus, we might view the relationship of autocorrelation function with power spectrum as analogous to a signal and its spectrum.

Finally, we note that not only deterministic power signals but also random power signals such as noise have power density spectra. This property of noise is discussed further in Chapter 4.

Power Density Spectrum of a Periodic Signal

We shall now show that the power density spectrum of a periodic signal is

$$\mathcal{S}_f(\omega) = 2\pi \sum_{n=-\infty}^{\infty} |C_n|^2\, \delta(\omega - n\omega_T), \tag{2.8-15}$$

where the Fourier series coefficients C_n are given by (2.1-13) and ω_T equals 2π divided by the period.

The proof* begins by writing the truncated signal $f_T(t)$ as the product $f(t)$ rect $(t/2T)$. Recall that rect $(t/2T)$ is a unit-amplitude function existing over $-T < t < T$. Now since

$$f(t) \leftrightarrow F(\omega), \tag{2.8-16}$$

$$\text{rect}\,(t/2T) \leftrightarrow 2T\,\text{Sa}\,(\omega T), \tag{2.8-17}$$

we may use these results in the frequency convolution integral (2.4-17) for finding the spectrum of a waveform which is the product of two signals. We get

$$F_T(\omega) = \frac{1}{2\pi}\int_{-\infty}^{\infty} F(\omega - u)2T\,\text{Sa}\,(uT)\, du. \tag{2.8-18}$$

Substituting (2.7-2) for $F(\omega)$ and forming the function $|F_T(\omega)|^2/2T$, we have

*Throughout the following proof T is *not* the signal period. To avoid any possible confusion on this point we everywhere use ω_T in place of 2π divided by the period for signal-related quantities, such as the spectrum $F(\omega)$.

ISBN 0-201-05758-1

$$\frac{|F_T(\omega)|^2}{2T} = \sum_{n=-\infty}^{\infty} \sum_{m=-\infty}^{\infty} 2TC_nC_m^* \text{ Sa }[(\omega - n\omega_T)T] \text{ Sa }[(\omega - m\omega_T)T]. \tag{2.8-19}$$

When the limit as $T \to \infty$ is formed to obtain $\mathscr{S}_f(\omega)$, functions of the form Sa $[(\omega - n\omega_T)T]$ will exist only at $\omega = n\omega_T$. Therefore, for a particular value of n, only the term for $m = n$ will produce a nonzero product. Hence, forming the limit using (2.8-19) will produce the same results as the following:

$$\mathscr{S}_f(\omega) = \lim_{T\to\infty} \frac{|F_T(\omega)|^2}{2T} = 2\pi \sum_{n=-\infty}^{\infty} |C_n|^2 \lim_{T\to\infty} \left\{ \frac{T}{\pi} \text{ Sa}^2 \, [(\omega - n\omega_T)T] \right\}. \tag{2.8-20}$$

Although we shall not prove it here, it may be shown that (see problem 2-25)

$$\lim_{T\to\infty} \left\{ \frac{T}{\pi} \text{ Sa}^2 \, (\omega T) \right\} = \delta(\omega). \tag{2.8-21}$$

The power density spectrum of a periodic signal now becomes

$$\mathscr{S}_f(\omega) = 2\pi \sum_{n=-\infty}^{\infty} |C_n|^2 \, \delta(\omega - n\omega_T) \tag{2.8-22}$$

from (2.8-20).

2.9. THE ENERGY DENSITY SPECTRUM

As noted in Section 2.8, an energy signal is one having finite energy and zero infinite-time average power. Bounded waveforms that exist only for a finite time are examples of energy signals.

The energy signal spectral function that is the analogy of the power density spectrum of the power signal is the *energy density spectrum.* Using Parseval's theorem the energy of a signal is

$$E = \int_{-\infty}^{\infty} |f(t)|^2 \, dt = \frac{1}{2\pi} \int_{-\infty}^{\infty} |F(\omega)|^2 \, d\omega. \tag{2.9-1}$$

We define the energy density spectrum $\mathscr{E}(\omega)$, to be

$$\mathscr{E}(\omega) = |F(\omega)|^2, \tag{2.9-2}$$

so that energy is given by the integral

$$E = \frac{1}{2\pi} \int_{-\infty}^{\infty} \mathscr{E}(\omega) \, d\omega$$

ISBN 0-201-05758-1

$$= \int_{-\infty}^{\infty} \mathcal{E}(2\pi f)\, df. \tag{2.9-3}$$

Clearly, $\mathcal{E}(\omega)$ is the distribution of the energy in the waveform over frequency.

For real waveforms $\mathcal{E}(\omega)$ is an even function of ω, and

$$E = 2\int_{0}^{\infty} \mathcal{E}(2\pi f)\, df. \tag{2.9-4}$$

2.10. TABLE OF USEFUL FOURIER TRANSFORMS

Most of the Fourier transform pairs developed in this chapter as well as other useful pairs are listed in Table 2.10-1. We shall often have occasion to use pairs from this table in the chapters to follow. A more extensive table of pairs is given in Campbell and Foster [3].

TABLE 2.10-1. Useful Fourier Transform Pairs

Pair No.	$f(t)$	$F(\omega)$
1	$f_n(t)$	$F_n(\omega)$
2	$\sum_{n=1}^{N} \alpha_n f_n(t)$	$\sum_{n=1}^{N} \alpha_n F_n(\omega)$
3	$f(t - t_0)$	$F(\omega)e^{-j\omega t_0}$
4	$f(t)e^{j\omega_0 t}$	$F(\omega - \omega_0)$
5	$f(\alpha t)$	$\frac{1}{\lvert\alpha\rvert} F\left(\frac{\omega}{\alpha}\right)$
6	$F(t)$	$2\pi f(-\omega)$
7	$\frac{d^n f(t)}{dt^n}$	$(j\omega)^n F(\omega)$
8	$(-jt)^n f(t)$	$\frac{d^n F(\omega)}{d\omega^n}$
9	$\int_{-\infty}^{t} f(\tau)\, d\tau$	$\frac{F(\omega)}{j\omega} + \pi F(0)\delta(\omega)$
10	$\delta(t)$	1
11	1	$2\pi\delta(\omega)$
12	$e^{j\omega_0 t}$	$2\pi\delta(\omega - \omega_0)$
13	sgn (t)	$\frac{2}{j\omega}$

ISBN 0-201-05758-1

Pair No.	$f(t)$	$F(\omega)$
14	$j\,\dfrac{1}{\pi t}$	$\mathrm{sgn}\,(\omega)$
15	$u(t)$	$\pi\delta(\omega) + \dfrac{1}{j\omega}$
16	$\sum_{n=-\infty}^{\infty} C_n e^{jn\omega_0 t}$	$2\pi \sum_{n=-\infty}^{\infty} C_n \delta(\omega - n\omega_0)$
17	$\mathrm{rect}\,(t/\tau)$	$\tau\,\mathrm{Sa}\,(\omega\tau/2)$
18	$\dfrac{W}{2\pi}\,\mathrm{Sa}\left(\dfrac{Wt}{2}\right)$	$\mathrm{rect}\left(\dfrac{\omega}{W}\right)$
19	$\mathrm{tri}\,(t)$	$\mathrm{Sa}^2\left(\dfrac{\omega}{2}\right)$
20	$A\cos\left(\dfrac{\pi t}{2\tau}\right)\mathrm{rect}\left(\dfrac{t}{2\tau}\right)$	$\dfrac{A\pi}{\tau}\;\dfrac{\cos(\omega\tau)}{(\pi/2\tau)^2 - \omega^2}$
21	$\cos(\omega_0 t)$	$\pi[\delta(\omega - \omega_0) + \delta(\omega + \omega_0)]$
22	$\sin(\omega_0 t)$	$\dfrac{\pi}{j}[\delta(\omega - \omega_0) - \delta(\omega + \omega_0)]$
23	$u(t)\cos(\omega_0 t)$	$\dfrac{\pi}{2}[\delta(\omega - \omega_0) + \delta(\omega + \omega_0)] + \dfrac{j\omega}{\omega_0{}^2 - \omega^2}$
24	$u(t)\sin(\omega_0 t)$	$\dfrac{\pi}{2j}[\delta(\omega - \omega_0) - \delta(\omega + \omega_0)] + \dfrac{\omega_0}{\omega_0{}^2 - \omega^2}$
25	$u(t)e^{-\alpha t}\cos(\omega_0 t)$	$\dfrac{(\alpha + j\omega)}{\omega_0{}^2 + (\alpha + j\omega)^2}$
26	$u(t)e^{-\alpha t}\sin(\omega_0 t)$	$\dfrac{\omega_0}{\omega_0{}^2 + (\alpha + j\omega)^2}$
27	$e^{-\alpha\|t\|}$	$\dfrac{2\alpha}{\alpha^2 + \omega^2}$
28	$e^{-t^2/(2\sigma^2)}$	$\sigma\sqrt{2\pi}\,e^{-\sigma^2\omega^2/2}$
29	$u(t)e^{-\alpha t}$	$\dfrac{1}{\alpha + j\omega}$
30	$u(t)te^{-\alpha t}$	$\dfrac{1}{(\alpha + j\omega)^2}$

ISBN 0-201-05758-1

PROBLEMS

2-1. Obtain the Fourier series of a periodic train of pulses of the type shown in Fig. 2.3-3(*a*). Assume $2\tau \leqslant T$.

2-2. (*a*) Plot the first term of the Fourier series for the pulse train of Fig. 2.1-2 and compare to the function. Use $T = 4\tau$. (*b*) Repeat (*a*) for two- and three-term approximations.

2-3. Find the percentage of the total average power that exists in the first (*a*) one, (*b*) two, and (*c*) three components of the Fourier series of the waveform of Fig. 2.1-3. Use $T = 6\tau$.

2-4. Let $T = 2\tau$ and form a train of pulses as shown in Fig. 2.3-3(*a*), which will represent a full-wave rectified cosine wave. Find the ratio of the peak value of the ac component at frequency $f = 1/(2\tau)$ to the dc component.

★ 2-5. (*a*) Find the real-form Fourier series of the pulse train shown in Fig. P2-5. (*b*) Decompose the pulse train into the sum of two trains having even and odd symmetry. Sketch the two trains. (*c*) Find the Fourier series of the two trains of part (*b*). Is their sum equal to the series found in (*a*)?

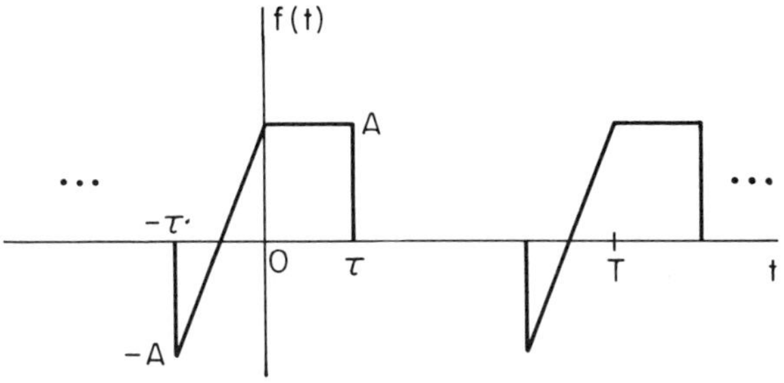

Fig. P2-5

2-6. (*a*) Find the Fourier series in its complex form for the pulse train of Fig. P2-6. In each period it is a cosine wave of peak amplitude *A* multiplied by a positive pulse of unit amplitude having duration 2τ. (*b*) sketch the Fourier coefficient spectrum.

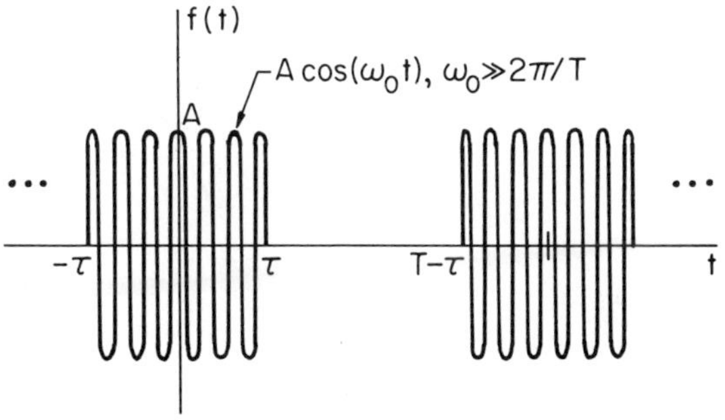

Fig. P2-6

ISBN 0-201-05758-1

★ 2-7. (*a*) Obtain the Fourier series of the signal of Fig. P2-7(*a*). (*b*) Find the series for the signal's derivative shown in (*b*). Show that the series also results from term by term differentiation of the series found in part (*a*).

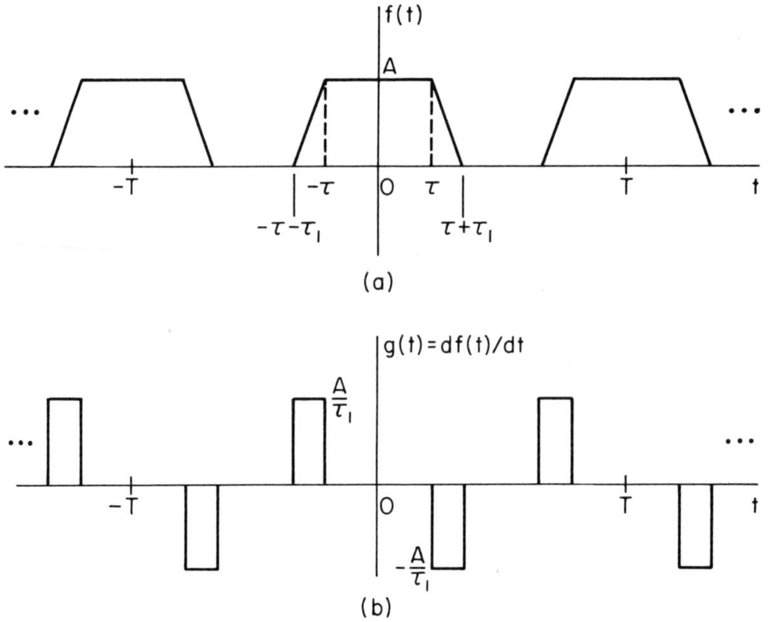

Fig. P2-7

2-8. Let A = 2 volts, τ = 5 seconds, τ_1 = 3 seconds, and T = 20 seconds for the signal of Fig. P2-7(*a*). (*a*) What will the reading be if the signal is applied to a meter that measures total power? (*b*) What would the power reading be for a dc power meter? Are the two power readings related? (*c*) What would the power reading be for an ac power meter?

2-9. The periodic waveform of Fig. P2-9 is applied to the input of an ideal integrator. Find the Fourier series of the output. (Hint: use results of example 2.1-4 and shift the time origin.)

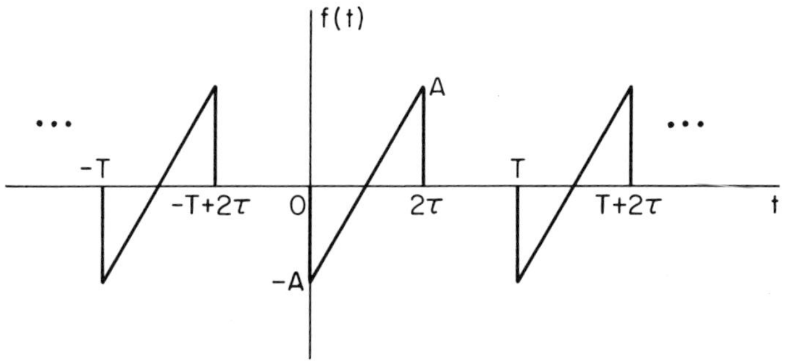

Fig. P2-9

2-10. A periodic signal with period T is defined by

$$f(t) = A\cos\left(\frac{2\pi}{t}\right), \qquad |t| < \tau$$

$$= 0, \qquad \tau < |t| \leqslant T/2.$$

Does it satisfy the Dirichlet conditions? Discuss.

ISBN 0-201-05758-1

2-11. Obtain a finite-range Fourier series expansion to represent the periodic signal consisting of the waveform of Fig. P2-5 in the intervals $nT < t < (n+1)T - \tau$, $n = 0, \pm 1, \pm 2, \ldots$. Assume an even symmetry case, and use results of an example of the text.

2-12. (*a*) Show that the Fourier transform of the nth derivative of the delta function is $(j\omega)^n$. (*b*) Show also that

$$t^n \leftrightarrow 2\pi j^n \frac{d^n \delta(\omega)}{d\omega^n}.$$

2-13. Prove eqs. (2.4-3) and (2.4-4).

2-14. Prove eq. (2.4-5).

2-15. Show that the Fourier transform of a periodic train of impulses is an infinite sum of spectral impulses; i.e.,

$$\sum_{n=-\infty}^{\infty} \delta(t - nT) \leftrightarrow 2\pi \sum_{n=-\infty}^{\infty} \frac{1}{T} \delta(\omega - n\omega_T),$$

where $\omega_T = 2\pi/T$.

2-16. Use the Fourier series representation of a periodic train of impulses to prove that

$$\sum_{n=-\infty}^{\infty} \delta(t - nT) = \frac{1}{T} \sum_{n=-\infty}^{\infty} e^{jn\omega_T t},$$

where $\omega_T = 2\pi/T$.

★ 2-17. The waveform shown in Fig. P2-17 is approximated by three linear line segments. Find the approximate spectrum of $f(t)$ by application of (2.4-8).

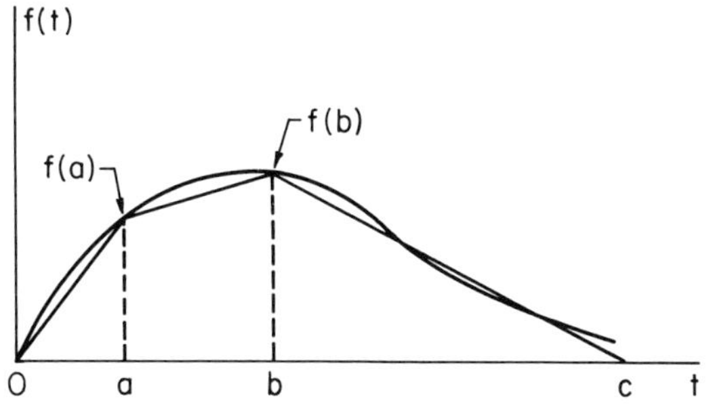

Fig. P2-17

2-18. Prove pair No. 23 of Table 2.10-1.

2-19. Find the Fourier transform of the signal

$$f(t) = u(t)e^{j\omega_0 t}.$$

2-20. Find the Fourier transform $F(\omega)$ of the signal

$$f(t) = [A + f_m(t)] \cos(\omega_0 t)$$

if $f_m(t)$ has a spectrum $F_m(\omega)$.

ISBN 0-201-05758-1

2-21. Let $f_m(t) = A_m \cos(\omega_m t)$ in problem 2.20. (*a*) Plot the spectrum $F(\omega)/\pi A$ for $\omega_m < \omega_0$ with $A_m/A = 0.5$ and 1.0. (*b*) How much of the total signal power is in the "sidebands" around $\pm f_0$?

2-22. Find the Fourier transform of the train of pulses shown in Fig. P2-22.

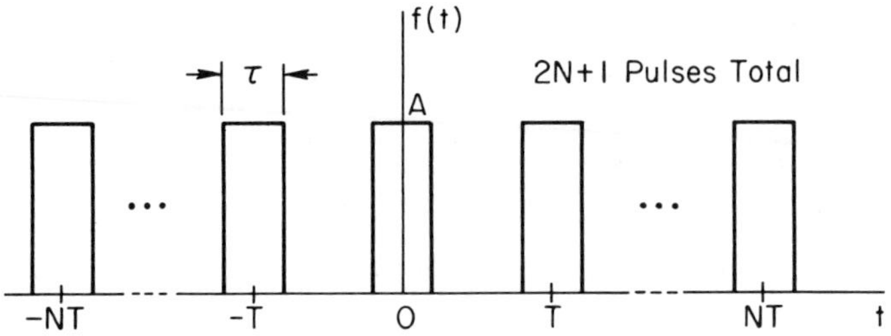

Fig. P2-22

2-23. (*a*) If $\cos(\omega_0 t)$ is multiplied by the pulse train of problem 2-22, what will the spectrum of the resulting signal be? (*b*) Sketch the magnitude of the spectrum if $\omega_0 >> 2\pi/\tau$.

2-24. Find the Fourier transforms of the waveforms of Fig. P2-24.

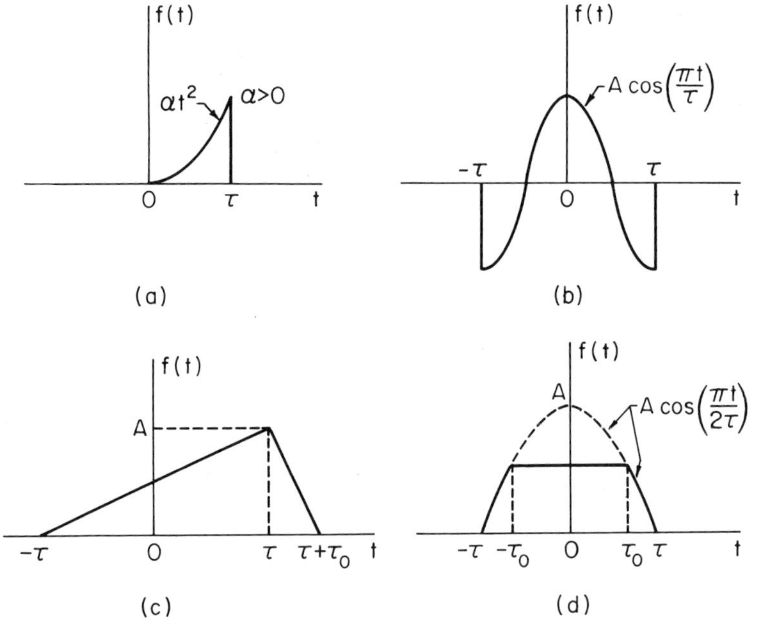

Fig. P2-24

2-25. Determine what the coefficient k must be in each of the following signals such that, if $\tau \to 0$, then $f(t) \to \delta(t)$ in the limit. (Hint: use tables for finding the area in (*a*), (*b*), and (*f*), and assume $\tau > 0$).

$$(a)\ f(t) = ke^{-(t/\tau)^2}.$$

ISBN 0-201-05758-1

(*b*) $f(t) = k\dfrac{\sin(t/\tau)}{(t/\tau)}$.

(*c*) $f(t) = k\left(1 - \dfrac{|t|}{\tau}\right), \quad |t| \leqslant \tau$

$= 0, \quad |t| > \tau.$

(*d*) $f(t) = ke^{-|t|/\tau}$.

(*e*) $f(t) = k \operatorname{rect}(t/\tau)$.

(*f*) $f(t) = k \operatorname{Sa}^2(t/\tau)$.

2-26. Find an expression for the total power in a periodic signal in terms of the complex Fourier series coefficients of (2.8-22). Check your result against (2.2-7).

2-27. Use the definition of the impulse function to evaluate the following integrals:

(*a*) $\displaystyle\int_{-\infty}^{\infty} \cos(9t)\delta(t - 2)\,dt.$

(*b*) $\displaystyle\int_{-\infty}^{\infty} \delta(t + 2)(1 + 4t + 8t^2)\,dt.$

(*c*) $\displaystyle\int_{-\infty}^{\infty} \delta(t - 3)e^{-8t^2}\,dt.$

(*d*) $\displaystyle\int_{-\infty}^{\infty} \delta(t)(1 + t)^{-1}\,dt.$

2-28. A signal's Fourier transform is shown in Fig. P2-28. Find the signal.

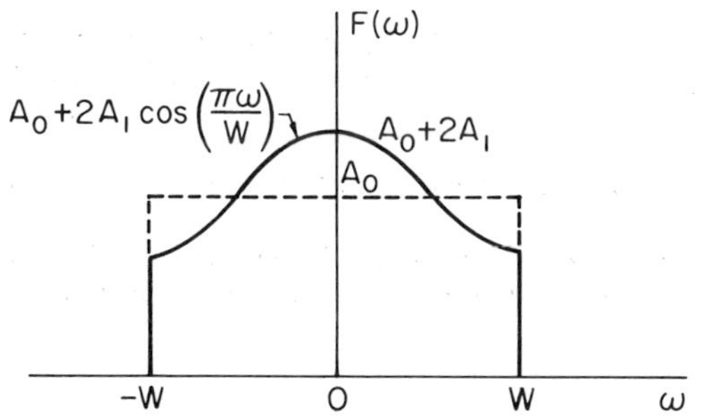

Fig. P2-28

ISBN 0-201-05758-1

2-29. Problem 2-28 may be generalized by assuming N terms. Thus, if the spectrum is

$$F(\omega) = A_0 + 2 \sum_{n=1}^{N} A_n \cos(n\pi\omega/W), \qquad -W < \omega < W$$

$$= 0, \qquad |\omega| > W,$$

find the signal $f(t)$.

2-30. Using the properties of the Fourier transform let

$$f(t) \leftrightarrow F(\omega)$$

and find the transforms of the following:

(*a*) $(t - 6)f(t - 3)$.

(*b*) $(t - 2)f(t)e^{j\omega_0 (t-3)}$.

(*c*) $t \dfrac{df(t)}{dt}$.

(*d*) $\dfrac{df(t)}{dt} e^{-j\omega_0 t}$.

(*e*) $\displaystyle\int_{-\infty}^{t+6} f(x)\, dx$.

(*f*) $3f(t) + 6 \dfrac{df(t)}{dt} + 2f(t - 1)$.

REFERENCES

[1] Wylie, C. R., Jr., *Advanced Engineering Mathematics,* McGraw-Hill Book Co., New York, 1960.
[2] Papoulis, A., *The Fourier Integral and Its Applications,* McGraw-Hill Book Co., New York, 1962.
[3] Campbell, G. A., and Foster, R. M., *Fourier Integrals for Practical Applications,* D. Van Nostrand Co., Princeton, New Jersey, 1948.
[4] Woodward, P. M., *Probability and Information Theory with Applications to Radar,* Pergamon Press, Oxford, 1960.
[5] Burdic, W. S., *Radar Signal Analysis,* Prentice-Hall, Englewood Cliffs, New Jersey, 1968.
[6] Hildebrand, F. B., *Introduction to Numerical Analysis,* McGraw-Hill Book Co., New York, 1956.

ISBN 0-201-05758-1

Chapter 3

Deterministic Signal Transfer through Networks

3.0. INTRODUCTION

In the previous chapter useful approaches were developed for characterizing signals through either time-domain or frequency-domain descriptions. In this chapter we shall see how these same principles can describe the response of a network to an applied signal. After all, the response is just another signal, so that the rules still apply. We shall consider only linear time-invariant networks—that is, those containing parameters that do not change with time. Since the networks are assumed to be linear, superposition will hold.

Just as there are two basic ways of describing a signal (time and frequency), there are two ways of approaching the analysis of network response. We shall present both of these. First will be the frequency-domain (spectral) method, since it is probably the simplest to understand. Second will be the time-domain (temporal) method, where the concept of convolution will be introduced. The general principles of both of these methods may be summarized with the help of Fig. 3.0-1.

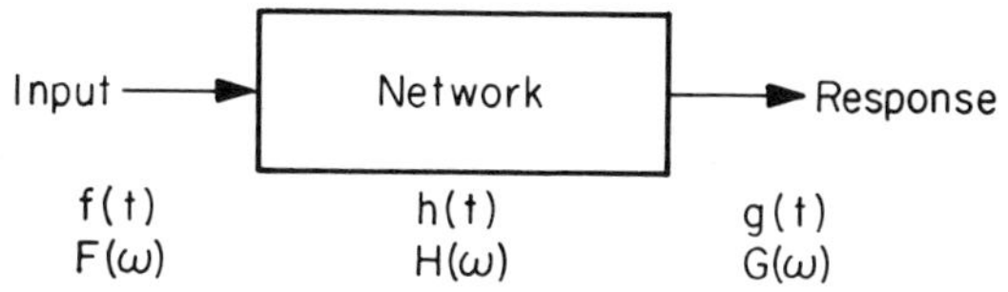

Fig. 3.0-1. General network.

At the input one may consider either the signal $f(t)$ applied (temporal) or its spectrum applied to the network. The use of temporal analysis will yield the output $g(t)$, and in the process we shall define $h(t)$, the characteristic or *impulse response* of the network. When we consider the input as a spectrum of frequencies we shall define a *network transfer function* $H(\omega)$, which will determine the output spectrum $G(\omega)$.

3.1. NETWORK RESPONSE BY SPECTRAL ANALYSIS

It will be convenient to approach the problem by first defining the transfer function, $H(\omega)$, and its symmetry for negative frequencies, since we have found it convenient to describe signals in terms of both positive and negative frequencies through the Fourier transform.

References start on p. 86.

ISBN 0-201-05758-1

Network Transfer Function and Symmetry

Consider a signal $f(t) = A\cos(\omega_0 t)$ applied to a network. At the output, the signal amplitude will be attenuated (or increased if gain is involved) by a factor α. The output signal's phase may also be shifted by an amount β. Both α and β are functions of frequency ω_0 for a general network, so that the output is

$$g(t) = A\alpha(\omega_0)\cos[\omega_0 t + \beta(\omega_0)]$$

$$= \frac{A}{2} e^{j\omega_0 t}\alpha(\omega_0)e^{j\beta(\omega_0)} + \frac{A}{2} e^{-j\omega_0 t}\alpha(\omega_0)e^{-j\beta(\omega_0)}. \tag{3.1-1}$$

In the last form of (3.1-1) we have expanded the cosine into its exponentials. If the same is done to represent $f(t)$, we may consider (3.1-1) as the response to the sum of two exponential signals, one, $f_1(t)$, at frequency ω_0 and the other $f_2(t)$, at the negative frequency $-\omega_0$. If we define the respective responses of (3.1-1) as $g_1(t)$ and $g_2(t)$, the network transfer function, $H(\omega_0)$, is the ratio of $g_1(t)$ to $f_1(t)$, or

$$H(\omega_0) = \frac{g_1(t)}{f_1(t)} = \alpha(\omega_0)e^{j\beta(\omega_0)}. \tag{3.1-2}$$

Now the second term in (3.1-1) can be used to find the behavior of $H(\omega_0)$ at the negative frequency $-\omega_0$. The transfer function for negative frequencies is given by

$$H(-\omega_0) = \frac{g_2(t)}{f_2(t)} = \alpha(\omega_0)e^{-j\beta(\omega_0)} = H^*(\omega_0). \tag{3.1-3}$$

Because ω_0 may have any arbitrary value, we may drop the subscripts to give the transfer function and its symmetry for any frequency ω.

$$H(\omega) = \alpha(\omega)e^{j\beta(\omega)}, \tag{3.1-4}$$

$$H(-\omega) = H^*(\omega). \tag{3.1-5}$$

Thus a network transfer function may be considered to exist at negative frequencies where its characteristics are defined by (3.1-5). It is worth noting two points concerning (3.1-5) that will come up later. First, $H(\omega)$, treated as a Fourier transform, satisfies the condition for being the spectrum of a real signal [see (2.3-26)]. Second, all the transmission characteristics at negative frequencies are known if they are known for positive frequencies. The latter result becomes important later in the definition of analytic signals.

Periodic Signal Input

The periodic signal, in its complex Fourier series form, is defined by

$$f(t) = \sum_{n=-\infty}^{\infty} C_n e^{jn\omega_T t} \leftrightarrow 2\pi \sum_{n=-\infty}^{\infty} C_n \delta(\omega - n\omega_T) = F(\omega). \tag{3.1-6}$$

ISBN 0-201-05758-1

Since this is just a sum of exponential inputs, the network output is

$$g(t) = \sum_{n=-\infty}^{\infty} C_n H(n\omega_T) e^{jn\omega_T t}. \tag{3.1-7}$$

The spectrum of the output signal, from use of (2.7-2), is

$$G(\omega) = 2\pi \sum_{n=-\infty}^{\infty} C_n H(n\omega_T)\delta(\omega - n\omega_T). \tag{3.1-8}$$

If we interpret $\delta(\omega - n\omega_T)$ as a function which "exists" only at $\omega = n\omega_T$, we may write (3.1-8) as

$$G(\omega) = F(\omega)H(\omega), \tag{3.1-9}$$

where $F(\omega)$ is the signal's spectrum.

Equation (3.1-9) is a basic result for the spectral analysis method. We show next that it applies equally well if the signal has a continuous spectrum.

Arbitrary Signal Input

When the Fourier transform of an arbitrary nonperiodic signal was developed in Chapter 2, the procedure was to obtain the limiting spectrum form of a periodic signal as the period $T = 2\pi/\omega_T$ was allowed to become infinite. We adopt the approach here and apply it to finding the spectrum of the output signal $g(t)$ in Fig. 3.0-1.

The output signal is given by (3.1-7). The Fourier series coefficients of the input signal are

$$C_n = \frac{1}{T}\int_{-T/2}^{T/2} f(t)e^{-jn\omega_T t}\, dt. \tag{3.1-10}$$

Placing (3.1-10) in (3.1-7) and rearranging terms gives

$$g(t) = \frac{1}{2\pi} \sum_{n=-\infty}^{\infty} \left\{ H(n\omega_T) \int_{-T/2}^{T/2} f(t)e^{-jn\omega_T t}\, dt \right\} e^{jn\omega_T t}\omega_T. \tag{3.1-11}$$

As before, we let $T \to \infty$, so that $n\omega_T \to \omega$ and $\omega_T \to d\omega$. Thus

$$g(t) = \frac{1}{2\pi}\int_{-\infty}^{\infty} \left\{H(\omega)F(\omega)\right\} e^{j\omega t}\, d\omega. \tag{3.1-12}$$

The braced quantity must be the spectrum of $g(t)$, so that (3.1-9) applies to any* signal of interest.

*We shall, throughout this chapter, be interested only in those signals that have Fourier transforms. See Section 2.3.

ISBN 0-201-05758-1

Example 3.1-1

As an example of the application of (3.1-9) we shall find the response of the network of Fig. 3.1-1 to a rectangular pulse of width τ. It can be shown (problem 3-5) that the network's transfer function is

$$H(\omega) = \frac{\sin(\omega\tau/2)}{\omega\tau/2} e^{-j\omega\tau/2}.$$

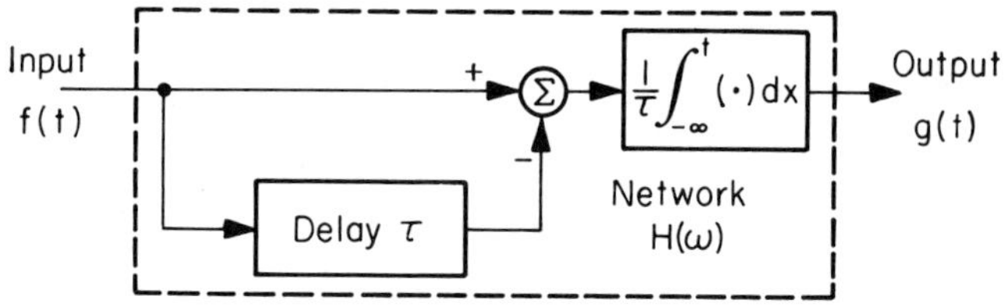

Fig. 3.1-1. A network.

The spectrum of a pulse of duration τ and amplitude A which is centered at the origin was found following example 2.3-1 of Chapter 2. It is

$$F(\omega) = A\tau \frac{\sin(\omega\tau/2)}{\omega\tau/2}.$$

The output spectrum is then

$$G(\omega) = A\tau \left[\frac{\sin(\omega\tau/2)}{\omega\tau/2}\right]^2 e^{-j\omega\tau/2} = A\tau \,\mathrm{Sa}^2\,(\omega\tau/2)e^{-j\omega\tau/2}.$$

The inverse Fourier transform of $A\tau\,\mathrm{Sa}^2\,(\omega\tau/2)$ was found earlier in example 2.3-2. The result was

$$A \,\mathrm{tri}\,(t/\tau) \leftrightarrow A\tau\,\mathrm{Sa}^2\,(\omega\tau/2).$$

Finally, we may use the time shifting property of Fourier transforms on this last result to find

$$g(t) = A \,\mathrm{tri}\left(\frac{t - \tau/2}{\tau}\right).$$

3.2. NETWORK RESPONSE BY TEMPORAL ANALYSIS

A network response that is fundamental to time-domain analysis is the impulse response.

Impulse Response

If an impulse is applied to a network, the output spectrum from (3.1-9) is

$$G(\omega) = H(\omega), \tag{3.2-1}$$

ISBN 0-201-05758-1

since the spectrum of the input impulse is unity—i.e., since

$$\delta(t) \leftrightarrow 1. \tag{3.2.2}$$

The output signal or *impulse response* $h(t)$ is the inverse Fourier transform of the output spectrum, giving

$$h(t) \leftrightarrow H(\omega). \tag{3.2-3}$$

We have obtained a very important result here. We have earlier shown (Section 2.5) that any signal may be represented by an infinite sum of impulses at times $n\,\Delta\tau$, where $\Delta\tau$ is the (small) time separation between impulses:

$$f(t) = \sum_{n=-\infty}^{\infty} f(n\,\Delta\tau)\delta(t - n\,\Delta\tau)\,\Delta\tau, \tag{3.2-4}$$

so that the response of the network to any signal may be found from the limit of the response to the sum of an infinite number of impulses.

General Response—Convolution

A component impulse, $\delta(t - n\,\Delta\tau)$, of (3.2-4) will cause an output response $h(t - n\,\Delta\tau)$. The response to $f(t)$ is then

$$g(t) = \sum_{n=-\infty}^{\infty} f(n\,\Delta\tau)h(t - n\,\Delta\tau)\,\Delta\tau. \tag{3.2-5}$$

Allowing $\Delta\tau \to 0$ will make the sum go over to an integral where we may let $n\,\Delta\tau \to \tau$, an arbitrary delay. The resulting output is

$$g(t) = \int_{-\infty}^{\infty} f(\tau)h(t - \tau)\,d\tau. \tag{3.2-6}$$

Equation (3.2-6) is known as the *convolution integral.* It is the fundamental analytical tool in the temporal analysis of linear time-invariant networks.

Another method may be used to derive (3.2-6) starting with the spectral method as a base. Since

$$g(t) = \frac{1}{2\pi}\int_{-\infty}^{\infty} F(\omega)H(\omega)e^{j\omega t}\,d\omega, \tag{3.2-7}$$

we may substitute the Fourier transforms,

$$F(\omega) = \int_{-\infty}^{\infty} f(\tau)e^{-j\omega\tau}\,d\tau, \tag{3.2-8}$$

ISBN 0-201-05758-1

$$H(\omega) = \int_{-\infty}^{\infty} h(x)\, e^{-j\omega x}\, dx, \tag{3.2-9}$$

and rearrange terms to obtain

$$g(t) = \int_{-\infty}^{\infty} f(\tau) \int_{-\infty}^{\infty} h(x) \left\{ \frac{1}{2\pi} \int_{-\infty}^{\infty} e^{j\omega(t-\tau-x)}\, d\omega \right\} dx\, d\tau. \tag{3.2-10}$$

The braced term is simply the inverse transform of the spectrum of an impulse; it is $\delta(t - \tau - x)$. The integration over x reduces, using the definition of the impulse function, to

$$g(t) = \int_{-\infty}^{\infty} f(\tau) \int_{-\infty}^{\infty} h(x)\delta(t - \tau - x)\, dx\, d\tau$$

$$= \int_{-\infty}^{\infty} f(\tau)h(t - \tau)\, d\tau, \tag{3.2-11}$$

leaving (3.2-6).

As a third and final way of obtaining (3.2-6) we could simply have used the convolution property of Fourier transforms for two signals as given by (2.4-15). By letting $f_1(t)$ represent $f(t)$, the input waveform, and letting $f_2(t)$ be the system impulse response $h(t)$, we obtain (3.2-6).

The convolution of two waveforms is sometimes represented symbolically by using an asterisk:

$$g(t) = f(t) * h(t), \tag{3.2-12}$$

where we define

$$f(t) * h(t) = \int_{-\infty}^{\infty} f(\tau)h(t - \tau)\, d\tau. \tag{3.2-13}$$

We may also write (3.2-11) as

$$g(t) = \int_{-\infty}^{\infty} h(\tau)f(t - \tau)\, d\tau. \tag{3.2-14}$$

Two simple changes in variable will show this fact.

Graphical Convolution

The convolution integral requires that one signal be (1) folded, (2) shifted, and (3) used to multiply the second signal; the product is then integrated to find the area, which equals $g(t)$ for a given shift t. Figure 3.2-1 illustrates the steps involved in graphically finding $g(t)$ for

ISBN 0-201-05758-1

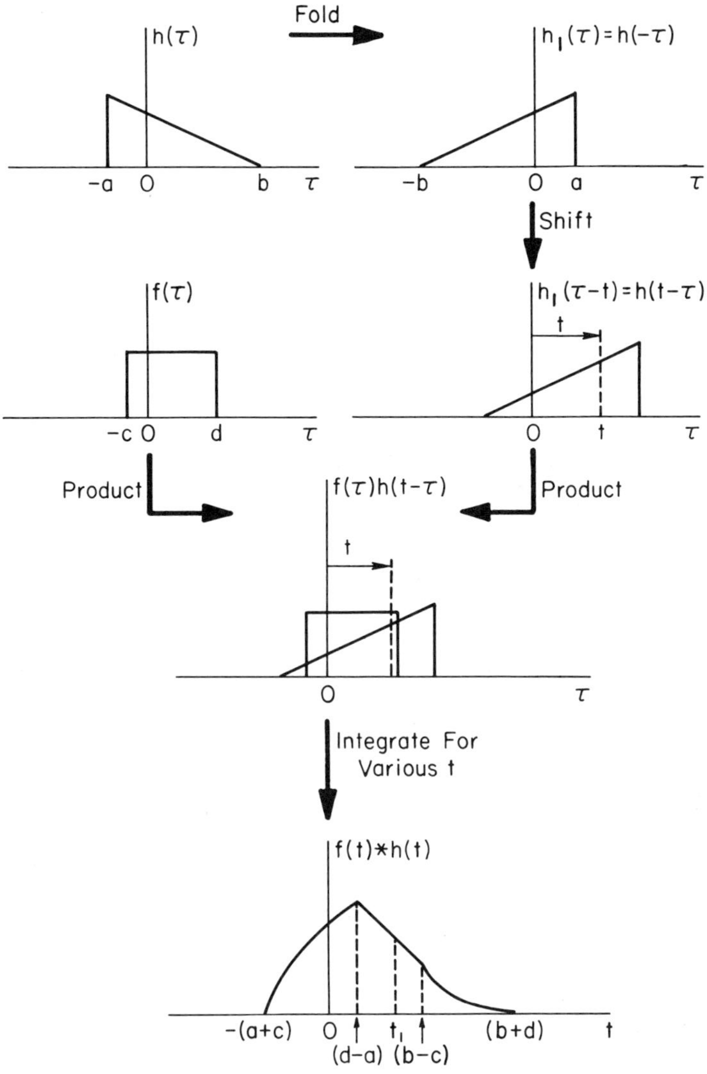

Fig. 3.2-1. Graphical convolution steps.

two waveforms. As a function of time, t, one may consider the folded waveform as sliding past the other waveform with the instantaneous area equaling $g(t)$.

Example 3.2-1

As an example which combines graphical interpretation with analytical approach let us find the convolution of two rectangular pulses as shown in Fig. 3.2-2(*a*) and defined by:

$$f_1(t) = A \text{ rect } (t/\tau_1),$$

$$f_2(t) = B \text{ rect } (t/\tau_2), \qquad \tau_2 < \tau_1.$$

We are interested, as found from graphical analysis, in three regions of time because of the

ISBN 0-201-05758-1

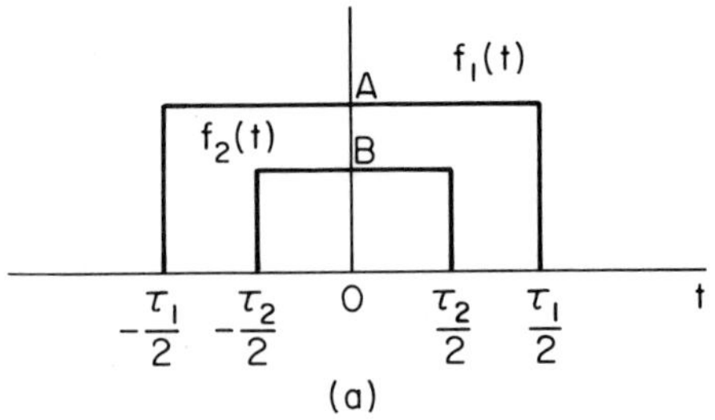

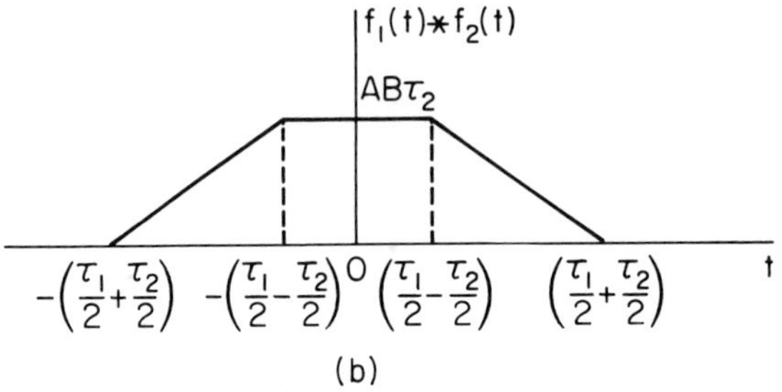

Fig. 3.2-2. Convolution of two rectangular pulses.

discontinuities in the waveforms. They are

$$\text{I:}\quad -\frac{\tau_1}{2} - \frac{\tau_2}{2} < t \leqslant -\frac{\tau_1}{2} + \frac{\tau_2}{2},$$

$$\text{II:}\quad -\frac{\tau_1}{2} + \frac{\tau_2}{2} < t \leqslant \frac{\tau_1}{2} - \frac{\tau_2}{2},$$

$$\text{III:}\quad \frac{\tau_1}{2} - \frac{\tau_2}{2} < t \leqslant \frac{\tau_1}{2} + \frac{\tau_2}{2}.$$

Because of symmetries we need analyze only region I and half of region II. In region I we have

$$\int_{-\infty}^{\infty} A \operatorname{rect}\left(\frac{x}{\tau_1}\right) B \operatorname{rect}\left(\frac{t-x}{\tau_2}\right) dx = \int_{-\tau_1/2}^{t+\tau_2/2} AB\, dx$$

$$= AB\left(t + \frac{\tau_1}{2} + \frac{\tau_2}{2}\right),$$

which increases linearly with t until the start of region II where graphically we see that $f_1(t)$ embraces all of $f_2(t)$, giving

$$\int_{-\infty}^{\infty} A \operatorname{rect}\left(\frac{x}{\tau_1}\right) B \operatorname{rect}\left(\frac{t-x}{\tau_2}\right) dx = \int_{t-\tau_2/2}^{t+\tau_2/2} AB\, dx = AB\tau_2,$$

a constant. The final plot of $f_1(t) * f_2(t)$ is shown in Fig. 3.2-2(b).

ISBN 0-201-05758-1

Algebraic Laws of Convolution

The usual commutative, distributive, and associative laws of algebra for multiplication apply to convolution. The commutative laws states that

$$f_1(t) * f_2(t) = f_2(t) * f_1(t). \tag{3.2-15}$$

This law may be proved by formally writing each side in integral form and using a suitable variable change.

For the distributive law:

$$f_1(t) * [f_2(t) + f_3(t)] = f_1(t) * f_2(t) + f_1(t) * f_3(t). \tag{3.2-16}$$

For the associative law:

$$f_1(t) * [f_2(t) * f_3(t)] = [f_1(t) * f_2(t)] * f_3(t). \tag{3.2-17}$$

Proof of these two laws is suggested as a reader exercise.

Even and Odd Symmetry in Convolution

Let $f_1(t)$ and $f_2(t)$ be real waveforms which may each be separated into even and odd parts according to the method of Section 2.1, which applies to any signal. The convolution $g(t)$ of $f_1(t)$ and $f_2(t)$ is

$$\begin{aligned} g(t) = f_1(t) * f_2(t) &= [f_{1e}(t) + f_{1o}(t)] * [f_{2e}(t) + f_{2o}(t)] \\ &= [f_{1e}(t) * f_{2e}(t)] + [f_{1o}(t) * f_{2o}(t)] \\ &+ [f_{1e}(t) * f_{2o}(t)] + [f_{1o}(t) * f_{2e}(t)], \end{aligned} \tag{3.2-18}$$

where

$$f_1(t) = f_{1e}(t) + f_{1o}(t), \tag{3.2-19}$$

$$f_2(t) = f_{2e}(t) + f_{2o}(t). \tag{3.2-20}$$

Examination of (3.2-18) will verify that if $f_1(t)$ and $f_2(t)$ both have even symmetry their convolution $g(t)$ will have even symmetry. If both have odd symmetry $g(t)$ will have even symmetry. However, if one of the two signals, $f_1(t)$ or $f_2(t)$, has even symmetry while the other is odd, then $g(t)$ will be odd. Similar statements apply to $g(t)$ when $f_1(t)$ and $f_2(t)$ are complex.

3.3. SUMMARY OF NETWORK RESPONSES

In summary, two basic ways were given for obtaining the response $g(t)$ of a network defined as in Fig. 3.0-1. One may arrive at $g(t)$ directly by using convolution (temporal analysis) if the input signal $f(t)$ and the network's impulse response $h(t)$ are given. On the other hand, if

ISBN 0-201-05758-1

the signal spectrum $F(\omega)$ and network transfer function $H(\omega)$ are specified instead, $g(t)$ is most easily found by first finding $G(\omega)$ by spectral analysis [product of $F(\omega)$ and $H(\omega)$] and then inverse Fourier transforming. In most cases these procedures are the easiest ones to follow in a given problem, since only one integration is involved in each case.

If *both* signal and network are not similarly specified, the solution method will not always be clear. For example, if $f(t)$ and $H(\omega)$ are given, the spectral method requires first transforming $f(t)$ to obtain $F(\omega)$ and then inverse transforming $G(\omega)$ to obtain $g(t)$. Two integrations (transformations) are involved with this approach. Alternatively, we could use temporal analysis. Here we would first inverse transform $H(\omega)$ to find $h(t)$ and then use convolution to obtain $g(t)$. Again two integrations are involved. Which method is the easiest may depend on the form of the given functions.

3.4. SPECIAL SYSTEM AND SIGNAL TOPICS

Distortionless System

If a system is required to pass a signal $f(t)$ without distortion, the output must be an exact replica of the input. Of course, the response may have a different magnitude than the input and may even occur at a different instant in time. In other words, the filter may scale the input by a constant K and shift it in time by an amount τ, but *it cannot change its form.* Figure 3.4-1 illustrates a distortionless system.

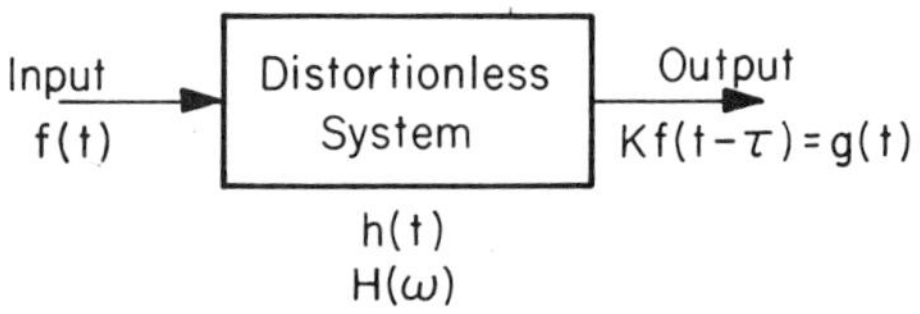

Fig. 3.4-1. Distortionless system.

If the system input is described by

$$f(t) \leftrightarrow F(\omega), \tag{3.4-1}$$

the time shifting property of Fourier transforms will give us the output spectrum $G(\omega)$:

$$Kf(t - \tau) = g(t) \leftrightarrow G(\omega) = KF(\omega)e^{-j\omega\tau}. \tag{3.4-2}$$

By spectral analysis $G(\omega) = F(\omega)H(\omega)$. Thus the filter's transfer function must be

$$H(\omega) = Ke^{-j\omega\tau} = |H(\omega)|e^{j\theta(\omega)} \tag{3.4-3}$$

for a distortion-free output. This filter characteristic has a constant magnitude K at all frequencies and a linear phase, $\theta(\omega) = -\omega\tau$, as shown in Fig. 3.4-2. The constant K may be positive or negative without changing the distortion-free transmission of the filter.

Well-known examples of practical systems that approach a distortionless transmission are waveguides, transmission lines, quality delay lines, and some amplifiers.

ISBN 0-201-05758-1

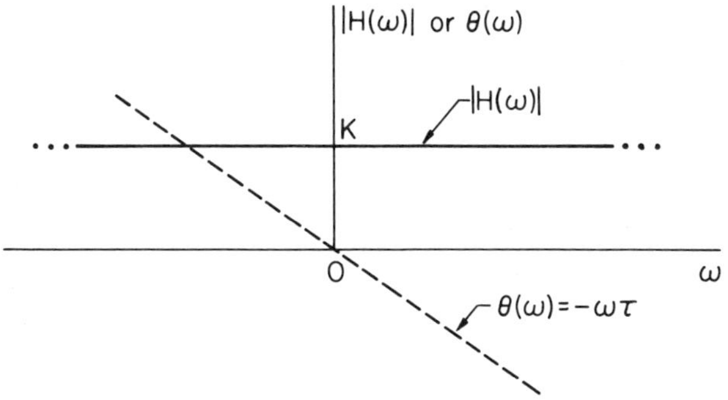

Fig. 3.4-2. Distortionless filter transfer function.

3-Decibel Bandwidth

In most cases when we speak of system bandwidth we mean 3-dB bandwidth. It is defined as the frequency interval over which the system transfer function magnitude $|H(\omega)|$ remains above $1/\sqrt{2}$ times its mid-band value. Figure 3.4-3 illustrates 3-dB bandwidth definitions for low-pass and band-pass systems. We use W for bandwidth in radians per second or B if in hertz. Thus,

$$W = 2\pi B. \tag{3.4-4}$$

The 3-dB bandwidth of a signal is defined on a similar basis using the magnitude of the signal spectrum $|F(\omega)|$.

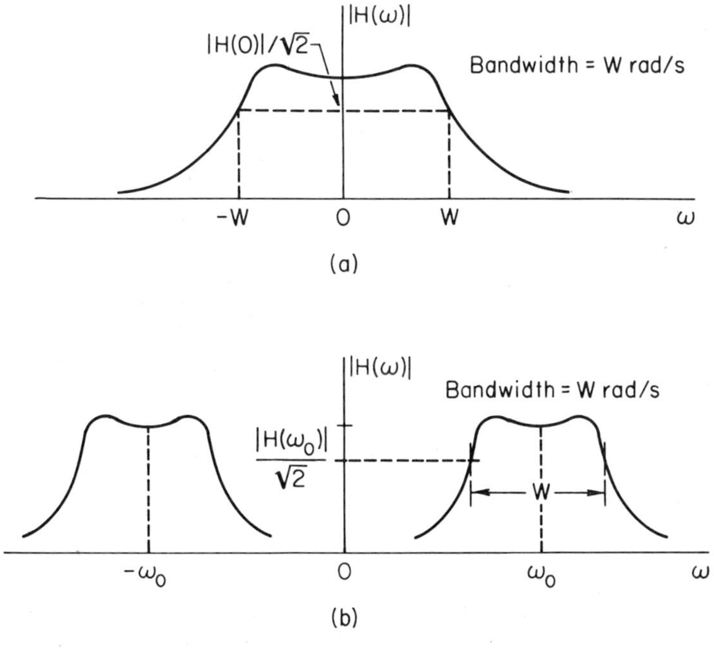

Fig. 3.4-3. Filters with bandwidth W: (a) Low-pass, and (b) band-pass.

ISBN 0-201-05758-1

For many applications more quantitative definitions of bandwidth are required; these are discussed in later paragraphs.

Ideal Filters

To simplify analysis and facilitate solution of some otherwise difficult problems, it is often convenient to use the concept of an *ideal filter.* We define an ideal filter as one that has unity gain and a perfectly flat response at frequencies inside a desired interval but zero response at frequencies outside the interval. This definition is visualized by use of Fig. 3.4-4, which illustrates low-pass (*a*), high-pass (*b*), and band-pass (*c*) ideal filters. In each case the filter may have a linear phase component $\theta(\omega) = -\omega\tau$, where τ is the *filter's delay constant.*

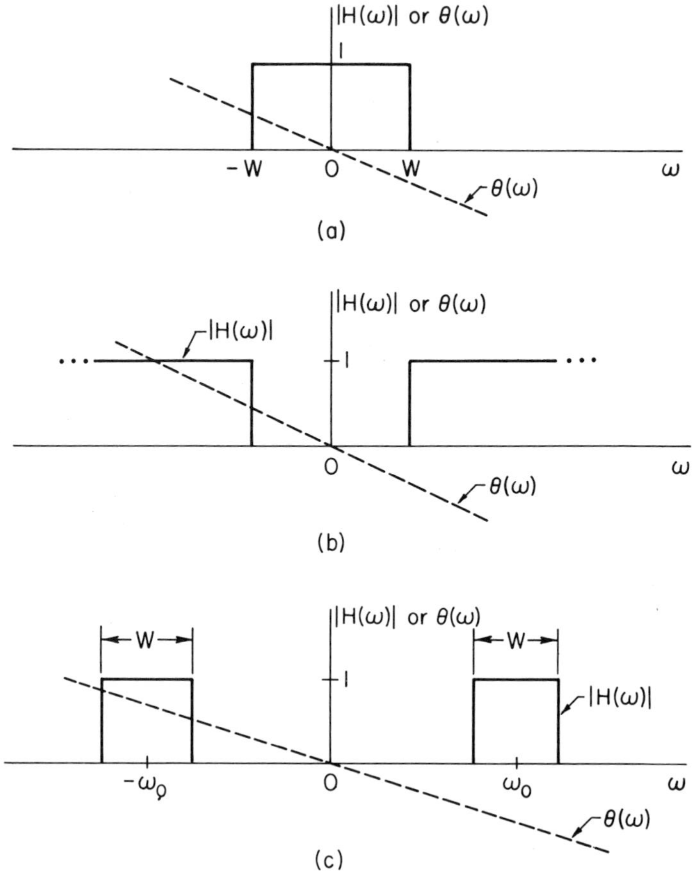

Fig. 3.4-4. Ideal filter transfer functions. (a) Low-pass, (b) high-pass, and (c) band-pass.

System Rise Time

A quantity called system *rise time* will sometimes be encountered in later work. Although we shall define the meaning of rise time precisely for our later use, it should be noted that there is no universally acceptable definition. Indeed, several are in use, and it may be helpful to place rise time in perspective by discussing some of these definitions.

ISBN 0-201-05758-1

Broadly stated, rise time relates to how rapidly a system can respond to an input signal, usually taken as a step function in the case of a low-pass system (for a band-pass system it would be a step of carrier at the center-band frequency).

Suppose we consider how fast the simplest low-pass system can respond to a step input by means of an example.

Example 3.4-1

Let the response $g(t)$ of a simple low-pass filter be found for a unit step input signal. The network transfer function is taken as

$$H(\omega) = \frac{1}{1 + j(\omega/W)},$$

where W is the 3-dB bandwidth. Using temporal analysis we need $h(t)$ the impulse response. From pair 29 of Table 2.10-1,

$$h(t) = u(t) We^{-Wt}.$$

The convolution integral now gives

$$g(t) = \int_{-\infty}^{\infty} u(x) We^{-Wx}\, u(t - x)\, dx$$

$$= \int_0^t We^{-Wx}\, dx = 1 - e^{-Wt}, \qquad 0 < t.$$

The response of the low-pass system is sketched in Fig. 3.4-5. The time required to attain 90% of its final value is often called rise time t_r. An easy calculation shows that

$$t_r = 2.303/W \tag{3.4-5}$$

or

$$t_r = 0.366/B. \tag{3.4-6}$$

Incidentally, the time required for the response to attain half its final value is sometimes called the system delay t_d. For the simple low-pass system

$$t_d = 0.693/W = 0.110/B. \tag{3.4-7}$$

If the system is more complicated than a simple low-pass case (but having the same bandwidth), 90% rise time will tend to be somewhat larger than that given by (3.4-5). However, regardless of the complexity of the low-pass system, t_r does not differ too greatly from π/W. To illustrate this fact, we observe that the slope of $|H(\omega)|$ for a more complicated system becomes steeper* in the vicinity of the 3-dB points. Such a system is said to have a sharp

*From network theory the fall-off rate with frequency increases with the number of network poles—that is, with increasing network complexity.

ISBN 0-201-05758-1

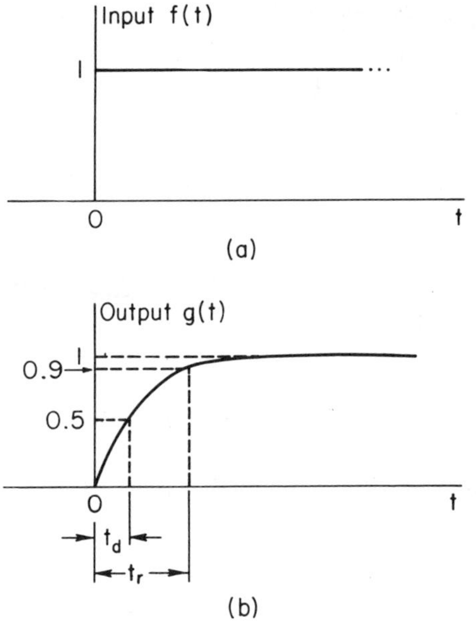

Fig. 3.4-5. Response (b) of a simple low-pass system to a unit step input signal (a).

cutoff transfer function, and the ideal low-pass filter may be used as a reasonable approximation to it. We proceed then to find the step response of an ideal low-pass filter. Its transfer function is

$$H(\omega) = Ke^{-j\omega t_d}, \qquad |\omega| < W$$
$$= 0, \qquad |\omega| > W. \tag{3.4-8}$$

K and t_d are chosen to approximate the center-band gain and phase slope of the actual filter, respectively.

The response of the approximate filter to a unit step input is (using spectral analysis)

$$g(t) = \frac{1}{2\pi}\int_{-\infty}^{\infty} H(\omega)\left[\pi\delta(\omega) + \frac{1}{j\omega}\right] e^{j\omega t}\, d\omega$$

$$= \frac{1}{2}\int_{-W}^{W} Ke^{-j\omega t_d + j\omega t}\,\delta(\omega)\, d\omega$$

$$+ \frac{1}{2\pi}\int_{-W}^{W} K\,\frac{e^{-j\omega t_d + j\omega t}}{j\omega}\, d\omega. \tag{3.4-9}$$

The first integral easily reduces to $K/2$ using (2.5-1). The second, after using the trigonometric expansion of the exponential, will reduce such that $g(t)$ may be written

$$g(t) = \frac{K}{2} + \frac{K}{\pi}\int_{0}^{W(t-t_d)} \frac{\sin(u)}{u}\, du. \tag{3.4-10}$$

ISBN 0-201-05758-1

The integral has no known closed form but is tabulated [1] as the *sine integral* in many places, which is defined as

$$\mathrm{Si}\,(x) = \int_0^x \frac{\sin\,(u)}{u}\,du. \tag{3.4-11}$$

In terms of the sine integral, which is plotted in Fig. 3.4-6(*a*), we have

$$g(t) = K\left\{\frac{1}{2} + \frac{1}{\pi}\,\mathrm{Si}\,[W(t - t_d)]\right\}, \tag{3.4-12}$$

which is shown in (*b*).

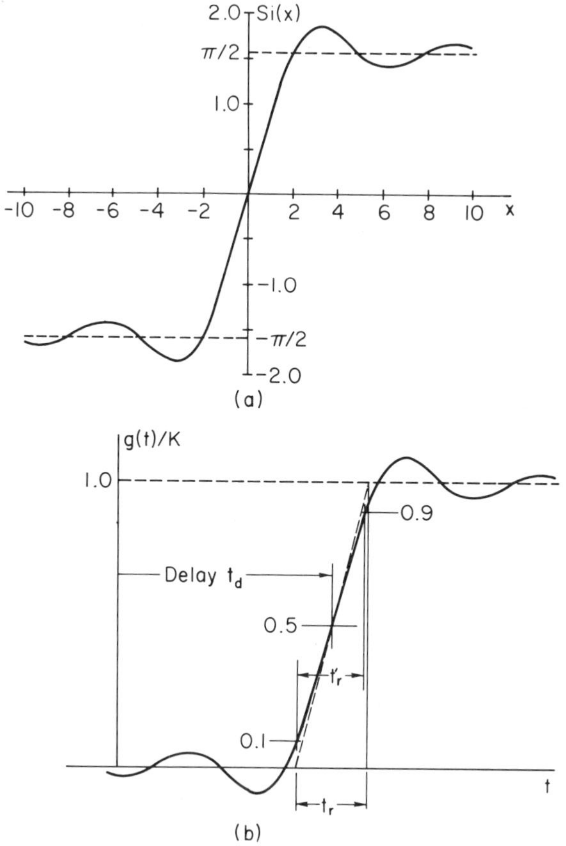

Fig. 3.4-6. Response of an ideal low-pass filter (b) as related to the sine integral (a).

Using Fig. 3.4-6(*b*), rise time may be found in many ways. In one way we could call it the time between where the response crosses zero for the last time and where it rises to unity for the first time. This time is $3.877/W = 0.617/B$. Another way might be the 10 to 90% time* t'_r as shown. It equals $2.802/W = 0.446/B$. All these definitions are not too different from $0.5/B$, and all are proportional to $1/B$. The definition we shall use in this text is the reciprocal of the signal's slope at its half-amplitude point. This time is

*The 0 to 90% time is $3.343/W = 0.532/B$.

ISBN 0-201-05758-1

$$t_r = \pi/W = 0.5/B. \tag{3.4-13}$$

From the above discussions we see that network rise time is proportional to the reciprocal of the 3-dB bandwidth. The proportionality constant may sometimes be less than 0.5, such as with simple systems, or larger in complicated systems, but always near 0.5. Thus, choice of (3.4-13) to represent rise time is a reasonable value for most systems.

Mean Frequency, RMS Bandwidth and Duration

Let $F(\omega)$ represent either the spectrum of a low-pass signal or the transfer function of a low-pass system, and assume $|F(\omega)|$ has even symmetry† about $\omega = 0$. We define root-mean-square (rms) bandwidth W_{rms} by

$$W_{rms}^2 = \frac{\int_{-\infty}^{\infty} \omega^2 |F(\omega)|^2 \, d\omega}{\int_{-\infty}^{\infty} |F(\omega)|^2 \, d\omega} = \frac{\int_{0}^{\infty} \omega^2 |F(\omega)|^2 \, d\omega}{\int_{0}^{\infty} |F(\omega)|^2 \, d\omega}. \tag{3.4-14}$$

Similarly, the rms duration of a signal $f(t)$ for which $|f(t)|$ has even symmetry about $t = 0$ is‡

$$\tau_{rms}^2 = \frac{\int_{-\infty}^{\infty} t^2 |f(t)|^2 \, dt}{\int_{-\infty}^{\infty} |f(t)|^2 \, dt}. \tag{3.4-15}$$

The reader may recognize these expressions as normalized second moments about the origins. They are measures of the spread of the spectrum or time function.

If $F(\omega)$ represents either a band-pass signal spectrum or a band-pass transfer function for which $|F(\omega)|$ has even symmetry about $\omega = 0$, we define

$$W_{rms}^2 = \frac{4\int_{0}^{\infty} (\omega - \bar{\omega}_0)^2 |F(\omega)|^2 \, d\omega}{\int_{0}^{\infty} |F(\omega)|^2 \, d\omega}, \tag{3.4-16}$$

where the *mean frequency* $\bar{\omega}_0$ is defined as

$$\bar{\omega}_0 = \frac{\int_{0}^{\infty} \omega |F(\omega)|^2 \, d\omega}{\int_{0}^{\infty} |F(\omega)|^2 \, d\omega}. \tag{3.4-17}$$

†All real low-pass signals and all realizable low-pass networks satisfy this condition.

‡Equation (3.4-15) also applies for arbitrary symmetry if the time origin is at the centroid of $|f(t)|^2$.

ISBN 0-201-05758-1

For most narrowband signals $\bar{\omega}_0$ will be very nearly the same as the frequency of the carrier signal involved in modulation.

Rms bandwidth is important in analysis of accuracy with which the time of occurrence of certain communication signals may be measured. Rms duration enters into accuracy analysis where one wishes to measure the carrier frequency of a waveform.

Uncertainty Principle for RMS Quantities

We state an uncertainty principle for rms bandwidth and time duration: The product of rms bandwidth and rms time duration for a low-pass waveform is never less than ½. That is,

$$W_{\mathrm{rms}}\tau_{\mathrm{rms}} \geqslant \frac{1}{2}. \tag{3.4-18}$$

The proof of this principle begins by using the *Schwarz inequality* which states that,* for two possibly complex waveforms $f_1(t)$ and $f_2(t)$,

$$\left[\mathrm{Re} \int_{-\infty}^{\infty} f_1(t) f_2^*(t)\, dt\right]^2 \leqslant \int_{-\infty}^{\infty} |f_1(t)|^2\, dt \int_{-\infty}^{\infty} |f_2(t)|^2\, dt, \tag{3.4-19}$$

where Re $(\cdot)$ stands for real part. The equality holds only if $f_1(t)$ is proportional to $f_2(t)$. Now if we substitute

$$f_1(t) = tf(t), \tag{3.4-20}$$

$$f_2(t) = \frac{df(t)}{dt} \tag{3.4-21}$$

into (3.4-19), we find that the left side may be integrated by parts. Carrying out the integration and assuming that $|f(t)| \to 0$ faster than $1/\sqrt{|t|}$ as $|t| \to \infty$, we have

$$\frac{1}{4}\left[\int_{-\infty}^{\infty} |f(t)|^2\, dt\right]^2 \leqslant \int_{-\infty}^{\infty} t^2 |f(t)|^2\, dt \int_{-\infty}^{\infty} \left|\frac{df(t)}{dt}\right|^2 dt. \tag{3.4-22}$$

Next we recognize that $df(t)/dt \leftrightarrow j\omega F(\omega)$ and replace the second right-side integral by its frequency-domain equivalent using Parseval's theorem (2.4-25). Equation (3.4-22) may now be written as

$$\left[\frac{\displaystyle\int_{-\infty}^{\infty} t^2 |f(t)|^2\, dt}{\displaystyle\int_{-\infty}^{\infty} |f(t)|^2\, dt}\right]\left[\frac{\displaystyle\frac{1}{2\pi}\int_{-\infty}^{\infty} \omega^2 |F(\omega)|^2\, d\omega}{\displaystyle\int_{-\infty}^{\infty} |f(t)|^2\, dt}\right] \geqslant \frac{1}{4}. \tag{3.4-23}$$

The last step in the proof is to apply Parseval's theorem again to the denominator of the

*An alternative form is given in (4.12-6).

ISBN 0-201-05758-1

second left-side factor to get

$$\frac{\int_{-\infty}^{\infty} t^2 |f(t)|^2 \, dt}{\int_{-\infty}^{\infty} |f(t)|^2 \, dt} \cdot \frac{\int_{-\infty}^{\infty} \omega^2 |F(\omega)|^2 \, d\omega}{\int_{-\infty}^{\infty} |F(\omega)|^2 \, d\omega} \geqslant \frac{1}{4}, \tag{3.4-24}$$

which proves (3.4-18), since the left side is simply $W_{\text{rms}}^2 \tau_{\text{rms}}^2$.

From steps in the above development the reader may wish to verify the following alternative definitions for W_{rms} and τ_{rms}:

$$W_{\text{rms}}^2 = \frac{\int_{-\infty}^{\infty} |df(t)/dt|^2 \, dt}{\int_{-\infty}^{\infty} |f(t)|^2 \, dt}, \tag{3.4-25}$$

$$\tau_{\text{rms}}^2 = \frac{\int_{-\infty}^{\infty} |dF(\omega)/d\omega|^2 \, d\omega}{\int_{-\infty}^{\infty} |F(\omega)|^2 \, d\omega}. \tag{3.4-26}$$

Equivalent Bandwidth and Duration

In Chapter 7 we shall have occasion to use a quantity called *equivalent duration* τ_{eq} for a pulsed waveform $f(t)$. It is defined by

$$\tau_{\text{eq}} = \frac{\int_{-\infty}^{\infty} f(t) \, dt}{f_{\text{max}}}, \tag{3.4-27}$$

where f_{max} represents the maximum magnitude of $f(t)$.

The analogous spectral quantity *equivalent bandwidth* W_{eq} is defined by

$$W_{\text{eq}} = \frac{\int_{-\infty}^{\infty} |F(\omega)| \, d\omega}{2F(0)}. \tag{3.4-28}$$

ISBN 0-201-05758-1

Uncertainty Principle for Equivalent Quantities

An uncertainty principle may also be stated in terms of equivalent bandwidth and time duration. It is

$$W_{eq}\tau_{eq} \geqslant \pi. \tag{3.4-29}$$

The proof results from working with the right side of (3.4-27). From the Fourier transform for $\omega = 0$,

$$F(0) = \int_{-\infty}^{\infty} f(t)\, dt. \tag{3.4-30}$$

The right-side numerator then becomes simply $F(0)$. Next, we apply the general integral relationship

$$\left| \int f(x)\, dx \right| \leqslant \int |f(x)|\, dx \tag{3.4-31}$$

to the inverse Fourier transform expression to obtain

$$|f(t)| \leqslant \frac{1}{2\pi} \int_{-\infty}^{\infty} |F(\omega)|\, d\omega, \tag{3.4-32}$$

which is true for any $|f(t)|$ including f_{max}. Thus, since $|f(t)| \leqslant f_{max}$, we use (3.4-32) and (3.4-30) in (3.4-27) to get

$$\tau_{eq} = \frac{F(0)}{f_{max}} \geqslant \frac{F(0)}{(1/2\pi) \int_{-\infty}^{\infty} |F(\omega)|\, d\omega} = \frac{\pi}{W_{eq}}, \tag{3.4-33}$$

which proves (3.4-29).

Effective Bandwidth and Duration

We complete our discussions of the various types of bandwidth and time definitions by including *effective bandwidth* W_e and *effective duration* τ_e applicable to a low-pass function $f(t)$ having a Fourier transform $F(\omega)$. They are [2]

$$W_e = \frac{\left[\int_{-\infty}^{\infty} |F(\omega)|^2\, d\omega \right]^2}{\int_{-\infty}^{\infty} |F(\omega)|^4\, d\omega}, \tag{3.4-34}$$

ISBN 0-201-05758-1

$$\tau_e = \frac{\left[\int_{-\infty}^{\infty} |f(t)|^2 \, dt\right]^2}{\int_{-\infty}^{\infty} |f(t)|^4 \, dt}. \tag{3.4-35}$$

W_e and τ_e arise in analyses related to the resolution (separation ability) of waveforms in time or frequency, respectively. In fact, a *time resolution constant* T_r and a *frequency resolution constant* W_r (radians per second) are sometimes used and are defined by

$$T_r = 2\pi/W_e, \tag{3.4-36}$$

$$W_r = 2\pi/\tau_e. \tag{3.4-37}$$

We shall actually have little occasion to use effective quantities. They are included to emphasize that bandwidth has many definitions. In fact, we shall consider still another definition in Chapter 4 called *noise bandwidth.*

Realizability

Let an impulse be applied at time $t = 0$ to a system with transfer function $H(\omega)$. The system response is $h(t)$. If $h(t) = 0$ for $t < 0$, $h(t)$ is said to be a *causal* impulse response. For a system to be realizable it must have a causal impulse response.

The Paley-Wiener [3] criterion provides us with a means of knowing whether or not the magnitude function, $|H(\omega)|$, corresponds to some causal signal. We summarize the criterion: If

$$\int_{-\infty}^{\infty} |H(\omega)|^2 \, d\omega < \infty, \tag{3.4-38}$$

then the Paley-Wiener criterion,

$$\int_{-\infty}^{\infty} \frac{\big|\ln|H(\omega)|\big|}{1 + \omega^2} \, d\omega < \infty, \tag{3.4-39}$$

is a necessary and sufficient condition for $|H(\omega)|$ to be the spectrum magnitude of a causal signal. The criterion only assures us that some phase may be associated with $|H(\omega)|$ such that a causal inverse Fourier transform exists [4]; it does not give the phase.

Some meaningful results [5] may be obtained from (3.4-39). First, $|H(\omega)|$ may be zero at discrete points in frequency but cannot be zero over a finite *band* of frequencies. This result means that no filter having a stop-band with infinite attenuation can ever be constructed. All the ideal filters discussed above fall in this category and consequently are not realizable.

Finally, $|H(\omega)|$ may not fall off faster than a function of exponential order [5]. This is observed by noting that

$$|H(\omega)| = Ke^{-a|\omega|} \tag{3.4-40}$$

ISBN 0-201-05758-1

is allowed in (3.4-39), where a is a real constant, but

$$|H(\omega)| = Ke^{-a\omega^2} \qquad (3.4\text{-}41)$$

will cause the integral to become infinite.

Although ideal filters cannot be realized, their spectrum magnitude may be approached by use of a high-order maximally flat *Butterworth filter** design. As the order becomes very large, the approximation becomes quite excellent. Unfortunately, as the order increases so does the time delay, so that in the limit we may approach an ideal filter but the signal never comes out.

3.5. ENERGY AND POWER DENSITY SPECTRA OF NETWORK RESPONSE

In Sections 2.9 and 2.8 we found the energy and power density spectra of energy and power signals, respectively. We now find the same quantities for the response signal of a network.

Energy Density Spectrum

Since the energy density spectrum of a Fourier-transformable signal is defined as the squared spectrum magnitude, the energy density spectra of the input signal $f(t)$ and output signal $g(t)$ are, respectively,

$$\mathcal{E}_f(\omega) = |F(\omega)|^2, \qquad (3.5\text{-}1)$$

$$\mathcal{E}_g(\omega) = |G(\omega)|^2. \qquad (3.5\text{-}2)$$

However, since

$$G(\omega) = F(\omega)H(\omega), \qquad (3.5\text{-}3)$$

where $H(\omega)$ is the network transfer function, we have

$$\mathcal{E}_g(\omega) = |F(\omega)H(\omega)|^2 = \mathcal{E}_f(\omega)|H(\omega)|^2. \qquad (3.5\text{-}4)$$

Equation (3.5-4) states that the energy density spectrum of the output signal is the product of the input signal energy density spectrum and the squared magnitude of the network transfer function.

For real output signals the total energy E in the response is

$$E = \frac{1}{2\pi}\int_{-\infty}^{\infty} \mathcal{E}_f(\omega)|H(\omega)|^2\, d\omega$$

$$= 2\int_0^{\infty} \mathcal{E}_f(2\pi f)|H(2\pi f)|^2\, df. \qquad (3.5\text{-}5)$$

*A Butterworth filter is one for which $|H(\omega)|^2$ is the nearest possible to a constant at band center. (See problem 3-9.) For a filter with N poles, the first $(N-1)$ derivatives of $|H(\omega)|^2$ with respect to ω^2 are all zero at band center.

ISBN 0-201-05758-1

Power Density Spectrum

Suppose a signal $f(t)$, applied to a network having a transfer function $H(\omega)$, invokes a response $g(t)$. If $f_T(t)$ and $g_T(t)$ represent $f(t)$ and $g(t)$ truncated to the time interval $-T < t < T$ and if $F_T(\omega)$ and $G_T(\omega)$ are their Fourier transforms, the power density spectra of $f(t)$ and $g(t)$ are

$$\mathcal{S}_f(\omega) = \lim_{T\to\infty} \left\{ \frac{|F_T(\omega)|^2}{2T} \right\}, \tag{3.5-6}$$

$$\mathcal{S}_g(\omega) = \lim_{T\to\infty} \left\{ \frac{|G_T(\omega)|^2}{2T} \right\}, \tag{3.5-7}$$

as defined in Section 2.8. Now $f_T(t)$ may be taken as applied to the network and the response found by letting $T \to \infty$. In general $g_T(t)$ is *not* the response to $f_T(t)$. However, for a stable network the response for $t > T$ will decay and become negligible in relation to $g_T(t)$ for $T \to \infty$. A similar argument allows the response for $-T > t$ relative to the portion of $f(t)$ following $t = -T$ to be neglected. Thus, for large enough T we may assume that $g_T(t)$ is the response to $f_T(t)$ with negligible error. This fact allows us to write (3.5-3) in the form

$$G_T(\omega) = F_T(\omega)H(\omega). \tag{3.5-8}$$

Equation (3.5-7) becomes

$$\begin{aligned} \mathcal{S}_g(\omega) &= \lim_{T\to\infty} \left\{ \frac{|F_T(\omega)H(\omega)|^2}{2T} \right\} \\ &= \lim_{T\to\infty} \left\{ \frac{|F_T(\omega)|^2}{2T} \right\} |H(\omega)|^2 \end{aligned} \tag{3.5-9}$$

or

$$\mathcal{S}_g(\omega) = \mathcal{S}_f(\omega)\ |H(\omega)|^2, \tag{3.5-10}$$

since $H(\omega)$ doesn't depend on T. Our result shows that the power spectrum of the output response is the product of the input signal power spectrum and the squared magnitude of the network transfer function.

For real output signals the total average power P is

$$\begin{aligned} P &= \frac{1}{2\pi} \int_{-\infty}^{\infty} \mathcal{S}_f(\omega)|H(\omega)|^2\, d\omega \\ &= 2 \int_0^{\infty} \mathcal{S}_f(2\pi f)|H(2\pi f)|^2\, df. \end{aligned} \tag{3.5-11}$$

As noted in Chapter 2, a noise signal possesses a power density spectrum. Therefore,

ISBN 0-201-05758-1

(3.5-11) is an important result, since it allows us to find an average (power) response of networks to noise* as well as to deterministic signals.

3.6. SPECTRUM AND SIGNAL BOUNDS

In some cases we may not need to know (or possibly cannot obtain) the spectrum of a signal. It may be sufficient to know bounds on the spectrum magnitude. Using the fact that the nth derivative, $n = 0, 1, \ldots$, of a signal $f(t)$ has the transform $(j\omega)^n F(\omega)$, or

$$\frac{d^n f(t)}{dt^n} \leftrightarrow (j\omega)^n F(\omega), \tag{3.6-1}$$

if such a transform exists, we may find bounds on $|F(\omega)|$.

Spectrum Bounds

A result [2] which may be obtained from the Schwarz inequality is

$$\left| \int_{-\infty}^{\infty} q(x)\, dx \right| \leqslant \int_{-\infty}^{\infty} |q(x)|\, dx, \tag{3.6-2}$$

where $q(x)$ is an arbitrary function. Writing (3.6-1) as

$$(j\omega)^n F(\omega) = \int_{-\infty}^{\infty} \frac{d^n f(t)}{dt^n} e^{-j\omega t}\, dt \tag{3.6-3}$$

and using (3.6-2) produces

$$|F(\omega)| \leqslant \frac{1}{|\omega|^n} \int_{-\infty}^{\infty} \left| \frac{d^n f(t)}{dt^n} \right| dt, \qquad n = 0, 1, \ldots . \tag{3.6-4}$$

Equation (3.6-4) may be considered as the $(n + 1)$th bound on $|F(\omega)|$.

Example 3.6-1

Suppose that $f(t)$ is a rectangular pulse defined by (assume $A > 0$ for convenience)

$$f(t) = A \operatorname{rect}\left(\frac{t}{\tau}\right) \leftrightarrow A\tau \operatorname{Sa}\left(\frac{\omega\tau}{2}\right) = F(\omega).$$

The first bound occurs for $n = 0$ and is

ISBN 0-201-05758-1

*Even though noise is a power signal, the analysis leading to (3.5-11) is different from that of Section 2.8 for deterministic signals.

$$\text{bound 1} = \int_{-\infty}^{\infty} A \operatorname{rect}\left(\frac{t}{\tau}\right) dt = A\tau.$$

The second bound is

$$\text{bound 2} = \frac{1}{|\omega|} \int_{-\infty}^{\infty} \left| \frac{d[A \operatorname{rect}(t/\tau)]}{dt} \right| dt$$

$$= \frac{A}{|\omega|} \int_{-\infty}^{\infty} |\delta(t + \tau/2) - \delta(t - \tau/2)| \, dt$$

$$= \frac{2A}{|\omega|}.$$

Plots of $|F(\omega)|$ and the two bounds are shown in Fig. 3.6-1. Only two bounds may be found, since rect (t/τ) can be differentiated only once [2].

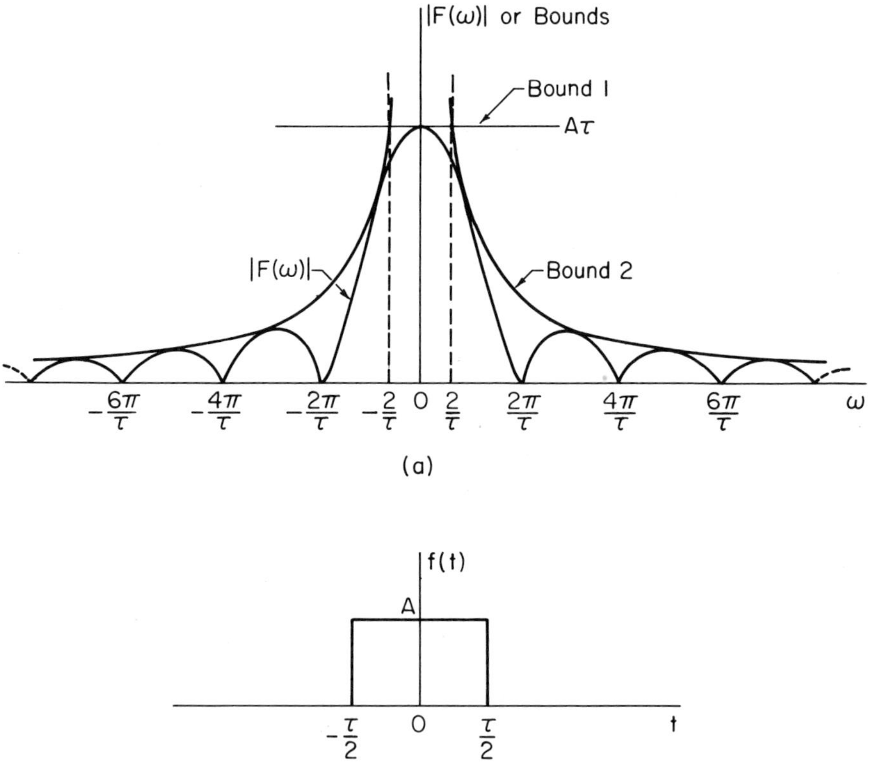

Fig. 3.6-1. (a) Spectrum and its bounds for the rectangular pulse of (b).

ISBN 0-201-05758-1

Example 3.6-2

As a second example we consider the triangular signal

$$f(t) = A \operatorname{tri}(t/\tau) \leftrightarrow A\tau \operatorname{Sa}^2(\omega\tau/2) = F(\omega).$$

Computing bounds:

$$\text{bound 1} = \int_{-\infty}^{\infty} A \operatorname{tri}(t/\tau)\, dt = A\tau,$$

$$\text{bound 2} = \frac{1}{|\omega|} \int_{-\infty}^{\infty} \frac{A}{\tau} \operatorname{rect}\left(\frac{t}{2\tau}\right) dt = \frac{2A}{|\omega|},$$

$$\text{bound 3} = \frac{1}{|\omega|^2} \int_{-\infty}^{\infty} \frac{A}{\tau} |\delta(t+\tau) - 2\delta(t) + \delta(t-\tau)|\, dt$$

$$= \frac{4A}{\tau|\omega|^2}.$$

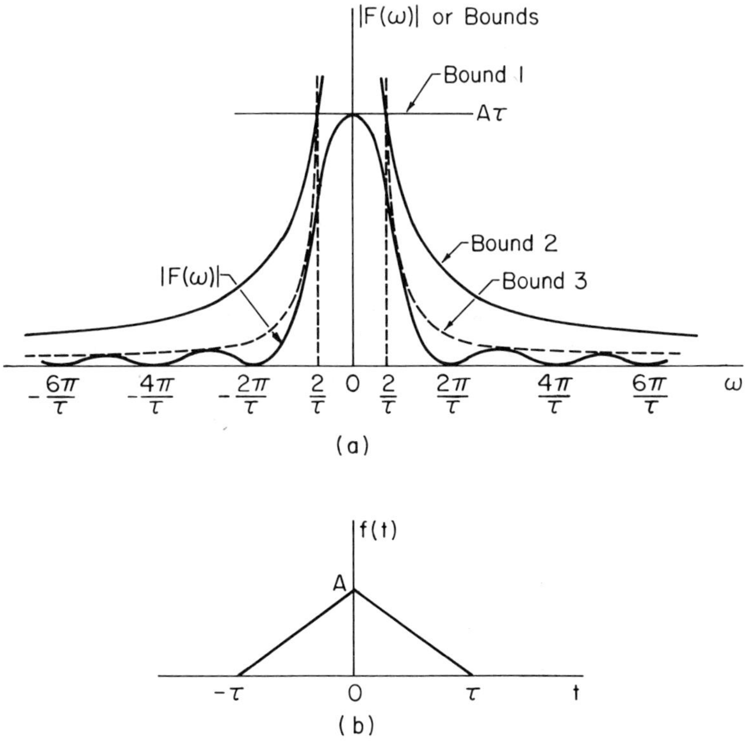

Fig. 3.6-2. Spectrum and its bounds (a) for the triangular pulse of (b).

ISBN 0-201-05758-1

Figure 3.6-2 illustrates $|F(\omega)|$ and the three bounds. Both bounds 2 and 3 cross bound 1 when $|\omega| = 2/\tau$. However, for $|\omega| > 2/\tau$, bound 3 is a better choice to make in approximating $|F(\omega)|$. The highest ordered bound will always be the best for higher $|\omega|$, since the bounds decrease as $1/|\omega|^n$.

Signal Bounds

In a manner similar to that used for spectrum magnitude bounds we may find bounds on $|f(t)|$. Starting with

$$(-jt)^n f(t) \leftrightarrow \frac{d^n F(\omega)}{d\omega^n}, \tag{3.6-5}$$

which means that

$$f(t) = \frac{1}{(-jt)^n} \frac{1}{2\pi} \int_{-\infty}^{\infty} \frac{d^n F(\omega)}{d\omega^n} e^{j\omega t} \, d\omega, \tag{3.6-6}$$

and using (3.6-2) gives

$$|f(t)| \leqslant \frac{1}{2\pi |t|^n} \int_{-\infty}^{\infty} \left| \frac{d^n F(\omega)}{d\omega^n} \right| d\omega = \text{bound } (n + 1) \tag{3.6-7}$$

for $n = 0, 1, 2, \ldots$.

★ 3.7. THE ANALYTIC SIGNAL

From our background in circuit analysis we are familiar with how the introduction of complex notation greatly simplifies many problems. A common example of the use of such notation is to specify the transfer function of a network as a complex (spectral) function rather than give the defining differential equation in time. If the network is realizable, then its transfer function will be an analytic* function (see [6] for introductory material). The impulse response is well known to exist only for $0 \leqslant t$ [4, p. 214]. We thus have an analytic spectral function that is characterized by a time function existing only for $0 \leqslant t$.

Definition

An analogous case can exist where we have an analytic time function characterized by a spectrum which exists only for $0 \leqslant \omega$. Such a signal is called the *analytic signal,* which appears to have been introduced initially by Gabor [7] and later by others [8–11]. An extremely lucid development has been given by Rihaczek [12].

The above and other useful and interesting properties of the analytic signal are listed by Bedrosian [13]:

*The term analytic does not mean a function which can be analyzed, merely one that satisfies certain conditions in the theory of complex variables. For our purposes the term is used only to describe a function having an inverse transform which is zero for negative argument.

ISBN 0-201-05758-1

(1) The Fourier transform of the analytic signal vanishes for negative frequencies (also, the real and imaginary parts have identical power spectra).
(2) The magnitude of the analytic signal is the envelope of the actual waveform. Though defined purely mathematically, the envelope has physical significance for narrow-band signals.
(3) The phase of the analytic signal is equal to the phase of the actual waveform.

Consider a *real* signal $f(t)$ having a continuous Fourier transform $F(\omega)$. Then

$$f(t) = \frac{1}{2\pi} \int_{-\infty}^{\infty} F(\omega) e^{j\omega t}\, d\omega. \tag{3.7-1}$$

We have previously shown, for real signals, that

$$F(-\omega) = F^*(\omega). \tag{3.7-2}$$

Splitting (3.7-1) into two integrals and using (3.7-2) we develop

$$f(t) = \frac{1}{2\pi} \int_{0}^{\infty} F(\omega) e^{j\omega t}\, d\omega + \frac{1}{2\pi} \int_{-\infty}^{0} F(\omega) e^{j\omega t}\, d\omega$$

$$= \left[\frac{1}{2\pi} \int_{0}^{\infty} F(\omega) e^{j\omega t}\, d\omega\right] + \left[\frac{1}{2\pi} \int_{0}^{\infty} F(\omega) e^{j\omega t}\, d\omega\right]^*$$

$$= \text{Re}\left[\frac{1}{2\pi} \int_{0}^{\infty} 2F(\omega) e^{j\omega t}\, d\omega\right], \tag{3.7-3}$$

which says that $f(t)$ can also be represented by the real part of a *complex* signal containing only the positive frequencies–i.e., a complex signal $\psi(t)$ having a spectrum

$$\Psi(\omega) = 2F(\omega), \qquad 0 \leqslant \omega$$

$$= 0, \qquad \omega < 0. \tag{3.7-4}$$

The signal $\psi(t)$ is the *analytic signal* given by

$$\psi(t) = \frac{1}{2\pi} \int_{0}^{\infty} 2F(\omega) e^{j\omega t}\, d\omega. \tag{3.7-5}$$

The Hilbert Transform

The analytic signal will be complex. Let us define it by

$$\psi(t) = f(t) + j\hat{f}(t), \tag{3.7-6}$$

ISBN 0-201-05758-1

where $\hat{f}(t)$ is a function that must be selected so that (3.7-4) is true. It is clearly related to $f(t)$ through $F(\omega)$ in (3.7-5).

Transforming (3.7-6) gives

$$\Psi(\omega) = F(\omega) + j\hat{F}(\omega), \tag{3.7-7}$$

where we define

$$\hat{f}(t) \leftrightarrow \hat{F}(\omega). \tag{3.7-8}$$

Since $\Psi(\omega)$ must be zero for $\omega < 0$ from (3.7-4), we have from (3.7-7) the requirement that

$$\hat{F}(\omega) = jF(\omega), \qquad \omega < 0. \tag{3.7-9}$$

For $\omega > 0$ we again find from (3.7-4) that

$$\hat{F}(\omega) = -jF(\omega), \qquad \omega > 0. \tag{3.7-10}$$

Combining these two results we have

$$\hat{F}(\omega) = -jF(\omega)[2U(\omega) - 1], \tag{3.7-11}$$

where $U(\omega)$ is the unit step function having the value 1.0 for $0 < \omega$ and zero for $\omega < 0$.

It is known from Section 2.6 that

$$\text{sgn}\,(t) = 2U(t) - 1 \leftrightarrow 2/j\omega, \tag{3.7-12}$$

so that

$$-1/j\pi t \leftrightarrow 2U(\omega) - 1 \tag{3.7-13}$$

from our symmetry property for transforms. Since the inverse transform of a product in the frequency domain is the same as convolution in the time domain, the inverse transform of (3.7-11), using (3.7-13), is

$$\hat{f}(t) = -j\int_{-\infty}^{\infty} f(\tau)\left[\frac{-1}{j\pi(t-\tau)}\right] d\tau$$

$$= \frac{1}{\pi}\int_{-\infty}^{\infty} \frac{f(\tau)}{t-\tau}\, d\tau. \tag{3.7-14}$$

The last form of (3.7-14) is called the *Hilbert transform* of $f(t)$. We see that $\hat{f}(t)$ is the Hilbert transform of $f(t)$. The inverse Hilbert transform applies if we know $\hat{f}(t)$ and must find $f(t)$ to make (3.7-4) hold:

$$f(t) = -\frac{1}{\pi}\int_{-\infty}^{\infty} \frac{\hat{f}(\tau)}{t-\tau}\, d\tau. \tag{3.7-15}$$

ISBN 0-201-05758-1

Relation to Modulated Carrier Signals

A modulated carrier signal consists of a carrier of frequency ω_0, which may have envelope modulation $a(t)$, or phase modulation $\phi(t)$, or both, imparted upon it [we assume that $a(t)$ and $\phi(t)$ are slowly varying with respect to ω_0 so that the terms amplitude and phase modulation have meaning]:

$$f(t) = a(t) \cos [\omega_0 t + \phi(t)]. \tag{3.7-16}$$

The analytic signal corresponding to this signal can be found from use of (3.7-6) and (3.7-14) in some cases. However, in many cases the required integration of (3.7-14) cannot be easily achieved. This leads us to ask: Under what conditions can the analytic signal corresponding to (3.7-16) be put in the simple form

$$\psi(t) = m(t)e^{j\omega_0 t} = a(t)\, e^{j\omega_0 t + j\phi(t)}\ ? \tag{3.7-17}$$

Here we define a complex modulation function $m(t)$ as

$$m(t) = a(t)e^{j\phi(t)}. \tag{3.7-18}$$

In answer to this question, it has been pointed out by Rihaczek [12] that (3.7-17) will be an analytic signal only if both $a(t)$ and $\cos\, [\omega_0 t + \phi(t)]$ have bandlimited and nonoverlapping spectra. Basically this requires $m(t)$ to be a bandlimited waveform. That is, $m(t)$ must have a spectrum $M(\omega)$ satisfying

$$M(\omega) = 0, \qquad 2\pi B < |\omega| \tag{3.7-19}$$

for some bandwidth B less than f_0.

Many current-day systems satisfy the narrowband approximation so that, with minor error, (3.7-17) represents the analytic signal. In a wideband system the approximation may not hold and computational difficulties may arise.

Waveform Energy

Let E be the energy of the actual real waveform $f(t)$ and E_c be the energy in the analytic signal $\psi(t)$. We have

$$E_c = \int_{-\infty}^{\infty} \psi(t)\psi^*(t)\, dt = 2\frac{1}{2\pi}\int_0^{\infty} 2|F(\omega)|^2\, d\omega = 2E \tag{3.7-20}$$

by use of Parseval's theorem [see eq. (2.4-24)]. Since the total energy is twice that existing in $f(t)$ alone, we conclude that the energy in $\hat{f}(t)$ must also equal E.

★ 3.8. NETWORK RESPONSE TO ANALYTIC SIGNALS

We now consider the passage of an analytic signal $\psi(t)$ through a linear time-invariant network having a real impulse response $h(t)$. It is permissible to consider that $\psi(t)$ comprises two distinct components being passed through the network (real and imaginary parts). From the

ISBN 0-201-05758-1

convolution theorem the output complex signal $\eta(t)$ is

$$\eta(t) = \int_{-\infty}^{\infty} h(\tau)\psi(t - \tau)\, d\tau. \tag{3.8-1}$$

Since the network passes both positive and negative frequencies while the analytic signal has only positive frequencies, we may simplify calculations and make the writing of (3.8-1) more consistent (signal is complex but impulse response is real) by using the *analytic impulse response.* That is, we replace $h(\tau)$, which has the spectrum $H(\omega)$, by the analytic signal $\psi_h(\tau)$ having the spectrum $\Psi_h(\omega)$:

$$\begin{aligned} \Psi_h(\omega) &= 2H(\omega), \qquad 0 \leqslant \omega \\ &= 0, \qquad \omega < 0 \end{aligned} \tag{3.8-2}$$

and include a factor of $1/2$ to account for the spectrum doubling. Thus, the output analytic signal is

$$\begin{aligned} \eta(t) &= \frac{1}{2}\int_{-\infty}^{\infty} \psi_h(\tau)\psi(t - \tau)\, d\tau \\ &= \frac{1}{2}\int_{-\infty}^{\infty} \psi(\tau)\psi_h(t - \tau)\, d\tau. \end{aligned} \tag{3.8-3}$$

One simplification resulting from this change in notation is that the convolution of an analytic signal $\psi_1(t)$ and the complex conjugate of an analytic signal $\psi_2(t)$ is zero [12]. That is,

$$\int_{-\infty}^{\infty} \psi_1(\tau)\, \psi_2^*(t - \tau)\, d\tau = 0 \tag{3.8-4}$$

or equivalently

$$\int_{-\infty}^{\infty} \psi_1^*(\tau)\psi_2(t - \tau)\, d\tau = 0. \tag{3.8-5}$$

This can be seen from the fact that going from $\psi(t)$ to $\psi^*(t)$ changes the spectrum $\Psi(\omega)$ to $\Psi^*(-\omega)$, which is zero.

Similarly, the correlation integral for the product of two analytic signals is zero:

$$\int_{-\infty}^{\infty} \psi_1(\tau)\psi_2(\tau + t)\, d\tau = 0 \tag{3.8-6}$$

ISBN 0-201-05758-1

or

$$\int_{-\infty}^{\infty} \psi_1^*(\tau)\psi_2^*(\tau + t)\, d\tau = 0. \tag{3.8-7}$$

We may readily verify that the real part of the analytic signal at the system output, as given by (3.8-3), is the correct response [12]. The true real response $g(t)$ is the real part of (3.8-1):

$$g(t) = \text{Re}\ [\eta(t)] = \int_{-\infty}^{\infty} h(\tau)f(t - \tau)\, d\tau. \tag{3.8-8}$$

Now substituting

$$h(\tau) = \frac{1}{2}\,[\psi_h(\tau) + \psi_h^*(\tau)], \tag{3.8-9}$$

$$f(t - \tau) = \frac{1}{2}\,[\psi(t - \tau) + \psi^*(t - \tau)] \tag{3.8-10}$$

into (3.8-8) and invoking the help of (3.8-4) will give

$$g(t) = \frac{1}{2}\left\{\frac{1}{2}\int_{-\infty}^{\infty} \psi_h(\tau)\psi(t - \tau)\, d\tau + \frac{1}{2}\int_{-\infty}^{\infty} \psi_h^*(\tau)\psi^*(t - \tau)\, d\tau\right\}$$

$$= \frac{1}{2}\,\text{Re}\left\{\int_{-\infty}^{\infty} \psi_h(\tau)\,\psi(t - \tau)\, d\tau\right\}. \tag{3.8-11}$$

This result is the real part of (3.8-3), which verifies that (3.8-3) provides the correct network response.

For a band-pass analytic signal, mean frequency $\bar{\omega}_0$ and rms bandwidth W_{rms} are given by (3.4-17) and (3.4-16) if $|F(\omega)|^2$ is replaced by the energy density spectrum $|\Psi(\omega)|^2$. Rms bandwidth is often called effective bandwidth in the literature on analytic signals.

PROBLEMS

3-1. (*a*) Obtain and sketch the magnitude and phase of the transfer function of the filter of Fig. P3-1. (*b*) Is the filter low-pass or high-pass? (*c*) If the positions of L and R are reversed, what kind of filter results?

ISBN 0-201-05758-1

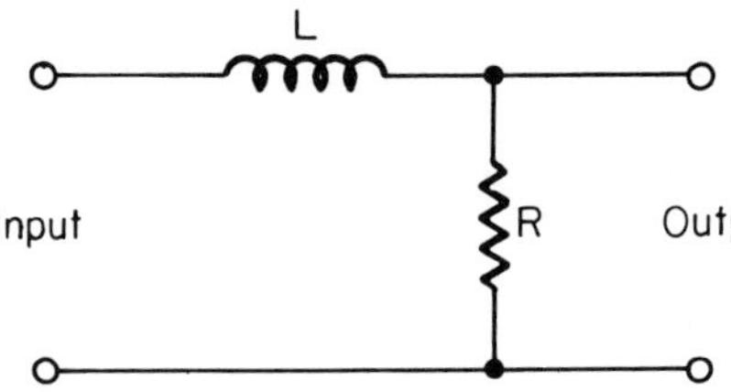

Fig. P3-1

3-2. A low-pass network has the transfer function

$$H(\omega) = \frac{K}{1 + j(\omega/W)},$$

where K and W are positive constants. Find the response of the network to a unit impulse applied at time τ_0. Use the spectral analysis method.

3-3. Solve problem 3-2 by the temporal analysis method using the convolution integral.

3-4. A network has an impulse response

$$\begin{aligned} h(t) &= \exp(-\alpha t), & 0 < t \\ &= 0, & t < 0, \end{aligned}$$

where α is a positive real constant. Use the convolution integral to find the network's response to a unit-amplitude pulse of width τ centered on the time origin.

3-5. Show that the transfer function of the network of Fig. 3.1-1 is

$$H(\omega) = \frac{\sin(\omega\tau/2)}{\omega\tau/2} e^{-j\omega\tau/2}.$$

3-6. By taking the Fourier transform of both sides of (3.2-11) show that $G(\omega) = F(\omega)H(\omega)$.

3-7. The output of the filter of problem 3-2 is applied to a second filter identical to the first. Find and plot the output of the second filter.

3-8. A filter has an impulse response

$$\begin{aligned} h(t) &= \beta t e^{-\alpha t}, & 0 < t \\ &= 0, & t < 0, \end{aligned}$$

where α and β are positive real constants. (*a*) Find the response of the filter to a unit-amplitude rectangular pulse of width τ centered on the time origin. (*b*) At what time will the response reach a maximum?

3-9. A certain class of filters called *Butterworth filters* has a transfer function magnitude defined by

$$|H(\omega)|^2 = \frac{1}{1 + (\omega/W)^{2n}},$$

ISBN 0-201-05758-1

where $n = 1, 2, \ldots$, is the number of filter poles, and W is the 3-dB bandwidth. Sketch $|H(\omega)|^2$ for $n = 1, 2, 4$, and 8, and note the behavior. As $n \to \infty$, what does $|H(\omega)|^2$ become?

3-10. A network is sketched in Fig. P3-10. (*a*) Under what conditions will the network behave approximately as a low-pass filter? (*b*) Under what conditions will it behave approximately as a high-pass filter? (*c*) Find a relationship between R_1, C_1, R_2, and C_2 such that the network behaves as a distortionless attenuator.

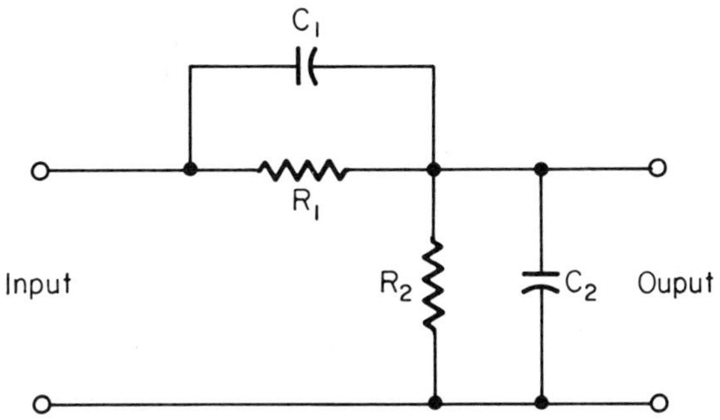

Fig. P3-10

3-11. The impulse response of the filter of Fig. 3.1-1 is an ideal rectangular pulse. If the integrator is replaced by the filter of problem 3-2 obtain and sketch the network impulse response for $W = \pi/\tau$, $W = \pi/(4\tau)$, and $W = \pi/(16\tau)$, assuming $K = \tau/W$ in each case. Which would be the better case to use in practice?

3-12. A low-pass filter is called *Gaussian* if it has a transfer function given by

$$H(\omega) = \frac{1}{\sqrt{2\pi}\, W_{\text{rms}}} \exp\left(\frac{-\omega^2}{2W_{\text{rms}}^2}\right),$$

where W_{rms} is the root-mean-square (rms) bandwidth. (*a*) Sketch $H(\omega)$. (*b*) How is the 3-dB bandwidth related to W_{rms}?

3-13. A nonlinear network produces an output

$$g(t) = a_0 + a_1 f(t) + a_2 f^2(t) + a_3 f^3(t)$$

with $f(t) = A \cos(\omega_0 t)$ applied as its input. (*a*) Find the amplitudes and frequencies of all components of $g(t)$. (*b*) If an ideal low-pass filter with bandwidth ω_0 follows the nonlinear device, what is its output? (*c*) If the filter is band-pass centered at frequency $2\omega_0$, find the output assuming a bandwidth less than $2\omega_0$. Note that the overall system behaves as a frequency doubler. (*d*) Can a frequency tripler be constructed? If so, explain how.

3-14. (*a*) Find the response of the network in problem 3-8 to a unit voltage step input signal. (*b*) Determine the output signal rise time as given by the time between 10% and 90% points. Compare to $t_r = \pi/W$ as given in the text if $\alpha = W$.

3-15. A waveform $g(t)$ is the time derivative of the signal $f(t)$ given in Fig. 3.6-1(*b*). (*a*) Find the spectrum $G(\omega)$ of $g(t)$. (*b*) Find all possible bounds on $G(\omega)$ using (3.6-4). How many are possible? Is agreement achieved when compared to the actual spectrum $G(\omega)$?

ISBN 0-201-05758-1

3-16. Find and plot the spectrum and its three bounds for the signal of Fig. P3-16. (Hint: Use results of example 3.6-2 to obtain the spectrum.)

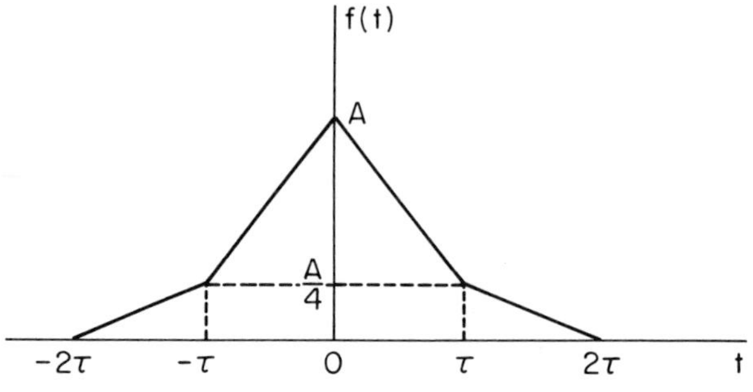

Fig. P3-16

3-17. Work problem 3-16 for the two bounds of the signal of Fig. P3-17.

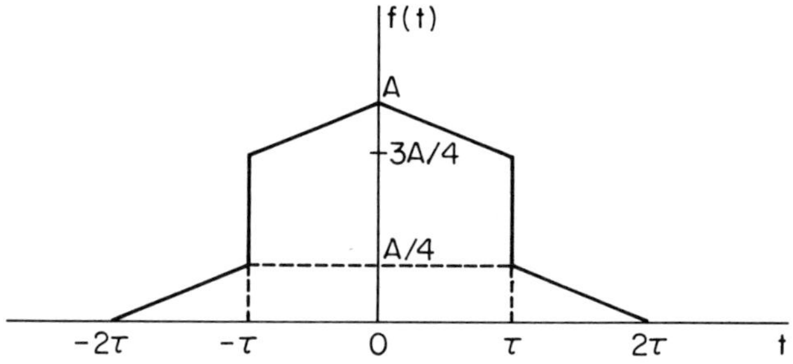

Fig. P3-17

★3-18. (*a*) What is the relationship between ω_0 and τ if

$$\psi(t) = \text{Sa}\,(\pi t/\tau)e^{j\omega_0 t}$$

is to be an analytic signal? (*b*) Sketch the spectra of Sa $(\pi t/\tau)$ and $\psi(t)$.

★3-19. Find the mean frequency and rms bandwidth of the analytic signal $\psi(t)$ of problem 3-18.

3-20. Calculate rms bandwidth W_{rms} for the filter of problem 3-9 assuming $n = 1$ and 2.

REFERENCES

[1] Abramowitz, M., and Stegun, I. A. (Editors), *Handbook of Mathematical Functions with Formulas, Graphs, and Mathematical Tables,* National Bureau of Standards Applied Mathematics Series 55, U.S. Government Printing Office, Washington, D.C., 1964.

[2] Burdic, W. S., *Radar Signal Analysis,* Prentice-Hall, Englewood Cliffs, New Jersey, 1968.

[3] Paley, R. E. A. C., and Wiener, N., *Fourier Transforms in the Complex Domain,* American Mathematical Society Colloquium Publication 19, New York, 1934.

[4] Papoulis, A., *The Fourier Integral and Its Applications,* McGraw-Hill Book Co., New York, 1962.

ISBN 0-201-05758-1

[5] Lathi, B. P., *Signals, Systems and Communication,* John Wiley & Sons, New York, 1965.
[6] Wylie, C. R., Jr., *Advanced Engineering Mathematics,* McGraw-Hill Book Co., New York, 1960.
[7] Gabor, D., Theory of Communication, *Journal of the IEE* (London), Part III, 1946, pp. 429–457.
[8] Ville, J., Theorie et Application de la Notion de Signal Analytique, *Cables et Trans.,* Vol. 2, January 1948, pp. 61–74. (Also RAND Translation T-92, Aug. 1, 1958, by I. Selin.)
[9] Oswald, J. R. V., The Theory of Analytic Band Limited Signals Applied to Carrier Systems, *Transactions of the IRE,* Vol. CT-3, No. 4, December 1956, pp. 244–251.
[10] Woodward, P. M., *Probability and Information Theory with Applications to Radar,* McGraw-Hill Book Co., New York, 1953.
[11] Dugundji, J., Envelopes and Pre-Envelopes of Real Waveforms, *Transactions of the IRE,* Vol. IT-4, No. 4, March 1958, pp. 53–57.
[12] Rihaczek, A. W., Delay-Doppler Ambiguity Function for Wideband Signals, *IEEE Transactions on AES,* Vol. AES-3, No. 4, July 1967, pp. 705–711.
[13] Bedrosian, E., The Analytic Signal Representation of Modulated Waveforms, RAND Corp. Memorandum RM-3080-PR, March, 1962.

ISBN 0-201-05758-1

Chapter 4

Statistical Concepts and the Description of Random Signals and Noise

4.0. INTRODUCTION

At this point the reader has become very familiar with deterministic signals—that is, those that may be well defined. Many signals important in communication systems are not well defined in the usual sense. These waveforms may be lumped into two broad categories: *random signals* and *random noise.* The former is a desired quantity such as an information-bearing signal, while the latter is undesirable. Both are *random,* however, and the same principles may be applied to their description. The purpose of this chapter is to develop these principles.

Almost everyone has some idea of what a random quantity is. These ideas may differ greatly in detail but in a broad sense would likely be similar to the definition to be used here. We define a *random quantity* as one having values which may not be precisely determined but are regulated in some probabilistic way. Thus, all our work with random waveforms must begin with, and be based on, probability theory.

4.1. PROBABILITY CONCEPT

Basic to the discussion of the concept of probability is the idea of an *experiment.* A single performance of the experiment is a *trial* for which there is an *outcome.* Although there exists a precise mathematical definition for an experiment, it will be sufficient for our purposes to simply resort to reason and example. To illustrate, the experiment might consist in tossing a die. A single toss would be a trial which could produce any one of six outcomes, the numbers from one to six.

Another example of an experiment is flipping a coin. Here there are now only two possible trial outcomes, a head or a tail. Still another example is drawing a card from an ordinary 52-card deck. There are 52 outcomes possible on any trial if cards are replaced after each trial.

Often one may be more interested in some characteristic of the outcome of an experiment as opposed to the outcome itself. In the case of drawing a card from a deck, we may be more interested in whether we obtain an ace, or king, or spade, etc. Clearly, there are four outcomes that satisfy the desire for a king while thirteen outcomes satisfy the desire for a spade. To handle these situations we must define an *event.*

Events

An event is something which may or may not happen. In drawing a card, an event might be to draw a king. If the actual trial outcome is not a king, the event did not occur. If a coin is tossed, either the head or tail result may be considered an event.

ISBN 0-201-05758-1

References start on p. 168.

Relative Frequency Definition of Probability

Let us perform n total trials of a given experiment. For an event A, suppose we observe that outcomes satisfy the event n_A times. The *relative frequency* with which A occurs is n_A/n. We reason that, if we perform a large number of trials, a point will be reached where n_A and n increase proportionately such that the ratio n_A/n approaches a constant. Such a ratio becomes the fraction of trials which are expected to satisfy event A. In other words it is the *probability* $P(A)$ of event A occurring which may be defined by

$$P(A) = \lim_{n\to\infty} \left(\frac{n_A}{n}\right). \tag{4.1-1}$$

The definition of probability as a relative frequency limit is usually most satisfying to engineers and scientists and is preferred here because it helps give a physical picture. On the other hand, the reader should be aware that there are several other ways of defining probability [1]. In particular, the axiomatic approach* may be considered as more mathematically sound, but it gives little if any physical insight.

Probability Properties

Certain properties of probability may be made evident by presuming that all the possible events in an experiment are $A, B, \ldots, M$. If only one of the events is satisfied on any given trial, the events are said to be *mutually exclusive.* Assuming mutually exclusive events we may perform the experiment n times producing event A n_A times, event B n_B times, and so forth. Since some event must be satisfied on each trial, we have

$$n_A + n_B + \cdots + n_M = n. \tag{4.1-2}$$

Dividing by n, allowing $n \to \infty$, and observing that (4.1-1) applies to any event, we write (4.1-2) as

$$P(A) + P(B) + \cdots + P(M) = 1. \tag{4.1-3}$$

In words, the sum of probabilities of all mutually exclusive events in an experiment must equal unity.

From (4.1-1) the largest $P(A)$ can be is unity, for then $n_A \to n$. The smallest value is zero if $n_A/n \to 0$. A *certain event* is one that must occur and has probability one. On the other extreme an *impossible event* cannot occur and has probability zero. Thus, we have

$$0 \leqslant P(A) \leqslant 1, \tag{4.1-4}$$

$$P(\text{certain event}) = 1, \tag{4.1-5}$$

$$P(\text{impossible event}) = 0. \tag{4.1-6}$$

The converse of these situations is not necessarily true. That is, an event with probability one

*Here a number, called probability, is assigned to an event. The number is chosen to satisfy a set of conditions or *axioms* from which the entire theory of probability may be constructed.

ISBN 0-201-05758-1

may not be certain to occur (there might be a finite number of trials where it does not occur, but n_A/n may still approach unity for large n). Similarly, an event with probability zero may actually occur (again only a finite number of times while $n_A/n \to 0$ as $n \to \infty$).

Example 4.1-1

As a simple example of mutually exclusive events, consider drawing cards (and replacing each draw) from a 52-card deck. Define events A, B, and C as drawing an ace, drawing a 7, and drawing anything other than an ace or 7. Since for either aces or 7's four of the 52 cards satisfy the event, the ratio n_{ace}/n or n_7/n is expected to approach 1/13 in the limit. Indeed they do, and $P(A) = P(B) = 1/13$. By the same argument $P(C) = 11/13$. From (4.1-3) the sum $P(A) + P(B) + P(C) = (1 + 1 + 11)/13 = 1$, as it must.

Example 4.1-2

In a box there are 100 resistors. Of these, 40 are 10 ohms, 35 are 47 ohms, and 25 are 82 ohms. We wish to determine the probability of extracting either a 10-ohm or a 47-ohm resistor from the box on a random draw. There are 40 + 35 mutually exclusive outcomes which will satisfy this event. Thus, $P(10\ \Omega \text{ or } 47\ \Omega) = 0.75$.

Joint and Conditional Probability

It may be that some events are not mutually exclusive. That is, more than one event may be satisfied on any given trial of an experiment. Consider only two events A and B to be possible. If in n trials A and B occur simultaneously n_{AB} times, we may define *joint probability* $P(A, B)$ by the relative frequency*

$$P(A, B) = \lim_{n \to \infty} \left(\frac{n_{AB}}{n} \right). \tag{4.1-7}$$

Now suppose A occurs n_A times while B occurs n_B times. The sum $n_A + n_B$ does *not* now equal n because n_{AB} of the outcomes satisfy both events A and B. What then equals n? To answer this question let n_1 be the number of times event A occurs *alone,* while event B occurs n_2 times alone. Thus,

$$n_A = n_1 + n_{AB}, \tag{4.1-8}$$

$$n_B = n_2 + n_{AB}, \tag{4.1-9}$$

$$n = n_1 + n_2 + n_{AB}. \tag{4.1-10}$$

Using the first two equations in the third gives $(n_A + n_B - n_{AB})/n = 1$. In the limit

$$P(A) + P(B) - P(A, B) = 1. \tag{4.1-11}$$

Comparing this result with (4.1-3) we may observe the effect of jointly probable events.

*We use $P(A, B)$ instead of the frequently seen notation $P(AB)$ to make the later transition to multidimensional random variables easier.

ISBN 0-201-05758-1

Returning to (4.1-7) let us write the ratio n_{AB}/n as $n_{AB}n_B/n_Bn$ to get

$$P(A, B) = \lim_{n\to\infty} \left(\frac{n_B}{n} \cdot \frac{n_{AB}}{n_B}\right). \tag{4.1-12}$$

The limit of n_B/n is obviously $P(B)$. The limit of n_{AB}/n_B is not obvious. We define it as the *conditional probability* $P(A\,|\,B)$:

$$P(A\,|\,B) = \lim_{n\to\infty} \left(\frac{n_{AB}}{n_B}\right), \tag{4.1-13}$$

so that

$$P(A, B) = P(A\,|\,B)P(B). \tag{4.1-14}$$

$P(A\,|\,B)$ is read "the probability of event A occurring conditional on the fact that event B has occurred." The meaning becomes clear by study of the ratio n_{AB}/n_B. It is a frequency of occurrence of A relative to n_B. By limiting the denominator to only those outcomes for which B occurs, we are placing the constraint "that event B has occurred."

By a similar analysis to that yielding (4.1-14) one may also derive

$$P(A, B) = P(B|A)P(A), \tag{4.1-15}$$

where

$$P(B|A) = \lim_{n\to\infty} \left(\frac{n_{AB}}{n_A}\right). \tag{4.1-16}$$

Equating (4.1-14) and (4.1-15) we obtain *Bayes' rule:*

$$P(A\,|\,B) = \frac{P(B|A)P(A)}{P(B)}. \tag{4.1-17}$$

For more than two events we may successively apply (4.1-14). For example, with three events A, B, and C,

$$P(A, B, C) = P(A\,|\,B, C)P(B, C) = P(A\,|\,B, C)P(B|C)P(C). \tag{4.1-18}$$

Example 4.1-3

Example 4.1-2 may be broadened to illustrate joint and conditional probability. In addition to resistance value let the resistors be graded by wattage according to the table below.

We shall find the various probabilities associated with defining events A and B as drawing a 47-ohm resistor and drawing a 2-watt resistor, respectively. From the table

$$P(A) = P(47\ \Omega) = 35/100,$$

$$P(B) = P(2\ \text{W}) = 43/100.$$

ISBN 0-201-05758-1

TABLE 4.1-1. Number of Resistors in Box Having Given Resistance and Wattage

Resistance	Wattage			Total
	1 W	2 W	5 W	
10 Ω	10	14	16	40
47 Ω	19	11	5	35
82 Ω	4	18	3	25
Total	33	43	24	100

The probability of (simultaneously) drawing a 47-ohm, 2-watt resistor is

$$P(A, B) = P(47\ \Omega, 2\ \mathrm{W}) = 11/100.$$

From (4.1-13) and (4.1-16) the conditional probabilities are

$$P(A|B) = P(47\ \Omega|2\ \mathrm{W}) = 11/43,$$

$$P(B|A) = P(2\ \mathrm{W}|47\ \Omega) = 11/35.$$

As a check on these results we may apply (4.1-14), (4.1-15), or (4.1-17).

Statistical Independence

Consider two events A and B. If $P(A|B)$ is actually independent of the occurrence of B, then $P(A|B) = P(A)$ and events A and B are said to be *statistically independent.* This fact allows

$$P(A, B) = P(A)P(B) \tag{4.1-19}$$

from (4.1-15). For statistically independent events $P(B|A) = P(B)$ also, as can be proved using (4.1-17). Equation (4.1-19) could also have been used as the definition of statistical independence and often is.

For N events A_i to be independent we must have [2]

$$\left.\begin{aligned} P(A_i, A_j) &= P(A_i)P(A_j), \\ P(A_i, A_j, A_k) &= P(A_i)P(A_j)P(A_k), \\ &\vdots \\ P(A_1, A_2, \ldots, A_N) &= P(A_1)P(A_2)\ldots P(A_N), \end{aligned}\right\} \tag{4.1-20}$$

for $1 \leqslant i < j < k < \ldots \leqslant N$.

Example 4.1-4

We return to example 4.1-3 to see if the two events A and B are statistically independent. In that example it was shown that $P(A, B) = 11/100$, $P(A) = 35/100$, and $P(B) = 43/100$. Thus,

ISBN 0-201-05758-1

from (4.1-19), $P(A, B) = 11/100 \neq P(A)P(B) = (35/100)(43/100)$, so A and B are not independent.

4.2. ELEMENTS OF SET THEORY

It is helpful to view probability theory from the standpoint of set theory. In this section the elementary aspects of set theory are introduced and used to further develop probability theory.

Basic Definitions

A *set* is a collection of *elements.* The elements may be anything whatsoever. We may have a set of voltages, a set of trees, or even a set of sets, called a *class* of sets. A set is usually denoted with an upper-case letter, while an element is denoted by a lower-case letter. If a is an element of a set A, then we write

$$a \in A. \tag{4.2-1}$$

If a is *not* an element of A, we write

$$a \notin A. \tag{4.2-2}$$

A set is specified by the contents of two braces $\{\cdot\}$. Two methods are usually used. In one, the elements are enumerated such as

$$A = \{2, 4, 6, 8\} \tag{4.2-3}$$

for the set of even integers I, $0 < I \leqslant 8$. A second method uses an explicit statement defining the elements such as

$$A = \{a \in I \mid 0 < a \leqslant 8\}, \tag{4.2-4}$$

which is read "A is the set of all integers a greater than 0 but not greater than 8."

If a set has a finite number of elements it is said to be *finite.* All other sets are *infinite* sets. An example of an infinite set would be the set of all numbers from zero to one.

If every element of a set A is contained in another set B, then A is a *subset* of B. That is, A is contained in B and we write

$$A \subseteq B. \tag{4.2-5}$$

If at least one element exists in B which is not in A, then A is a *proper subset* of B, and we write

$$A \subset B. \tag{4.2-6}$$

The largest or all-encompassing set is called the *universal set* or *sample space S.* All sets are subsets of S. On the other extreme a set having no elements is said to be *empty* and is called the *null set* ϕ. The null set is a subset of all other sets.

ISBN 0-201-05758-1

One may think of the universal set as having elements which are all the possible outcomes of an experiment. An event may be considered to be a set of elements which are the outcomes that satisfy the event. In the case of drawing a card from a 52-card deck the universal set would have 52 elements, one for each card. The event "draw a five" would have four elements, one for the five in each suit.

Two sets A and B are said to be *disjoint* or *mutually exclusive* if they have no elements in common.

Venn Diagram

One of the main advantages of introducing sets is that a geometrical interpretation is possible for many probability principles. The *Venn diagram,* illustrated in Fig. 4.2-1, provides this interpretation. It associates area with outcomes that correspond to a given set (event). The universal set is represented as a rectangular area. Other sets are areas inside the rectangle. In (*a*) the set of points B forms a subset of the set A. Neither A nor B is a subset of C. Furthermore, A and B are not disjoint, but A and C are, as well as B and C.

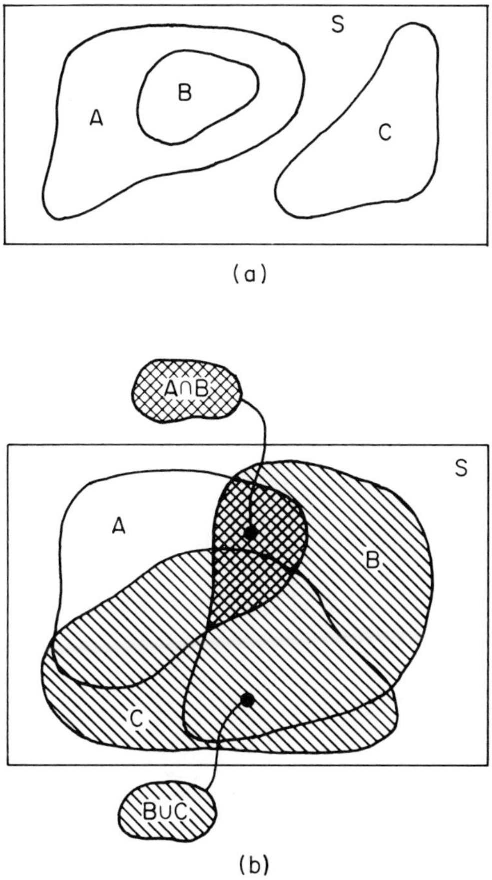

Fig. 4.2-1. Venn diagrams for three events A, B, and C. (a) Illustration of subsets and mutually exclusive sets, and (b) illustration of intersection and union of sets.

ISBN 0-201-05758-1

Set Algebra

Two sets A and B are *equal sets* if, and only if, every element of A is contained in B and every element of B is contained in A.

The *union* C of two sets A and B, sometimes called the *sum*, is the set of all elements in A or B or both. The symbol $\cup$ is commonly used to represent union. Hence, if C is the union of A and B,

$$C = A \cup B. \tag{4.2-7}$$

The *intersection* C of two sets A and B is the set of all elements common to both A and B. Intersection is usually written

$$C = A \cap B. \tag{4.2-8}$$

Intersection and union are illustrated in Fig. 4.2-1(b).

The usual algebraic commutative, associative, and distributive laws hold for union and intersection. The commutative law is

$$A \cup B = B \cup A, \tag{4.2-9}$$

$$A \cap B = B \cap A. \tag{4.2-10}$$

The associative law is

$$(A \cup B) \cup C = A \cup (B \cup C) = A \cup B \cup C, \tag{4.2-11}$$

$$(A \cap B) \cap C = A \cap (B \cap C) = A \cap B \cap C. \tag{4.2-12}$$

The distributive law is

$$A \cap (B \cup C) = (A \cap B) \cup (A \cap C), \tag{4.2-13}$$

$$A \cup (B \cap C) = (A \cup B) \cap (A \cup C). \tag{4.2-14}$$

The *complement* of a set A is denoted by A^c. It is the set of all elements not in A. Clearly,

$$A \cup A^c = S, \tag{4.2-15}$$

$$S^c = \phi, \tag{4.2-16}$$

$$\phi^c = S. \tag{4.2-17}$$

The *difference* of two sets A and B, denoted by $A - B$, is a set having elements of A which are not in B. It is easy to show by Venn diagrams that

$$A - B = A \cap B^c = A - (A \cap B). \tag{4.2-18}$$

Finally, the union and intersection may be extended to multiple sets by repeated application of the associative law [2].

ISBN 0-201-05758-1

Example 4.2-1

Let us show that the cross-hatched area in Fig. 4.2-2 is given by $(A \cap B) \cup (A^c \cap B^c)$. The center portion common to A and B is $(A \cap B)$. The portion not common to either A or B is $(A \cup B)^c$. By using *DeMorgan's law* (see problem 4-13) we know that $(A \cup B)^c = A^c \cap B^c$. The desired area is the union of the two portions which proves the area is $(A \cap B) \cup (A^c \cap B^c)$.

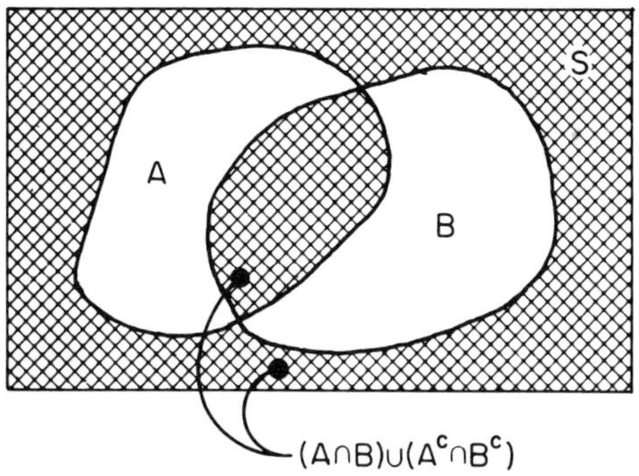

Fig. 4.2-2. Venn diagram applicable to example 4.2-1.

Relationship between Sets and Probability

Earlier probability discussions have already indicated that an event in a random experiment is satisfied by only a certain number of outcomes–in other words, by a *set* of outcomes. Thus, we may associate a set with an event having a certain probability of occurrence. Since this probability is that corresponding collectively to *all* the elements in the set, we may associate a probability with a set. Hence, the probability of a certain set corresponds to the probability of occurrence of the event defined by the set.

In the next section we combine sets and probability concepts via the above viewpoint. We use the notation $P\{X \leqslant x\}$ to denote the probability associated with the set $\{X \leqslant x\}$ of all points on the real line X that do not exceed some specific value x. Similar notations are also adopted for joint and conditional probabilities.

4.3. THE RANDOM VARIABLE AND ITS CHARACTERIZATION

A *random variable* is neither random nor a variable in the usual sense. *It is a function,* one which maps elements in the sample space into the real line. Thus, for every possible outcome of a random experiment (which does not actually have to be conducted) we may assign a real number to the outcome which becomes the value of the random variable.

It is conventional to represent a random variable by a capital letter (typically, X, Y, or Z) and a particular *value* of the variable by a lower-case letter (x, y, or z). The probability that a random variable will have a particular value is just the probability corresponding to those sample space elements that correspond to the particular value (an event). In terms of *domain,** the domain for probability is all subsets (events) of the sample space S. The domain for the random variable is just S.

*Recall from calculus that the domain of an arbitrary real function $f(x)$ is the set of values which x may take on, while the *range* is the set of corresponding values of f.

ISBN 0-201-05758-1

Random variables may be either discrete or continuous. We next discuss both cases and consider examples which will help illustrate the concept of a random variable.

Discrete and Continuous Random Variables

A *discrete random variable* is one that has only discrete values. To illustrate the discrete case we take two examples based on tossing a single ordinary six-faced die.

Example 4.3-1

We toss a single die and observe the number showing on the upper face. There are six possible outcomes (call them $\zeta_i = i$, $i = 1, 2, \ldots, 6$) which form the sample space $S = \{1, 2, 3, 4, 5, 6\}$. Now suppose we arbitrarily define the random variable X such that

$$X(\zeta_i) = 2\zeta_i.$$

X will then have values $X = \{2, 4, 6, 8, 10, 12\}$. The probability of each sample space element is 1/6. The probabilities are the same for the random variable values $x_i = X(\zeta_i) = 2\zeta_i$.

Example 4.3-2

As another, less obvious example, we again toss a single die. The sample space is defined by the set $S = \{1, 2, 3, 4, 5, 6\}$. Now, however, we define X by

$$x_1 = -4, \qquad i = 1, 2,$$

$$x_2 = -1, \qquad i = 3, 4,$$

$$x_3 = 25, \qquad i = 5, 6.$$

The values of X form the set $\{-4, -1, 25\}$, which has fewer elements than there are in S. It is not necessary that they have the same number. This point is emphasized by passing to the extreme.

In the extreme we assign only one value (say 10) to *every* element in S. Thus, X has only one possible value, 10, corresponding to a certain event with probability one. In this case six sample points map into the single value of X.

A *continuous random variable* is one having a continuous range. One of the simplest examples of the continuous case is spinning the pointer on a wheel of chance such as that shown in Fig. 4.3-1. Let the wheel be numbered like a clock. The sample space S now consists of *all* real numbers from 0 to 12. Such a space is different from any encountered thus far but should cause no problem* at this point. If the random variable is simply pointer position, it may likewise have values from 0 to 12. On the other hand, we could equally well define a random variable with values equal to the *square* of pointer values. X would then range from 0 to 144. This freedom of assignment emphasizes the important fact that a random variable is a function.

*The careful reader may ask what probability is associated with any one of an infinity of possible numbers in S? We subsequently deal with this problem in detail. However, the answer will be found to be zero, as can be justified from careful consideration of (4.1-1).

ISBN 0-201-05758-1

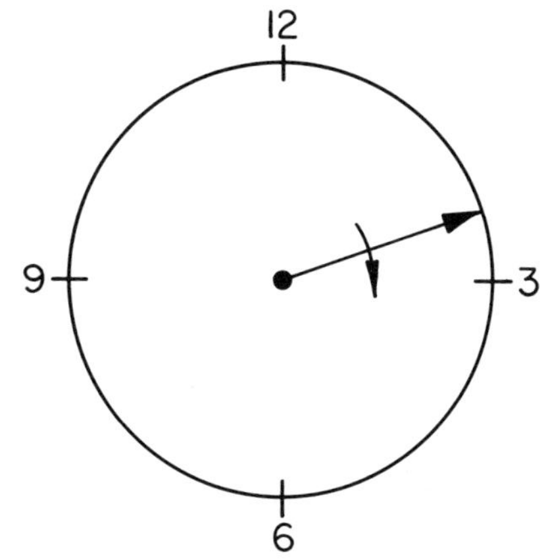

Fig. 4.3-1. A wheel of chance.

A random variable may in general be discrete, continuous, or a combination of both. The case of a mixture is usually the least important of the three but may sometimes be encountered.

Distribution Function

The above discussions have dealt with the definition of a random variable X in terms of its values. We now turn to a consideration of how probability is distributed over the range of X.

Suppose we divide the range of X into two parts, one with points $\{X \leqslant x\}$ and another with points $\{X > x\}$, where x is any arbitrary real number from $-\infty$ to $+\infty$. The *cumulative distribution function* $P_X(x)$ of X is defined as the probability that X has a value less than or equal to x:

$$P_X(x) = P\{X \leqslant x\}. \tag{4.3-1}$$

Regardless of whether X is discrete, continuous, or mixed, $P_X(x)$ has certain properties which may easily be deduced. These are:

$$(1)\ 0 \leqslant P_X(x) \leqslant 1, \tag{4.3-2}$$

$$(2)\ P_X(-\infty) = 0, \tag{4.3-3}$$

$$(3)\ P_X(\infty) = 1, \tag{4.3-4}$$

$$(4)\ P_X(x_1) \leqslant P_X(x_2) \text{ if } x_1 < x_2, \tag{4.3-5}$$

$$(5)\ P_X(x) \text{ is a nondecreasing function of } x. \tag{4.3-6}$$

Property 1 results because $P_X(x)$ is a probability. Properties 2 and 3 result from the fact that the event $\{X \leqslant x\}$ is less likely for small x (fewer points in S are mapped into the set $\{X \leqslant x\}$) and more likely for large x. Proof of (4.3-5) begins by writing

$$P\{X \leqslant x_2\} = P\{X \leqslant x_1\} + P\{x_1 < X \leqslant x_2\}; \tag{4.3-7}$$

it is left for the reader as a problem (problem 4-18). Property 5 follows from property 4.

ISBN 0-201-05758-1

With these properties in mind let us illustrate cumulative distribution function (often abbreviated cdf) for both a discrete and a continuous random variable.

Example 4.3-3

Assume that a discrete variable X has values $\{-1, -0.5, 0.7, 1.5, 3\}$ which occur with probabilities $\{0.1, 0.2, 0.1, 0.4, 0.2\}$, respectively. The probability that $X \leqslant x$ is zero if $x < -1$. Only when $x = -1$ does $P_X(x)$ assume the value 0.1. For $-1 \leqslant x < -0.5, P_X(x)$ remains constant at 0.1 because $P_X(x)$ cannot change until $x = -0.5$. Thus, there is a step in $P_X(x)$ at -1 equal to the probability of occurrence of $x = -1$. Other steps occur at the values of X as x increases. Each step equals the probability of occurrence of the particular value of X. The points are illustrated in Fig. 4.3-2(a).

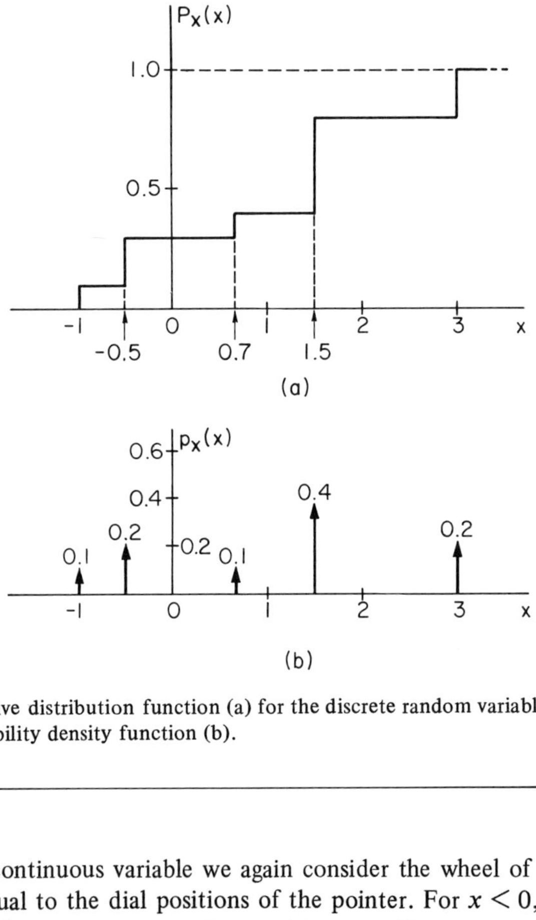

Fig. 4.3-2. The cumulative distribution function (a) for the discrete random variable of example 4.3-3, and the corresponding probability density function (b).

Example 4.3-4

As an example of a continuous variable we again consider the wheel of chance of Fig. 4.3-1. Let X have values equal to the dial positions of the pointer. For $x < 0$, $P_X(x) = 0$, for there are no sample space elements corresponding to this event. Now suppose $x = 3$. We desire the probability that X is 3 or less. Assuming a fair wheel, reason tells us this is 0.25. In fact, as x increases linearly around the wheel, we expect the probability of the pointer's falling between 0 and x to increase linearly. It reaches unity when $x = 12$. These facts allow $P_X(x)$ to be constructed as shown in Fig. 4.3-3(a).

From the above examples we see that the cdf of a continuous random variable is continuous while that of a discrete variable is discontinuous, having "stair-step" discontinuities.

ISBN 0-201-05758-1

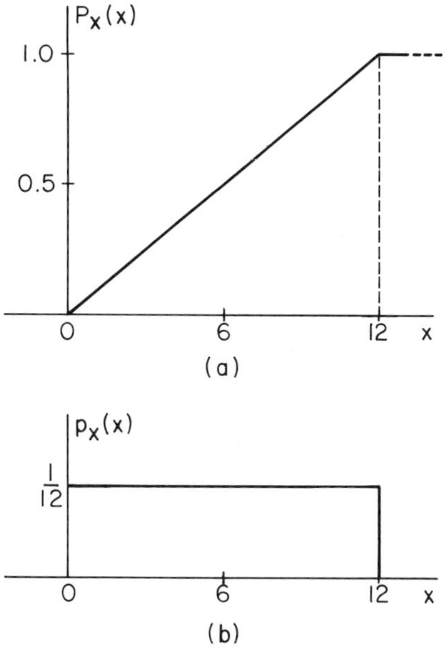

Fig. 4.3-3. The cumulative distribution function (a), and the corresponding probability density function (b), for the random variable of example 4.3-4.

Density Function

In many problems it is more convenient to know the *density* of probability rather than the distribution of probability. We define the *probability density function* $p_X(x)$ as the derivative of the cumulative distribution function:

$$p_X(x) = \frac{dP_X(x)}{dx}. \tag{4.3-8}$$

The probability density function (abbreviated pdf) may have step discontinuities for continuous random variables (see Fig. 4.3-3). For discrete variables it will be a sum of impulse functions (see Fig. 4.3-2).

Several properties for pdf's may be proved from the properties of cdf's:

$$(1)\ p_X(x) \geqslant 0, \qquad \text{all } x, \tag{4.3-9}$$

$$(2)\ \int_{-\infty}^{\infty} p_X(x)\,dx = 1, \tag{4.3-10}$$

$$(3)\ P_X(x) = \int_{-\infty}^{x} p_X(u)\,du, \tag{4.3-11}$$

$$(4)\ P\{x_1 < X \leqslant x_2\} = \int_{x_1}^{x_2} p_X(u)\,du. \tag{4.3-12}$$

ISBN 0-201-05758-1

Proofs of these properties are left to the reader as exercises. Properties 1 and 2 may be used as tests to see if a given function may be a valid pdf. Both tests must be satisfied for validity.

Example 4.3-5

Given the function $g(x) = e^{-x/2}$, for $0 < x$ (and zero for $x < 0$), we determine if it is a valid pdf. Since it satisfies (4.3-9) we test using (4.3-10):

$$\int_{-\infty}^{\infty} g(x)\,dx = \int_{0}^{\infty} e^{-x/2}\,dx = -2e^{-x/2}\Big|_{0}^{\infty} = 2.$$

Thus $g(x)$ is not a valid pdf.

4.4. OPERATIONS ON A RANDOM VARIABLE

There are many operations which may be performed on a random variable. Most of these are essential to the fundamental theory of random variables and are discussed in this section. Other, somewhat specialized operations, involve transforming one random variable into another through arbitrary transformations; these are discussed in Section 4.6.

Expected (Mean) Value

Let a random variable X be characterized by its pdf, $p_X(x)$. The *mean* or *expected value* of X, denoted as $E[X]$, is*

$$E[X] = \int_{-\infty}^{\infty} x p_X(x)\,dx, \tag{4.4-1}$$

which holds for a discrete, continuous, or mixed random variable X. We shall often use the symbol m_X to represent the mean value of X. For a discrete variable the pdf will have the form

$$p_X(x) = \sum_i P(x_i)\delta(x - x_i), \tag{4.4-2}$$

where the sum is indexed over all possible values of i and $P(x_i)$ is the probability of occurrence of a particular discrete value x_i of X. Using (4.4-2) in (4.4-1) gives

$$m_X = E[X] = \sum_i P(x_i)x_i. \tag{4.4-3}$$

The expected value operation will be illustrated by two examples.

*The expected value of a random variable X is often represented by an overbar $\overline{X} = E[X]$. This representation will be used in many places in later work because of its simplicity.

ISBN 0-201-05758-1

Example 4.4-1

We find the mean value of the continuous random variable X having the density function of Fig. 4.3-3(b). Using (4.4-1), we obtain

$$E[X] = \int_0^{12} x\left(\frac{1}{12}\right) dx = 6.$$

Example 4.4-2

For the discrete random variable having the pdf of Fig. 4.3-2(b), the expected value is

$$E[X] = -0.1 - 0.1 + 0.07 + 0.6 + 0.6 = 1.07$$

by using (4.4-3).

Expected Value of a Function of a Random Variable

Let $g(X)$ be any real function of the random variable X. The expected value of this function is defined as

$$E[g(X)] = \int_{-\infty}^{\infty} g(x)p_X(x)\, dx. \tag{4.4-4}$$

This operation is used to develop several important topics in the remainder of this section.

Moments about the Origin

The moments of a random variable X taken about the origin may be obtained from (4.4-4) by letting

$$g(X) = X^n. \tag{4.4-5}$$

Thus, defining these moments as m_n, we have

$$m_n = E[X^n] = \int_{-\infty}^{\infty} x^n p_X(x)\, dx \tag{4.4-6}$$

for $n = 0, 1, 2, \ldots$.

For a discrete random variable, where the pdf of (4.4-2) applies, (4.4-6) reduces to

$$m_n = \sum_i P(x_i)x_i^n, \qquad \text{discrete variable.} \tag{4.4-7}$$

ISBN 0-201-05758-1

Of course $m_0 = 1$ from (4.3-10) and m_1 is the expected value of X. The second moment m_2 is of special interest:

$$m_2 = E[X^2] = \int_{-\infty}^{\infty} x^2 p_X(x)\, dx. \tag{4.4-8}$$

It may often be interpreted as the total average power in a random waveform. More is said about this fact when *random processes* (particularly *ergodic processes*) are discussed later.

Central Moments

A central moment is defined as a moment taken about the mean value of the random variable. Labeling these moments u_n, $n = 0, 1, \ldots$, we have

$$u_n = E[(X - m_1)^n] = \int_{-\infty}^{\infty} (x - m_1)^n p_X(x)\, dx. \tag{4.4-9}$$

Calculation shows $u_0 = 1$ and $u_1 = 0$. The second central moment is so important it is given the special name *variance* and symbol $\sigma_X{}^2$ (subscript changes according to the random variable symbol):

$$\sigma_X{}^2 = E[(X - m_1)^2] = \int_{-\infty}^{\infty} (x - m_1)^2 p_X(x)\, dx. \tag{4.4-10}$$

Variance may be related to the second moment by direct expansion:

$$\sigma_X{}^2 = E[(X - m_1)^2] = E[X^2 - 2m_1 X + m_1{}^2]. \tag{4.4-11}$$

The expected value of a sum can be proved to equal the sum of expected values. Hence

$$\sigma_X{}^2 = E[X^2] - m_1{}^2 = m_2 - m_1{}^2 \tag{4.4-12}$$

on use of (4.4-6) in (4.4-11). Alternatively

$$m_2 = \sigma_X{}^2 + m_1{}^2. \tag{4.4-13}$$

The positive square root of variance is called *standard deviation.* It is a measure of the spread of $p_X(x)$ about the mean value. Variance may often be interpreted as the average power in the fluctuating (ac) part of a random waveform. As previously noted, m_2 may sometimes carry the interpretation of total average power. Thus, (4.4-13) implies that total power is the sum of dc and ac powers in a random waveform.

The third central moment u_3 is a measure of the asymmetry of $p_X(x)$ about m_1. It is here called *skew.** A pdf with even symmetry about m_1 would have zero skew; that is, $u_3 = 0$. In fact, $u_n = 0$ for odd values of n (why?).

*The normalized third central moment, given by $u_3/\sigma_X{}^3$, is known as the *skewness* of the distribution or as the *coefficient of skewness.*

ISBN 0-201-05758-1

For the discrete variable we use (4.4-2) with (4.4-9) to obtain

$$u_n = \sum_i P(x_i)(x_i - m_1)^n, \qquad \text{discrete variable.} \tag{4.4-14}$$

Central moments may be related to moments about the origin by use of the *binomial expansion*

$$(X - m_1)^n = \sum_{k=0}^{n} (-1)^k \binom{n}{k} X^{n-k} m_1{}^k, \tag{4.4-15}$$

where the *binomial coefficient* $\binom{n}{k}$ is given by

$$\binom{n}{k} = \frac{n!}{(n-k)!k!}. \tag{4.4-16}$$

Taking the expected value of both sides of (4.4-15) we have

$$u_n = \sum_{k=0}^{n} (-1)^k \binom{n}{k} m_1{}^k m_{n-k}. \tag{4.4-17}$$

Example 4.4-3

We find the moments and central moments for the continuous random variable of example 4.4-1. The given density function [Fig. 4.3-3(*b*)] is

$$p_X(x) = \frac{1}{12}, \qquad 0 < x < 12$$

$$= 0, \qquad \text{elsewhere.}$$

From (4.4-6),

$$m_n = \int_0^{12} x^n \frac{1}{12}\, dx = \left. \frac{x^{n+1}}{12(n+1)} \right|_0^{12} = \frac{12^n}{n+1}.$$

Central moments result from (4.4-9) with $m_1 = 6$:

$$u_n = \int_0^{12} (x-6)^n \frac{1}{12}\, dx = \left. \frac{(x-6)^{n+1}}{12(n+1)} \right|_0^{12} = \frac{6^n[1 - (-1)^{n+1}]}{2(n+1)}.$$

Using (4.4-12) the variance of X is

$$\sigma_X{}^2 = \frac{12^2}{3} - \left(\frac{12}{2}\right)^2 = 12,$$

which checks with $u_2 = 12$ found from the above equation for u_n.

ISBN 0-201-05758-1

★ Characteristic and Moment-Generating Functions

Two useful functions, both capable of yielding the moments of a random variable about the origin, are the *characteristic function* $F_X(jv)$:

$$F_X(jv) = E[e^{jvX}] = \int_{-\infty}^{\infty} p_X(x)e^{jvx}\,dx, \tag{4.4-18}$$

and the *moment-generating function* $M_X(u)$:

$$M_X(u) = E[e^{uX}] = \int_{-\infty}^{\infty} p_X(x)e^{ux}\,dx. \tag{4.4-19}$$

In these equations u and v are real numbers and $j = \sqrt{-1}$ as usual in this book. The characteristic function is especially useful. It is recognized as the Fourier transform (with sign of j reversed) of $p_X(x)$. Thus, if $F_X(jv)$ is known, $p_X(x)$ may be found by the inverse transform:

$$p_X(x) = \frac{1}{2\pi}\int_{-\infty}^{\infty} F_X(jv)e^{-jvx}\,dv. \tag{4.4-20}$$

Because Fourier transform pairs are unique, $p_X(x)$ uniquely determines $F_X(jv)$ through (4.4-18), and $p_X(x)$ is conversely determined by $F_X(jv)$ using (4.4-20). Similarly, $M_X(u)$ uniquely specifies $p_X(x)$; that is, it is the only pdf corresponding to the given function $M_X(u)$.

An important advantage of the characteristic function, as compared to the moment-generating function, is that it always exists, whereas the latter exists only if all the moments exist [2].

If $F_X(jv)$ is known for a random variable X, the moments m_n may be obtained as follows:

$$m_n = (-j)^n \left.\frac{d^n F_X(jv)}{dv^n}\right|_{v=0}. \tag{4.4-21}$$

Similarly, if $M_X(u)$ is known, moments are

$$m_n = \left.\frac{d^n M_X(u)}{du^n}\right|_{u=0}. \tag{4.4-22}$$

We proceed to give the proof of (4.4-21). The proof of (4.4-22) is quite similar. We use the series expansion

$$e^{jvX} = \sum_{n=0}^{\infty} \frac{(jvX)^n}{n!} \tag{4.4-23}$$

and take the expected value of both sides,

ISBN 0-201-05758-1

$$F_X(j\nu) = E[e^{j\nu X}] = \sum_{n=0}^{\infty} \frac{(j\nu)^n}{n!} E[X^n] = \sum_{n=0}^{\infty} \frac{(j\nu)^n}{n!} m_n. \quad (4.4\text{-}24)$$

Now we differentiate n times and then set $\nu = 0$:

$$\left.\frac{d^n F_X(j\nu)}{d\nu^n}\right|_{\nu=0} = (j)^n m_n. \quad (4.4\text{-}25)$$

Multiplying both sides by $(-j)^n$ now gives the desired result (4.4-21).

We close the discussion of $F_X(j\nu)$ and $M_X(u)$ by using (4.4-2) for discrete random variables to obtain

$$F_X(j\nu) = \sum_i P(x_i)e^{j\nu x_i}, \qquad \text{discrete variable,} \quad (4.4\text{-}26)$$

$$M_X(u) = \sum_i P(x_i)e^{ux_i}, \qquad \text{discrete variable,} \quad (4.4\text{-}27)$$

from (4.4-18) and (4.4-19), respectively. Substitution of these results into (4.4-21) and (4.4-22), respectively, will form a check on, and verify, (4.4-7).

4.5. EXAMPLE DISTRIBUTIONS AND DENSITIES [3, 4]

In this section we list a number of probability density functions which are used enough to have been given names. In each case we give the mean, variance, characteristic function, and cumulative distribution function where available. The first one is for a discrete random variable X, while the remaining ones are for a continuous variable X.

Binomial

For $0 < p < 1$, $q = 1 - p$ and $n = 1, 2, \ldots$:

$$p_X(x) = \sum_{k=0}^{n} \binom{n}{k} p^k q^{n-k} \delta(x - k), \quad (4.5\text{-}1)$$

$$m_X = np, \quad (4.5\text{-}2)$$

$$\sigma_X^2 = np(1 - p), \quad (4.5\text{-}3)$$

$$F_X(j\nu) = (1 - p + pe^{j\nu})^n, \quad (4.5\text{-}4)$$

$$P_X(x) = \sum_{k=0}^{n} \binom{n}{k} p^k q^{n-k} u(x - k). \quad (4.5\text{-}5)$$

Here $u(x)$ is the unit step function and $\binom{n}{k}$ is the *binomial coefficient* of (4.4-16).

ISBN 0-201-05758-1

The binomial distribution is useful in games of chance, certain detection systems such as radar and sonar, and almost any type of problem where only two outcomes are possible on any one trial, one with probability p and the other with probability q. By using the proper limiting operations the binomial distribution yields either the *normal* (defined below) or *Poisson* distribution [4].

Uniform

For $-\infty < a < \infty$, $-\infty < b < \infty$, and $a < h$:

$$p_X(x) = 1/(b - a), \qquad a < x < b$$

$$= 0, \qquad \text{otherwise}, \tag{4.5-6}$$

$$m_X = (a + b)/2, \tag{4.5-7}$$

$$\sigma_X{}^2 = (b - a)^2/12, \tag{4.5-8}$$

$$F_X(jv) = \frac{e^{jbv} - e^{jav}}{j(b - a)v}, \tag{4.5-9}$$

$$P_X(x) = 0, \qquad x < a$$

$$= \frac{x - a}{b - a}, \qquad a \leqslant x \leqslant b$$

$$= 1, \qquad b < x. \tag{4.5-10}$$

The uniform distribution is useful in describing quantization errors in some communication systems, and it will be used later. It is also useful in accuracy analyses of some measurement systems.

Exponential

For $0 < b$:

$$p_X(x) = \frac{1}{b} e^{-x/b}, \qquad 0 < x$$

$$= 0, \qquad \text{otherwise}, \tag{4.5-11}$$

$$m_X = b, \tag{4.5-12}$$

$$\sigma_X{}^2 = b^2, \tag{4.5-13}$$

$$F_X(jv) = 1/(1 - jbv). \tag{4.5-14}$$

$$P_X(x) = 1 - e^{-x/b}, \qquad 0 < x$$

$$= 0, \qquad x < 0. \tag{4.5-15}$$

ISBN 0-201-05758-1

The exponential distribution has broad application. It is used to describe size distribution of raindrops when a large number of rain storms is averaged. It is also known to approximately describe fluctuations of signal strength received by radar from certain types of aircraft.

Gaussian or Normal

For $-\infty < a < \infty$ and $0 < b$:

$$p_X(x) = \frac{1}{\sqrt{\pi b}} e^{-(x-a)^2/b}, \qquad -\infty < x < \infty, \tag{4.5-16}$$

$$m_X = a, \tag{4.5-17}$$

$$\sigma_X^2 = b/2, \tag{4.5-18}$$

$$F_X(jv) = \exp\,[jav - (bv^2/4)], \tag{4.5-19}$$

$$P_X(x) = \frac{1}{2} + \frac{1}{2}\,\text{erf}\left(\frac{x-a}{\sqrt{b}}\right), \qquad -\infty < x < \infty. \tag{4.5-20}$$

Here the *error function*, erf $(\cdot)$, is defined by

$$\text{erf}\,(\alpha) = \frac{2}{\sqrt{\pi}} \int_0^{\alpha} e^{-\xi^2}\, d\xi, \tag{4.5-21}$$

$$\text{erf}\,(-\alpha) = -\text{erf}\,(\alpha). \tag{4.5-22}$$

It is an integral which cannot be evaluated in closed form but is extensively tabulated [5].

The Gaussian distribution is the most important of all distributions, since it enters into nearly all areas of engineering and science. We shall encounter it frequently, since many types of noise are Gaussian. Recognizing that a and b are related to m_X and σ_X^2 through (4.5-17) and (4.5-18), the density function is usually expressed in the form

$$p_X(x) = \frac{1}{\sqrt{2\pi\sigma_X^2}} e^{-(x-m_X)^2/2\sigma_X^2}, \qquad -\infty < x < \infty. \tag{4.5-23}$$

Rayleigh

For $0 < b$:

$$p_X(x) = \frac{2x}{b} e^{-x^2/b}, \qquad 0 < x$$

$$= 0, \qquad \text{otherwise}, \tag{4.5-24}$$

$$m_X = \sqrt{\pi b/4}, \tag{4.5-25}$$

$$\sigma_X^2 = b\left(1 - \frac{\pi}{4}\right), \tag{4.5-26}$$

ISBN 0-201-05758-1

$$P_X(x) = 1 - e^{-x^2/b}, \qquad 0 < x$$

$$= 0, \qquad \text{otherwise.} \tag{4.5-27}$$

The Rayleigh distribution describes the envelope of a band-pass noise waveform. As a consequence, it is quite valuable in noise analysis of various communication systems.

Rice [6, p. 161]

For $0 < a$ and $0 < b$:

$$p_X(x) = \frac{x}{b} e^{-(x^2+a^2)/2b} I_0\left(\frac{ax}{b}\right), \qquad 0 < x$$

$$= 0, \qquad \text{otherwise,} \tag{4.5-28}$$

$$m_X = \sqrt{\frac{\pi b}{2}}\, e^{-a^2/4b}\left[\left(1 + \frac{a^2}{2b}\right) I_0\left(\frac{a^2}{4b}\right) + \frac{a^2}{2b} I_1\left(\frac{a^2}{4b}\right)\right]. \tag{4.5-29}$$

Here $I_0(\cdot)$ and $I_1(\cdot)$ are the modified Bessel functions of the first kind of orders zero and one. They are tabulated in reference [5]. When ax/b is large, $I_0(\cdot)$ approaches an exponential form, and

$$p_X(x) \approx \sqrt{\frac{x}{2\pi ab}}\, e^{-(x-a)^2/2b}, \qquad 1 \ll \frac{ax}{b}. \tag{4.5-30}$$

The Rician distribution is encountered when narrowband Gaussian noise is added to a constant amplitude sinusoidal signal. It describes the envelope of the sum.

★ 4.6. TRANSFORMATIONS OF A RANDOM VARIABLE

It is sometimes necessary to transform a variable X into a new variable Y by a functional transformation T; that is,

$$Y = T(X). \tag{4.6-1}$$

For example, X might represent a noise voltage at some time instant which is applied to a square-law envelope (magnitude) detector. The detector output would be

$$Y = cX^2, \tag{4.6-2}$$

where c is a real constant. The question of interest becomes, "What probability density function should be associated with the random variable Y, assuming that of X is known?" The answer to this question is developed in this section as applied to the general problem (4.6-1).

Monotonic Transformations

A transformation T is said to monotonically increase if $T(x_2) > T(x_1)$ for $x_2 > x_1$. It monotonically decreases if $T(x_2) < T(x_1)$ for $x_2 > x_1$. Figure 4.6-1 illustrates these transformations.

ISBN 0-201-05758-1

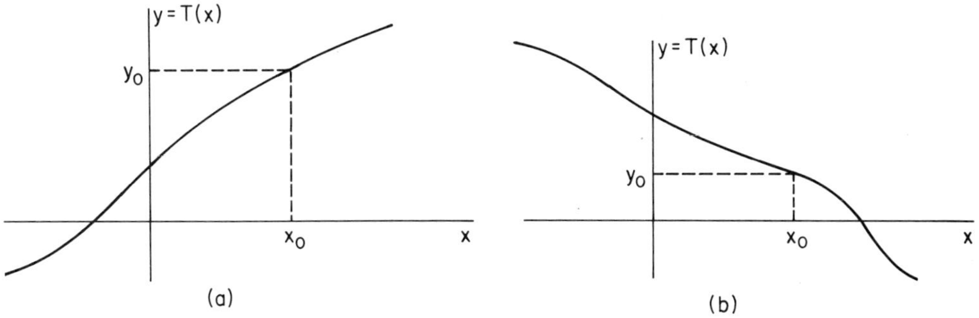

Fig. 4.6-1. Monotonic functions, (a) increasing, and (b) decreasing.

Let us discuss first the increasing transformation. Assume X has a pdf $p_X(x)$ and a cdf $P_X(x)$, while those of Y are $p_Y(y)$ and $P_Y(y)$. Let Y have a particular value y_0 corresponding to the value x_0 for X as seen in Fig. 4.6-1(a). The two are related by

$$x_0 = T^{-1}(y_0), \tag{4.6-3}$$

where $T^{-1}(\cdot)$ is the function associated with the inverse transformation T^{-1} of T. The probability that Y will have a value in the interval $Y \leqslant y_0$ must be the same as the probability that X has a value in the interval $X \leqslant x_0$, since there is a one-to-one correspondence between X and Y. Thus,

$$P_Y(y_0) = P\{Y \leqslant y_0\} = P\{X \leqslant x_0\} = P_X(x_0) \tag{4.6-4}$$

or

$$\int_{-\infty}^{y_0} p_Y(y)\, dy = \int_{-\infty}^{x_0 = T^{-1}(y_0)} p_X(x)\, dx. \tag{4.6-5}$$

Differentiating both sides with respect to y_0 using *Leibnitz's rule** gives

$$p_Y(y_0) = p_X[T^{-1}(y_0)] \frac{dT^{-1}(y_0)}{dy_0}. \tag{4.6-6}$$

The derivative will be positive here, since $T^{-1}(y_0)$ has a positive slope. We may now drop the subscripts and write this result in either of the forms

*Leibnitz's rule says that, if $H(x, u)$ is continuous in x and u and

$$G(u) = \int_{\alpha(u)}^{\beta(u)} H(x, u)\, dx,$$

then

$$\frac{dG(u)}{du} = \int_{\alpha(u)}^{\beta(u)} \frac{\partial H(x, u)}{\partial u}\, dx + H[\beta(u), u] \frac{d\beta(u)}{du} - H[\alpha(u), u] \frac{d\alpha(u)}{du}.$$

ISBN 0-201-05758-1

$$p_Y(y) = p_X[T^{-1}(y)] \frac{dT^{-1}(y)}{dy}, \tag{4.6-7}$$

$$p_Y(y) = p_X(x) \frac{dx}{dy}. \tag{4.6-8}$$

It is understood in the latter form that x is a function of y through (4.6-3).

By repeating the above arguments for the decreasing transformation using Fig. 4.6-1(b), the result analogous to (4.6-4) is

$$P_Y(y_0) = P\{Y \leqslant y_0\} = P\{X > x_0\} = 1 - P_X(x_0). \tag{4.6-9}$$

Development now results in (4.6-6) but with the right side negative. However, since the slope of $T^{-1}(y_0)$ is also negative, we conclude that for either type of monotonic transformation

$$p_Y(y) = p_X[T^{-1}(y)] \left| \frac{dT^{-1}(y)}{dy} \right| \tag{4.6-10}$$

or simply

$$p_Y(y) = p_X(x) \left| \frac{dx}{dy} \right|. \tag{4.6-11}$$

Nonmonotonic Transformations

In the more general case a transformation may not be monotonic. Figure 4.6-2 illustrates such a case. Now, since there is no one-to-one correspondence between X and Y, the probability $P_Y(y_0)$ that $Y \leqslant y_0$ must now equal the probability that X has values corresponding to $Y \leqslant y_0$. We may write

$$P_Y(y_0) = P(\text{set of } x \text{ yielding } y \leqslant y_0)$$

$$= \int_{\{x|y \leqslant y_0\}} p_X(x)\, dx. \tag{4.6-12}$$

In terms of the function illustrated in Fig. 4.6-2, (4.6-12) would be

$$P_Y(y_0) = \int_{-\infty}^{x_1} p_X(x)\, dx + \int_{x_2}^{x_3} p_X(x)\, dx. \tag{4.6-13}$$

Having computed (4.6-12), one is still left with the problem of differentiating to find $p_Y(y_0)$. The formal result is

$$p_Y(y_0) = \frac{d}{dy_0} \int_{\{x|y \leqslant y_0\}} p_X(x)\, dx. \tag{4.6-14}$$

ISBN 0-201-05758-1

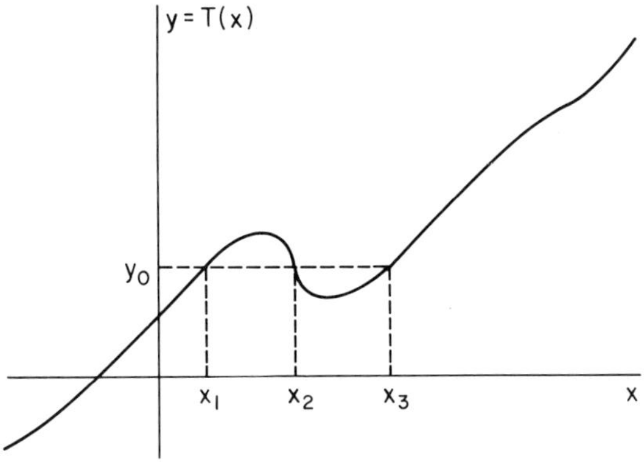

Fig. 4.6-2. A nonmonotonic function.

Examples

Several examples are next developed to illustrate both monotonic and nonmonotonic transformations.

Example 4.6-1

We find $p_Y(y)$ for the linear transformation $Y = aX + b = T(X)$. Clearly $X = (Y - b)/a = T^{-1}(Y)$, and (4.6-10) becomes

$$p_Y(y) = p_X\left(\frac{y-b}{a}\right)\left|\frac{1}{a}\right| .$$

If X has the Gaussian pdf of (4.5-23), then we have

$$p_Y(y) = \frac{1}{\sqrt{2\pi a^2 \sigma_X{}^2}} \exp\left\{-\frac{[y-(b+am_X)]^2}{2a^2\sigma_X{}^2}\right\},$$

which is another Gaussian pdf with mean m_Y and variance $\sigma_Y{}^2$ defined by

$$m_Y = b + am_X,$$

$$\sigma_Y{}^2 = a^2\sigma_X{}^2.$$

Thus, a linear transformation of a Gaussian random variable produces another Gaussian random variable.

Example 4.6-2

Suppose we now develop the square-law detector problem mentioned earlier for $c > 0$. Equation (4.6-2) applies. Although this is a nonmonotonic transformation, the integration required by (4.6-14) is straightforward, since the set of x corresponding to $y \leqslant y_0$ is $\{x \mid -x_0 \leqslant x \leqslant x_0\}$ as seen from Fig. 4.6-3. Using (4.6-14) gives (dropping the subscript on y_0)

ISBN 0-201-05758-1

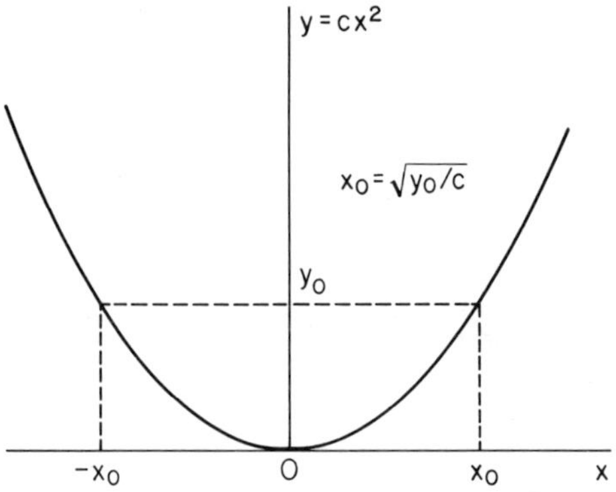

Fig. 4.6-3. Square-law transformation.

$$p_Y(y) = \frac{d}{dy}\int_{-\sqrt{y/c}}^{\sqrt{y/c}} p_X(x)\,dx$$

$$= \left[p_X\left(\sqrt{\frac{y}{c}}\right) + p_X\left(-\sqrt{\frac{y}{c}}\right)\right]\frac{1}{2\sqrt{cy}}$$

from Leibnitz's rule. Of course, this result applies only for $0 < y$.

If $p_X(x) = 1/(2a)$, $-a < x < a$, we reduce this result for a uniform density function:

$$p_Y(y) = \frac{1}{2a\sqrt{cy}}, \qquad 0 < y < ca^2$$

$$= 0, \qquad \text{elsewhere.}$$

This function and $p_X(x)$ are sketched in Fig. 4.6-4.

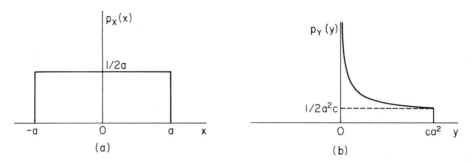

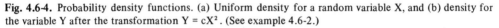

Fig. 4.6-4. Probability density functions. (a) Uniform density for a random variable X, and (b) density for the variable Y after the transformation Y = cX². (See example 4.6-2.)

Example 4.6-3

As another application of the square-law result found in example 4.6-2 we find $p_Y(y)$ for X a zero-mean Gaussian variable, as given in (4.5-23) with $m_X = 0$. We develop

ISBN 0-201-05758-1

$$p_Y(y) = \frac{1}{\sqrt{2\pi\sigma_X{}^2}} \left\{ e^{-(\sqrt{y/c})^2/2\sigma_X{}^2} + e^{-(-\sqrt{y/c})^2/2\sigma_X{}^2} \right\} \frac{1}{2\sqrt{cy}}$$

$$= \frac{1}{\sqrt{2\pi c\sigma_X{}^2 y}} e^{-y/2c\sigma_X{}^2}, \qquad 0 < y \quad \text{(zero elsewhere).}$$

This is sometimes called the *distribution of power* [4] because the transformation $Y = cX^2$ is analogous to power if X is taken analogous to current and c represents resistance. Hence we have, in effect, developed the distribution of power Y in a resistor c if X is a zero-mean Gaussian current.

4.7. SEVERAL RANDOM VARIABLES AND THEIR CHARACTERIZATION

In most problems the theory of a single random variable is inadequate. Therefore, it is necessary that the theory be extended to include multiple variables. It does result, however, that a great many problems–indeed, nearly all of interest in practical situations–may be handled using two-variable theory. Ultimately we develop the concept of a random process (Section 4.8) to characterize the behavior of noise and random signals and of networks with these waveforms present. Nearly all random processes of practical interest may be handled by the methods of two-variable theory. Thus, in the following we emphasize the case of two random variables although we do state the more general results.

Joint Distribution and Density Functions

Consider two random variables X and Y defined on the same sample space. It is convenient to think of the two random variables as being components of a *random point* in the two-dimensional XY plane. The point may be called a *vector random variable* or simply a *random vector.* From this viewpoint the XY plane becomes a two-dimensional or joint sample space. In this space we define two events A and B by the sets $\{X \leqslant x\}$ and $\{Y \leqslant y\}$, respectively, where x and y are any specific values of X and Y. The joint event (A, B) is therefore defined by the set of points $\{X \leqslant x, Y \leqslant y\}$ and the probability of this joint event is $P(A, B) = P\{X \leqslant x, Y \leqslant y\}$. Thus, analogous to cumulative distribution function (cdf) for a single random variable, we define cdf $P_{XY}(x, y)$ for two variables by

$$P_{XY}(x, y) = P\{X \leqslant x, Y \leqslant y\}. \tag{4.7-1}$$

For convenience, we drop the subscripts in following work and write

$$P(x, y) = P\{X \leqslant x, Y \leqslant y\}. \tag{4.7-2}$$

Such a distribution function may be defined whether or not X and Y are random variables on the same sample space [7, p. 21] as originally assumed.

We state without proof the properties of (4.7-2) which are the extensions of (4.3-2) through (4.3-6).

$$(1)\ 0 \leqslant P(x, y) \leqslant 1, \qquad \text{all } x \text{ and } y, \tag{4.7-3}$$

$$(2)\ P(-\infty, y) = 0, P(x, -\infty) = 0, P(-\infty, -\infty) = 0, \tag{4.7-4}$$

ISBN 0-201-05758-1

(3) $P(\infty, \infty) = 1,$ (4.7-5)

(4) $P(x, y)$ is a nondecreasing function of increasing either x or y or both, (4.7-6)

(5) $P(x, \infty) = P_X(x), P(\infty, y) = P_Y(y).$ (4.7-7)

We shall say more about property 5 when we discuss marginal probability density function (pdf). Briefly, $P_X(x)$ and $P_Y(y)$ are *marginal distribution functions,* and subscripts are chosen to indicate that they are not necessarily the same functions.

The joint pdf is defined as the two-variable extension of (4.3-8):

$$p(x, y) = \frac{\partial^2 P(x, y)}{\partial x \, \partial y}. \tag{4.7-8}$$

Hence,

$$P(x, y) = \int_{-\infty}^{y} \int_{-\infty}^{x} p(u, v) \, du \, dv, \tag{4.7-9}$$

which is property 3 of the properties of the density function listed below. The properties are the two-variable extensions of (4.3-9) through (4.3-12).

(1) $p(x, y) \geqslant 0, \quad$ all x and y, (4.7-10)

(2) $\displaystyle\int_{-\infty}^{\infty} \int_{-\infty}^{\infty} p(x, y) \, dx \, dy = 1,$ (4.7-11)

(3) $\displaystyle P(x, y) = \int_{-\infty}^{y} \int_{-\infty}^{x} p(u, v) \, du \, dv,$ (4.7-12)

(4) $P\{x_1 < X \leqslant x_2, y_1 < Y \leqslant y_2\}$

$$= \int_{y_1}^{y_2} \int_{x_1}^{x_2} p(x, y) \, dx \, dy, \tag{4.7-13}$$

(5) $\displaystyle p_X(x) = \int_{-\infty}^{\infty} p(x, y) \, dy,$ (4.7-14)

$$p_Y(y) = \int_{-\infty}^{\infty} p(x, y) \, dx. \tag{4.7-15}$$

Property 5 is a definition of *marginal density functions.* They are discussed further subsequently.

ISBN 0-201-05758-1

If the random variables X and Y are both discrete, their density function will be a sum of impulses. The two-variable extension of (4.4-2) is

$$p(x, y) = \sum_i \sum_j P(x_i, y_j)\delta(x - x_i, y - y_j). \tag{4.7-16}$$

However, for two-dimensional impulse functions

$$\delta(x - x_i, y - y_j) = \delta(x - x_i)\delta(y - y_j), \tag{4.7-17}$$

so

$$p(x, y) = \sum_i \sum_j P(x_i, y_j)\delta(x - x_i)\delta(y - y_j). \tag{4.7-18}$$

Here $P(x_i, y_j)$ is the probability of the joint occurrence of discrete values x_i and y_j. The sums are taken over all possible values of i and j.

Integrating (4.7-18) using (4.7-9) we have the cdf for two discrete random variables:

$$P(x, y) = \sum_i \sum_j P(x_i, y_j)u(x - x_i)u(y - y_j). \tag{4.7-19}$$

As usual, $u(\cdot)$ is the unit step function.

For the more general problem of N random variables $X_1, X_2, \ldots, X_N$ we define

$$P(x_1, x_2, \ldots, x_N) = P\{X_1 \leqslant x_1, X_2 \leqslant x_2, \ldots, X_N \leqslant x_N\}, \tag{4.7-20}$$

$$p(x_1, x_2, \ldots, x_N) = \frac{\partial^N P(x_1, x_2, \ldots, x_N)}{\partial x_1 \, \partial x_2 \ldots \partial x_N}. \tag{4.7-21}$$

Thus,

$$P(x_1, x_2, \ldots, x_N) = \int_{-\infty}^{x_N} \cdots \int_{-\infty}^{x_2} \int_{-\infty}^{x_1} p(u_1, u_2, \ldots, u_N)\, du_1\, du_2 \ldots du_N. \tag{4.7-22}$$

Marginal Distribution and Density Functions

We have already briefly touched on these topics. In a given problem a two-variable pdf may arise when one wants to know the pdf of only *one* of the variables. This may be obtained by simply integrating out one variable such as:

$$p_X(x) = \int_{-\infty}^{\infty} p(x, y)\, dy, \tag{4.7-23}$$

ISBN 0-201-05758-1

$$p_Y(y) = \int_{-\infty}^{\infty} p(x, y)\, dx. \tag{4-7.24}$$

The result is called a *marginal density function.*

We have already stated that the Rayleigh pdf represents the envelope (magnitude) of narrowband Gaussian noise. In the derivation the noise is actually described by two random variables, magnitude and phase. The Rayleigh pdf is a marginal pdf obtained by integrating out the phase variable.

Marginal distribution functions $P_X(x)$ and $P_Y(y)$ follow, in a somewhat similar manner, from (4.7-9). If $y \to \infty$,

$$P_X(x) = P(x, \infty) = \int_{-\infty}^{\infty} \int_{-\infty}^{x} p(u, v)\, du\, dv. \tag{4.7-25}$$

Upon using (4.7-23) in (4.7-25),

$$P_X(x) = \int_{-\infty}^{x} p_X(u)\, du. \tag{4.7-26}$$

The marginal cdf is the integral of the marginal pdf which is consistent with one-variable theory. If $x \to \infty$ we may similarly develop

$$P_Y(y) = P(\infty, y) = \int_{-\infty}^{y} \int_{-\infty}^{\infty} p(u, v)\, du\, dv \tag{4.7-27}$$

or

$$P_Y(y) = \int_{-\infty}^{y} p_Y(v)\, dv. \tag{4.7-28}$$

Finally, on differentiation of (4.7-26) and (4.7-28) we have

$$p_X(x) = \frac{dP_X(x)}{dx}, \tag{4.7-29}$$

$$p_Y(y) = \frac{dP_Y(y)}{dy} \tag{4.7-30}$$

for marginal quantities.

Conditional Distribution and Density Functions

Recall from (4.1-15) that, for two events A and B, the conditional probability of B is

$$P(B|A) = \frac{P(A, B)}{P(A)}. \tag{4.7-31}$$

ISBN 0-201-05758-1

We use this as a starting point and define events A and B by the sets $\{x < X \leqslant x + \Delta x\}$ and $\{Y \leqslant y\}$, respectively, with Δx a small quantity. Equation (4.7-31) becomes

$$P\{Y \leqslant y \mid x < X \leqslant x + \Delta x\} = \frac{P\{x < X \leqslant x + \Delta x, Y \leqslant y\}}{P\{x < X \leqslant x + \Delta x\}}. \tag{4.7-32}$$

If $\Delta x \to 0$, the left side of this expression becomes the probability of $\{Y \leqslant y\}$ given that X has the particular value x; we define this as the *conditional distribution function* $P(y|x)$ of Y.

Working now on the right side of (4.7-32) we use (4.7-13) and (4.3-12) to express numerator and denominator, respectively, as follows

$$P(y|x) = \frac{\displaystyle\int_{-\infty}^{y} \int_{x}^{x+\Delta x} p(u, v)\, du\, dv}{\displaystyle\int_{x}^{x+\Delta x} p_X(u)\, du}. \tag{4.7-33}$$

Again, if $\Delta x \to 0$, the integrand in the denominator is approximately constant at $p_X(x)$. The integral becomes $p_X(x)\,\Delta x$. Similar consideration in the numerator gives

$$P(y|x) = \frac{\displaystyle\int_{-\infty}^{y} p(x, v)\, dv}{p_X(x)} \tag{4.7-34}$$

for the conditional distribution function of Y.

The *conditional density function* $p(y|x)$ of Y is defined as the derivative of $P(y|x)$:

$$p(y|x) = \frac{dP(y|x)}{dy}. \tag{4.7-35}$$

Differentiation of (4.7-34) leads to

$$p(y|x) = \frac{p(x, y)}{p_X(x)} \tag{4.7-36}$$

or

$$p(x, y) = p(y|x)p_X(x). \tag{4.7-37}$$

Similar analysis produces the conditional distribution and density of X:

$$P(x|y) = \frac{\displaystyle\int_{-\infty}^{x} p(u, y)\, du}{p_Y(y)}, \tag{4.7-38}$$

ISBN 0-201-05758-1

$$p(x|y) = \frac{dP(x|y)}{dx} = \frac{p(x, y)}{p_Y(y)}. \tag{4.7-39}$$

Properties of the conditional density may be summarized.

$$(1)\ p(y|x) \geqslant 0, \qquad \text{all } y \text{ and } x, \tag{4.7-40}$$

$$(2)\ \int_{-\infty}^{\infty} p(y|x)\, dy = 1, \tag{4.7-41}$$

$$(3)\ P(y|x) = \int_{-\infty}^{y} p(v|x)\, dv, \tag{4.7-42}$$

$$(4)\ P\{y_1 < Y \leqslant y_2 | x\} = \int_{y_1}^{y_2} p(v|x)\, dv. \tag{4.7-43}$$

Properties of the conditional distribution may be found from these properties. We give an example of conditional pdf after considering statistical independence.

Statistical Independence

Two events A and B are statistically independent if (4.1-19) holds. If we define these events by the sets $\{X \leqslant x\}$ and $\{Y \leqslant y\}$, respectively, for two random variables X and Y, then X and Y are statistically independent if (and only if)

$$P\{X \leqslant x, Y \leqslant y\} = P\{X \leqslant x\}\, P\{Y \leqslant y\}. \tag{4.7-44}$$

In other words, joint distribution must be factorable:

$$P(x, y) = P_X(x)P_Y(y). \tag{4.7-45}$$

By applying (4.7-8) this also means a factorable density function is necessary:

$$p(x, y) = p_X(x)p_Y(y). \tag{4.7-46}$$

Equation (4.7-46) is also a sufficient condition for the definition of statistical independence of two variables.

For multiple random variables X_i we may define events corresponding to $\{X_i \leqslant x_i\}$ and follow similar procedures to define statistical independence using (4.1-20).

Example 4.7-1

One of the most direct and useful applications of conditional quantities is in evaluating the sum of signal plus independent noise in a system.

Let noise, represented by the random variable N having the pdf $p_N(n)$, be added to a signal, represented by the random variable X having pdf $p_X(x)$. The sum is represented by the

ISBN 0-201-05758-1

random variable $Y = X + N$. The joint density function of N and X is, since X and N are statistically independent,

$$p(n, x) = p_N(n)p_X(x).$$

However, since $Y = X + N$ may be considered to be a linear transformation of N to Y, the joint density of Y and X becomes

$$p(y, x) = p_N(y - x)p_X(x).$$

Thus, we recognize $p_N(y - x)$ as the conditional density $p(y|x)$ from (4.7-36).

Often we are more interested in a "typical" pdf of Y alone. We would then develop the marginal pdf of Y with the aid of (4.7-24):

$$p_Y(y) = \int_{-\infty}^{\infty} p_N(y - x)p_X(x)\, dx,$$

which is the convolution of noise and signal densities. Our example also shows that the pdf of a sum of two random variables is the convolution of the two pdf's if the variables are statistically independent.

Mean Values, Moments, and Correlation

The two-variable extension of (4.4-4) is

$$E[g(X, Y)] = \int_{-\infty}^{\infty}\int_{-\infty}^{\infty} g(x, y)p(x, y)dx\, dy, \tag{4.7-47}$$

where $g(X, Y)$ is some real function of the random variables X and Y.

The moments of X and Y result by letting $g(X, Y) = X^n$ and $g(X, Y) = Y^n$, respectively, for $n = 0, 1, 2, \ldots$:

$$E[X^n] = \int_{-\infty}^{\infty}\int_{-\infty}^{\infty} x^n p(x, y)\, dx\, dy = \int_{-\infty}^{\infty} x^n p_X(x)\, dx, \tag{4.7-48}$$

$$E[Y^n] = \int_{-\infty}^{\infty}\int_{-\infty}^{\infty} y^n p(x, y)\, dx\, dy = \int_{-\infty}^{\infty} y^n p_Y(y)\, dy. \tag{4.7-49}$$

Thus, moments of any one variable involve only the marginal pdf of that variable [see (4.7-23)]. Mean values m_X and m_Y result for $n = 1$.

Central moments are similarly derived. For, say, the variable X,

$$E[(X - m_X)^n] = \int_{-\infty}^{\infty}\int_{-\infty}^{\infty} (x - m_X)^n p(x, y)\, dx\, dy. \tag{4.7-50}$$

ISBN 0-201-05758-1

Generalized moments m_{nk} are defined by letting $g(X, Y) = X^n Y^k$ in (4.7-47):

$$m_{nk} = E[X^n Y^k] = \int_{-\infty}^{\infty} \int_{-\infty}^{\infty} x^n y^k p(x, y)\, dx\, dy. \tag{4.7-51}$$

The sum $n + k$ is the *order* of the moments. The first-order moments are the mean values: $m_{10} = m_X$ and $m_{01} = m_Y$. The second-order moment is $m_{11} = E[XY]$; it is called *correlation.* If the correlation of X and Y is factorable, then

$$m_{11} = E[X]E[Y], \tag{4.7-52}$$

and X and Y are said to be *uncorrelated.* If $m_{11} = 0$ they are said to be *orthogonal.* Statistical independence is sufficient to make X and Y uncorrelated as is easily shown from (4.7-51).

Generalized central moments u_{nk} are defined by

$$u_{nk} = E[(X - m_X)^n (Y - m_Y)^k] = \int_{-\infty}^{\infty} \int_{-\infty}^{\infty} (x - m_X)^n (y - m_Y)^k p(x, y)\, dx\, dy. \tag{4.7-53}$$

Two second central moments are the variances: $u_{20} = \sigma_X{}^2$ and $u_{02} = \sigma_Y{}^2$. The second central moment u_{11} is extremely important and is called *covariance.* If the covariance is zero, X and Y are uncorrelated. Sometimes covariance is normalized so as to be independent of the absolute magnitude of either X or Y. This quantity, called the *correlation coefficient* ρ, is

$$\rho = E\left[\left(\frac{X - m_X}{\sigma_X}\right)\left(\frac{Y - m_Y}{\sigma_Y}\right)\right] = \int_{-\infty}^{\infty} \int_{-\infty}^{\infty} \left(\frac{x - m_X}{\sigma_X}\right)\left(\frac{y - m_Y}{\sigma_Y}\right) p(x, y)\, dx\, dy. \tag{4.7-54}$$

It can be shown that (problem 4-33)

$$-1 \leqslant \rho \leqslant 1. \tag{4.7-55}$$

It is straightforward to generalize (4.7-47) to N variables $X_1, X_2, \ldots, X_N$:

$$E[g(X_1, X_2, \ldots, X_N)] = \int_{-\infty}^{\infty} \cdots \int_{-\infty}^{\infty} g(x_1, x_2, \ldots, x_N) p(x_1, x_2, \ldots, x_N)\, dx_1\, dx_2 \ldots dx_N. \tag{4.7-56}$$

Example 4.7-2

Let us find the mean value of a sum Y of N random variables X_i:

$$Y = \sum_{i=1}^{N} a_i X_i.$$

ISBN 0-201-05758-1

The a_i are arbitrary real weighting coefficients. From (4.7-56)

$$m_Y = E[Y] = \int_{-\infty}^{\infty} \cdots \int_{-\infty}^{\infty} \sum_{i=1}^{N} a_i x_i p(x_1, x_2, \ldots, x_N)\, dx_1\, dx_2 \ldots dx_N$$

$$= \sum_{i=1}^{N} a_i E[X_i] = \sum_{i=1}^{N} a_i m_{X_i}.$$

This equation states that *the mean value of a weighted sum of random variables equals the weighted sum of mean values.*

Example 4.7-3

Suppose we find the variance of Y in example 4.7-2 if the variables are jointly statistically independent:

$$\sigma_Y^2 = E[(Y - m_Y)^2] = \int_{-\infty}^{\infty} \cdots \int_{-\infty}^{\infty} \left\{ \sum_{i=1}^{N} a_i(x_i - m_{X_i}) \cdot \sum_{j=1}^{N} a_j(x_j - m_{X_j}) \right\}$$

$$p(x_1, x_2, \ldots, x_N)\, dx_1\, dx_2 \ldots dx_N.$$

Because of statistical independence this reduces to

$$\sigma_Y^2 = \sum_{i=1}^{N} a_i^2 \int_{-\infty}^{\infty} (x_i - m_{X_i})^2 p_{X_i}(x_i)\, dx_i$$

$$+ \sum_{\substack{i=1 \\ i \neq j}}^{N} \sum_{j=1}^{N} a_i a_j \int_{-\infty}^{\infty} (x_i - m_{X_i}) p_{X_i}(x_i)\, dx_i \int_{-\infty}^{\infty} (x_j - m_{X_j}) p_{X_j}(x_j)\, dx_j.$$

Here $p_{X_i}(x_i)$ is the marginal pdf of X_i. Every term in the double sum is zero, while the elements of the first sum are a_i^2 times the variances $\sigma_{X_i}^2$ of the X_i. Thus,

$$\sigma_Y^2 = \sum_{i=1}^{N} a_i^2 \sigma_{X_i}^2$$

proves that *the variance of a weighted sum of statistically independent random variables (weights* a_i*) equals the weighted sum of the variances of the individual variables (weights* a_i^2*).*

ISBN 0-201-05758-1

★ Characteristic Function

The multidimensional characteristic function is defined as

$$F(jv_1, jv_2, \ldots, jv_N) = E\left[\exp\left(j \sum_{i=1}^{N} v_i x_i\right)\right] \tag{4.7-57}$$

for random variables X_i, $i = 1, 2, \ldots, N$. For two variables X and Y,

$$F(jv_1, jv_2) = \int_{-\infty}^{\infty} \int_{-\infty}^{\infty} p(x, y) e^{jv_1 x + jv_2 y}\, dx\, dy, \tag{4.7-58}$$

which is the *two-dimensional Fourier transform* (with sign of j changed) of $p(x, y)$. From the inverse transform,

$$p(x, y) = \frac{1}{(2\pi)^2} \int_{-\infty}^{\infty} \int_{-\infty}^{\infty} F(jv_1, jv_2) e^{-jv_1 x - jv_2 y}\, dv_1\, dv_2 \tag{4.7-59}$$

completes the Fourier transform pair. The generalization to N variables is more or less obvious.

As in the one-variable case, the characteristic function allows us to find generalized moments. For two variables the exponentials in (4.7-58) may be put in series form to get

$$F(jv_1, jv_2) = \sum_{n=0}^{\infty} \sum_{k=0}^{\infty} \frac{(jv_1)^n (jv_2)^k}{n!k!}\, m_{nk}. \tag{4.7-60}$$

By differentiation of this equation we have

$$m_{nk} = (-j)^{n+k} \left. \frac{\partial^{n+k} F(jv_1, jv_2)}{\partial v_1{}^n \partial v_2{}^k} \right|_{\substack{v_1 = 0 \\ v_2 = 0}} . \tag{4.7-61}$$

★ Complex Random Variables

A *complex random variable* Z is defined by

$$Z = X + jY, \tag{4.7-62}$$

where X and Y are real random variables. A formal extension of the various topics already discussed for single and multiple random variables could be made for complex variables. However, since we shall have only limited occasion to use complex random variables, the present discussion is limited to the most important topics.

Let $g(Z)$ be defined as a function (real or complex) of the complex random variable Z. The expected value of $g(Z)$ must now be taken with respect to *two* random variables; that is,

ISBN 0-201-05758-1

$$E[g(Z)] = \int_{-\infty}^{\infty} \int_{-\infty}^{\infty} g(z)p(x, y)\, dx\, dy, \tag{4.7-63}$$

where $p(x, y)$ is the joint probability density function of X and Y. If $g(Z) = Z$, we have the *mean value* m_Z of Z:

$$m_Z = E[Z] = E[X] + jE[Y]. \tag{4.7-64}$$

Clearly, the mean is a complex quantity, and

$$E[Z^*] = \{E[Z]\}^*, \tag{4.7-65}$$

where * represents the complex conjugate operation. The *variance* $\sigma_Z{}^2$ of Z is defined as the mean value of the function $g(Z) = \{Z - E[Z]\}\{Z^* - E[Z^*]\}$. Thus,

$$\sigma_Z{}^2 = E\left[\{Z - E[Z]\}\{Z^* - E[Z^*]\}\right]$$

$$= E[|Z|^2] - |E[Z]|^2. \tag{4.7-66}$$

Two complex random variables Z_i and Z_j are statistically independent if the joint probability density function of the random variables X_i, Y_i, X_j, and Y_j satisfies

$$p(x_i, y_i, x_j, y_j) = p(x_i, y_i)p(x_j, y_j). \tag{4.7-67}$$

The extension to multiple variables is accomplished through the use of (4.1-20).

The *correlation* m_{11} and *covariance* u_{11} of two complex random variables are defined by

$$m_{11} = E[Z_i Z_j{}^*] \tag{4.7-68}$$

$$u_{11} = E\left[\{Z_i - E[Z_i]\}\{Z_j^* - E[Z_j^*]\}\right]. \tag{4.7-69}$$

Z_i and Z_j are said to be *uncorrelated variables* if their covariance is zero. On setting (4.7-69) to zero we find that correlation must be factorable for uncorrelated variables:

$$E[Z_i Z_j^*] = E[Z_i]\, E[Z_j^*], \qquad i \neq j. \tag{4.7-70}$$

Statistical independence of Z_i and Z_j is sufficient to guarantee that the variables are uncorrelated. Finally, we note that, if the correlation m_{11} is zero, two complex random variables are said to be *orthogonal.*

4.8. RANDOM PROCESSES

Several times it has been suggested that a random variable can represent a random waveform at an instant in time. In this section we shall show this to be true as part of a complete characterization of such waveforms through introduction of the concept of a random process.

ISBN 0-201-05758-1

The Random Process Concept

A random process may be considered as a generalization of a random variable to include time. Thus, if the random variable X becomes time-dependent, $X(t)$ is called a *random process.* If $X(t)$ is complex we have a *complex random process.* We shall deal almost entirely with real processes, and, unless otherwise indicated, all discussions imply $X(t)$ real. Just as x is a sample value of X, so is $x(t)$ a *sample function* of the process $X(t)$. Similarly, as the collection of all possible sample values x defines X, so the collection of all possible sample functions $x(t)$ defines the process $X(t)$. The collection of all sample functions is called an *ensemble,* and any one member $x(t)$ is an *ensemble member.* At a fixed instant in time, say t_1, averaging $X(t_1)$ over its values $x(t_1)$ amounts to averaging over the ensemble at time t_1. However, at the given instant in time this amounts to taking the expected value of X so expected values are also referred to as *ensemble averages.* They are also often called *statistical averages.*

To fix the concept of random process more firmly let us presume that we have an infinity of similar noise generators at our disposal. Each generator produces a noise waveform. The waveforms will all be different (but still noise) because no two generators are identical, owing to the mechanisms of noise generation as well as to random tolerance and selection of component parts in the manufacturing process. The various output waveforms are illustrated in Fig. 4.8-1. The whole set (infinite number) is the ensemble. Any one generator produces an ensemble member. At some time, say t_1, some probability density function [$p_X(x)$ at t_1] will define the statistical distribution of the voltage random variable X. Observe that, consistent with the random variable concept, the random experiment corresponds to choosing a generator. The experiment outcome is a generator (number $n = 1, 2, \ldots$); the outcome gives rise to the value $x_n(t_1)$ of the random variable $X(t_1)$. By finding the statistical average $E[X(t_1)]$ we find the average output at time t_1 *when all generators are considered.* We may then think of $E[X(t_1)]$ as the "typical" generator output for the generators being manufactured.

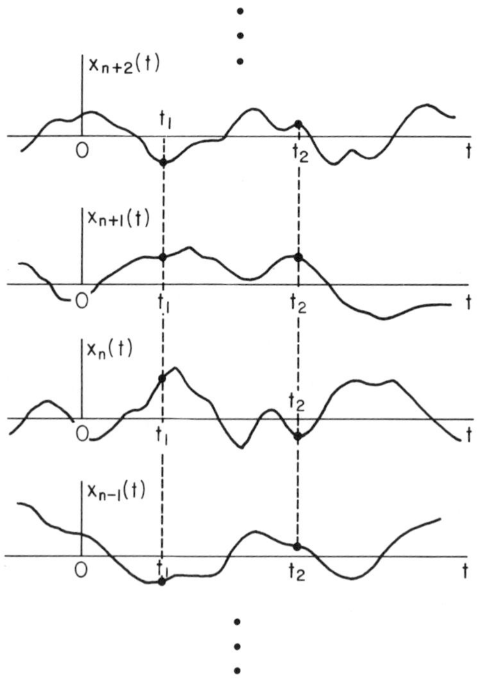

Fig. 4.8-1. A continuous random process.

ISBN 0-201-05758-1

The reader may ask, Why bring time into the problem at all? The answer hinges on the fact that a system must respond to *time* waveforms. Thus, if we are to determine its response to a random signal, where amplitude is random, we must combine time with a random variable description of amplitude. Hence, the random process. At this point we should comment on how the random process concept ties in with system analysis.

Consider the situation of a random signal $x(t)$ applied to a system to cause an output $y(t)$. The problem is to describe $y(t)$ in some way. Since $x(t)$ is random and cannot be described exactly, we cannot describe $y(t)$ exactly. However, it may be useful to obtain a "typical" or *expected* response. This expected value of the response is, of course, something that can be found if $x(t)$ is a sample function of a random process $X(t)$. The response $y(t)$ will then be a sample function of another (output) random process $Y(t)$. We may then find $E[Y(t)]$ in terms of characteristics of the network and input process (such as $E[X(t)]$).

Naturally, this discussion only scratches the surface, for one may want to know more about the system output than just the expected value. These refinements will become obvious through subsequent discussion. At this point we should only seek to understand the meaning of a random process and to become familiar with the idea of a random process being applied to a network rather than a single random waveform.

Types of Random Processes

Broadly speaking, there are four types of processes which correspond to whether the random variable or time is continuous or discrete.

A *continuous random process* has both a continuous random variable and continuous time. An example has already been given in the process of Fig. 4.8-1. Most noise problems arising in communication receivers involve this type of process.

If the random variable has a continuum of possible values but the values occur only at discrete points in time, we have a *continuous random sequence.* An example is illustrated in Fig. 4.8-2, where the process of Fig. 4.8-1 has been sampled periodically in time. Sample

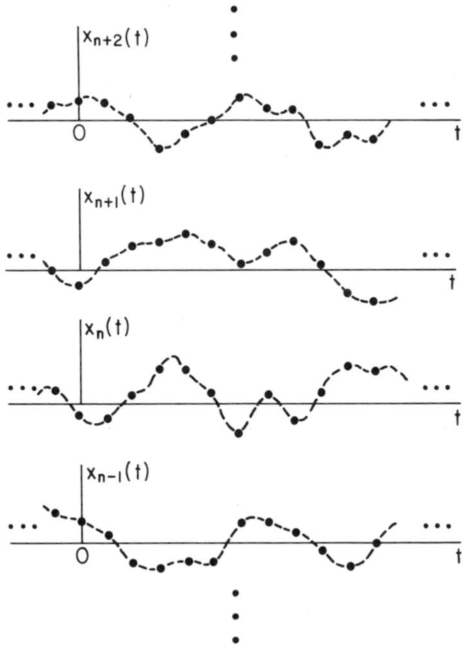

Fig. 4.8-2. A continuous random sequence.

ISBN 0-201-05758-1

values are indicated by the heavy dots. This situation is encountered if one transmits samples of some noisy message, such as over a pulsed communication system. The noise part of the sampled waveform is represented by the continuous random sequence.

A *discrete random process* occurs when the random variable has only discrete values but time is continuous. An example derives from severely limiting or clipping the process of Fig. 4.8-1 such that the resulting sample functions equal either a positive level or a negative level between zero crossings. Such a process is illustrated in Fig. 4.8-3.

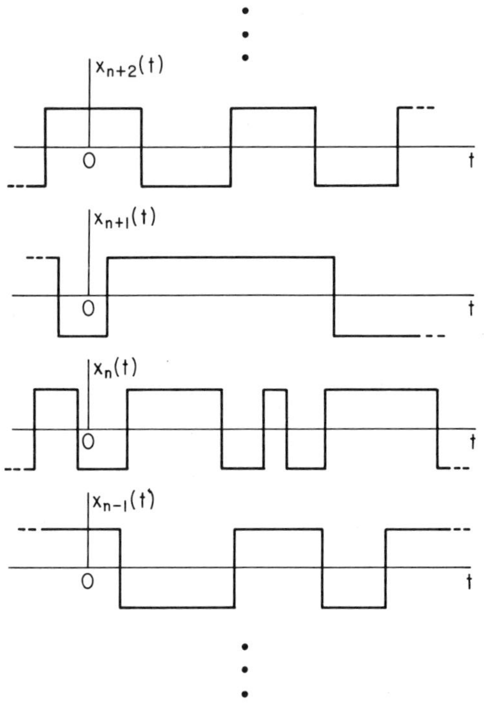

Fig. 4.8-3. A discrete random process.

The last type, a *discrete random sequence,* corresponds to both the random variable and time having only discrete values. As an example, suppose at uniformly separated time instants we toss a die and let values of the random variable be the numbers one to six on the die. The example discrete random sequence is sketched in Fig. 4.8-4.

Perhaps it should be pointed out at this point that random processes do not always have to involve the parameter time. Any other parameter will do. In the last example we could just as well have replaced time by, say, the number of the toss; that is, we could have used a numerical index. In this book we select time as the parameter of interest purely because engineers are most often interested in random time waveforms.

Stationarity

Broadly speaking, a random process is stationary if its statistical properties do not change under a shift in time origin. To be more specific we shall need to define *autocorrelation function* and consider multiple random variables defined by the process at multiple instants in time.

ISBN 0-201-05758-1

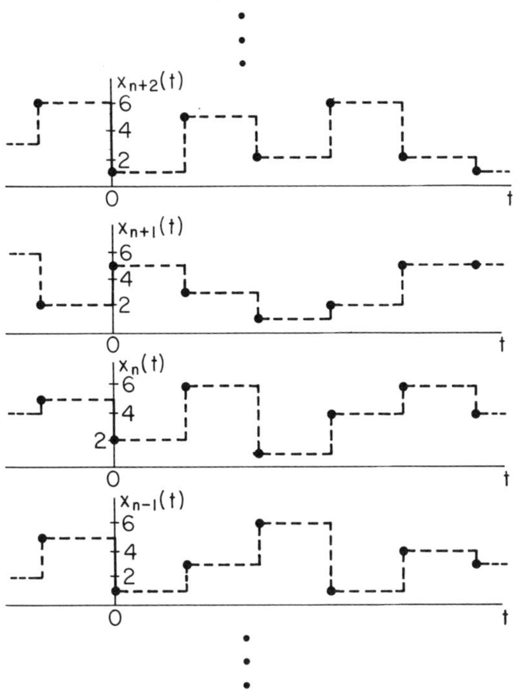

Fig. 4.8-4. A discrete random sequence.

Consider first two instants in time, t_1 and t_2, such as shown in Fig. 4.8-1. Random variables at these times are $X(t_1)$ and $X(t_2)$. With these definitions autocorrelation function is defined as

$$R_X(t_1, t_2) = E[X(t_1)X(t_2)]. \tag{4.8-1}$$

Generalizing to N random variables $X(t_i)$ at N time instants t_i, $i = 1, 2, \ldots, N$, then a random process is said to be stationary to order N on the time interval $(0, T)$ if, for all t_i and Δ such that t_i and $t_i + \Delta$ are in $(0, T)$,

$$p[x(t_1), x(t_2), \ldots, x(t_N)] = p[x(t_1 + \Delta), x(t_2 + \Delta), \ldots, x(t_N + \Delta)]. \tag{4.8-2}$$

That is, the joint pdf of the $X(t_i)$ is independent of an arbitrary shift Δ in the time origin. If $X(t)$ is stationary to *all* orders it is said to be stationary in the *strict sense* or *strictly stationary*. If $X(t)$ is stationary to order N it is also stationary to orders $n < N$.

For a first-order stationary process the *first-order pdf* is independent of absolute time:

$$p[x(t_1)] = p[x(t_1 + \Delta)] = p(x) \tag{4.8-3}$$

(let arbitrary Δ equal $-t_1$). This means that the mean value is constant:

$$E[X(t)] = m_X \text{ (constant)}. \tag{4.8-4}$$

For a second-order stationary process a consequence of (4.8-2) is that the autocorrelation function depends only on time *differences* $t_2 - t_1$. To show this fact we write (4.8-2) as

ISBN 0-201-05758-1

$$p[x(t_1), x(t_2)] = p[x(t_1 + \Delta), x(t_2 + \Delta)] \tag{4.8-5}$$

and form the autocorrelation function on both sides using (4.8-1). We have

$$E[X(t_1)X(t_2)] = E[X(t_1 + \Delta)X(t_2 + \Delta)] \tag{4.8-6}$$

or

$$R_X(t_1, t_2) = R_X(t_1 + \Delta, t_2 + \Delta). \tag{4.8-7}$$

Letting $\Delta = -t_1$, $R_X(t_1, t_2)$ becomes a function of time differences only:

$$R_X(t_1, t_2) = R_X(0, t_2 - t_1) = R_X(\tau), \tag{4.8-8}$$

where $\tau = t_2 - t_1$. In terms of the definition of autocorrelation function this result may be written in the very useful form

$$R_X(\tau) = E[X(t)X(t + \tau)] \tag{4.8-9}$$

for a second-order stationary process.

The assumption of strict stationarity is much more restrictive in practical problems than is necessary. Indeed, the extremely important class of problems involving Gaussian noise involves only first- and second-order moments. Even with other types of noise the most useful analytical results come from use of only first- and second-order moments. Thus, a more relaxed form of stationarity is often defined which is adequate for most problems.

We define a *wide-sense stationary process* as one that satisfies two conditions:

$$(1)\ E[X(t)] = m_X = \text{constant}, \tag{4.8-10}$$

$$(2)\ E[X(t)X(t + \tau)] = R_X(\tau). \tag{4.8-11}$$

A process stationary to order two is clearly wide-sense stationary. The converse is not necessarily true [6].

Example 4.8-1

We determine if the random process $X(t) = A \cos(\omega_0 t + \theta)$ is wide-sense stationary if θ is a uniformly distributed random *variable* with $0 \leqslant \theta \leqslant 2\pi$. Clearly,

$$E[X(t)] = \int_0^{2\pi} A \cos(\omega_0 t + \theta) \frac{1}{2\pi} d\theta = 0.$$

The autocorrelation function becomes

$$\begin{aligned} R_X(t, t + \tau) &= E[X(t)X(t + \tau)] \\ &= E[A^2 \cos(\omega_0 t + \theta) \cos(\omega_0 t + \omega_0 \tau + \theta)] \end{aligned}$$

ISBN 0-201-05758-1

$$= \frac{A^2}{2} E[\cos(\omega_0 \tau) + \cos(2\omega_0 t + \omega_0 \tau + 2\theta)]$$

$$= \frac{A^2}{2} \cos(\omega_0 \tau) + \frac{A^2}{2} E[\cos(2\omega_0 t + \omega_0 \tau + 2\theta)]$$

$$= \frac{A^2}{2} \cos(\omega_0 \tau).$$

Since the autocorrelation function does not depend on t and $E[X(t)]$ is constant (zero), then $X(t)$ is wide-sense stationary.

Sometimes it is necessary to define stationarity of one process $X(t)$ with respect to another $Y(t)$. We deal only with the wide-sense case. Two processes are *jointly* wide-sense stationary if each is separately wide-sense stationary and their *cross-correlation function* depends only on time difference $t_2 - t_1 = \tau$:

$$E[X(t)Y(t + \tau)] = R_{XY}(\tau), \tag{4.8-12}$$

where the cross-correlation function is defined generally by

$$R_{XY}(t_1, t_2) = E[X(t_1)Y(t_2)]. \tag{4.8-13}$$

Operations, Definitions, and Correlation Functions

In the above discussion of stationarity some of the main operations on random processes were introduced. The mean value operation is the same, for example, as with random variables and, as long as the process is at least wide-sense stationary, gives a constant (4.8-10). Auto- and cross-correlation operations, (4.8-1) and (4.8-13), are taken in a formal sense the same way as was done with moments in random variables. However, the results were just constant numbers for random variables, while with processes that are at least wide-sense stationary the numbers become functions of $\tau = t_2 - t_1$. This discussion will consider other operations and some definitions, and will develop properties of correlation functions. We limit ourselves to wide-sense processes.

The *autocovariance function* of $X(t)$ is the covariance of the random variables $X(t)$ and $X(t + \tau)$:

$$C_X(\tau) = E[\{X(t) - m_X\}\{X(t + \tau) - m_X\}]. \tag{4.8-14}$$

By direct expansion of the product,

$$C_X(\tau) = R_X(\tau) - m_X{}^2. \tag{4.8-15}$$

With $\tau = 0$ the variance $\sigma_X{}^2$ of $X(t)$ is seen to equal $C_X(0)$:

$$\sigma_X{}^2 = C_X(0) = R_X(0) - m_X{}^2. \tag{4.8-16}$$

ISBN 0-201-05758-1

This result agrees with earlier random variable theory [see (4.4-12)] by recognizing $R_X(0)$ as the second moment of $X(t)$.

Another second-order central moment of the processes $X(t)$ and $Y(t)$ is the *cross-covariance function*:

$$C_{XY}(\tau) = E[\{X(t) - m_X\}\{Y(t+\tau) - m_Y\}] \tag{4.8-17}$$

or

$$C_{XY}(\tau) = R_{XY}(\tau) - m_X m_Y. \tag{4.8-18}$$

By analogy with random variables, two processes are *uncorrelated* if

$$C_{XY}(\tau) = 0, \tag{4.8-19}$$

or, from (4.8-18), this requires a cross-correlation function which is a constant:

$$R_{XY}(\tau) = m_X m_Y = E[X(t)]E[Y(t)]. \tag{4.8-20}$$

They are *orthogonal* processes if

$$R_{XY}(\tau) = 0. \tag{4.8-21}$$

$X(t)$ and $Y(t)$ are *statistically independent* processes [1] if the random variable group

$$X(t_1), X(t_2), \ldots, X(t_n) \tag{4.8-22a}$$

is independent of the group

$$Y(t_1'), Y(t_2'), \ldots, Y(t_m') \tag{4.8-22b}$$

for any $t_1, t_2, \ldots, t_n, t_1', t_2', \ldots, t_m'$. This independence requires that the joint density be factorable by groups. That is,

$$p[X(t_1), \ldots, X(t_n), Y(t_1'), \ldots, Y(t_m')] = p[X(t_1), \ldots, X(t_n)]\,p[Y(t_1'), \ldots, Y(t_m')]. \tag{4.8-23}$$

From this definition it follows from (4.8-20) that two statistically independent processes are also uncorrelated. The converse situation is not necessarily true, although it is true for two jointly Gaussian processes [7, p. 59].

Some important properties of the autocorrelation function are:

$$(1)\ R_X(0) = E[X^2(t)], \quad \text{(mean-squared value)}, \tag{4.8-24}$$

$$(2)\ R_X(\tau) = R_X(-\tau) \quad \text{(symmetry)}, \tag{4.8-25}$$

$$(3)\ |R_X(\tau)| \leqslant R_X(0) \quad \text{(bound)}. \tag{4.8-26}$$

The first two follow easily from direct substitutions into the definition (4.8-9). The third

ISBN 0-201-05758-1

results from expansion of the quantity [6]

$$E[\{X(t) \pm X(t+\tau)\}^2] \geqslant 0. \qquad (4.8\text{-}27)$$

Some important properties of the cross-correlation function are:

$$(1)\ R_{XY}(-\tau) = R_{YX}(\tau) \qquad \text{(symmetry)}, \qquad (4.8\text{-}28)$$

$$(2)\ |R_{XY}(\tau)| \leqslant \sqrt{R_X(0)R_Y(0)} \qquad \text{(bound)}, \qquad (4.8\text{-}29)$$

$$(3)\ |R_{XY}(\tau)| \leqslant \frac{1}{2}[R_X(0) + R_Y(0)] \qquad \text{(bound)}. \qquad (4.8\text{-}30)$$

Property 1 follows from (4.8-12). Property 2 follows from expansion of the quantity

$$E[\{Y(t+\tau) + \alpha X(t)\}^2] \geqslant 0 \qquad (4.8\text{-}31)$$

and observing that the discriminant of the quadratic in α is nonpositive for any α. The proof of property 3 uses property 2 and the fact that the geometric mean of two positive numbers does not exceed their arithmetic mean; that is,

$$|R_{XY}(\tau)| \leqslant \sqrt{R_X(0)R_Y(0)} \leqslant \frac{1}{2}[R_X(0) + R_Y(0)]. \qquad (4.8\text{-}32)$$

Thus, property 2 constitutes a tighter bound on $|R_{XY}(\tau)|$ than does property 3.

Time Averages and Ergodicity

The *time average* of a quantity is defined as follows:

$$A[\cdot] = \lim_{T\to\infty} \frac{1}{2T} \int_{-T}^{T} [\cdot]\, dt. \qquad (4.8\text{-}33)$$

Here A is used to denote time average in a manner analogous to E for the statistical average. Time average is taken over the infinite interval because, as applied to a random process,* a sample function is a power signal existing for all time.

Specific time averages of interest are: the mean value m_x of a sample function (we use a lower-case subscript for a sample function time average):

$$m_x = A[x(t)] = \lim_{T\to\infty} \frac{1}{2T} \int_{-T}^{T} x(t)\, dt, \qquad (4.8\text{-}34)$$

and the *time autocorrelation function* $\mathcal{R}_x(\tau)$, already introduced in Section 2.8:

$$\mathcal{R}_x(\tau) = A[x(t)x(t+\tau)] = \lim_{T\to\infty} \frac{1}{2T} \int_{-T}^{T} x(t)x(t+\tau)\, dt. \qquad (4.8\text{-}35)$$

*In this discussion we consider only stationary processes.

ISBN 0-201-05758-1

For any one sample function of a random process these last two integrals simply give two numbers.† However, if we consider that we are dealing strictly with random processes we must consider the two numbers as values of *random variables.* Taking the mean values of both sides of these results and assuming the expectation can be brought inside the integrals, we have

$$E[m_x] = m_X, \tag{4.8-36}$$

$$E[\mathcal{R}_x(\tau)] = R_X(\tau). \tag{4.8-37}$$

Now it would be nice if, by some theorem, the random variables m_x and $\mathcal{R}_x(\tau)$ could be made to have zero variances. This would make the integrals of (4.8-34) and (4.8-35) equal to constants instead of random variables. Thus, time averages would equal statistical averages as follows:

$$m_x = m_X, \tag{4.8-38}$$

$$\mathcal{R}_x(\tau) = R_X(\tau). \tag{4.8-39}$$

The *ergodic theorem* stated in very loose terms allows the validity of (4.8-38) and (4.8.39), and, in a broader sense, allows time averages to equal statistical averages. Processes that satisfy the ergodic theorem are called *ergodic processes.*

Although the details of ergodic processes are too advanced to be treated in this text, we may develop a mental justification of the theorem. It seems reasonable that over infinite time each sample function of a random process may take on, at some time or other, all the statistically possible values of the process. If such is true, it seems reasonable that the time average of every sample function would be the same and (4.8-38) would be valid. A similar argument can be presented for (4.8-39).

It will often be assumed in this book that a random process describing a random waveform is ergodic. Thus, time averages equal statistical averages. We shall further presume that for *jointly ergodic processes*

$$\mathcal{R}_{xy}(\tau) = \lim_{T\to\infty} \frac{1}{2T} \int_{-T}^{T} x(t)y(t+\tau)\,dt = R_{XY}(\tau). \tag{4.8-40}$$

Measurement of Correlation Functions

Obviously we can never measure the true correlation functions $R_X(\tau)$ or $R_{XY}(\tau)$ for random processes because we do not have access in the real world to *all* sample functions or even to *one* complete sample function. The only recourse we have is to determine time averages over a finite time interval taken large enough to approximate the true result for ergodic processes. Because of our capability of only working in time, we are forced, like it or not, to presume that a given problem is ergodic. Fortunately, such as assumption is valid, or at least approximately so, in many practical cases.

Correlation functions of an ergodic random process may be approximately determined using the circuits of Fig. 4.8-5. A simple, practical means of achieving the integrator is to use

†Of course, we are considering (4.8-35) for a particular value of τ.

ISBN 0-201-05758-1

a low-pass filter having a rectangular impulse response over the time interval $(0, 2T)$, where $2T$ is the integration interval used. T is chosen to be large.

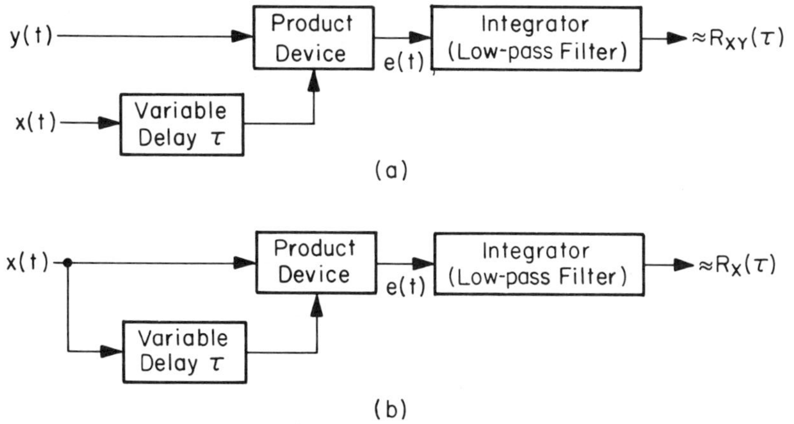

Fig. 4.8-5. Block diagrams of measurement systems for (a) cross-correlation, and (b) autocorrelation functions.

Let us now consider the circuit of (*a*) and show that the output approximately equals $R_{XY}(\tau)$. The development also shows that, if the low-pass filter transfer function is

$$H(\omega) = \frac{\sin(\pi\omega/W)}{\pi\omega/W} e^{-j\pi\omega/W}, \tag{4.8-41}$$

the integration indicated is realized if $W = \pi/T$. The required filter impulse response is

$$h(t) = \frac{W}{2\pi}\left[u(t) - u\left(t - \frac{2\pi}{W}\right)\right] \tag{4.8-42}$$

from transform pairs 3, 5, and 17 of Table 2.10-1. If the response of the network is denoted by $\hat{\mathcal{R}}_{xy}(t)$, then

$$\hat{\mathcal{R}}_{xy}(t) = \int_{-\infty}^{\infty} h(\zeta)e(t - \zeta)\, d\zeta$$

$$= \frac{W}{2\pi}\int_{0}^{2\pi/W} y(t - \zeta)x(t - \zeta - \tau)\, d\zeta. \tag{4.8-43}$$

Next, if we change variables by letting $\eta = t - \zeta - \tau$ and use the fact that $T = \pi/W$, (4.8-43) reduces to

$$\hat{\mathcal{R}}_{xy}(t) = \frac{1}{2T}\int_{t-\tau-2T}^{t-\tau} x(\eta)y(\eta + \tau)\, d\eta. \tag{4.8-44}$$

Now if we formulate our measurement as the value of $\hat{\mathcal{R}}_{xy}(t)$ when $t = \tau + T$, then

ISBN 0-201-05758-1

$$\hat{\mathcal{R}}_{xy}(\tau + T) = \frac{1}{2T}\int_{-T}^{T} x(\eta)y(\eta + \tau)\,d\eta \approx R_{XY}(\tau) \tag{4.8-45}$$

for $T \to \infty$. This corresponds to sampling the output at time $t = T + \tau$. In reality the integral (4.8-44) will tend to be independent of t if T is large so that the main requirement is only to wait $2T + \tau$ seconds after $x(t)$ is applied so that transients can die out.

It should be clear that our derivation applies to autocorrelation functions if $y(\eta + \tau)$ in (4.8-45) is replaced by $x(\eta + \tau)$.

★ Complex Random Processes

A *complex random process* $Z(t)$ may be considered as the generalization of a complex random variable Z to include time. Formally, we define $Z(t)$ by [8]

$$Z(t) = X(t) + jY(t), \tag{4.8-46}$$

where $X(t)$ and $Y(t)$ are real random processes. The complex process is said to be strictly stationary if $X(t)$ and $Y(t)$ are jointly stationary in the strict sense. Similarly, $Z(t)$ is wide-sense stationary if $X(t)$ and $Y(t)$ are jointly wide-sense stationary.

Mean value m_Z, *autocorrelation function* $R_Z(t, t+\tau)$ and *autocovariance function* $C_Z(t, t + \tau)$ are important quantities describing characteristics of a complex process. Mean value is defined by

$$m_Z = E[Z(t)] = E[X(t)] + jE[Y(t)]. \tag{4.8-47}$$

Clearly, we may also write

$$E[Z^*(t)] = \{E[Z(t)]\}^*. \tag{4.8-48}$$

Autocorrelation function is defined by

$$R_Z(t, t + \tau) = E[Z(t)Z^*(t + \tau)], \tag{4.8-49}$$

while autocovariance is defined as

$$C_Z(t, t + \tau) = E\left[\{Z(t) - E[Z(t)]\}\,\{Z^*(t + \tau) - E[Z^*(t + \tau)]\}\right]. \tag{4.8-50}$$

If $Z(t)$ is at least wide-sense stationary, these last two expressions are functions of τ and not absolute time:

$$R_Z(t, t + \tau) = R_Z(\tau), \tag{4.8-51}$$

$$C_Z(t, t + \tau) = C_Z(\tau). \tag{4.8-52}$$

From the definition of autocorrelation function we may easily determine its symmetry:

$$R_Z(-\tau) = R_Z^*(\tau). \tag{4.8-53}$$

ISBN 0-201-05758-1

For two complex random processes $Z_i(t)$ and $Z_j(t)$ the cross-correlation function is defined as

$$R_{Z_iZ_j}(t, t + \tau) = E[Z_i(t)Z_j^*(t + \tau)]. \tag{4.8-54}$$

Similarly, the cross-covariance function is defined by

$$C_{Z_iZ_j}(t, t + \tau) = E\Big[\{Z_i(t) - E[Z_i(t)]\}\ \{Z_j^*(t + \tau) - E[Z_j^*(t + \tau)]\}\Big]. \tag{4.8-55}$$

For processes which are at least wide-sense stationary these two expressions become functions of τ only:

$$R_{Z_iZ_j}(t, t + \tau) = R_{Z_iZ_j}(\tau), \tag{4.8-56}$$

$$C_{Z_iZ_j}(t, t + \tau) = C_{Z_iZ_j}(\tau). \tag{4.8-57}$$

The two processes are said to be *uncorrelated complex processes* if the cross-covariance function is zero. For uncorrelated processes

$$R_{Z_iZ_j}(t, t + \tau) = E[Z_i(t)]\,E[Z_j^*(t + \tau)]. \tag{4.8-58}$$

The two processes are called *orthogonal complex processes* if the cross-correlation function is zero.

4.9. SPECTRAL CHARACTERISTICS OF RANDOM PROCESSES

We know from working with Fourier transforms that any time waveform will have some type of spectral characterization. For deterministic signals it was just the Fourier transform or spectrum. But what about a random signal? Is there such a thing as a Fourier spectrum? If not, what then? We seek the answers to these questions in this section.

Power Density Spectrum

In Section 2.8 the power density spectrum was derived for deterministic power signals. Recall that it was a function describing how average power in the signal was distributed over frequency. Noise and random signals are also power signals, since we model them as sample functions from a random process. Our intuition tells us that a power density spectrum should then be possible for a random process. Indeed, this is true, and it is the basis for the spectral characterization of a process.

We present a derivation of the power density spectrum in order to gain insight into its underlying logic. We assume stationary processes. Let us begin by truncating the process $X(t)$ to the interval $(-T, T)$ to form $X_T(t)$; that is,

$$X_T(t) = X(t), \qquad -T < t < T$$
$$= 0, \qquad \text{elsewhere.} \tag{4.9-1}$$

ISBN 0-201-05758-1

Truncation allows a finite energy, so

$$\int_{-\infty}^{\infty} |x_T(t)|^2 \, dt < \infty \tag{4.9-2}$$

for a typical truncated sample function $x_T(t)$. Thus, the Fourier transform of $x_T(t)$ will exist,† and we may "transform" the random process $X_T(t)$ to a new *random process* $F_T(\omega)$:

$$F_T(\omega) = \int_{-\infty}^{\infty} X_T(t) e^{-j\omega t} \, dt \tag{4.9-3}$$

or

$$X_T(t) \leftrightarrow F_T(\omega). \tag{4.9-4}$$

The inverse, of course, holds:

$$X_T(t) = \frac{1}{2\pi} \int_{-\infty}^{\infty} F_T(\omega) e^{j\omega t} \, d\omega. \tag{4.9-5}$$

Using Parseval's theorem we may write the energy* $E_T(T)$ of the process over the interval $(-T, T)$:

$$E_T(T) = \int_{-T}^{T} X_T^2(t) \, dt = \frac{1}{2\pi} \int_{-\infty}^{\infty} |F_T(\omega)|^2 \, d\omega. \tag{4.9-6}$$

The average power $P_T(T)$ over the interval of time $2T$ is

$$P_T(T) = \frac{1}{2T} \int_{-T}^{T} X^2(t) \, dt = \frac{1}{2\pi} \int_{-\infty}^{\infty} \frac{|F_T(\omega)|^2}{2T} \, d\omega, \tag{4.9-7}$$

where over the region of the left integral (4.9-1) has been used. Now $P_T(T)$ is a random quantity. The mean power in the process over the interval $(-T, T)$ is found by taking the expected value

$$E[P_T(T)] = \frac{1}{2T} \int_{-T}^{T} E[X^2(t)] \, dt = \frac{1}{2\pi} \int_{-\infty}^{\infty} \frac{E[|F_T(\omega)|^2]}{2T} \, d\omega. \tag{4.9-8}$$

Finally, we obtain the total mean (average) power P by forming the limit to obtain

$$P = \lim_{T\to\infty} \{E[P_T(T)]\} = A\{E[X^2(t)]\} = \frac{1}{2\pi} \int_{-\infty}^{\infty} \lim_{T\to\infty} \frac{E[|F_T(\omega)|^2]}{2T} \, d\omega. \tag{4.9-9}$$

†Equation (4.9-2) is a more stringent condition for existence than is (2.3-17) [4, p. 133].
*We interpret $X(t)$ as a voltage process.

ISBN 0-201-05758-1

For stationary processes the time average of $E[X^2(t)]$ is just $E[X^2(t)]$, giving

$$P = E[X^2(t)] = \frac{1}{2\pi}\int_{-\infty}^{\infty} \mathcal{S}_X(\omega)\, d\omega, \tag{4.9-10}$$

where we define the power density spectrum $\mathcal{S}_X(\omega)$ as

$$\mathcal{S}_X(\omega) = \lim_{T\to\infty} \frac{E[|F_T(\omega)|^2]}{2T}. \tag{4.9-11}$$

The above discussion has shown two things. First, the average power in a stationary random process equals the mean-squared value $E[X^2(t)]$. This, of course, also equals the autocorrelation function for $\tau = 0$. Second, power may be obtained by integrating the power density spectrum, which is defined by (4.9-11).

Power Spectrum as the Fourier Transform of the Autocorrelation Function

It will now be shown that the power density spectrum is the Fourier transform of the autocorrelation function. That is,

$$\mathcal{S}_X(\omega) = \int_{-\infty}^{\infty} R_X(\tau) e^{-j\omega\tau}\, d\tau. \tag{4.9-12}$$

The inverse also holds:

$$R_X(\tau) = \frac{1}{2\pi}\int_{-\infty}^{\infty} \mathcal{S}_X(\omega) e^{j\omega\tau}\, d\omega. \tag{4.9-13}$$

We shall only show the proof of (4.9-12). For a proof of (4.9-13) the reader is referred to Thomas [6]. Our procedure is to develop the right side of (4.9-11). Using (4.9-3) and (4.9-1), we obtain

$$\begin{aligned}\frac{E[|F_T(\omega)|^2]}{2T} &= E\left\{\frac{1}{2T}\int_{-T}^{T} X(t)e^{-j\omega t}\, dt \int_{-T}^{T} X(\zeta)e^{j\omega\zeta}\, d\zeta\right\}\\ &= \frac{1}{2T}\int_{-T}^{T}\int_{-T}^{T} R_X(t-\zeta)e^{-j\omega(t-\zeta)}\, dt\, d\zeta. \end{aligned}\tag{4.9-14}$$

If next we make the changes of variable $\tau = t - \zeta$ and $\nu = \zeta$ in (4.9-14) and observe carefully the regions of integration, we may write

$$\frac{E[|F_T(\omega)|^2]}{2T} = \frac{1}{2T}\int_{-2T}^{0} R_X(\tau)e^{-j\omega\tau} \int_{-T-\tau}^{T} d\nu\, d\tau$$

ISBN 0-201-05758-1

$$+ \frac{1}{2T} \int_0^{2T} R_X(\tau) e^{-j\omega\tau} \int_{-T}^{T-\tau} dv \, d\tau$$

$$= \int_{-2T}^{2T} \left(1 - \frac{|\tau|}{2T}\right) R_X(\tau) e^{-j\omega\tau} \, d\tau. \tag{4.9-15}$$

Under the assumption that

$$\int_{-\infty}^{\infty} |\tau R_X(\tau)| \, d\tau < \infty \tag{4.9-16}$$

the second right-side term in (4.9-15) vanishes if we take the limit as $T \to \infty$. The remaining term shows that

$$\lim_{T\to\infty} \frac{E[|F_T(\omega)|^2]}{2T} = \mathcal{S}_X(\omega) = \int_{-\infty}^{\infty} R_X(\tau) e^{-j\omega\tau} \, d\tau \tag{4.9-17}$$

as we desired to prove. Equation (4.9-17) may also be generalized to include nonstationary processes [4, p. 147]. $R_X(\tau)$ is replaced by $A\,[R_X(t, t+\tau)]$.

Example 4.9-1

Let the power density spectrum be found for the random process $X(t)$ of example 4.8-1. It was previously shown that

$$R_X(\tau) = \frac{A^2}{2} \cos(\omega_0 \tau).$$

Fourier transformation results in

$$\mathcal{S}_X(\omega) = \int_{-\infty}^{\infty} R_X(\tau) e^{-j\omega\tau} \, d\tau$$

$$= \frac{A^2}{4} \int_{-\infty}^{\infty} [e^{-j(\omega-\omega_0)\tau} + e^{-j(\omega+\omega_0)\tau}] \, d\tau$$

$$= \frac{A^2 \pi}{2} [\delta(\omega - \omega_0) + \delta(\omega + \omega_0)].$$

Example 4.9-2

The autocorrelation function for a given noise process $N(t)$ is

$$R_N(\tau) = \frac{\alpha}{2} e^{-\alpha|\tau|},$$

ISBN 0-201-05758-1

where α is a real positive constant. (This function is also a *Laplace* probability density function having zero-mean value and variance $\sigma^2 = 2/\alpha^2$.) We find the power density spectrum of $N(t)$. From pair 27 of Table 2.10-1 it is

$$\mathscr{S}_N(\omega) = \frac{1}{1 + (\omega/\alpha)^2}.$$

This result can be shown to apply to broadband noise, with uniform spectral density, passed through a simple low-pass filter. Figure 4.9-1 illustrates $R_N(\tau)$ and $\mathscr{S}_N(\omega)$.

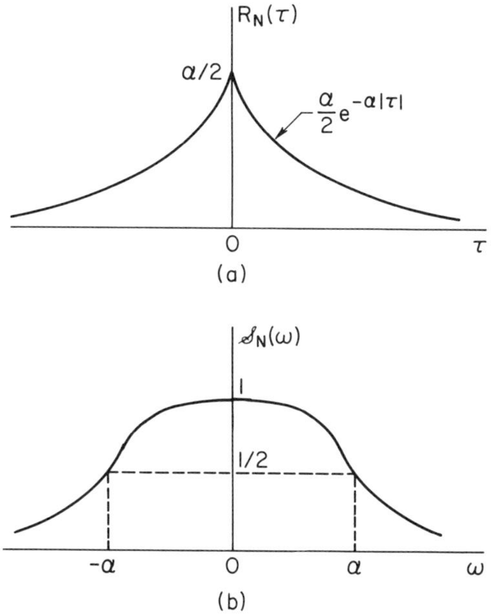

Fig. 4.9-1. Autocorrelation function (a), and power density spectrum (b), for low-pass filtered broadband noise. (See example 4.9-2.)

Example 4.9-3

A noise signal has the power density spectrum shown in Fig. 4.9-2(*b*). We shall find its autocorrelation function. The power spectrum is

$$\mathscr{S}_N(\omega) = \frac{\pi P_0}{W}\left[\text{rect}\left(\frac{\omega - \omega_0}{W}\right) + \text{rect}\left(\frac{\omega + \omega_0}{W}\right)\right].$$

Using pairs from Table 2.10-1 we easily find the inverse Fourier transform of this result to be

$$R_N(\tau) = P_0 \text{ Sa } (W\tau/2) \cos (\omega_0 \tau).$$

$R_N(\tau)$ is plotted in Fig. 4.9-2(*a*). Our example corresponds to broadband noise filtered by an ideal band-pass filter.

ISBN 0-201-05758-1

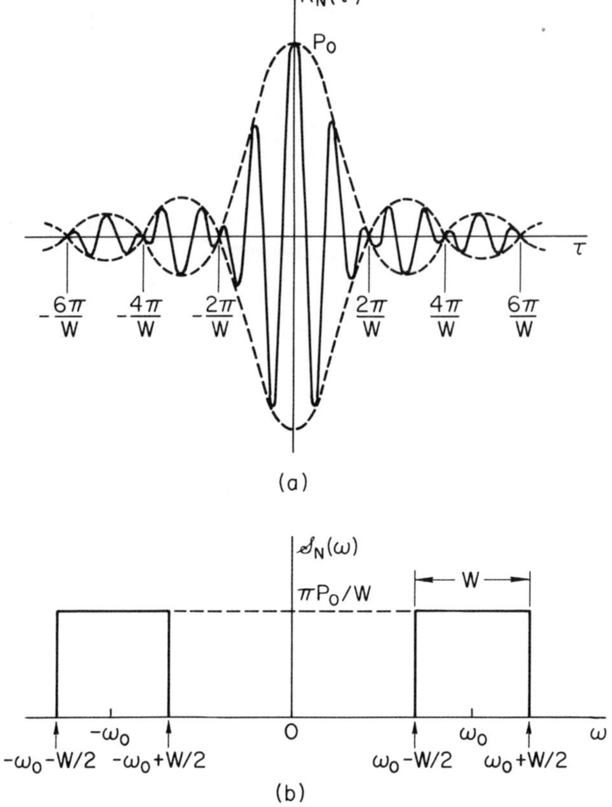

Fig. 4.9-2. Autocorrelation function (a), and power density spectrum (b), of band-pass filtered broadband noise.

Properties of the Power Density Spectrum

Some properties of $\mathscr{S}_X(\omega)$ are listed as follows:

$$\text{(1)}\quad \mathscr{S}_X(\omega) \geqslant 0 \qquad \text{(nonnegative function),} \tag{4.9-18}$$

$$\text{(2)}\quad \mathscr{S}_X(-\omega) = \mathscr{S}_X(\omega) \qquad \text{(symmetry),} \tag{4.9-19}$$

$$\text{(3)}\quad \mathscr{S}_X(\omega) \text{ is real,} \tag{4.9-20}$$

$$\text{(4)}\quad \frac{1}{2\pi}\int_{-\infty}^{\infty} \mathscr{S}_X(\omega)\, d\omega = E[X^2(t)] \qquad \text{(power expression).} \tag{4.9-21}$$

Property 1 derives from (4.9-11) by observing that the expected value of a nonnegative function must be nonnegative and T is positive. The second and third properties are proved from (4.9-12) by using the fact that $R_X(\tau)$ is a real, even function. Property 4 has already been shown from the derivation of (4.9-10).

Now since $\mathscr{S}_X(\omega)$ is real and even, we also have

ISBN 0-201-05758-1

$$E[X^2(t)] = \frac{1}{\pi}\int_0^\infty \mathcal{S}_X(\omega)\, d\omega$$

$$= 2\int_0^\infty \mathcal{S}_X(2\pi f)\, df. \tag{4.9-22}$$

Sometimes [4, p. 140] another property is given. If $\dot{X}(t) = dX(t)/dt$ is a process with power density spectrum $\mathcal{S}_{\dot{X}}(\omega)$, then a fifth property is

$$(5)\quad \mathcal{S}_{\dot{X}}(\omega) = \omega^2 \mathcal{S}_X(\omega) \qquad \text{(differentiation).} \tag{4.9-23}$$

The reader may wish to prove this property as an exercise. (See problem 4-42.)

Cross-Power Density Spectrum

Analogous to the power density spectrum for a single process we may define a cross-power density spectrum $\mathcal{S}_{XY}(\omega)$ for two processes $X(t)$ and $Y(t)$ as the Fourier transform of the cross-correlation function:

$$\mathcal{S}_{XY}(\omega) = \int_{-\infty}^{\infty} R_{XY}(\tau)e^{-j\omega\tau}\, d\tau. \tag{4.9-24}$$

The inverse holds:

$$R_{XY}(\tau) = \frac{1}{2\pi}\int_{-\infty}^{\infty} \mathcal{S}_{XY}(\omega)e^{j\omega\tau}\, d\omega. \tag{4.9-25}$$

The two results assume the processes are at least jointly wide-sense stationary.

For the cross-correlation function $R_{YX}(\tau)$ we have

$$R_{YX}(\tau) \leftrightarrow \mathcal{S}_{YX}(\omega). \tag{4.9-26}$$

Properties of the Cross-Power Density Spectrum

Several properties of $\mathcal{S}_{XY}(\omega)$ are listed below:

$$(1)\quad \mathcal{S}_{XY}(\omega) = \mathcal{S}_{YX}(-\omega) = \mathcal{S}^*_{YX}(\omega), \tag{4.9-27}$$

$$(2)\quad \text{Re}\,[\mathcal{S}_{XY}(\omega)] \text{ is an even function,} \tag{4.9-28}$$

$$(3)\quad \text{Im}\,[\mathcal{S}_{XY}(\omega)] \text{ is an odd function,} \tag{4.9-29}$$

$$(4)\quad \mathcal{S}_{XY}(\omega) = 0 \text{ if } X(t) \text{ and } Y(t) \text{ are orthogonal,} \tag{4.9-30}$$

$$(5)\quad \mathcal{S}_{XY}(\omega) = 2\pi m_X m_Y \delta(\omega) \text{ if } X(t) \text{ and } Y(t) \text{ are uncorrelated.} \tag{4.9-31}$$

ISBN 0-201-05758-1

These properties also hold for $\mathscr{S}_{YX}(\omega)$. Since $R_{XY}(\tau)$ is not necessarily an even function, $\mathscr{S}_{XY}(\omega)$ is not necessarily real and property 1 reflects this fact. It is proved by direct substitution into (4.9-24). Properties 2 and 3 follow from the first. The fourth property comes from the definition of orthogonal processes—that is, that $R_{XY}(\tau) = 0$ [see (4.8-21)]. Property 5 states that the cross-power density spectrum of uncorrelated processes consists of at most a dc component; it is proved from use of (4.8-20).

Measurement of Power Density Spectrum

A simple measurement system is shown in Fig. 4.9-3(*a*). The input is a sample function of a random process $X(t)$ having a power density spectrum $\mathscr{S}_X(\omega)$, such as that sketched in (*b*). We may think of $\mathscr{S}_X(\omega)$ as being impressed on the network. The input is multiplied by

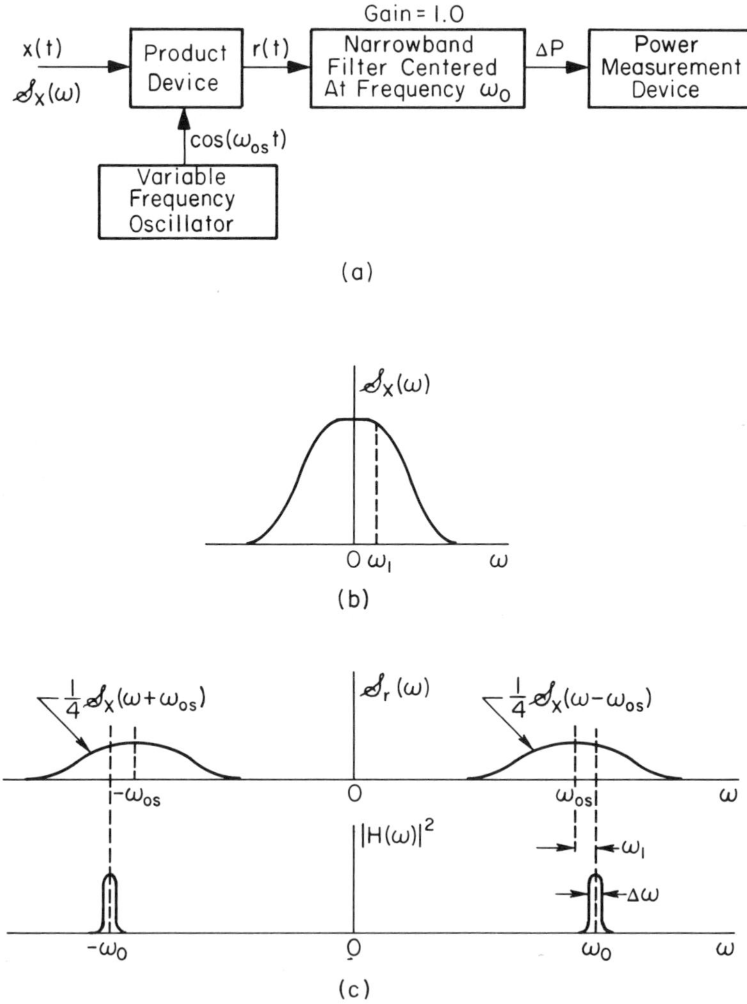

Fig. 4.9-3. A power density measurement system (a). (b) A typical power density spectrum at the input, and (c) power spectrum at product device output along with the narrowband filter's power transfer function.

ISBN 0-201-05758-1

cos $(\omega_{os}t)$, the output of a variable frequency oscillator, in a product device which serves to translate* the power spectrum to new center frequencies ω_{os} and $-\omega_{os}$ as shown in (*c*).

If the bandpass filter is centered at a frequency ω_0 and has a narrow bandwidth $\Delta\omega$ as shown in (*c*), then the average power ΔP emerging† from the filter will be approximately

$$\Delta P = \frac{1}{4\pi} \mathscr{S}_X(\omega_1)\, \Delta\omega, \tag{4.9-32}$$

where

$$\omega_1 = \omega_0 - \omega_{os}. \tag{4.9-33}$$

Since ω_0 is fixed, we may vary ω_{os} to give various values of ω_1. Thus, by measuring the average filter output power, the power density spectrum is found for the frequency ω_1:

$$\mathscr{S}_X(\omega_1) = 4\pi\, \Delta P/\Delta\omega. \tag{4.9-34}$$

In a practical system $\Delta\omega$ should be much smaller than the bandwidth of the power spectrum $\mathscr{S}_X(\omega)$ and would be the noise bandwidth of the network (see Section 4.11).

Simpler methods exist for measuring a power density spectrum. For example, the product device could be eliminated and the narrowband filter could have a variable center frequency. The main disadvantage of such a method is that the filter bandwidth $\Delta\omega$ changes as the center frequency changes. This disadvantage is not present in the system having the product device.

Bandwidth of the Power Density Spectrum

When we speak of the *bandwidth* of a power density spectrum $\mathscr{S}_X(\omega)$ we usually mean the frequency interval between points 3 dB down from the maximum (at $\omega = 0$ for low-pass and at ω_0 for band-pass processes). Hence, we imply the 3-dB bandwidth. However, rms bandwidth is very useful in many problems and is defined by

$$W_{\text{rms}}^2 = \frac{\displaystyle\int_{-\infty}^{\infty} \omega^2 \mathscr{S}_X(\omega)\, d\omega}{\displaystyle\int_{-\infty}^{\infty} \mathscr{S}_X(\omega)\, d\omega} \tag{4.9-35}$$

for low-pass processes, which is an extension of (3.4-14). A similar extension of (3.4-16) could also be made.

★ Power Spectra of Complex Random Processes

For complex processes that are at least wide-sense stationary, power density spectrum and cross-power density spectrum are given by (4.9-12) and (4.9-24). Thus, a complex random

*This operation is considered further in Section 4.11.

†The negative-frequency contribution doubles the positive-frequency power contribution. The bandwidth in the following expression should be the noise bandwidth of Section 4.11.

ISBN 0-201-05758-1

process $Z(t)$ has the power density spectrum

$$\mathscr{S}_Z(\omega) = \int_{-\infty}^{\infty} R_Z(\tau)e^{-j\omega\tau}\, d\tau, \tag{4.9-36}$$

while the cross-power density spectrum for two processes $Z_i(t)$ and $Z_j(t)$ is

$$\mathscr{S}_{Z_iZ_j}(\omega) = \int_{-\infty}^{\infty} R_{Z_iZ_j}(\tau)e^{-j\omega\tau}\, d\tau. \tag{4.9-37}$$

The Fourier inversion formulas may be used to recover the autocorrelation function $R_Z(\tau)$ if $\mathscr{S}_Z(\omega)$ is known. Similarly, the cross-correlation function $R_{Z_iZ_j}(\tau)$ may be recovered from knowledge of $\mathscr{S}_{Z_iZ_j}(\omega)$.

4.10. NETWORK TRANSMISSION OF RANDOM SIGNALS AND NOISE

All the theory described so far has little meaning unless it can be applied in some practical manner. To the engineer or applied scientist this means that it should help him better describe and understand his systems. We must then proceed, in a more or less natural way, to investigate the response of systems to random inputs. The usual situation calls for the input (random process or processes) to be specified or known in some manner, such as through its correlation functions or power density spectrum or both. The items we seek to find are the characteristics of the response, such as the correlation functions or power spectra or all of these.

We consider only stable linear time-invariant networks and wide-sense stationary processes as illustrated in Fig. 4.10-1. For such networks we need only one basic relationship from which we may derive all desired response characteristics. It is the convolution integral relating the network output $Y(t)$ to network input $X(t)$. Thus,

$$Y(t) = \int_{-\infty}^{\infty} X(t-u)h(u)\, du, \tag{4.10-1}$$

where $h(t)$ is the impulse response (assumed real) of the network. The input process power density spectrum is $\mathscr{S}_X(\omega)$, while that of the response is $\mathscr{S}_Y(\omega)$. The filter transfer function is $H(\omega)$. In some of our developments we let $X(t)$ represent a sum of jointly wide-sense stationary processes.

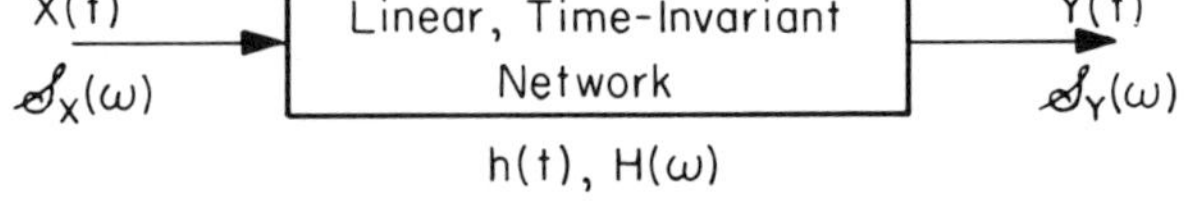

Fig. 4.10-1. A network illustrating definitions of input and output random processes.

Output Correlation Functions

The mean value of the response is found from (4.10-1):

ISBN 0-201-05758-1

$$m_Y = E[Y(t)] = \int_{-\infty}^{\infty} E[X(t-u)]\,h(u)\,du$$

$$= m_X \int_{-\infty}^{\infty} h(u)\,du. \tag{4.10-2}$$

If the input process has zero mean the output process also has zero mean.

The output autocorrelation function becomes

$$R_Y(t, t+\tau) = E[Y(t)Y(t+\tau)] = E\left[\int_{-\infty}^{\infty} X(t-u)h(u)\,du \int_{-\infty}^{\infty} X(t+\tau-v)h(v)\,dv\right], \tag{4.10-3}$$

or, since $E[X(t-u)X(t+\tau-v)] = R_X(\tau + u - v)$,

$$R_Y(\tau) = \int_{-\infty}^{\infty}\int_{-\infty}^{\infty} R_X(\tau + u - v)h(u)h(v)\,du\,dv. \tag{4.10-4}$$

This result indicates that the output autocorrelation function is the two-fold convolution of the input autocorrelation function with the system impulse response.

Observe that, since $E[Y(t)]$ is a constant and the output autocorrelation function depends only on τ, a result of $X(t)$ being wide-sense stationary, the output $Y(t)$ is wide-sense stationary.

The average power P_Y in the output process is

$$P_Y = R_Y(0) = \int_{-\infty}^{\infty}\int_{-\infty}^{\infty} R_X(u-v)h(u)h(v)\,du\,dv. \tag{4.10-5}$$

The cross-correlation function for input and output processes becomes

$$R_{XY}(\tau) = E[X(t)Y(t+\tau)] = E\left[X(t)\int_{-\infty}^{\infty} X(t+\tau-u)h(u)\,du\right] \tag{4.10-6}$$

or

$$R_{XY}(\tau) = \int_{-\infty}^{\infty} R_X(\tau-u)h(u)\,du. \tag{4.10-7}$$

In words, the cross-correlation function of the input with the output is the single convolution of the input autocorrelation function and the system impulse response.

ISBN 0-201-05758-1

In a practical problem, even if $R_X(\tau)$ is known, it may not be simple to solve the integrals associated with the output mean value, autocorrelation function, or cross-correlation functions. In many cases it is easier to resort to spectral analysis.

Output Spectral Characteristics

The output power density spectrum is perhaps most easily obtained by direct Fourier transformation of the autocorrelation function:

$$\mathscr{S}_Y(\omega) = \int_{-\infty}^{\infty} R_Y(\tau)e^{-j\omega\tau}\,d\tau. \tag{4.10-8}$$

Using (4.10-4)

$$\mathscr{S}_Y(\omega) = \int_{-\infty}^{\infty} h(u) \int_{-\infty}^{\infty} h(v) \int_{-\infty}^{\infty} R_X(\tau + u - v)e^{-j\omega\tau}\,d\tau\,dv\,du. \tag{4.10-9}$$

By introducing the variable change $w = \tau + u - v$ we get

$$\mathscr{S}_Y(\omega) = \int_{-\infty}^{\infty} h(u)e^{j\omega u}\,du \int_{-\infty}^{\infty} h(v)e^{-j\omega v}\,dv \int_{-\infty}^{\infty} R_X(w)e^{-j\omega w}\,dw. \tag{4.10-10}$$

These three integrals are recognized as $H^*(\omega)$, $H(\omega)$, and $\mathscr{S}_X(\omega)$, respectively, owing to our assumption of real system impulse response. As a result, the output power density spectrum becomes

$$\mathscr{S}_Y(\omega) = |H(\omega)|^2\mathscr{S}_X(\omega), \tag{4.10-11}$$

which is an extremely important result. It states that the output power spectrum is the product of the input power spectrum and the *power transfer function* $|H(\omega)|^2$ of the network.

Development of the cross-power density spectrum of the system output is similar to the above. By operating on (4.10-7) we derive

$$\mathscr{S}_{XY}(\omega) = H(\omega)\mathscr{S}_X(\omega). \tag{4.10-12}$$

From (4.9-27) we also have

$$\mathscr{S}_{YX}(\omega) = H(-\omega)\mathscr{S}_X(\omega). \tag{4.10-13}$$

System Response to a Sum of Inputs

The behavior of a system when several processes are simultaneously applied to its input is of great interest. The obvious example comes to mind where, of two processes, one represents a random signal while the second represents a noise. We shall develop the two-process case in the following discussion. Generalization to more inputs is straightforward.

ISBN 0-201-05758-1

The input $X(t)$ is now the sum of two processes $X_1(t)$ and $X_2(t)$, which are jointly wide-sense stationary:

$$X(t) = X_1(t) + X_2(t). \tag{4.10-14}$$

Because the network is linear, each input gives rise to a corresponding output, so

$$Y(t) = Y_1(t) + Y_2(t). \tag{4.10-15}$$

The mean value of $Y(t)$ is easily found from an application of (4.10-2):

$$E[Y(t)] = m_Y = (m_{X_1} + m_{X_2}) \int_{-\infty}^{\infty} h(u)\, du. \tag{4.10-16}$$

To evaluate the output autocorrelation function we must evaluate the terms of

$$\begin{aligned} R_Y(\tau) &= E[\{Y_1(t) + Y_2(t)\}\{Y_1(t + \tau) + Y_2(t + \tau)\}] \\ &= R_{Y_1}(\tau) + R_{Y_2}(\tau) + R_{Y_1Y_2}(\tau) + R_{Y_2Y_1}(\tau). \end{aligned} \tag{4.10-17}$$

The first two terms present no problem as they are direct applications of (4.10-4). However, the remaining terms are new. They are the cross-correlations between the *output components* $Y_1(t)$ and $Y_2(t)$. By direct evaluation we may show that

$$R_{Y_1Y_2}(\tau) = \int_{-\infty}^{\infty} \int_{-\infty}^{\infty} R_{X_1X_2}(\tau + u - v) h(u) h(v)\, du\, dv, \tag{4.10-18}$$

which relates output component correlation to input component correlation. Substituting (4.10-18) and (4.10-4) for each component into (4.10-17) and noting that $R_Y(\tau)$ must also be given by (4.10-4), we find that $R_X(\tau)$ has the same form as $R_Y(\tau)$. That is,

$$R_X(\tau) = R_{X_1}(\tau) + R_{X_2}(\tau) + R_{X_1X_2}(\tau) + R_{X_2X_1}(\tau). \tag{4.10-19}$$

By a similar analysis it can be shown that the cross-correlation function between input and output is given by

$$R_{XY}(\tau) = R_{X_1Y_1}(\tau) + R_{X_2Y_2}(\tau) + R_{X_1Y_2}(\tau) + R_{X_2Y_1}(\tau). \tag{4.10-20}$$

Each term is given by proper application of (4.10-7). An alternative development will show that if (4.10-19) is used in (4.10-7) we again obtain (4.10-20).

By Fourier transforming (4.10-19) and using (4.10-11) we have the output power spectrum:

$$\mathcal{S}_Y(\omega) = |H(\omega)|^2 [\mathcal{S}_{X_1}(\omega) + \mathcal{S}_{X_2}(\omega) + \mathcal{S}_{X_1X_2}(\omega) + \mathcal{S}_{X_2X_1}(\omega)]. \tag{4.10-21}$$

ISBN 0-201-05758-1

A special case of interest allows the two inputs to be statistically independent (alternatively we allow uncorrelated inputs). Applying (4.8-20) to (4.10-19) gives

$$R_X(\tau) = R_{X_1}(\tau) + R_{X_2}(\tau) + 2m_{X_1}m_{X_2}. \tag{4.10-22}$$

The Fourier transform of (4.10-22) reveals that

$$\mathscr{S}_X(\omega) = \mathscr{S}_{X_1}(\omega) + \mathscr{S}_{X_2}(\omega) + 4\pi m_{X_1}m_{X_2}\delta(\omega). \tag{4.10-23}$$

If either input is zero mean,

$$R_X(\tau) = R_{X_1}(\tau) + R_{X_2}(\tau), \tag{4.10-24}$$

$$\mathscr{S}_X(\omega) = \mathscr{S}_{X_1}(\omega) + \mathscr{S}_{X_2}(\omega). \tag{4.10-25}$$

In other words, the autocorrelation function or power spectrum of the sum of two statistically independent (or uncorrelated), zero-mean, random processes equals the sum of individual autocorrelation functions or power spectra.

System Evaluation Using Random Waveforms

Suppose a system with unknown impulse response $h(t)$ is to be evaluated by finding $h(t)$. A procedure using a random waveform can be evolved from (4.10-7). If we assume that an available source random process $X(t)$ is very broadband and has uniform power density spectrum, $h(t)$ may be found using the setup of Fig. 4.10-2. The cross-correlation measurement system would be as shown in Fig. 4.8-5.

Since $\mathscr{S}_X(\omega)$ is assumed approximately constant (at a level $\mathscr{N}_0/2$, say), its autocorrelation function will be approximately

$$R_X(\tau) = \frac{\mathscr{N}_0}{2}\delta(\tau). \tag{4.10-26}$$

Using this fact in (4.10-7) we have

$$\begin{aligned} R_{XY}(\tau) &= \int_{-\infty}^{\infty} \frac{\mathscr{N}_0}{2}\delta(\tau - u)h(u)\,du \\ &= \frac{\mathscr{N}_0}{2}h(\tau). \end{aligned} \tag{4.10-27}$$

Now the measurement $\hat{R}_{XY}(\tau) \approx R_{XY}(\tau)$, so

$$h(\tau) \approx \frac{2}{\mathscr{N}_0}\hat{R}_{XY}(\tau). \tag{4.10-28}$$

In other words, the output $\hat{R}_{XY}(\tau)$ of Fig. 4.10-2, when scaled by the constant $2/\mathscr{N}_0$, is approximately the value of $h(\tau)$. Since τ is a variable controlled within the cross-correlation measurement system (see Fig. 4.8-5), we may vary τ and obtain a complete record of $h(\tau)$.

ISBN 0-201-05758-1

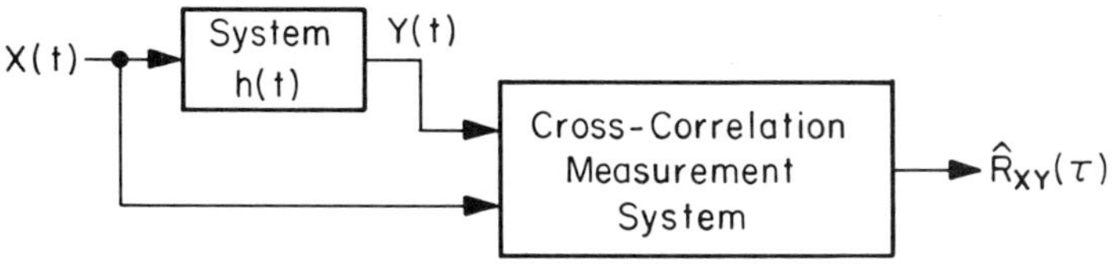

Fig. 4.10-2. Method for determining a system's impulse response.

4.11. SOME NOISE DEFINITIONS AND TOPICS

It will be helpful in later discussions to make certain definitions of types of noise. The noise process is labeled $N(t)$. In addition we consider certain noise topics used in several places throughout the rest of the book.

White Noise

Earlier we found that a random (noise) waveform could be used to find the impulse response of a network. The waveform's power density spectrum was required to be constant over a broad frequency range. *White noise* satisfies this requirement. We define white noise as having a constant power density spectrum at *all* frequencies. Thus,

$$\mathcal{S}_N(\omega) = \mathcal{N}_0/2, \tag{4.11-1}$$

where $\mathcal{N}_0$ is a constant as illustrated in Fig. 4.11-1(*b*). The autocorrelation function of white noise is shown in Fig. 4.11-1(*a*). The term "white" derives from an analogy to white light which contains all visible light frequencies in its spectrum.

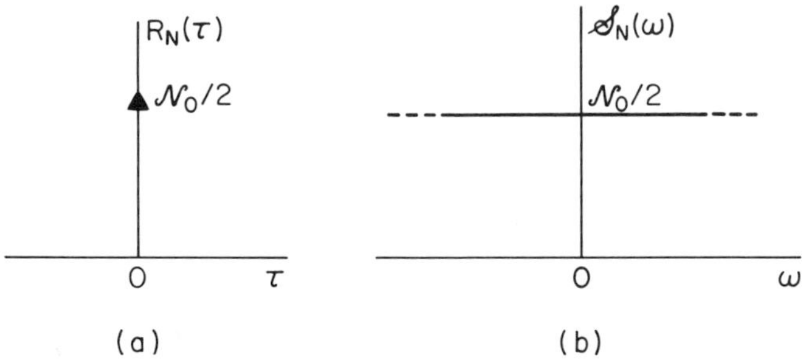

Fig. 4.11-1. Characteristics of white noise. (a) Autocorrelation function, and (b) power density spectrum.

In reality there is no truly white noise. However, thermal noise, such as that derived from a resistive network (see Chapter 8), approximates white noise out to about 10^{13} Hz, which is well beyond the frequency of nearly every communication system of interest.* Other noise types may also be considered as white noise for analysis purposes when the bandwidth of the noise is much larger than the bandwidth of the network or system to which it is applied.

*Our comments neglect systems at optical frequencies where lasers are important.

ISBN 0-201-05758-1

Colored Noise

Noise characterized by a power density spectrum which is either bandlimited or nonuniform can be classified as *colored noise.* As an example, white noise passed through a network having a transfer function $H(\omega)$ will emerge as colored noise $N_o(t)$ at the output with power density spectrum

$$\mathscr{S}_{N_o}(\omega) = \frac{\mathscr{N}_0}{2} |H(\omega)|^2. \tag{4.11-2}$$

Colored noise derives its name from analogy with colored light which possesses only a band of frequencies in the visible spectrum.

Noise Response of a Product Device

Product devices have already been encountered in connection with correlation measurement methods (Fig. 4.8-5) and power density spectrum measurement methods (Fig. 4.9-3). A particular product operation which is used in nearly every communication receiver involving a modulated carrier signal is illustrated in Fig. 4.11-2. We shall determine the power density spectrum of the response $Y(t)$ to an input wide-sense stationary random process $X(t)$.

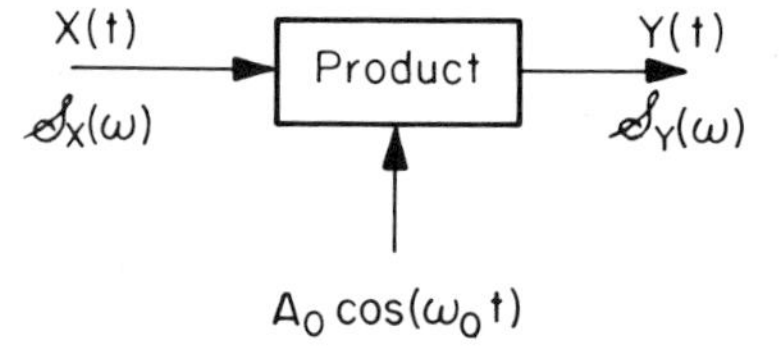

Fig. 4.11-2. A product of interest in communication receivers.

We elect to find $\mathscr{S}_Y(\omega)$ by first finding the output autocorrelation function and Fourier transforming to get the power spectrum. The definition leads to

$$\begin{aligned} R_Y(t, t+\tau) &= E[Y(t)Y(t+\tau)] \\ &= E[A_0{}^2 X(t)X(t+\tau)\cos(\omega_0 t)\cos\{\omega_0(t+\tau)\}] \\ &= \frac{A_0{}^2 R_X(\tau)}{2}[\cos(\omega_0\tau) + \cos\{2\omega_0 t + \omega_0\tau\}], \end{aligned} \tag{4.11-3}$$

which is seen to be nonstationary, since it is time-dependent. As mentioned following (4.9-17), the power density can still be obtained for nonstationary process as follows:

$$\mathscr{S}_Y(\omega) = \int_{-\infty}^{\infty} A[R_Y(t, t+\tau)]\, e^{-j\omega\tau}\, d\tau. \tag{4.11-4}$$

The time average of (4.11-3) is

ISBN 0-201-05758-1

$$A\left[R_Y(t, t+\tau)\right] = \frac{A_0{}^2 R_X(\tau)}{2}\cos(\omega_0 \tau). \tag{4.11-5}$$

Fourier transformation leads to

$$\mathscr{S}_Y(\omega) = \frac{A_0{}^2}{4}\left[\mathscr{S}_X(\omega - \omega_0) + \mathscr{S}_X(\omega + \omega_0)\right]. \tag{4.11-6}$$

Retracing the procedure will show that this same result is obtained if $\cos(\omega_0 t)$ is replaced by $\sin(\omega_0 t)$.

The power density spectrum for $X(t)$ and that given by (4.11-6) are illustrated in Fig. 4.11-3. The figure presumes that $X(t)$ has a power spectrum clustered around $\omega = 0$. This is not a constraint, as $\mathscr{S}_X(\omega)$ may also be band-pass as next discussed.

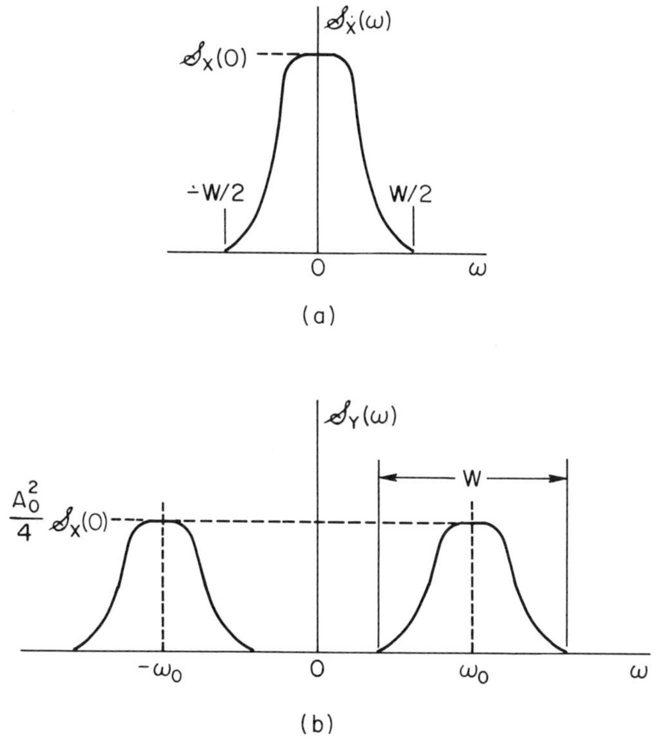

Fig. 4.11-3. Power density spectra appropriate to Fig. 4.11-2. (a) For the input, and (b) for the output.

Band-Pass and Narrowband Gaussian Noise

Communication systems using carrier modulation all involve band-pass waveforms. In practically all cases the systems are narrowband. The networks are usually designed with large enough bandwidth to pass the information-bearing signals of interest but not so large as to admit excessive noise. Thus, system bandwidth is usually on the order of the signal bandwidth which is typically much smaller than the carrier frequency. System noise will occupy the sys-

ISBN 0-201-05758-1

tem bandwidth. These facts lead us to define *band-pass noise* as that having a nonzero power density spectrum confined to a frequency band W centered about some frequency ω_0, while *narrowband noise* is a band-pass noise having the additional constraint of small bandwidth. That is,

$$\omega_0 >> W. \tag{4.11-7}$$

Because all the spectral components of narrowband noise have frequencies of approximately ω_0, the noise, as viewed on an oscilloscope, would tend to look much like a sinusoid of frequency ω_0 which slowly undulates in magnitude and phase. Figure 4.11-4 illustrates these points.

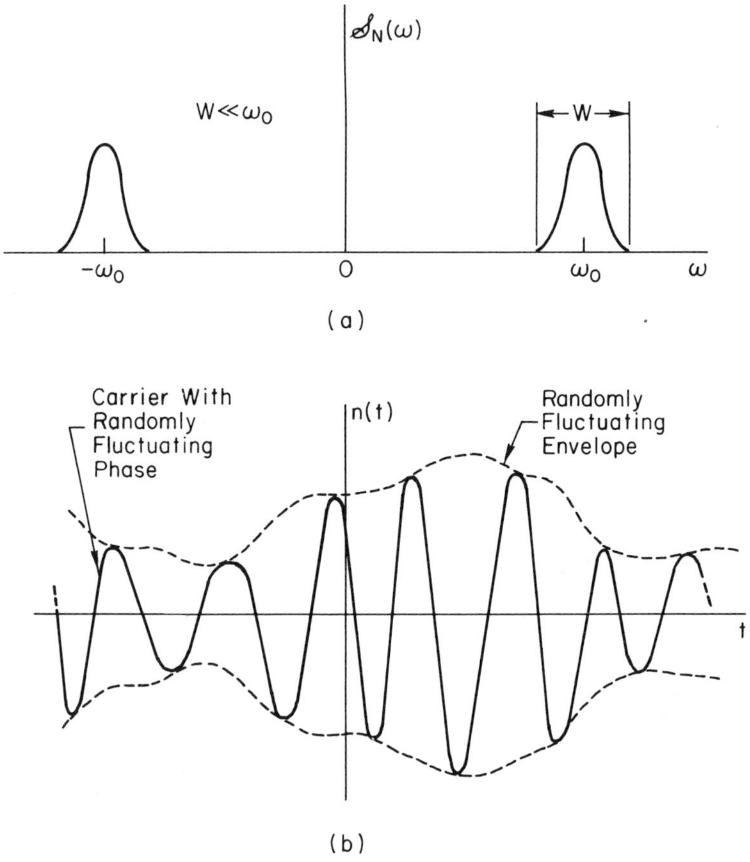

Fig. 4.11-4. Narrowband noise. (a) Power density spectrum, and (b) the waveform.

The appearance of narrowband noise suggests that it may be represented by a random process:

$$N(t) = A(t) \cos [\omega_0 t + \Theta(t)], \tag{4.11-8}$$

where $A(t)$ is a random process representing the slowly varying amplitude and $\Theta(t)$ is a process representing the slowly varying phase. Indeed this is the case, and it results that $A(t)$ and $\Theta(t)$ have Rayleigh and uniform first-order probability density functions, respectively,

ISBN 0-201-05758-1

for Gaussian noise $N(t)$. The *processes* are not statistically independent [7], but at any one instant in time the process *random variables* are independent.

In some problems (4.11-8) is a preferred representation for $N(t)$. For others it is convenient to use the equivalent form:

$$N(t) = X_c(t) \cos(\omega_0 t) - X_s(t) \sin(\omega_0 t), \tag{4.11-9}$$

where

$$X_c(t) = A(t) \cos[\Theta(t)], \tag{4.11-10}$$

$$X_s(t) = A(t) \sin[\Theta(t)]. \tag{4.11-11}$$

Expressions relating $A(t)$ and $\Theta(t)$ to $X_c(t)$ and $X_s(t)$ are

$$A(t) = \sqrt{X_c^2(t) + X_s^2(t)}, \tag{4.11-12}$$

$$\Theta(t) = \tan^{-1}[X_s(t)/X_c(t)]. \tag{4.11-13}$$

The representation of (4.11-9) is most important in describing narrowband, zero-mean, stationary, Gaussian (normal) noise processes. However, it is actually more general than implied, since it applies even to *wideband* normal noise. For the wideband case $X_c(t)$ and $X_s(t)$, postulated to be normal stationary zero-mean processes, are *not* statistically independent in general [9, p. 352], although the random variables $X_c(t)$ and $X_s(t)$ for a given time t *are* statistically independent. There are two cases, however, for which the processes $X_c(t)$ and $X_s(t)$ become either exactly or approximately independent. In the extreme case where the noise has infinite bandwidth, becoming white noise, the processes are statistically independent in an exact sense. On the other hand, $X_c(t)$ and $X_s(t)$ will be approximately independent processes if the noise $N(t)$ is narrowband and has a power density spectrum $\mathcal{S}_N(\omega)$ that is symmetric about ω_0. The approximation is reasonable enough for most practical applications [9].

Returning to the more general case, the following properties of $X_c(t)$ and $X_s(t)$ apply to either narrowband or band-pass Gaussian noise:

$$(1)\ \mathcal{S}_{X_c}(\omega) = \mathcal{S}_{X_s}(\omega), \tag{4.11-14}$$

$$(2)\ \mathcal{S}_{X_c}(\omega) = \begin{cases} \mathcal{S}_N(\omega - \omega_0) + \mathcal{S}_N(\omega + \omega_0), & |\omega| < W/2 \\ 0, & \text{elsewhere}, \end{cases} \tag{4.11-15}$$

$$(3)\ R_{X_c}(\tau) = \frac{1}{\pi}\int_0^\infty \mathcal{S}_N(\omega) \cos[(\omega - \omega_0)\tau]\, d\omega, \tag{4.11-16}$$

$$(4)\ R_{X_s}(\tau) = R_{X_c}(\tau), \tag{4.11-17}$$

$$(5)\ R_{X_c X_s}(\tau) = \frac{1}{\pi}\int_0^\infty \mathcal{S}_N(\omega) \sin[(\omega - \omega_0)\tau]\, d\omega, \tag{4.11-18}$$

ISBN 0-201-05758-1

$$(6)\ E[X_c^2(t)] = E[X_s^2(t)] = E[N^2(t)], \tag{4.11-19}$$

$$(7)\ E[X_c(t)X_s(t)] = 0. \tag{4.11-20}$$

Property 1 follows from 4; 2 is proved during development of the proof of 3; 6 follows from 4 and 3 with $\tau = 0$; and property 7 follows from 5 with $\tau = 0$. Thus, all these properties result easily if 3, 4, and 5 can be demonstrated. All three of these proofs depend on showing that $X_c(t)$ and $X_s(t)$ are produced by the networks of Fig. 4.11-5. Let us show this fact first and then prove 3. Proofs of 4 and 5 are similar to that of 3.

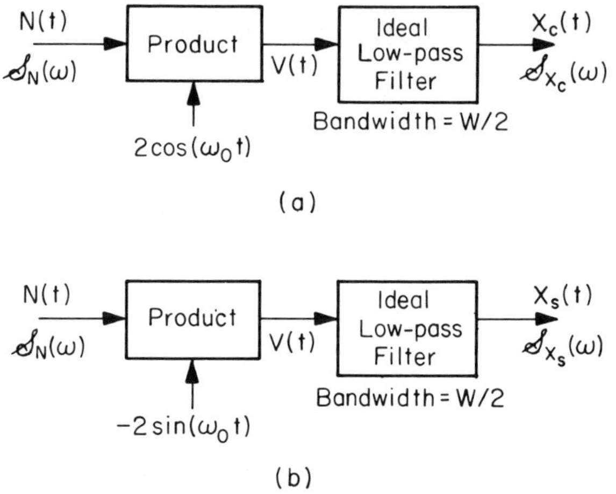

Fig. 4.11-5. Networks which give (a) $X_c(t)$, and (b) $X_s(t)$.

For the network in (*a*) the process $V(t)$ is

$$\begin{aligned} V(t) &= 2N(t)\cos(\omega_0 t) \\ &= 2X_c(t)\cos^2(\omega_0 t) - 2X_s(t)\sin(\omega_0 t)\cos(\omega_0 t) \\ &= X_c(t) + [X_c(t)\cos(2\omega_0 t) - X_s(t)\sin(2\omega_0 t)]. \end{aligned} \tag{4.11-21}$$

The term in brackets is just the input band-pass process shifted to a new center frequency $2\omega_0$, while $X_c(t)$ is a low-pass process centered about $\omega = 0$. The ideal filter will pass $X_c(t)$ unaltered but eliminates the band-pass component. Hence, the output is $X_c(t)$. A similar development yields $X_s(t)$ for the circuit of (*b*).

Having shown that Fig. 4.11-5(*a*) gives $X_c(t)$ as its output, we may calculate the autocorrelation function of $X_c(t)$ using the circuit. We apply the procedure of Davenport [2, p. 518]. The basic output autocorrelation function is

$$E[X_c(t)X_c(t+\tau)] = E\left[\int_{-\infty}^{\infty} h(u)V(t-u)\,du \int_{-\infty}^{\infty} h(v)V(t+\tau-v)\,dv\right]. \tag{4.11-22}$$

Substituting for $V(t)$,

ISBN 0-201-05758-1

$$R_{X_c}(t, t+\tau) = E[X_c(t)X_c(t+\tau)]$$

$$= \int_{-\infty}^{\infty}\int_{-\infty}^{\infty} h(u)h(v)E[N(t-u)N(t+\tau-v)]$$

$$\cdot\, 4\cos\{\omega_0(t-u)\}\cos\{\omega_0(t+\tau-v\}\, du\, dv. \qquad (4.11\text{-}23)$$

We recognize that $E[N(t-u)N(t+\tau-v)] = R_N(\tau+u-v)$. Replacing $R_N(\tau+u-v)$ by its equivalent, the inverse Fourier transform of the power density spectrum, and expanding the cosine terms into their exponential forms allow us to write (4.11-23) as four terms:

$$R_{X_c}(t, t+\tau) =$$

$$\frac{e^{j2\omega_0 t}}{2\pi}\int_{-\infty}^{\infty} \mathscr{S}_N(\omega)e^{j(\omega+\omega_0)\tau}\int_{-\infty}^{\infty} h(u)e^{j(\omega-\omega_0)u}\, du \int_{-\infty}^{\infty} h(v)e^{-j(\omega+\omega_0)v}\, dv\, d\omega$$

$$+\frac{1}{2\pi}\int_{-\infty}^{\infty} \mathscr{S}_N(\omega)e^{j(\omega+\omega_0)\tau}\int_{-\infty}^{\infty} h(u)e^{j(\omega+\omega_0)u}\, du \int_{-\infty}^{\infty} h(v)e^{-j(\omega+\omega_0)v}\, dv\, d\omega$$

$$+\frac{1}{2\pi}\int_{-\infty}^{\infty} \mathscr{S}_N(\omega)e^{j(\omega-\omega_0)\tau}\int_{-\infty}^{\infty} h(u)e^{j(\omega-\omega_0)u}\, du \int_{-\infty}^{\infty} h(v)e^{-j(\omega-\omega_0)v}\, dv\, d\omega$$

$$+\frac{e^{-j2\omega_0 t}}{2\pi}\int_{-\infty}^{\infty} \mathscr{S}_N(\omega)e^{j(\omega-\omega_0)\tau}\int_{-\infty}^{\infty} h(u)e^{j(\omega+\omega_0)u}\, du \int_{-\infty}^{\infty} h(v)e^{-j(\omega-\omega_0)v}\, dv\, d\omega.$$

$$(4.11\text{-}24)$$

This result is readily placed in the form

$$R_{X_c}(t, t+\tau) = \frac{e^{j2\omega_0 t}}{2\pi}\int_{-\infty}^{\infty} \mathscr{S}_N(\omega)[H^*(\omega-\omega_0)H(\omega+\omega_0)]\, e^{j(\omega+\omega_0)\tau}\, d\omega$$

$$+\frac{1}{2\pi}\int_{-\infty}^{\infty} \mathscr{S}_N(\omega)[|H(\omega+\omega_0)|^2 e^{j(\omega+\omega_0)\tau} + |H(\omega-\omega_0)|^2 e^{j(\omega-\omega_0)\tau}]\, d\omega$$

$$+\frac{e^{-j2\omega_0 t}}{2\pi}\int_{-\infty}^{\infty} \mathscr{S}_N(\omega)[H^*(\omega+\omega_0)H(\omega-\omega_0)]\, e^{j(\omega-\omega_0)\tau}\, d\omega. \qquad (4.11\text{-}25)$$

At this point observe that $H^*(\omega-\omega_0)H(\omega+\omega_0) = 0$ and $H^*(\omega+\omega_0)H(\omega-\omega_0) = 0$ because the shifted ideal transfer functions do not overlap. Thus, the first and third integrals are zero. In the second, let $\zeta_1 = \omega + \omega_0$ in the first term and $\zeta_2 = \omega - \omega_0$ in the second term.

ISBN 0-201-05758-1

Then

$$R_{X_c}(t, t+\tau) = \frac{1}{2\pi}\int_{-\infty}^{\infty} \mathscr{S}_N(\zeta_1 - \omega_0)|H(\zeta_1)|^2 e^{j\zeta_1\tau}\, d\zeta_1$$

$$+ \frac{1}{2\pi}\int_{-\infty}^{\infty} \mathscr{S}_N(\zeta_2 + \omega_0)|H(\zeta_2)|^2 e^{j\zeta_2\tau}\, d\zeta_2. \qquad (4.11\text{-}26)$$

Now on replacing the dummy variables by ω and noting that the right side is independent of t, we get

$$R_{X_c}(\tau) = \frac{1}{2\pi}\int_{-W/2}^{W/2} [\mathscr{S}_N(\omega + \omega_0) + \mathscr{S}_N(\omega - \omega_0)]\, e^{j\omega\tau}\, d\omega. \qquad (4.11\text{-}27)$$

The limits derive from $|H(\omega)|^2 = 1$, $|\omega| < W/2$, and $|H(\omega)|^2 = 0$, elsewhere. Equation (4.11-26) proves (4.11-15). It only remains to prove (4.11-16) using (4.11-27) or its equivalent, the middle right-side term of (4.11-25). The latter may be written as

$$R_{X_c}(\tau) = \frac{1}{2\pi}\int_{-\infty}^{0} \mathscr{S}_N(\omega)e^{j(\omega+\omega_0)\tau}\, d\omega + \frac{1}{2\pi}\int_{0}^{\infty} \mathscr{S}_N(\omega)e^{j(\omega-\omega_0)\tau}\, d\omega. \qquad (4.11\text{-}28)$$

In the first term we let $\zeta = -\omega$, use the fact that $\mathscr{S}_N(-\zeta) = \mathscr{S}_N(\zeta)$, and then replace ζ with ω. The result is

$$R_{X_c}(\tau) = \frac{1}{2\pi}\int_{0}^{\infty} \mathscr{S}_N(\omega)e^{-j(\omega-\omega_0)\tau}\, d\omega + \frac{1}{2\pi}\int_{0}^{\infty} \mathscr{S}_N(\omega)e^{j(\omega-\omega_0)\tau}\, d\omega. \qquad (4.11\text{-}29)$$

Finally, combining these terms proves (4.11-16). As already mentioned, proofs of (4.11-17) and (4.11-18) are similar to the above proof.

Noise Bandwidth

As a final topic of this section we define a quantity called *noise bandwidth,* W_N, of a network. It is the bandwidth of an equivalent ideal rectangular passband filter, having the same mid-band gain as the actual filter, and producing the same output noise power for white noise applied at the input.

If the actual filter transfer function is $H(\omega)$, the output average noise power P_N is

$$P_N = \frac{1}{2\pi}\int_{-\infty}^{\infty} \frac{\mathscr{N}_0}{2}|H(\omega)|^2\, d\omega, \qquad (4.11\text{-}30)$$

ISBN 0-201-05758-1

where $\mathscr{N}_0/2$ is the density of white noise power at the input. The average noise power emerging from the equivalent filter with noise bandwidth W_N is

$$P_N = \frac{1}{2\pi}\frac{\mathscr{N}_0}{2} 2W_N |H(\omega_0)|^2, \tag{4.11-31}$$

where ω_0 is the center frequency of the actual filter. The noise bandwidth is found from equating these two expressions:

$$W_N = \frac{\int_{-\infty}^{\infty} |H(\omega)|^2 \, d\omega}{2|H(\omega_0)|^2} = \frac{\int_{0}^{\infty} |H(\omega)|^2 \, d\omega}{|H(\omega_0)|^2}. \tag{4.11-32}$$

The second form results from the fact that $|H(\omega)|^2$ is an even function of ω.* If $H(\omega)$ is a low-pass filter transfer function, $\omega_0 = 0$ and

$$W_N = \frac{\int_{-\infty}^{\infty} |H(\omega)|^2 \, d\omega}{2|H(0)|^2} = \frac{\int_{0}^{\infty} |H(\omega)|^2 \, d\omega}{|H(0)|^2}. \tag{4.11-33}$$

Noise bandwidth is also discussed in Chapter 9.

4.12. MATCHED FILTERS

In modern digital data transfer systems a common signal might be a stream of ones (pulses) and zeros (absence of pulses) mixed with noise as illustrated in Fig. 4.12-1. The problem of the receiving system is to determine if, at an instant in time, a pulse or just noise is present. For such problems reason indicates that the ability to correctly determine which is true will be better for a large ratio of peak signal power (as measured at the signal's maximum) to average noise power. For a given input signal, it is conceivable that some filter may exist that will enhance the signal as much as possible while reducing noise as much as possible until the ratio of peak signal power to average noise power is a maximum. Such a filter is called a *matched filter* because, as we shall see, its transfer function must be "matched" to the signal spectrum.

Filter for Colored Noise

Let the more general problem of signal plus colored noise be investigated first. Let the sum $q(t)$ of a signal $f(t)$ and noise $N(t)$, such that

$$q(t) = f(t) + N(t), \tag{4.12-1}$$

be applied to a network with transfer function $H(\omega)$. The signal component, $f_o(t)$, of the filter output is

*This fact is always true for networks having a real impulse response. Networks which may be constructed always satisfy this condition.

ISBN 0-201-05758-1

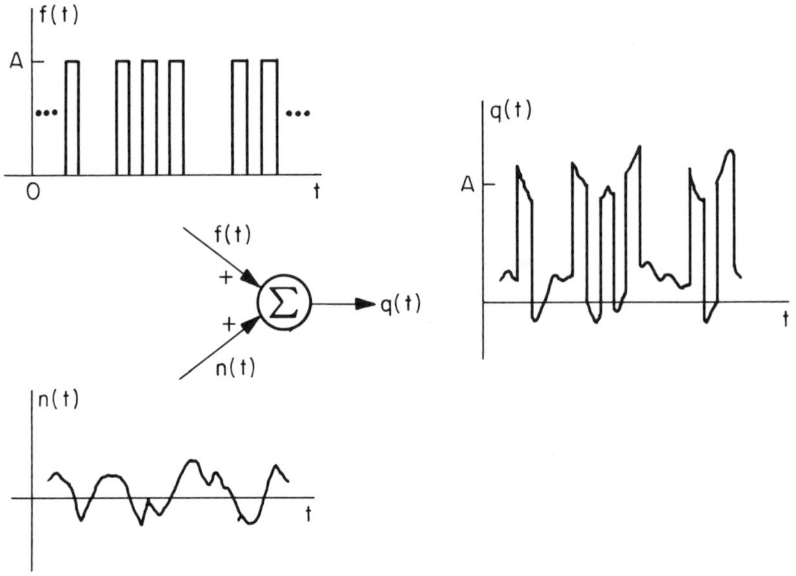

Fig. 4.12-1. A digital signal plus noise.

$$f_o(t) = \frac{1}{2\pi} \int_{-\infty}^{\infty} F(\omega)H(\omega)e^{j\omega t}\, d\omega, \tag{4.12-2}$$

where $F(\omega)$ is the Fourier spectrum of $f(t)$. Let t_0 be the time that the output signal component is maximum. The peak signal power, S_o, of the output signal at time t_0 is

$$S_o = |f_o(t_0)|^2 = \left| \frac{1}{2\pi} \int_{-\infty}^{\infty} F(\omega)H(\omega)e^{j\omega t_0}\, d\omega \right|^2. \tag{4.12-3}$$

The average output noise power, N_o, is given by the output noise autocorrelation function when $\tau = 0$. Using (4.10-11) the average output noise power is

$$N_o = \frac{1}{2\pi} \int_{-\infty}^{\infty} \mathscr{S}_N(\omega)|H(\omega)|^2\, d\omega, \tag{4.12-4}$$

where $\mathscr{S}_N(\omega)$ is the input colored noise power density spectrum. The ratio of powers to be maximized becomes

$$\left(\frac{S_o}{N_o}\right) = \frac{\left|(1/2\pi) \int_{-\infty}^{\infty} F(\omega)H(\omega)e^{j\omega t_0}\, d\omega \right|^2}{(1/2\pi) \int_{-\infty}^{\infty} \mathscr{S}_N(\omega)|H(\omega)|^2\, d\omega}. \tag{4.12-5}$$

To maximize (4.12-5) we may apply the *Schwarz inequality.* If $A(\omega)$ and $B(\omega)$ are in

ISBN 0-201-05758-1

general complex functions of the real variable ω, the inequality states that

$$\left| \int_{-\infty}^{\infty} A(\omega)B(\omega)\, d\omega \right|^2 \leqslant \int_{-\infty}^{\infty} |A(\omega)|^2\, d\omega \int_{-\infty}^{\infty} |B(\omega)|^2\, d\omega. \tag{4.12-6}$$

The equality holds only when $B(\omega)$ is proportional to $A(\omega)$—that is,

$$B(\omega) = CA^*(\omega), \tag{4.12-7}$$

where C is an arbitrary real constant.

If we make the substitutions

$$A(\omega) = \sqrt{\mathcal{S}_N(\omega)}\, H(\omega), \tag{4.12-8}$$

$$B(\omega) = \frac{F(\omega)e^{j\omega t_0}}{2\pi\sqrt{\mathcal{S}_N(\omega)}} \tag{4.12-9}$$

in the Schwarz inequality we obtain

$$\left(\frac{S_o}{N_o}\right) \leqslant \frac{1}{2\pi} \int_{-\infty}^{\infty} \frac{|F(\omega)|^2}{\mathcal{S}_N(\omega)}\, d\omega. \tag{4.12-10}$$

The maximum ratio occurs when the equality holds which implies that (4.12-7) is true. If $H_{opt}(\omega)$ denotes the optimum or matched filter, substitution of (4.12-8) and (4.12-9) in (4.12-7) produces

$$H_{opt}(\omega) = \frac{1}{2\pi C} \frac{F^*(\omega)}{\mathcal{S}_N(\omega)} e^{-j\omega t_0}. \tag{4.12-11}$$

In summary, if a signal having a spectrum $F(\omega)$ and noise having a power density spectrum $\mathcal{S}_N(\omega)$ are passed through a filter which maximizes the ratio of peak output signal power to average output noise power, the filter must have the transfer function defined in (4.12-11). Since C is arbitrary and since the exponential factor only represents a delay (which controls the time of occurrence of the output signal maximum), the important information in (4.12-11) is the factor $F^*(\omega)/\mathcal{S}_N(\omega)$. The form of the matched filter's transfer function must be given by the ratio of the complex conjugate of the signal spectrum to the noise power density spectrum.

For a given signal and noise the matched filter may not be realizable. Indeed, it may be hard to find colored noise situations where a realizable filter results. In practice one may only try to approximate nonrealizable filters. Some light is shed on these problems in Thomas [6].

Filter for White Noise

If the noise is white, $\mathcal{S}_N(\omega)$ is a constant at all frequencies. The matched filter is now given by

$$H_{opt}(\omega) = KF^*(\omega)e^{-j\omega t_0}, \tag{4.12-12}$$

ISBN 0-201-05758-1

where K is an arbitrary real constant. In this case we see that the filter form is given by the complex conjugate of the signal spectrum.

By inverse Fourier transforming (4.12-12) the impulse response $h_{\text{opt}}(t)$ of the matched filter is obtained. It is easily shown that

$$h_{\text{opt}}(t) = Kf^*(t_0 - t). \tag{4.12-13}$$

If $f(t)$ is real, then

$$h_{\text{opt}}(t) = Kf(t_0 - t). \tag{4.12-14}$$

Example 4.12-1

Let the matched filter be found for a pulse in white noise such as shown in Fig. 4.12-2(*a*). The signal was discussed in example 2.3-1 where its spectrum was found to be

$$F(\omega) = A\tau \frac{\sin(\omega\tau/2)}{\omega\tau/2} e^{j\omega[\tau_0 - (\tau/2)]}.$$

From (4.12-12) the matched filter transfer function is

$$H_{\text{opt}}(\omega) = KA\tau \frac{\sin(\omega\tau/2)}{\omega\tau/2} e^{-j\omega[t_0 + \tau_0 - (\tau/2)]}.$$

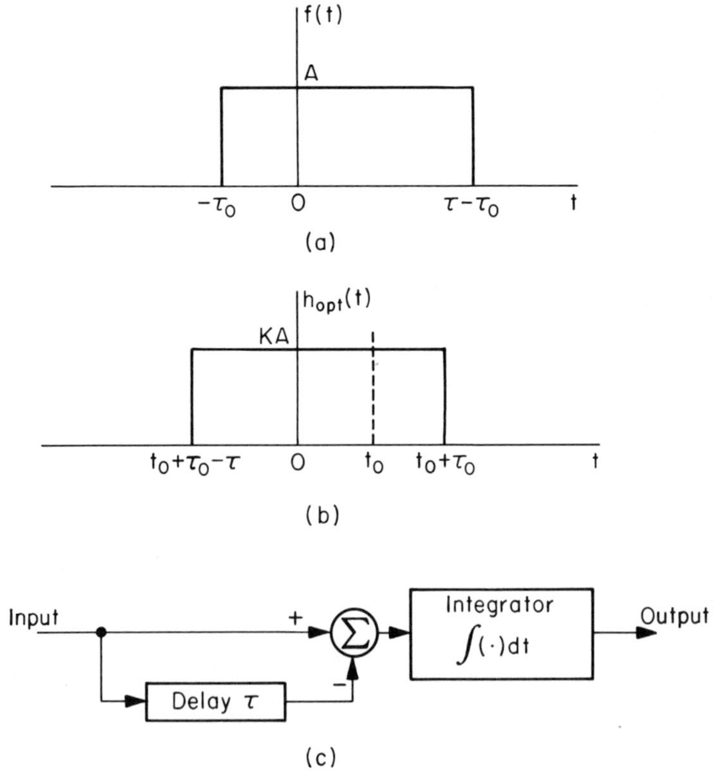

Fig. 4.12-2. A matched filter and its related signals. (a) Input signal, (b) impulse response, and (c) filter block diagram.

ISBN 0-201-05758-1

Whether any chance exists for such a filter to be realizable may be seen from the filter's impulse response obtained from (4.12-14) and sketched in Fig. 4.12-2(*b*). Clearly, to be realizable at all we require a delay of at least

$$t_0 \geqslant \tau - \tau_0$$

for $h_{\text{opt}}(t)$ to be causal. Such a filter could be realized by the network shown in (*c*) if perfect integrators could be built. Of course, they can't. However, very good approximations are possible using modern operational amplifiers with feedback, so for all practical purposes matched filters for pulses in white noise may be constructed.

Finally, we may demonstrate that the output signal $f_o(t)$ emerging from the optimum filter is proportional to the correlation of the input signal with itself and that the output signal magnitude is maximum when $t = t_0$. From the convolution integral and (4.12-13) we have

$$f_o(t) = \int_{-\infty}^{\infty} f(t-u)h_{\text{opt}}(u)\,du$$

$$= K\int_{-\infty}^{\infty} f(t-u)f^*(t_0-u)\,du. \tag{4.12-15}$$

When $t = t_0$, the integrand is $|f(t_0-u)|^2$, which is positive for all values of u. Such an integrand leads to the largest possible magnitude for the integral as can be seen from an application of the Schwarz inequality (4.12-6). It gives

$$|f_o(t)| = \left| K\int_{-\infty}^{\infty} f(t-u)\,f^*(t_0-u)\,du\right| \leqslant |K|\int_{-\infty}^{\infty} |f(t_0-u)|^2\,du. \tag{4.12-16}$$

The right side forms an upper bound for the magnitude of (4.12-15) which is realized on the left side when $t = t_0$.

Making the variable change $\tau = t_0 - u$ in (4.12-15) gives

$$f_o(t) = K\int_{-\infty}^{\infty} f^*(\tau)f(\tau + t - t_0)\,d\tau. \tag{4.12-17}$$

By comparing the integral with (2.4-19) we have shown that the output signal is proportional to the correlation of the input signal with itself.

PROBLEMS

4-1. In an experiment where a single die is rolled the outcomes are the numbers showing on the upper die face. Two events are defined as (1) a six occurs and (2) a one or a three occurs. (*a*) Find the probability for the two events. (*b*) Define a third event such that the sum of probabilities of the three events is unity.

ISBN 0-201-05758-1

4-2. In a box there are 100 balls—15 black, 25 white, 20 green, 30 red, and 10 blue. (*a*) What is the probability of getting a black ball on any draw? (*b*) What are the probabilities for the other colors?

4-3. A certain missile can be accidentally launched if two systems A and B simultaneously fail. A and B are known to have probabilities of failure of 0.01 and 0.05, respectively. Furthermore B fails with probability 0.15 (more rapidly) if A has already failed. (*a*) What is the probability of an accidental missile launch? (*b*) What is the probability of failure of A if B is known to have failed? (*c*) Are the events statistically independent?

4-4. Explain why the probabilities derived in example 4.1-3 do not satisfy (4.1-11).

4-5. Rework example 4.1-3 by defining events A and B as the drawing of a 10-ohm resistor and the drawing of a 5-watt resistor, respectively.

4-6. In an experiment two dice are thrown and the outcomes are the sums of numbers showing on the upper faces. Find the probabilities of (*a*) a twelve, (*b*) an even number, (*c*) an odd number, and (*d*) a six or larger.

★4-7. On a production line that manufactures capacitors to a 5% tolerance the probability of any given capacitor's being out of tolerance is 0.01. If three capacitors are selected at random: (*a*) What is the probability that they are all out of tolerance? (*b*) What is the probability of exactly two being out? (*c*) Of none?

4.8- Write a set of equations that must all be satisfied if three events A_1, A_2, and A_3 are to be statistically independent.

4-9. State whether the following sets are finite or infinite.

(*a*) $\{1, 8, 16, 44\}$.
(*b*) $\{a \in I \mid 16 < a \leqslant \infty\}$.
(*c*) $\{x \in X \mid 1.4 \leqslant x \leqslant 24.3\}$.
(*d*) $\{1.1, 1.11, 1.111, 1.1111\}$.
(*e*) $\{3, 4, 5, 6, \{x \in X \mid 6 < x \leqslant 12\}\}$.

4-10. A space $S = \{1, 3, 5, 7, 9, 11, 13, 15, 17\}$. If subsets A and B are defined as $A = \{1, 5, 9, 13, 17\}$ and $B = \{5, 7, 9, 11, 13\}$, determine the following:

(*a*) $A \cup B, A \cup B^c, (A \cup B)^c,$
$A \cap B, A \cap B^c, (A \cap B)^c.$
(*b*) Are A and B disjoint?

4-11. Using Venn diagrams for three events A, B, and C, cross-hatch the area corresponding to:

$$(A \cup B) \cap C, \quad A \cup B \cup C, \quad (A \cup B)^c \cup C,$$
$$(A \cap B) \cup C, \quad A^c \cup B^c \cup C^c, \quad (A \cap B)^c \cap C.$$

4-12. Use Venn diagrams to justify the validity of the following:

$$A \cap A = A, \qquad A \cup A = A,$$
$$A \cap \phi = \phi, \qquad A \cup \phi = A,$$
$$A \cap S = A, \qquad A \cup S = S,$$
$$A \cap A^c = \phi, \qquad A \cup A^c = S,$$
$$A - \phi = A, \qquad A - S = \phi,$$
$$(A \cup A) - A = \phi, \; A \cup (A - A) = A,$$
$$(A \cap A) - A = \phi, \; A \cap (A - A) = \phi.$$

4-13. Use Venn diagrams to prove *DeMorgan's laws* for two sets A and B:

$$(A \cup B)^c = A^c \cap B^c,$$
$$(A \cap B)^c = A^c \cup B^c.$$

4-14. Use Venn diagrams to prove that $P(A \cup B) = P(A) + P(B) - P(A \cap B)$.

ISBN 0-201-05758-1

4-15. Prove that $P(A \cup B \cup C) = P(A) + P(B) + P(C) - P(A \cap B) - P(A \cap C) - P(B \cap C) + P(A \cap B \cap C)$ for three events A, B, and C.

4-16. The possible outcomes of a random experiment form the set $X = \{1, 1.5, 2, 2.5, 3, 4, 6\}$. What is the range of each of the following random variables:

(*a*) $Y = 3X$.
(*b*) $Y = 10X^2$.
(*c*) $Y = 5X^2 - 6X - 3$.
(*d*) $Y = \cos(\pi X)$.
(*e*) $Y = \sin(\pi X)$.
(*f*) $Y = (1 - 2X)^{-1}$.

4-17. Repeat problem 4-16 with $X = \{x \in X \mid -3 < x \leqslant 4\}$.

4-18. Prove (4.3-5) using (4.3-7) as a starting point.

4-19. Assuming the outcomes of the experiment of problem 4-16 are equally probable, sketch the cumulative distribution and probability density functions for parts (*a*), (*d*), (*e*), and (*f*).

4-20. Determine which of the following are valid probability density functions:

(*a*) $f(x) = \dfrac{u(x)}{6} e^{-x/6}$.
(*b*) $f(x) = u(x + 4) - u(x - 16)$.
(*c*) $f(x) = [u(x + \pi) - u(x - \pi)] \cos(x/2)/4$.
(*d*) $f(x) = \dfrac{-u(x - 2)}{1 + 3x}$.
(*e*) $f(x) = (1 + x^2)^{-1}$.

4-21. Determine a (for arbitrary m and b) so that

$$p(x) = ae^{-|x-m|/b}$$

is a valid probability density function (called the *Laplace* density). Find the mean and variance.

4-22. Find the moments and central moments of the random variable defined in Fig. 4.3-2. What is the numerical value of the variance?

★ 4-23. Prove (4.4-22).

★ 4.24. Using the characteristic function given in (4.5-4) for the binomial distribution, find the first and second moments and verify m_X and $\sigma_X{}^2$ as given by (4.5-2) and (4.5-3).

4-25. A spacecraft is expected to land in a prescribed recovery zone 70% of the time. If five spacecraft land: (*a*) Find the probability of exactly none of the five hitting the recovery zone. (*b*) Find the probability for one of five. (*c*) Repeat for two, three, four, and five out of five. (Hint: Use the binomial distribution.)

4-26. Verify the validity of (4.5-25) and (4.5-26) for the Rayleigh probability density function.

4-27. A zero-mean Gaussian random variable X has a variance of $\sigma_X{}^2 = 4$. (*a*) Find the probability that $X > 2$. (*b*) Repeat for $|X| > 4$. (*c*) If the variable has a mean of $m_X = 1.5$, what is the probability that $X > 2$? Compare your result with that found in (*a*).

4-28. A Rayleigh distributed random variable X has a variance of 7. (*a*) What is the mean value of X? (*b*) What is the probability that X will be not more than 10 but greater than its mean value?

ISBN 0-201-05758-1

★ 4-29. A zero-mean Gaussian random variable X is changed to a new variable Y by the transformation $Y = bX$ for $0 < X$, and $Y = 0$ for $X \leqslant 0$, with b being a positive constant. Find and sketch the probability density function of Y.

★ 4-30. If the transformation of problem 4-29 is applied to a Rayleigh random variable, what is its effect?

★ 4.31. A random variable Θ is uniformly distributed on the interval $\theta_1 < \Theta < \theta_2$, where θ_2 is not greater than π and $0 \leqslant \theta_1$. Find and sketch the probability density function of $Y = \cos(\Theta)$.

★ 4-32. Show that the probability density function of the sum Z of two statistically independent random variables X and Y is

$$p_Z(z) = \int_{-\infty}^{\infty} p_X(z - y)p_Y(y)\, dy,$$

where $p_X(x)$ and $p_Y(y)$ are the density functions of X and Y. Start the proof by equating probabilities: $P_Z(z) = P\{Z \leqslant z\} = P\{X + Y \leqslant z\}$.

★ 4-33. Show that the magnitude of the normalized correlation coefficient ρ is not greater than unity for two random variables X and Y.

4-34. List the properties of the conditional distribution function which may be deduced from (4.7-40) through (4.7-43).

★ 4-35. Show that the N-dimensional joint characteristic function for N statistically independent random variables equals the N-fold product of individual characteristic functions.

★ 4-36. Find the characteristic function of the sum $Y = \sum_{i=1}^{N} X_i$ of N statistically independent random variables X_i, in terms of the individual characteristic functions, (*a*) when the variables X_i have different distributions and (*b*) when the X_i are identically distributed.

4-37. Sample functions of a random process $X(t)$ are periodic square waves with period T and random "timing" t_0. Timing is uniformly distributed over one period. Show that $X(t)$ is wide-sense stationary. (Hint: Expand $X(t)$ into a Fourier series.)

4-38. By expanding (4.8-27), prove (4.8-26).

4-39. By expanding (4.8-31), prove (4.8-29).

4-40. Determine the exact response of the network of Fig. 4.8-5(*a*) for any integrator having an impulse response $h(t) = 0$, $t < 0$. What properties should $h(t)$ have (for $t > 0$) for the result to closely approximate the cross-correlation function $R_{XY}(\tau)$?

4-41. A random process $Y(t) = X(t) \cos[\omega_0 t + \Theta]$ is the result of a wide-sense process $X(t)$ amplitude modulating a carrier with uniformly distributed (over 2π) phase Θ. Assume $X(t)$ and Θ are statistically independent. (*a*) Is $Y(t)$ wide-sense stationary? (*b*) Find and sketch the power density spectrum of $Y(t)$ if $X(t)$ has a low-pass power density spectrum.

4-42. Prove (4.9-23).

4-43. A random process $X(t)$ having an autocorrelation function $R_X(\tau) = \delta(\tau)$ is applied to a low-pass RC filter having a 3-dB bandwidth W and an impulse response $h(t) = e^{-Wt}$ for $t > 0$ and $h(t) = 0$ for $t < 0$. (*a*) Write an expression for the filter response process $Y(t)$. (*b*) What is the output mean value? (*c*) Find the output autocorrelation function.

4-44. In problem 4-43 find the cross-correlation function of the input and output processes.

ISBN 0-201-05758-1

4-45. Two orthogonal, stationary random processes $X_1(t)$ and $X_2(t)$ are applied to the network of Fig. P4-45. By superposition the response is the sum of individual responses $Y_1(t)$ and $Y_2(t)$. Show that $Y_1(t)$ and $Y_2(t)$ are orthogonal processes.

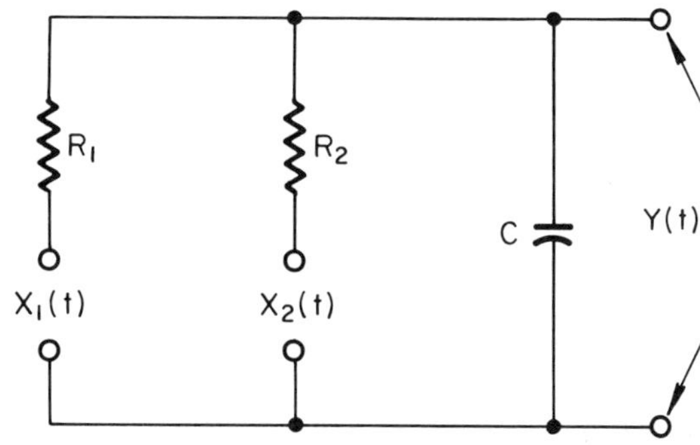

Fig. P4-45

4-46. In problem 4-45 show that $R_Y(\tau) = R_{Y_1}(\tau) + R_{Y_2}(\tau)$.

4-47. In problem 4-45 let the power density spectra of $X_1(t)$ and $X_2(t)$ be $\mathscr{S}_{X_1}(\omega) = \mathscr{N}_0/2$ and $\mathscr{S}_{X_2}(\omega) = K\mathscr{N}_0/2$, respectively. Write an expression for the power spectrum of $Y(t)$.

★ 4-48. In problem 4-45 let $Y_1(t)$ and $Y_2(t)$ be zero-mean Gaussian processes having variances of $\sigma_{Y_1}{}^2 = 4$ and $\sigma_{Y_2}{}^2 = 9$, respectively. Write an expression for the probability density function of $Y(t)$.

4-49. The power density spectrum of thermal noise can be written as $\mathscr{S}_N(\omega) = \mathscr{N}_0/2[1 + (\omega/W_N)^2]$, where $W_N \approx 10^{14}$. Sketch this function and determine at what frequencies (in hertz) it drops to $0.9(\mathscr{N}_0/2)$, $0.7(\mathscr{N}_0/2)$, and $0.5(\mathscr{N}_0/2)$. Over what frequency range does thermal noise approximate white noise?

4-50. Gaussian narrowband noise can be represented by either (4.11-8) or (4.11-9) where for any one instant in time the random variables $X_c(t)$ and $X_s(t)$ are statistically independent, have zero-mean values, and have equal variances σ^2. (*a*) Find the joint probability density function of $A(t)$ and $\Theta(t)$. (*b*) Show that the random variables $A(t)$ and $\Theta(t)$ in (4.11-8) are statistically independent. (*c*) Show that the probability density functions of $A(t)$ and $\Theta(t)$ are Rayleigh and uniform, respectively.

★ 4-51. Let a sinusoid $A_0 \cos(\omega_0 t)$ be added to the noise of problem 4-50. (*a*) Find the joint probability density function of the magnitude and phase of the sum for any instant in time. (*b*) Find the marginal density function for magnitude alone.

4-52. An amplifier has a power gain which varies with frequency as $G(\omega) = 10/[1 + (\omega/1000)^2]^2$. Find its noise bandwidth (in hertz).

4-53. If an energy signal having energy E plus white noise with power density $\mathscr{N}_0/2$ is applied to a matched filter, show that the output peak signal power to rms noise power ratio cannot exceed $2E/\mathscr{N}_0$.

ISBN 0-201-05758-1

REFERENCES

[1] Papoulis, A., *Probability, Random Variables, and Stochastic Processes,* McGraw-Hill Book Co., New York, 1965.

[2] Davenport, W. B., Jr., *Probability and Random Processes,* McGraw-Hill Book Co., New York, 1970.

[3] Drake, A. W., *Fundamentals of Applied Probability Theory,* McGraw-Hill Book Co., New York, 1967.

[4] Cooper, G. R., and McGillem, C. D., *Probabilistic Methods of Signal and System Analysis,* Holt, Rinehart and Winston, New York, 1971.

[5] Abramowitz, M., and Stegun, I. A. (Editors), *Handbook of Mathematical Functions with Formulas, Graphs, and Mathematical Tables,* National Bureau of Standards Applied Mathematics Series 55, U.S. Government Printing Office, Washington, D.C., 1964.

[6] Thomas, J. B., *An Introduction to Statistical Communication Theory,* John Wiley & Sons, New York, 1969.

[7] Davenport, W. B., Jr., and Root, W. L., *An Introduction to the Theory of Random Signals and Noise,* McGraw-Hill Book Co., New York, 1958.

[8] Miller, K. S., *Complex Stochastic Processes, An Introduction to Theory and Application,* Addison-Wesley Publishing Co., Reading, Massachusetts, 1974.

[9] Middleton, D., *An Introduction to Statistical Communication Theory,* McGraw-Hill Book Co., New York, 1960.

ISBN 0-201-05758-1

Chapter 5

Amplitude Modulation

5.0. INTRODUCTION

The goal of all communication systems is to convey information from one point to another. The form of the system may vary widely, from simple baseband information signals being transmitted over wires by telephone, to complicated space communication links through satellites using radio waves. In this latter case, a radio-frequency (RF) carrier is used to form an information-bearing signal which will propagate readily by electromagnetic radio waves. Recall from our discussion in Chapter 1 that a carrier is necessary because baseband information signals may not be propagated easily by radio waves. The process by which the carrier is made to become an information-bearing signal is called *carrier-wave modulation.*

A carrier is simply a signal of the form $A \cos [\theta(t)]$, where A is its peak amplitude and $\theta(t)$ is its instantaneous phase angle. There are several schemes by which information may be imparted onto the carrier by modulation. In one scheme, the instantaneous phase angle $\theta(t)$ is made to vary as some function of the information signal while A is held constant. This type of modulation, called *angle modulation,* is considered in detail in Chapter 6. In another type, called *amplitude modulation* (AM), A is made to vary as some function of the information signal $f(t)$ while the phase angle is independent of $f(t)$. In AM, $\theta(t)$ has the form $\omega_0 t + \theta_0$, where θ_0 is an arbitrary constant and ω_0 is the carrier radian frequency, a constant. In this chapter we consider the various types of AM which have these characteristics, such as *standard AM* and *double-sideband (DSB) AM.*

In addition, we consider two other modulation types which are classified as AM even though both amplitude *and* phase angle vary with $f(t)$. They are *single-sideband* (SSB) and *vestigial-sideband* (VSB) modulation. The reason that these may be considered as AM is that SSB and VSB are the result of linear operations on a carrier much in the same manner as for standard AM and DSB.

Deterministic Signal Representation in AM

At this point it is appropriate to define representations to be used for the information signal involved in our subsequent work. We shall consistently represent the information signal by the notation $f(t)$. However, $f(t)$ will be assumed to be either deterministic or random, according to which interpretation leads to the best explanation of the topic under discussion. For example, we shall most often interpret $f(t)$ as a deterministic signal. This interpretation will allow us to discuss the Fourier spectrum of either $f(t)$ or the carrier waveform modulated by $f(t)$. The spectral approach is a quite natural way to convey many of the principles involved in the various modulation processes.

References start on p. 217.

ISBN 0-201-05758-1

Random Signal Representation in AM

Even though the deterministic signal interpretation of $f(t)$ leads to rapid understanding of many principles, there are times when it is more desirable to consider $f(t)$ a random signal. Indeed, modeling the information signal as a random waveform is the correct approach, since one is never quite sure what signal will originate from most information sources. The times when the random interpretation will be used mainly involve calculations of either signal power or correlation functions. One departure from this rule will occur when $f(t)$ is presumed to be a simple sinusoidal waveform as discussed in Section 5.1. When the random representation of $f(t)$ is employed, the carrier phase θ_0 will be assumed to be a random variable having a uniform distribution on the interval $(0, 2\pi)$.

For the case where $f(t)$ is nonrandom, power or correlation functions are found by forming an appropriate time average $A(\cdot)$. When $f(t)$ is assumed to be random, the appropriate operation is the statistical expectation $E(\cdot)$. Thus, two operations are necessary. However, in order to have a consistent notation in the following work, we shall use an overbar $\overline{(\cdot)}$ to imply either a time average *or* statistical expectation. For example, the average power in a random signal is $E[f^2(t)]$, while the power in a nonrandom signal is $A[f^2(t)]$. In either case we represent the power by $\overline{f^2(t)}$ and let the context make clear which situation applies.

5.1. STANDARD AM

Ordinary broadcast radio using the frequency band from 535 kHz to 1605 kHz employs what will be called *standard AM.*

Waveform and Spectrum

Figure 5.1-1 illustrates the standard amplitude modulation method. An information signal $f(t)$, as shown in (*a*) existing only for $0 < t < T$, causes the instantaneous amplitude of the carrier to change as shown in (*b*). In general, the expression for the modulated signal $s_{AM}(t)$ is

$$s_{AM}(t) = [A_0 + f(t)] \cos(\omega_0 t + \theta_0), \tag{5.1-1}$$

where θ_0, ω_0, and A_0 are constants.

To obtain the spectrum of the modulated signal we may first write (5.1-1) in its exponential form:

$$s_{AM}(t) = [A_0 + f(t)] \left[\frac{e^{j(\omega_0 t+\theta_0)} + e^{-j(\omega_0 t+\theta_0)}}{2}\right]. \tag{5.1-2}$$

It is a simple matter to use results from Chapter 2 to obtain the Fourier transform of (5.1-2). Let $F(\omega)$ be defined as the spectrum of $f(t)$; that is,

$$f(t) \leftrightarrow F(\omega). \tag{5.1-3}$$

We now use the relationships

$$A_0 e^{j\omega_0 t} \leftrightarrow 2\pi A_0 \delta(\omega - \omega_0), \tag{5.1-4}$$

$$f(t)e^{j\omega_0 t} \leftrightarrow F(\omega - \omega_0) \tag{5.1-5}$$

ISBN 0-201-05758-1

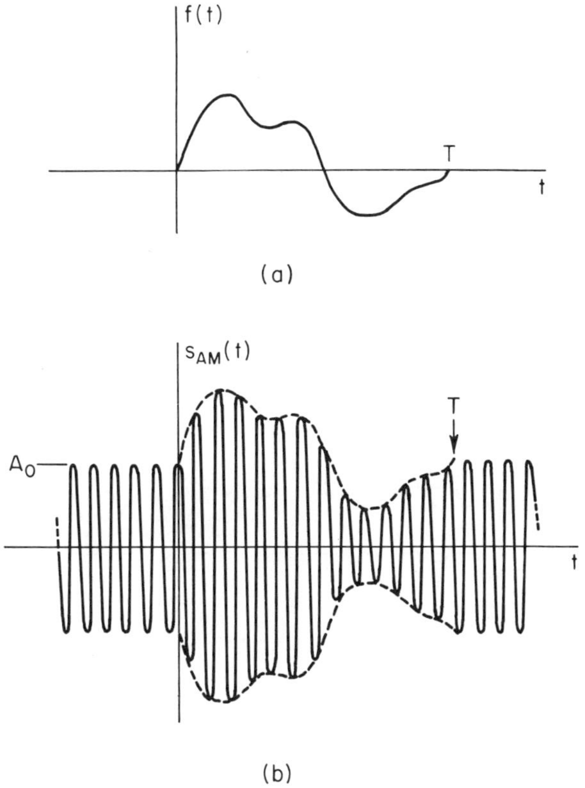

Fig. 5.1-1. Standard AM signal (b) corresponding to the information signal (a).

to obtain the spectrum $S_{AM}(\omega)$ through Fourier transformation of (5.1-2):

$$S_{AM}(\omega) = [2\pi A_0 \delta(\omega - \omega_0) + F(\omega - \omega_0)] \frac{e^{j\theta_0}}{2} + [2\pi A_0 \delta(\omega + \omega_0) + F(\omega + \omega_0)] \frac{e^{-j\theta_0}}{2}. \tag{5.1-6}$$

Figurc 5.1-2(*a*) illustrates the spectrum of a possible signal $f(t)$, while the spectrum (5.1-6) of the modulated carrier is sketched in (*b*) for $\theta_0 = 0$. By assuming $f(t) = 0$, we see that the impulses at $\omega = \omega_0$ and $\omega = -\omega_0$ are due to the carrier, whether modulated or not. Thus, the effect of AM is to shift half-amplitude replicas of the information signal spectrum out to frequencies of ω_0 and $-\omega_0$. If the maximum frequency extent of the information signal is W_f, we observe from this figure that the frequency extent of a standard AM signal is $2W_f$. The band of frequencies above ω_0 or less than $-\omega_0$ is called the *upper sideband* (USB). That on the opposing side of $\pm\omega_0$ is called the *lower sideband* (LSB).

Waveform Generation

A commonly used method of obtaining standard AM as used in broadcast radio is the plate-modulated class C amplifier. Most standard AM transmitters still employ tubes, so we discuss

ISBN 0-201-05758-1

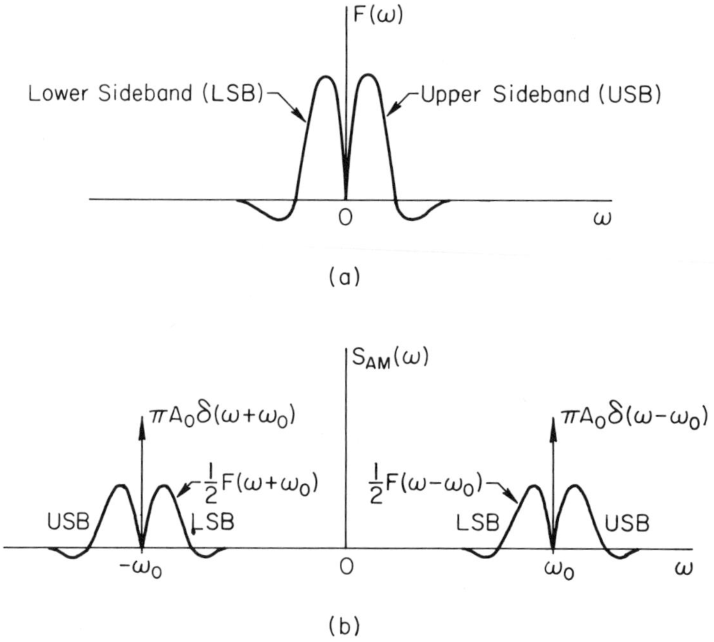

Fig. 5.1-2. Signal spectrum (b) of standard AM corresponding to the information signal spectrum of (a).

this case. A typical network using a triode tube is shown in Fig. 5.1-3. The tube is nominally biased beyond cutoff by the grid bias $-E_c$. In the absence of an information signal the RF carrier input waveform turns the tube on for a short time during each positive half-cycle of the grid voltage. During the times that the tube is on, pulses of current at the carrier frequency ω_0 flow in the transformer T2 which is tuned to the frequency ω_0 by capacitor C_2. The action of the tuned circuit is to remove higher harmonics of ω_0 present in the output plate current. As a result, only the fundamental component of plate current causes a significant excitation of the tuned circuit, and the output RF waveform amplitude becomes proportional to the current in T2 at the fundamental frequency ω_0. When a modulating signal is inserted

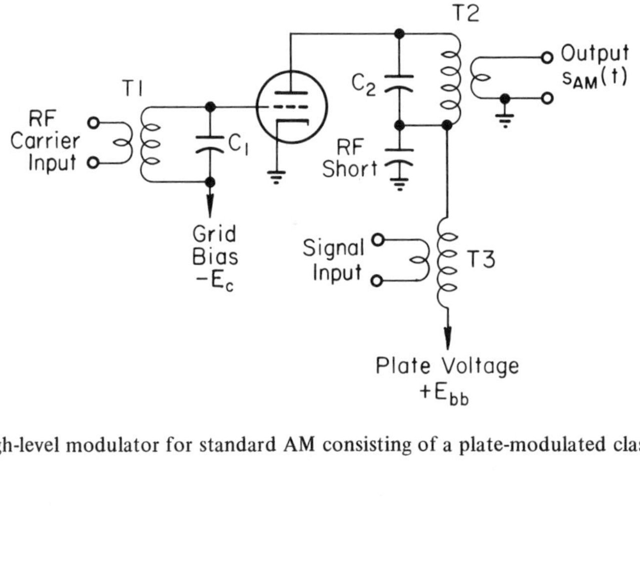

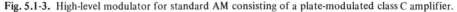

Fig. 5.1-3. High-level modulator for standard AM consisting of a plate-modulated class C amplifier.

ISBN 0-201-05758-1

into the plate circuit via T3, its level adds to the dc plate voltage, thereby altering the magnitude of the fundamental component of plate current and causing the output RF amplitude to vary approximately as a linear function of $f(t)$ as needed.

In using the plate-modulated amplifier care must be taken to assure that the maximum modulating signal level in series with the plate voltage E_{bb} does not exceed the plate voltage. If such a case occurs it is possible to reduce the effective supply voltage to zero, cutting off the tube. This is called *overmodulation.* One effect of overmodulation is to cause distortion in $f(t)$ when recovered by the receiver demodulator.

The plate-modulated amplifier is usually a high-level stage; that is, it is the power stage that drives the antenna. Other high-level modulators, such as the grid-modulated class C amplifier, are also possible but are not as attractive because of lower power efficiency, higher distortion, and poorer power-handling capabilities than in the plate-modulated unit [1].

It is also possible to use any number of low-level standard AM modulation methods and amplify the modulated wave to the desired power level. In general, such techniques are not as efficient as the high-level modulators. However, efficiency may be improved somewhat by use of class B push-pull power amplifiers, similar to many audio power output stages, rather than using strictly linear class A amplifiers.

One of the most attractive low-level modulation methods is illustrated in block diagram form in Fig. 5.1-4. This modulator implements the sum derived from writing $s_{AM}(t)$ in the form

$$s_{AM}(t) = A_0 \cos(\omega_0 t + \theta_0) + f(t) \cos(\omega_0 t + \theta_0). \tag{5.1-7}$$

The *balanced modulator* is nothing more than a product device that can be implemented easily using diodes and passive elements. It is said to be balanced because no component of output is ideally present that is proportional to either input signal. Practical methods of implementing balanced modulators are discussed in Section 5.2.

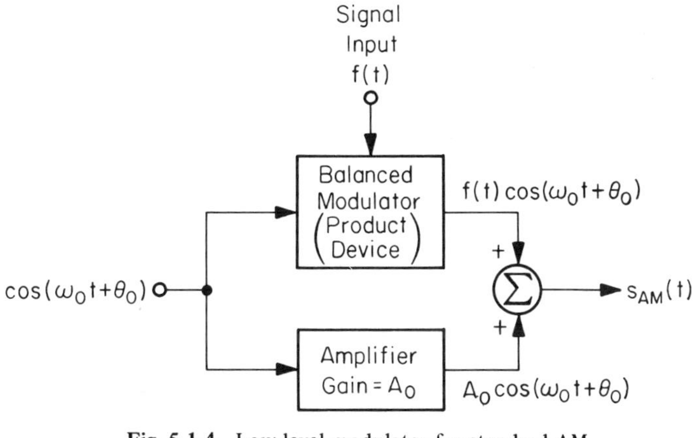

Fig. 5.1-4. Low-level modulator for standard AM.

Signal Power Distribution and Efficiency

The total average power P_{AM} in the waveform of (5.1-7) is given by the mean-squared value of the waveform. It can be written as

ISBN 0-201-05758-1

$$P_{AM} = \overline{A_0{}^2 \cos^2(\omega_0 t + \theta_0)} + \overline{f^2(t)\cos^2(\omega_0 t + \theta_0)}$$
$$+ \overline{2A_0 f(t)\cos^2(\omega_0 t + \theta_0)}. \tag{5.1-8}$$

The first term here is the power P_c due to the carrier, while the second term is the power P_f in the sidebands due to the information signal. The third term is a crosspower which is often zero because $f(t)$ usually has no average or dc component. Assuming that the crosspower is zero, we may solve (5.1-8) for the carrier and sideband powers (see problem 5-4):

$$P_c = A_0{}^2/2, \tag{5.1-9}$$

$$P_f = \overline{f^2(t)}/2. \tag{5.1-10}$$

The power in the AM signal becomes

$$P_{AM} = P_c + P_f = \frac{1}{2}[A_0{}^2 + \overline{f^2(t)}], \tag{5.1-11}$$

which is an expression involving time waveforms.

By finding the power density spectrum $\mathscr{S}_{AM}(\omega)$ of the modulated signal we can also illustrate how power is distributed as a function of frequency. $\mathscr{S}_{AM}(\omega)$ is found by Fourier transforming the autocorrelation function $R_{AM}(\tau)$ of the AM signal. From the definition of autocorrelation function given in Chapter 4 we have

$$R_{AM}(\tau) = \overline{s_{AM}(t)s_{AM}(t+\tau)}$$
$$= \overline{[A_0 + f(t)]\cos(\omega_0 t + \theta_0)[A_0 + f(t+\tau)]\cos(\omega_0 t + \omega_0\tau + \theta_0)}. \tag{5.1-12}$$

In problem 5-5 it is found that (5.1-12) will reduce to

$$R_{AM}(\tau) = \frac{A_0{}^2}{2}\cos(\omega_0\tau) + \frac{R_f(\tau)}{2}\cos(\omega_0\tau), \tag{5.1-13}$$

where $R_f(\tau)$ is the autocorrelation function of $f(t)$ which is assumed to have no average value. Fourier transformation of $R_{AM}(\tau)$ gives the desired power density spectrum:

$$\mathscr{S}_{AM}(\omega) = \frac{\pi A_0{}^2}{2}[\delta(\omega - \omega_0) + \delta(\omega + \omega_0)]$$
$$+ \frac{1}{4}[\mathscr{S}_f(\omega - \omega_0) + \mathscr{S}_f(\omega + \omega_0)]. \tag{5.1-14}$$

In writing this expression we have let $\mathscr{S}_f(\omega)$ represent the power density spectrum of $f(t)$; that is, it is the Fourier transform of the autocorrelation function $R_f(\tau)$.

The first two terms in (5.1-14) are due to the carrier, while the second two represent sideband power due to the information signal. The sidebands actually represent the useful part of the modulated signal, while the carrier component conveys no information. By integration

ISBN 0-201-05758-1

of the first two terms we again find that carrier power P_c is given by (5.1-9). Sideband power P_f is determined from an integration of the second two terms:

$$P_f = \frac{1}{2\pi}\int_{-\infty}^{\infty} \frac{1}{4}\left[\mathcal{S}_f(\omega - \omega_0) + \mathcal{S}_f(\omega + \omega_0)\right] d\omega$$

$$= \frac{1}{2\pi}\int_{-\infty}^{\infty} \frac{1}{2}\mathcal{S}_f(\omega - \omega_0)\, d\omega = \frac{1}{2\pi}\int_{-\infty}^{\infty} \frac{1}{2}\mathcal{S}_f(\omega)\, d\omega. \quad (5.1\text{-}15)$$

The last form of this equation shows that the total power in the sidebands of the standard AM signal is one-half the power in the modulating information signal $f(t)$.

Efficiency η_{AM} of the modulated signal may be defined as the ratio of sideband power to total power. Using (5.1-10) and (5.1-11) this ratio is

$$\eta_{AM} = \frac{P_f}{P_c + P_f} = \frac{\overline{f^2(t)}}{A_0^2 + \overline{f^2(t)}}. \quad (5.1\text{-}16)$$

In terms of (5.1-15), efficiency may also be expressed as

$$\eta_{AM} = \frac{(1/2\pi)\displaystyle\int_{-\infty}^{\infty} \mathcal{S}_f(\omega)\, d\omega}{A_0^2 + (1/2\pi)\displaystyle\int_{-\infty}^{\infty} \mathcal{S}_f(\omega)\, d\omega}. \quad (5.1\text{-}17)$$

An example will illustrate efficiency for standard AM.

Example 5.1-1

We find the efficiency η_{AM} for a carrier with peak amplitude A_0 modulated by a square wave having a peak-to-peak amplitude of $2A_m$ and no dc component. It is easy to see that

$$P_f = \frac{1}{2}\overline{f^2(t)} = A_m^2/2.$$

Thus, efficiency is

$$\eta_{AM} = \frac{(A_m/A_0)^2}{1 + (A_m/A_0)^2}.$$

The largest value of A_m for no overmodulation is A_0. Assuming $A_m = A_0$, the highest efficiency possible is 1/2 or 50%. For such a case, sideband power equals carrier power. On careful reflection, the reader may verify, through this example, that all other information signals having no dc component will give an efficiency less than 50% so long as no overmodulation occurs.

ISBN 0-201-05758-1

Sinusoidal Modulation

The special case of sinusoidal modulation is of interest because it helps develop a picture of the points discussed above. We assume that the information signal is the sinusoid

$$f(t) = A_m \cos(\omega_m t + \theta_m) \tag{5.1-18}$$

having amplitude A_m, frequency ω_m, and phase θ_m. The spectrum of $f(t)$, as determined by Fourier transformation, is found to be

$$F(\omega) = \pi A_m [\delta(\omega - \omega_m)e^{j\theta_m} + \delta(\omega + \omega_m)e^{-j\theta_m}]. \tag{5.1-19}$$

Thus, from (5.1-1) and (5.1-6), the modulated signal and its spectrum are given by

$$s_{AM}(t) = A_0[1 + \beta_{AM} \cos(\omega_m t + \theta_m)] \cos(\omega_0 t + \theta_0), \tag{5.1-20}$$

$$\begin{aligned} S_{AM}(\omega) = {} & \pi A_0[\delta(\omega - \omega_0)e^{j\theta_0} + \delta(\omega + \omega_0)e^{-j\theta_0}] \\ & + \frac{\pi A_m}{2}[\delta(\omega - \omega_0 - \omega_m)e^{j\theta_m} + \delta(\omega - \omega_0 + \omega_m)e^{-j\theta_m}]e^{j\theta_0}. \\ & + \frac{\pi A_m}{2}[\delta(\omega + \omega_0 - \omega_m)e^{j\theta_m} + \delta(\omega + \omega_0 + \omega_m)e^{-j\theta_m}]e^{-j\theta_0}. \end{aligned} \tag{5.1-21}$$

Here β_{AM} is called the *modulation index* defined as

$$\beta_{AM} = A_m/A_0. \tag{5.1-22}$$

Figure 5.1-5 illustrates the AM signal (*a*) and its spectrum (*b*) for $\theta_0 = 0$ and $\theta_m = 0$.

Efficiency is most easily found from (5.1-16). Since

$$\overline{f^2(t)} = A_m{}^2/2, \tag{5.1-23}$$

efficiency becomes

$$\eta_{AM} = \frac{(A_m/A_0)^2}{2 + (A_m/A_0)^2} = \frac{\beta_{AM}{}^2}{2 + \beta_{AM}{}^2}. \tag{5.1-24}$$

For 100% modulation $A_m = A_0$, and the maximum possible efficiency becomes 1/3 or 33.3% for modulation by a sinusoidal signal.

Example 5.1-2

We find the power density spectrum for sinusoidal modulation. The autocorrelation function of a signal of the form $f(t) = A_m \cos(\omega_m t + \theta_m)$ was found in example 4.8-1 of Chapter 4. If θ_m is assumed to be a uniformly distributed random variable over $-\pi < \theta_m < \pi$, then

$$R_f(\tau) = \frac{A_m{}^2}{2} \cos(\omega_m \tau).$$

ISBN 0-201-05758-1

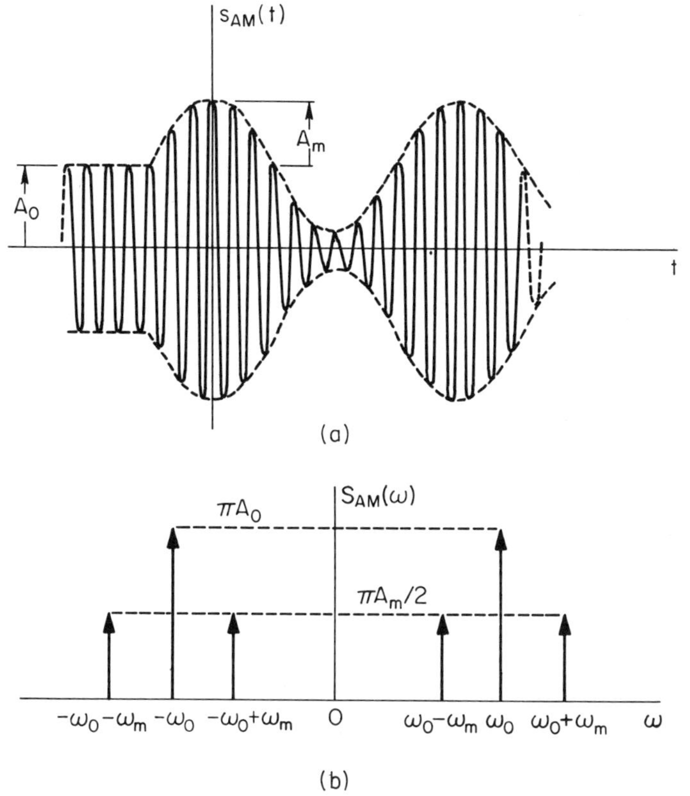

Fig. 5.1-5. Standard AM signal (a) and its spectrum (b) for sinusoidal modulation.

Fourier transforming this result produces

$$\mathscr{S}_f(\omega) = \frac{\pi A_m^{\,2}}{2} [\delta(\omega - \omega_m) + \delta(\omega + \omega_m)],$$

and on use in (5.1-14) we obtain the desired power density spectrum:

$$\begin{aligned}\mathscr{S}_{AM}(\omega) = \frac{\pi A_0^{\,2}}{2} [\delta(\omega - \omega_0) + \delta(\omega + \omega_0)] \\ + \frac{\pi A_m^{\,2}}{8} [\delta(\omega - \omega_0 - \omega_m) + \delta(\omega - \omega_0 + \omega_m) \\ + \delta(\omega + \omega_0 - \omega_m) + \delta(\omega + \omega_0 + \omega_m)].\end{aligned}$$

5.2. SUPPRESSED CARRIER AM

In the previous discussion of standard AM it was found that only a fraction of the total power in the modulated waveform is due to the spectral sidebands which carry the information. In fact, over half the power is due to the carrier and is wasted in the sense that the carrier conveys no useful information. In this section, two types of modulation are described which cor-

ISBN 0-201-05758-1

respond to suppressing the carrier in order to place all available power into the information sidebands. The first type, known as *double-sideband* (DSB) *suppressed carrier* AM, is achieved by simply letting $A_0 = 0$ in the standard AM signal given by (5.1-1). The efficiency of the DSB-modulated signal becomes 100%.

The second type of suppressed carrier modulation also has the advantage of an efficiency of 100%. It is called *single-sideband* (SSB) AM. However, SSB provides an additional advantage. It requires only half the bandwidth of a DSB signal. The saving is possible by observing that $F(-\omega) = F^*(\omega)$ for the spectrum of a real information signal $f(t)$. In other words, the spectrum for negative frequencies, which forms the lower sideband of the spectrum of $f(t)$, is directly related to the spectrum for positive frequencies, which forms the upper sideband. Thus, only one sideband is really required to characterize $f(t)$, and, because the spectrum of a DSB signal will subsequently be shown to consist of replicas of the spectrum of $f(t)$ shifted to frequencies of ω_0 and $-\omega_0$, the SSB waveform may be generated by simply eliminating one sideband of a DSB signal's spectrum.

Suppressed carrier AM is used in amateur radio among other applications, and, although in some cases it has power, bandwidth, and certain other advantages, it is not without certain disadvantages. The most serious of which is implementation complexity. Both transmitter and receiver are often more complex and expensive than for standard AM.

Double-Sideband AM

As already mentioned, the DSB signal is achieved by setting $A_0 = 0$ in (5.1-1). The DSB waveform becomes

$$s_{\text{DSB}}(t) = f(t) \cos(\omega_0 t + \theta_0). \tag{5.2-1}$$

The spectrum of the DSB signal is obtained either from (5.1-6) with $A_0 = 0$ or from the Fourier transform of (5.2-1):

$$S_{\text{DSB}}(\omega) = \frac{e^{j\theta_0}}{2} F(\omega - \omega_0) + \frac{e^{-j\theta_0}}{2} F(\omega + \omega_0). \tag{5.2-2}$$

Clearly, the spectrum of the DSB signal has no carrier, and all power is placed in the sidebands. Thus, from (5.1-10),

$$P_{\text{DSB}} = P_f = \overline{f^2(t)}/2. \tag{5.2-3}$$

A consequence of this fact is that efficiency becomes 100%. From (5.1-16), with $A_0 = 0$,

$$\eta_{\text{DSB}} = 1.0. \tag{5.2-4}$$

An example of a DSB waveform and its spectrum are illustrated in Fig. 5.2-1(*b*) and (*a*), respectively. It has been assumed that $\theta_0 = 0$ in the sketches. The information signal and its spectrum are shown in (*c*). For convenience in illustration $f(t)$ is assumed to exist only for a finite period of time.

DSB Modulation Methods

The most obvious method of generating a DSB waveform is to implement the product of a carrier $\cos(\omega_0 t + \theta_0)$ and the information signal $f(t)$ which is suggested by the form of

ISBN 0-201-05758-1

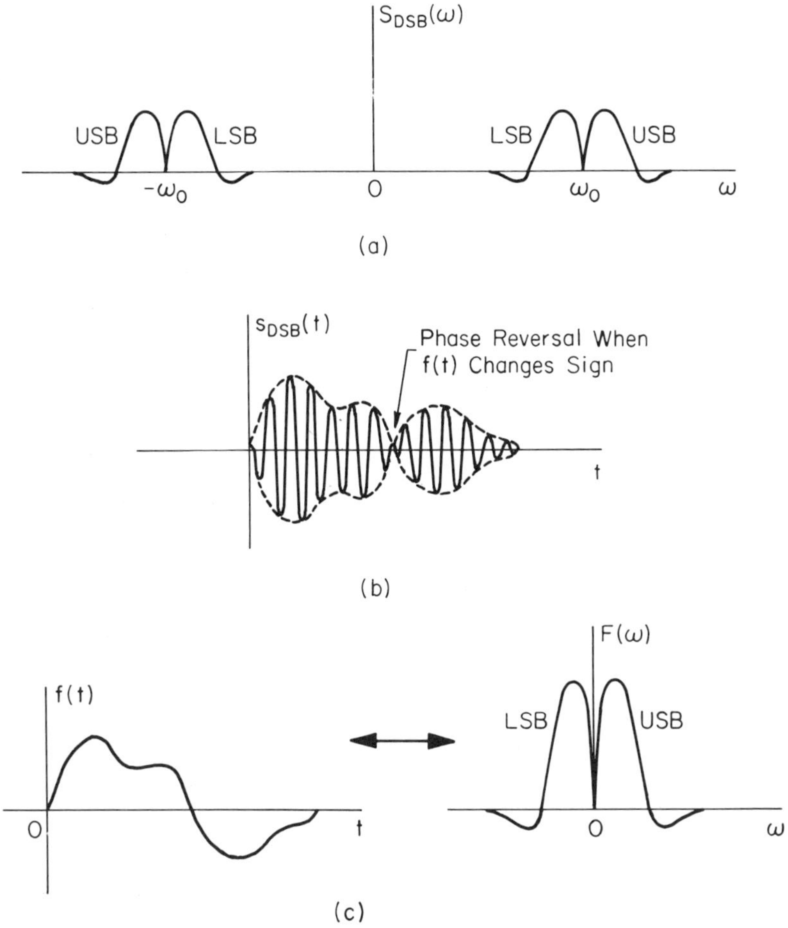

Fig. 5.2-1. Spectrum (a) of the DSB suppressed carrier signal (b) corresponding to the information signal and spectrum of (c).

(5.2-1). Many techniques exist for realizing the necessary product. Some of these techniques depend on control of an amplifier's gain, while others depend on the response of a nonlinear circuit element. In the following paragraphs we shall discuss only a third way of implementing the product by using a *balanced modulator*. The balanced modulator is one of the simplest, most used, and easily constructed product devices.

One form of balanced modulator is shown in Fig. 5.2-2(*a*). The carrier level A_0 is chosen large enough so that, when the carrier signal is positive as shown, the diodes conduct heavily during each positive half-cycle. During these times the two transformers are connected together and the output of T2 is proportional to $f(t)$. During negative half-cycles of the carrier the diodes are open-circuited and essentially no output occurs. The total output would appear as illustrated in (*d*). This response is equivalent to the product of $f(t)$ times a square-wave train of positive pulses. Using the Fourier series for such a train of pulses, we have

$$\text{unfiltered output} = Kf(t) \sum_{n=0}^{\infty} \frac{\sin(n\pi/2)}{(n\pi/2)} \cos(n\omega_0 t), \qquad (5.2\text{-}5)$$

ISBN 0-201-05758-1

where K is a proportionality constant. This output contains DSB signals at any "carrier"

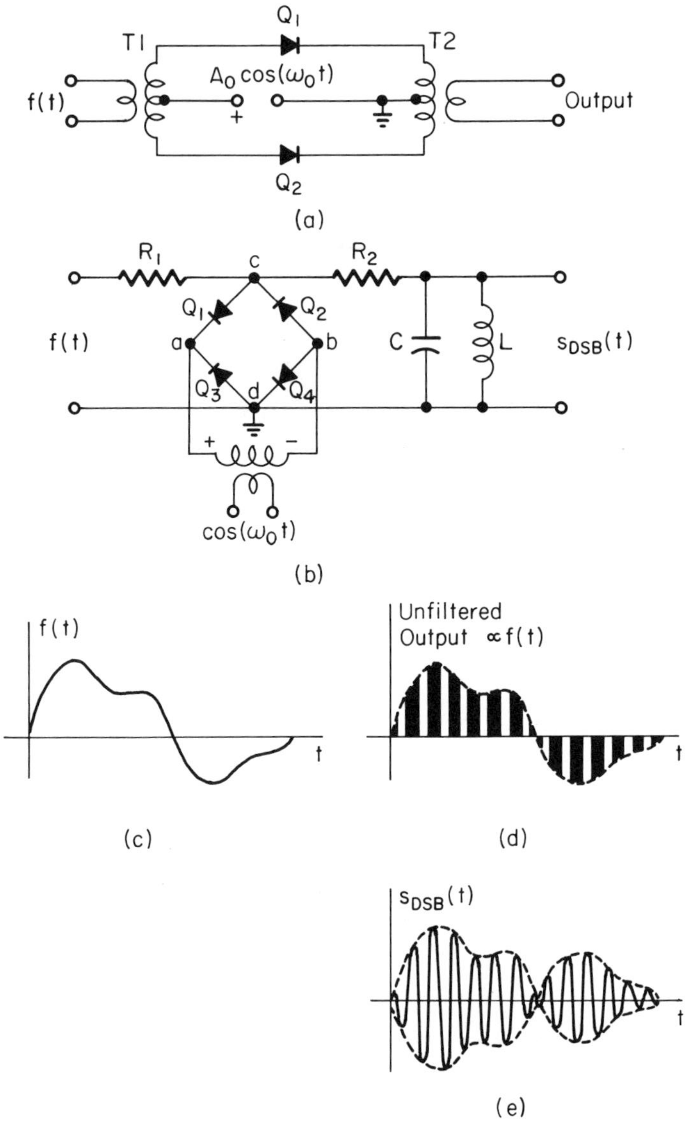

Fig. 5.2-2. Balanced modulators. (a) Using series arrangement of two diodes, and (b) with four diodes in shunt configuration. The input signal (c) gives rise to unfiltered output (d) or filtered output (e).

frequency which is an odd multiple of ω_0. Since the component at frequency ω_0 has the highest level, it is usually the one of interest. By using a band-pass filter centered at frequency ω_0, the "undesired" DSB signals may be filtered out. This filter must be considered part of the modulator. Thus, the filtered output is

$$s_{DSB}(t) = \frac{2K}{\pi} f(t) \cos(\omega_0 t), \tag{5.2-6}$$

which is shown in (*e*). The constant $2K/\pi$ is the overall gain of the balanced modulator.

Another form of balanced modulator is shown in Fig. 5.2-2(*b*). Operation is similar to that of the modulator in (*a*). When the carrier signal driving the diodes is positive as shown, the

ISBN 0-201-05758-1

diodes are approximately open circuits and essentially an open circuit exists between points c and d. On negative half-cycles the diodes conduct heavily and a virtual short between c and d exists. During the half-cycles that the diodes are open circuits, the output is again proportional to $f(t)$ as shown in (d) if no tuning of the output circuit were used. When the LC band-pass resonant circuit is added, the filtered response is again given by (5.2-6).

Two other forms of diode balanced modulators are shown in Fig. 5.2-3. The circuit of (a) is analyzed in problem 5-12. It works similar to the modulators of Fig. 5.2-2. By adjustment of R the output may be balanced; that is, any carrier frequency component in the output can be nulled. The circuit of Fig. 5.2-3(b) is similar to that of Fig. 5.2-2(a) except that four diodes are used. By adding the two extra diodes the unfiltered output will now exist also during negative half-cycles of the carrier. During these times diodes Q_3 and Q_4 conduct, connecting terminals a and c to d and b, respectively. The result is a sign reversal causing the unfiltered output to appear as a square-wave pulse train amplitude modulated by $f(t)$ as illustrated in

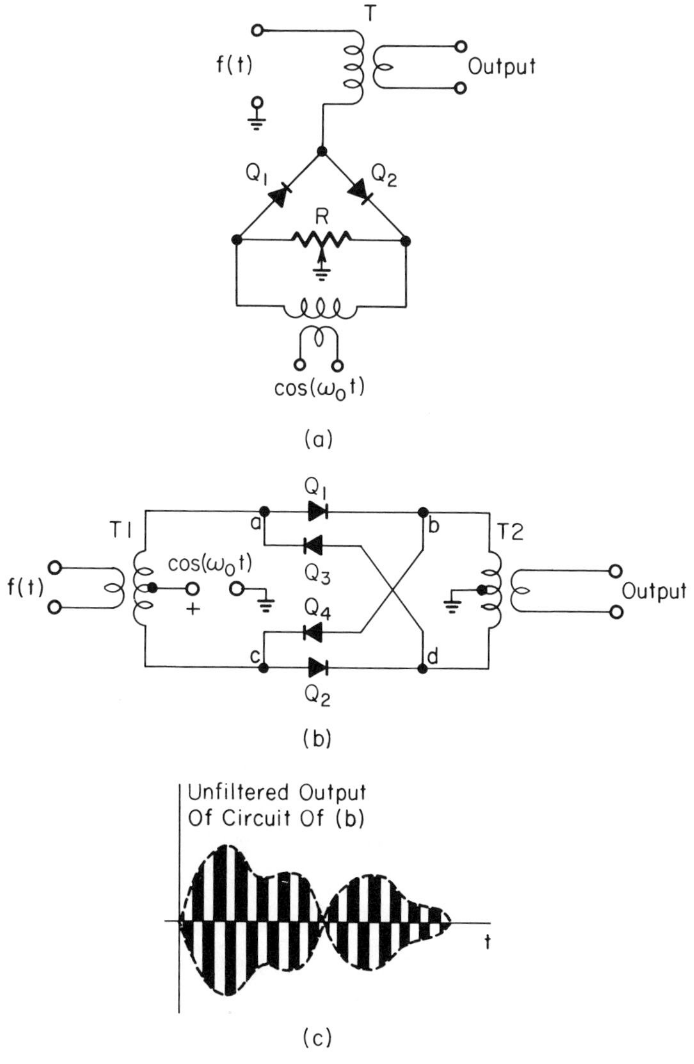

Fig. 5.2-3. Balanced modulators using two diodes (a) and four diodes (b). (c) The unfiltered output of the four-diode modulator.

ISBN 0-201-05758-1

(*c*). The filtered output will still be of the form of (5.2-6); however, the gain of the four-diode unit will be twice as large as that of the two-diode unit of Fig. 5.2-2(*a*).

Practical balanced modulators do not give perfect carrier cancellation. With good design, suppression to levels of −30 dB to −40 dB are possible.

DSB signals may be generated in ways other than the simple product with a cosine signal as indicated in (5.2-1). More generally, a product $f(t)q(t)$ will also produce a DSB modulated waveform, where $q(t)$ is almost arbitrary. One need only require that $q(t)$ be periodic and have a spectral component at the desired carrier frequency.* An example will help illustrate these points.

Example 5.2-1

Let $q(t) = A_0 \cos^2 (\omega_0 t/2) = A_0 [1 + \cos (\omega_0 t)]/2$ where the desired carrier frequency is ω_0. The product $f(t)q(t)$ is

$$\frac{A_0}{2} f(t) + \frac{A_0}{2} f(t) \cos (\omega_0 t).$$

By using a band-pass filter the baseband term $A_0 f(t)/2$ is removed. The second term is the desired DSB-modulated waveform which passes through the filter unaffected.

Single-Sideband AM

As previously noted, both sidebands in the spectrum of a DSB signal carry the same information. Thus, one is redundant, since full information may be transmitted by using only one sideband. The result of removing one sideband is called *single-sideband* (SSB) AM. The great advantage of SSB is the 2 : 1 bandwidth savings as compared to DSB.

There are several ways of generating SSB waveforms. Each of these ways will be discussed in the following paragraphs, and an explicit expression for the SSB waveform will be developed during the discussion of the second method.

Filter Method for SSB Generation

The most obvious way to eliminate one sideband is to pass a DSB signal through a filter that passes only the desired band. Figure 5.2-4 illustrates this procedure, assuming an ideal band-pass filter positioned to pass the upper sideband. A high-pass filter could also have been used. Alternatively, a band-pass or low-pass filter may be used to generate a lower sideband SSB spectrum. One of the chief disadvantages of the filter method is that sharp cutoff or steep-skirted filters are needed if no appreciable change from ideal is to be made in the sideband selected. A modified method of generating SSB which allows for more realistic filters is called *vestigial-sideband* AM; it is considered in Section 5.3.

Phase Method for SSB Generation

The *phase method* [2] is theoretically capable of generating a SSB signal without distortion. A block diagram of the necessary circuitry is shown in Fig. 5.2-5. The method is clearly more complex than the filter method, since more equipment is required.

*If the repetition frequency ω_q of $q(t)$ is not large we must also require $\omega_q \geqslant 2W_f$, where W_f is the maximum spectral extent of $f(t)$.

ISBN 0-201-05758-1

The manner in which the system generates a SSB waveform is most easily studied through a consideration of the spectra involved. The output SSB spectrum is obtained by either adding or subtracting a spectrum $Q(\omega)$, having odd symmetry about frequencies $\pm\omega_0$, from a DSB spectrum. As described previously, the DSB spectrum is generated by taking the product

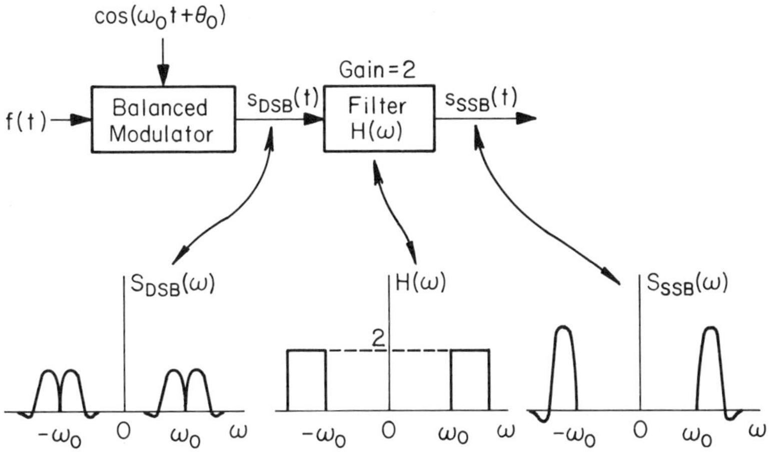

Fig. 5.2-4. Block diagram of the filter method for generating SSB.

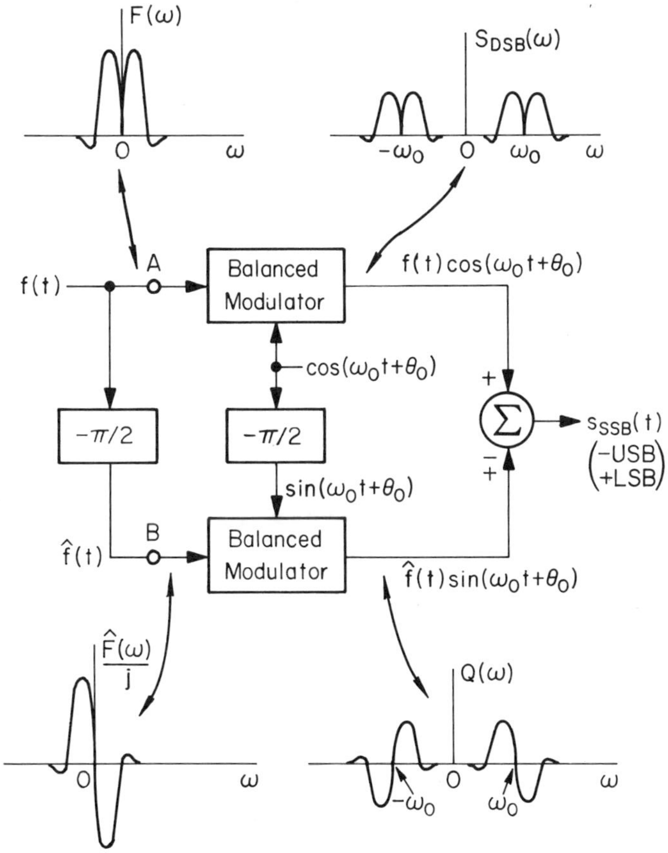

Fig. 5.2-5. Block diagram of the phase method for generating SSB.

ISBN 0-201-05758-1

of the information signal $f(t)$ and the carrier $\cos(\omega_0 t + \theta_0)$. The key to the phase method is the way in which $Q(\omega)$ is formed.

$Q(\omega)$ is the result of the product of a signal $\hat{f}(t)$ and $\sin(\omega_0 t + \theta_0)$ in a balanced modulator, where $\hat{f}(t)$ is produced by passing the signal $f(t)$ through a phase shift of $-\pi/2$ rad. Since phase shift is an odd function of ω, the action of the -90-degree phase shift is to multiply $F(\omega)$, the spectrum of $f(t)$, by $-j$ for $\omega > 0$ and by $+j$ for $\omega < 0$. If $\hat{F}(\omega)$ is defined as the spectrum of $\hat{f}(t)$, then

$$\hat{F}(\omega) = -jF(\omega)[2U(\omega) - 1]. \tag{5.2-7}$$

A sketch of $\hat{F}(\omega)/j$ is illustrated in Fig. 5.2-5.

By expanding $\sin(\omega_0 t + \theta_0)$ into its exponential form

$$\sin(\omega_0 t + \theta_0) = \frac{e^{j\omega_0 t + j\theta_0} - e^{-j\omega_0 t - j\theta_0}}{2j} \tag{5.2-8}$$

and using the Fourier transform frequency shifting property, we may obtain $Q(\omega)$ as

$$Q(\omega) = -\frac{1}{2}F(\omega - \omega_0)[2U(\omega - \omega_0) - 1]\, e^{j\theta_0}$$

$$+ \frac{1}{2}F(\omega + \omega_0)[2U(\omega + \omega_0) - 1]\, e^{-j\theta_0}. \tag{5.2-9}$$

The spectrum of the output SSB signal $s_{SSB}(t)$ is the difference between (5.2-2) and (5.2-9). It can be written as

$$S_{SSB}(\omega) = U(\omega - \omega_0)F(\omega - \omega_0)e^{j\theta_0} + U(-\omega - \omega_0)F(\omega + \omega_0)e^{-j\theta_0}, \tag{5.2-10}$$

which assumes that the upper sidebands are desired. If lower sidebands are desired instead, $Q(\omega)$ can be added to (5.2-2) to obtain the lower sideband result:

$$S_{SSB}(\omega) = U(\omega_0 - \omega)F(\omega - \omega_0)e^{j\theta_0} + U(\omega + \omega_0)F(\omega + \omega_0)e^{-j\theta_0}. \tag{5.2-11}$$

The output SSB signal, using Fig. 5.2-5, takes the form

$$s_{SSB}(t) = f(t)\cos(\omega_0 t + \theta_0) \mp \hat{f}(t)\sin(\omega_0 t + \theta_0), \tag{5.2-12}$$

where the upper sign corresponds to generation of the USB signal and the lower sign goes with LSB generation.

Since $\hat{f}(t)$ is a phase-shifted version of $f(t)$, it is clear that the two are related in some way. The relationship can be found by noting that the spectrum of $\hat{f}(t)$, given by (5.2-7), has already been encountered in Chapter 3 where the *analytic signal* was discussed. It was shown there that $\hat{f}(t)$ is related to $f(t)$ through the *Hilbert transform:*

ISBN 0-201-05758-1

$$\hat{f}(t) = \frac{1}{\pi}\int_{-\infty}^{\infty} \frac{f(\tau)}{t-\tau}\, d\tau. \tag{5.2-13}$$

One of the most serious problems with the phase method is stability. For good performance, stability and accuracy of the −90-degree carrier phase shifter must be quite good (held to within a few degrees). Another problem arises in the practical design of the signal phase shifter because all spectral components in $f(t)$ must be shifted by −90 degrees. This requirement is difficult to approximate in practice. On the other hand, the requirement is more stringent than is necessary. It can be shown that ideal performance is achievable even if the path containing $\hat{f}(t)$ does not have a constant −90-degree phase shift, so long as its phase *differs* from that over the path containing $f(t)$ by −90 degrees. In other words, if the phase shift of the path processing $f(t)$ is given by a function $\theta(\omega)$, then the phase shift over the path processing $\hat{f}(t)$ must be $\theta(\omega) - \pi/2$. In this case, the two paths simply must track in phase* as a function of ω with a −90-degree constant difference. This is a much easier situation to realize than a fixed −90-degree shift in the one path.

Example 5.2-2

Suppose $f(t) = A \cos(\omega_f t)$. We find $\hat{f}(t)$ and $s_{SSB}(t)$. Here $\hat{f}(t) = A \cos(\omega_f t - \pi/2) = A \sin(\omega_f t)$. Also $f(t) \cos(\omega_0 t + \theta_0) = A \cos(\omega_f t) \cos(\omega_0 t + \theta_0) = \frac{A}{2}\{\cos[(\omega_0 + \omega_f)t + \theta_0] + \cos[(\omega_0 - \omega_f)t + \theta_0]\}$ and $\hat{f}(t) \sin(\omega_0 t + \theta_0) = A \sin(\omega_f t) \sin(\omega_0 t + \theta_0) = \frac{A}{2}\{\cos[(\omega_0 - \omega_f)t + \theta_0] - \cos[(\omega_0 + \omega_f)t + \theta_0]\}$. The output becomes

$$s_{SSB}(t) = \frac{A}{2}\{(1 \mp 1)\cos[(\omega_0 - \omega_f)t + \theta_0] + (1 \pm 1)\cos[(\omega_0 + \omega_f)t + \theta_0]\},$$

where upper signs correspond to USB and lower to LSB.

★ Weaver's Method for SSB Generation

A third method [3] of generating SSB signals is actually a combination of the filter and phase methods. It has the advantages of the phase method without the disadvantage of needing a constant −90-degree signal-path phase shifter. A possible implementation of the method is illustrated in Fig. 5.2-6. To the right of terminals A and B the circuit is the same as in the phase method of Fig. 5.2-5. To the left, the −90-degree phase shifter of the phase method is replaced by a special circuit that performs an equivalent function by using the filter method.

Several spectrum sketches are shown in Fig. 5.2-6 to aid in understanding operation. As shown, the information signal spectrum $F(\omega)$ is assumed to have a low-frequency gap and exists only in the range $W_L < |\omega| < W_f$, where W_L is the low-frequency end of the spectrum and W_f is the maximum spectral extent of $f(t)$. The key to operation is the use of a preliminary carrier frequency

$$\omega_b = (W_L + W_f)/2 \tag{5.2-14}$$

*See problem 5-13.

ISBN 0-201-05758-1

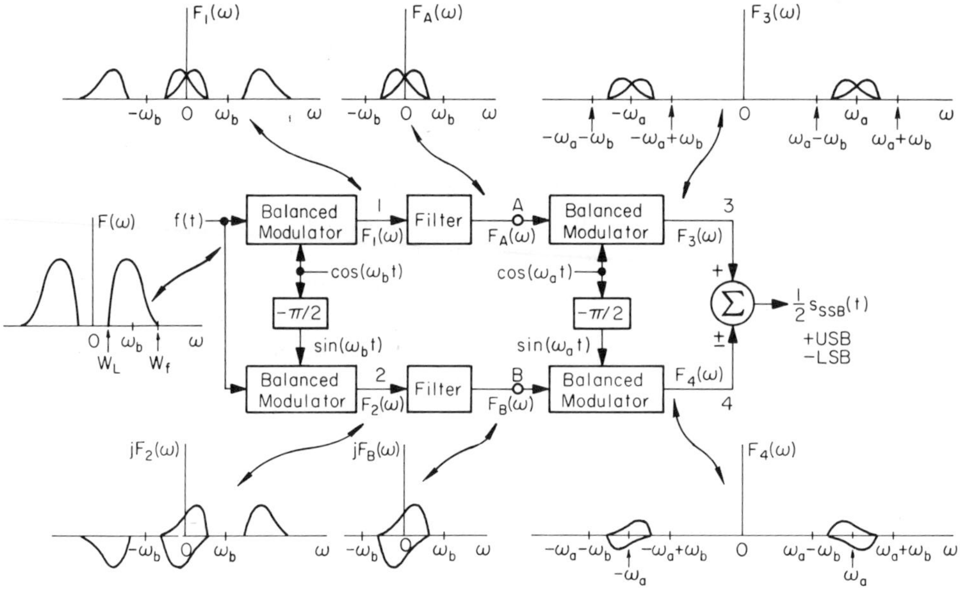

Fig. 5.2-6. Block diagram of a third method for generating SSB. Spectra shown assume a low-pass filter.

to generate double-sideband spectra having either even or odd symmetry about $\omega = 0$, depending on which balanced modulator is considered. These are the spectra at points 1 and 2, respectively, which, on being passed through low-pass filters with cutoff frequency $(W_f - W_L)/2$, become the spectra shown at points A and B. The remaining circuit performs as described in the phase method.

As in the phase method, either USB or LSB operation is possible depending on whether $F_3(\omega)$ and $F_4(\omega)$ are added or subtracted. Now, however, the carrier frequency will be different. The proper carrier frequency ω_0 will be $\omega_0 = \omega_a + \omega_b$ for a difference, giving the LSB, while a sum gives the USB and a carrier frequency $\omega_0 = \omega_a - \omega_b$.

The low-pass filter transfer function will have a region $2W_L$ wide beginning at $|\omega| = (W_f - W_L)/2$ in which to roll off to a negligible value. Thus, for signal spectra having a significant low-frequency gap, practical filters are possible. As $W_L \to 0$ an ideal filter is required with bandwidth $W_f/2$.

In the above discussion Weaver's method has been described as originally developed. The method may be applied, however, even if identical high-pass filters are used. Performing a spectral analysis of Fig. 5.2-6 for filters having a transfer function $H(\omega)$ gives

$$\frac{1}{2} S_{SSB}(\omega) = \frac{1}{2} [H(\omega - \omega_a)F(\omega + \omega_b - \omega_a) + H(\omega + \omega_a)F(\omega - \omega_b + \omega_a)] \quad \text{(sum)} \tag{5.2-15}$$

when the sum is taken at the summation point, or

$$\frac{1}{2} S_{SSB}(\omega) = \frac{1}{2} [H(\omega - \omega_a)F(\omega - \omega_b - \omega_a) + H(\omega + \omega_a)F(\omega + \omega_b + \omega_a)] \quad \text{(difference)} \tag{5.2-16}$$

ISBN 0-201-05758-1

when a difference is taken. When $H(\omega)$ corresponds to a high-pass filter, the sum will produce a LSB spectrum at a carrier frequency of $\omega_a - \omega_b$. A difference produces USB at a carrier frequency $\omega_a + \omega_b$.

5.3. VESTIGIAL-SIDEBAND AM

Signal Generation by the Filter Method

In the last section it was found that a single-sideband spectrum could be obtained by using a filter to pass only the desired sideband of the spectrum of a double-sideband suppressed carrier signal. The major disadvantage of the method was the need for a filter having a rapid cutoff characteristic in the vicinity of the carrier frequency. *Vestigial-sideband* (VSB) modulation is generated by a very similar procedure except that the filter is not required to have a sharp cutoff characteristic. Indeed, instead of completely eliminating one sideband, a "vestige" of the band is left remaining while a portion of the desired sideband is attenuated. In other words, a more realistic filter transfer function is allowed having a finite frequency region over which the transition is made in a gradual way, from full transmission to complete attenuation.

Figure 5.3-1(b) illustrates the magnitude of a suitable filter transfer function $H_v(\omega)$ for selecting the upper sideband when used in the circuit of (a). The USB region lies from $|\omega|$

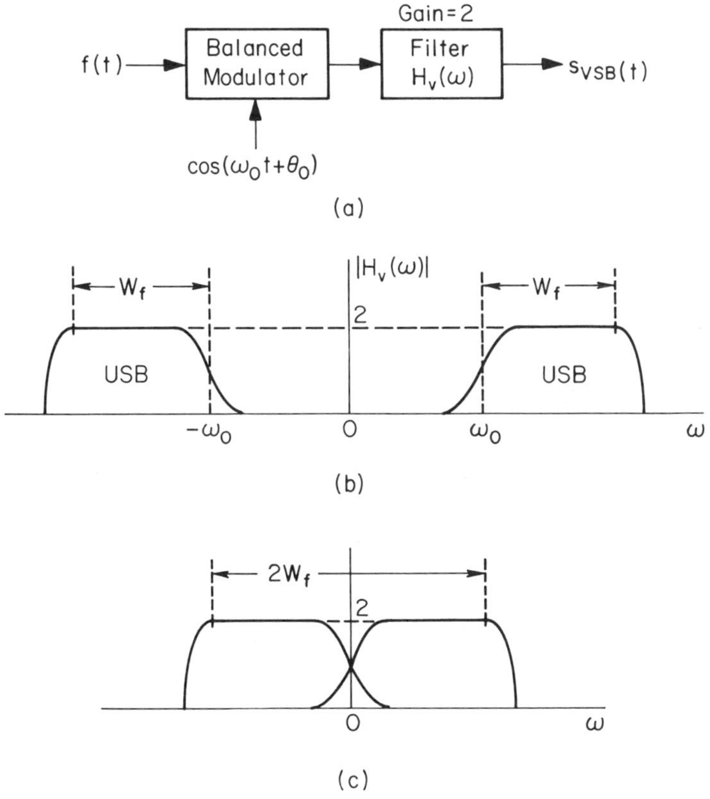

Fig. 5.3-1. Block diagram of the filter method for VSB generation (a). Characteristics (b) and (c) of the required filter.

ISBN 0-201-05758-1

$= \omega_0$ to $|\omega| = \omega_0 + W_f$. The transfer function magnitude is flat over this region except in the vicinity of $|\omega| = \omega_0$ where the roll-off characteristic has odd symmetry with respect to the half-amplitude level at ω_0. Such a behavior, also called *complementary symmetry,* means that if two shifted VSB filter transfer functions, $H_v(\omega - \omega_0)$ and $H_v(\omega + \omega_0)$, are added together, the result for frequencies near zero will be a constant.* Thus,

$$H_v(\omega - \omega_0) + H_v(\omega + \omega_0) = \text{constant} = 2 \tag{5.3-1}$$

for $|\omega| < W_f$ as illustrated in (*c*). In a subsequent discussion of the demodulation of the VSB signal it will be found that (5.3-1) is a *required* condition for the recovery of an undistorted information signal. In the frequency region above the desired band, the filter roll-off behavior is arbitrary and may be chosen for convenience.

Since the output of the modulator of Fig. 5.3-1(*a*) is a DSB spectrum as given by (5.2-2), the VSB signal spectrum $S_{\text{VSB}}(\omega)$ at the filter output is easily stated:

$$S_{\text{VSB}}(\omega) = \frac{H_v(\omega)}{2} [e^{j\theta_0} F(\omega - \omega_0) + e^{-j\theta_0} F(\omega + \omega_0)]. \tag{5.3-2}$$

Commercial broadcast television in the United States uses a form of VSB modulation and is its principal application.

Relationship of VSB to SSB—The Phase Method

Since VSB is basically a form of SSB it is instructive to develop the relationship between the two modulation methods. In doing so, we shall find that a VSB signal can also be generated by a phase method similar to that of Fig. 5.2-5 for a SSB waveform. As a matter of fact, the VSB phase method may be considered more general than the SSB phase method, since the latter is a special case of the former when the VSB filter becomes an ideal filter as a limiting case.

The development begins by observing from Fig. 5.3-1(*a*) that, using the convolution theorem, the output signal may be stated as

$$s_{\text{VSB}}(t) = \int_{-\infty}^{\infty} h_v(\xi) f(t - \xi) \cos [\omega_0(t - \xi) + \theta_0]\, d\xi, \tag{5.3-3}$$

where $h_v(t)$ is the impulse response of the VSB filter. By expansion of the cosine term, (5.3-3) may be put in an alternative form:

$$s_{\text{VSB}}(t) = \left[\int_{-\infty}^{\infty} h_v(\xi) \cos (\omega_0 \xi) f(t - \xi)\, d\xi\right] \cos (\omega_0 t + \theta_0)$$
$$- \left[-\int_{-\infty}^{\infty} h_v(\xi) \sin (\omega_0 \xi) f(t - \xi)\, d\xi\right] \sin (\omega_0 t + \theta_0). \tag{5.3-4}$$

*The filter has been assigned a nominal maximum gain of 2 so that the desired spectral components of the VSB signal in the flat region of $H_v(\omega)$ will have the same magnitude as those of the SSB signal previously defined.

ISBN 0-201-05758-1

The bracketed factors represent time functions which DSB modulate "carriers" $\cos(\omega_0 t + \theta_0)$ and $\sin(\omega_0 t + \theta_0)$. The time functions both have the form of a convolution integral. Thus, they may each be viewed as the response of a suitably defined network where the input is $f(t)$. Let the networks be defined by their impulse responses denoted $h_I(t)$ and $h_Q(t)$. From (5.3-4) these are recognized as

$$h_I(t) = h_v(t)\cos(\omega_0 t), \tag{5.3-5}$$

$$h_Q(t) = -h_v(t)\sin(\omega_0 t). \tag{5.3-6}$$

The corresponding transfer functions, defined as $H_I(\omega)$ and $H_Q(\omega)$, are the Fourier transforms of $h_I(t)$ and $h_Q(t)$, respectively. They are

$$H_I(\omega) = \frac{1}{2}[H_v(\omega - \omega_0) + H_v(\omega + \omega_0)], \tag{5.3-7}$$

$$H_Q(\omega) = \frac{-1}{2j}[H_v(\omega - \omega_0) - H_v(\omega + \omega_0)]. \tag{5.3-8}$$

With these considerations in mind, it is easily seen that the network illustrated in Fig. 5.3-2 will provide the VSB output signal defined by (5.3-4).

The network of Fig. 5.3-2 is very similar to that of Fig. 5.2-5 for SSB. In fact, we may demonstrate exact correspondence in the limit as the VSB filter becomes ideal. Using the requirement (5.3-1) in the above two equations we have

$$H_I(\omega) = 1, \tag{5.3-9}$$

$$H_Q(\omega) = -j[H_v(\omega + \omega_0) - 1], \tag{5.3-10}$$

which must hold over the band occupied by the signal spectrum—that is, over $|\omega| < W_f$. If $H_v(\omega)$ defines a high-pass (or appropriate band-pass) filter, corresponding to selecting the USB, then in the limit as the filter becomes ideal

$$\begin{aligned} H_Q(\omega) = -j[2U(\omega) - 1] &= -j, \quad 0 < \omega \\ &= j, \quad \omega < 0, \end{aligned} \tag{5.3-11}$$

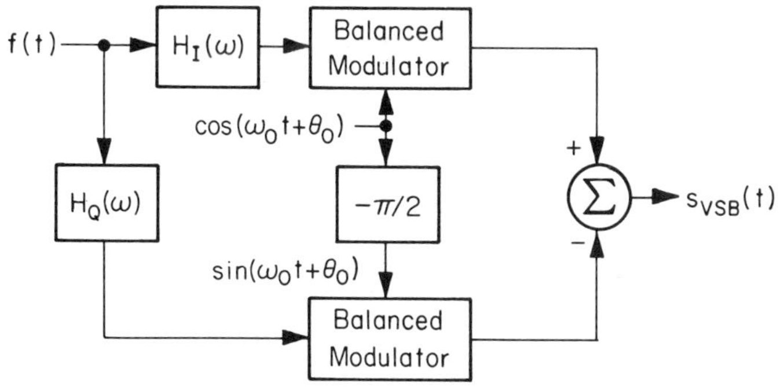

Fig. 5.3-2. Block diagram of the phase method for generating VSB.

ISBN 0-201-05758-1

for $|\omega| < W_f$. Here $U(\omega)$ is the unit step function of ω. From use of (5.3-9) and (5.3-11) with Fig. 5.3-2, we see that the VSB system becomes exactly the SSB system of Fig. 5.2-5 for selection of the upper sideband.

Finally, if $H_v(\omega)$ defines a low-pass (or appropriate band-pass) filter for LSB selection, (5.3-11) is again obtained except without the negative sign. The sign change corresponds to reversing the negative sign at the summing point in Fig. 5.3-2. The same change was found to be required in the SSB system when the LSB was desired.

5.4. COHERENT DEMODULATION OF AM SIGNALS

The generation of any of the various AM signals previously discussed involved use of a product device. In this section we shall demonstrate that using a product device in the receiver will also allow the demodulation (recovery) of the information signal without distortion, regardless of the type of AM used. That is, it will work for standard AM, DSB, SSB, and VSB waveforms.

Figure 5.4-1 illustrates the appropriate receiver demodulation operation. The input AM signal is multiplied by a *local carrier* which must be generated within the receiver and must be *phase-coherent* or *synchronous* with the transmitted carrier for best performance. This requirement means that, if the transmitted carrier phase is $(\omega_0 t + \theta_0)$, with ω_0 and θ_0 being constants, the local carrier phase must also be $(\omega_0 t + \theta_0)$. In the following developments, however, we shall assume a local carrier phase of $(\omega_0 t + \phi)$, where ϕ is not necessarily equal to θ_0 or a constant,* in order to demonstrate the effect of nonsynchronism of the local carrier.

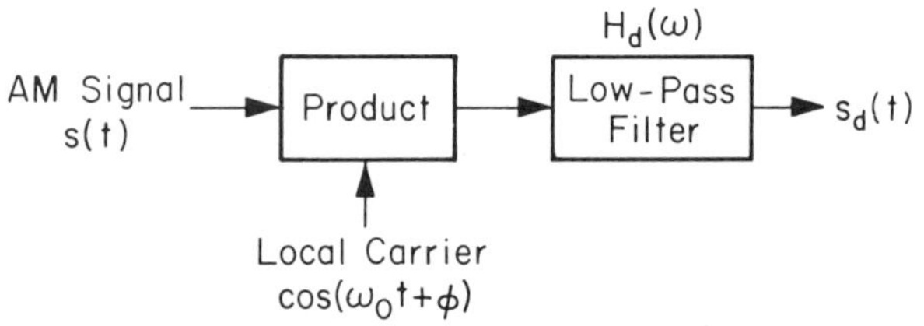

Fig. 5.4-1. Coherent demodulator applicable to AM, DSB, SSB, and VSB signal demodulation.

The output of the product device in Fig. 5.4-1 is passed through a low-pass filter. The purpose of the filter is to allow only those spectral components falling in the band of the information signal to reach the final output. The filter is assumed to be an ideal filter approximation to a real filter. Assuming the signal's spectrum extends to a frequency W_f, the filter bandwidth is also W_f.

A demodulator of the form of Fig. 5.4-1 is called a *coherent* or *synchronous demodulator* because of the need for a coherent local carrier. Generation of the local carrier within the receiving system is not always a straightforward task. Some methods for accomplishing this task are given in Section 5.7.

Standard AM and DSB Signal Demodulation

Let the input to the receiver be a standard AM signal:

*We allow ϕ to be a function of time in some following work to illustrate the effect of both frequency and phase errors in the local carrier.

ISBN 0-201-05758-1

$$s_{AM}(t) = [A_0 + f(t)] \cos(\omega_0 t + \theta_0). \tag{5.4-1}$$

The product in Fig. 5.4-1 becomes

$$s_p(t) = [A_0 + f(t)] \cos(\omega_0 t + \theta_0) \cos(\omega_0 t + \phi), \tag{5.4-2}$$

which reduces to

$$s_p(t) = \frac{1}{2} [A_0 + f(t)] [\cos(\theta_0 - \phi) + \cos(2\omega_0 t + \theta_0 + \phi)]. \tag{5.4-3}$$

The low-pass filter will remove the components involving the frequency $2\omega_0$ so that the output signal is

$$s_d(t) = \frac{1}{2} \cos(\theta_0 - \phi)[A_0 + f(t)]. \tag{5.4-4}$$

The factor $\cos(\theta_0 - \phi)$ determines whether or not the information signal is recovered without distortion. If ϕ is a constant phase angle, not equal to θ_0, then $\cos(\theta_0 - \phi)$ simply scales the magnitude of the term involving $f(t)$ and no distortion occurs. However, if $\theta_0 - \phi = \pi/2$ or $-\pi/2$ there is *no* output, while if $\theta_0 - \phi = \pi$ or $-\pi$ the sign of $f(t)$ changes. A sign change is of no significance if $f(t)$ is a voice or music signal but could be disastrous if $f(t)$ represents digital data.

Next let ϕ represent a time-dependent angle, such as

$$\phi = \omega_\epsilon t + \phi_\epsilon + \theta_0, \tag{5.4-5}$$

where ω_ϵ is a frequency error, while ϕ_ϵ is a constant phase error. Equation (5.4-4) may be written as

$$s_d(t) = \frac{1}{2} [A_0 + f(t)] \cos(\omega_\epsilon t + \phi_\epsilon). \tag{5.4-6}$$

This result is a standard AM signal with a "small" carrier frequency ω_ϵ and phase ϕ_ϵ. Distortion is now clearly present in the demodulated waveform. For speech signals and small frequency errors the output will sound as though slow but periodic changes in volume are occurring.

With a properly phased local carrier $\phi = \theta_0$ and the demodulator output is

$$s_d(t) = \frac{1}{2} [A_0 + f(t)] \qquad \text{(standard AM)}. \tag{5.4-7}$$

The problem of demodulating a DSB-modulated waveform is handled by using the above standard AM results with A_0 set equal to zero. For a properly phased local carrier the demodulated signal becomes

$$s_d(t) = \frac{1}{2} f(t) \qquad \text{(DSB)}. \tag{5.4-8}$$

ISBN 0-201-05758-1

SSB Signal Demodulation

Let us next examine the effect of a general single-sideband signal

$$s_{SSB}(t) = f(t)\cos(\omega_0 t + \theta_0) \mp \hat{f}(t)\sin(\omega_0 t + \theta_0) \tag{5.4-9}$$

applied to the demodulator of Fig. 5.4-1. According to the sign convention established earlier, the upper sign applies to an upper sideband SSB waveform while the lower sign is for the lower sideband case. The signal at the product device output becomes

$$s_p(t) = \frac{1}{2}[f(t)\cos(\theta_0 - \phi) \mp \hat{f}(t)\sin(\theta_0 - \phi)]$$

$$+ \frac{1}{2}[f(t)\cos(2\omega_0 t + \theta_0 + \phi) \mp \hat{f}(t)\sin(2\omega_0 t + \theta_0 + \phi)]. \tag{5.4-10}$$

Terms in the second brackets form a SSB signal at a carrier frequency of $2\omega_0$ having a phase angle of $\theta_0 + \phi$. This component is removed by the low-pass filter while the first bracketed expression forms the demodulator output:

$$s_d(t) = \frac{1}{2}[f(t)\cos(\theta_0 - \phi) \mp \hat{f}(t)\sin(\theta_0 - \phi)]. \tag{5.4-11}$$

Regarding distortion in the output signal, developments similar to those given previously for demodulation of standard AM apply. Again allowing ϕ to be given by (5.4-5) we have

$$s_d(t) = \frac{1}{2}[f(t)\cos(\omega_\epsilon t + \phi_\epsilon) \pm \hat{f}(t)\sin(\omega_\epsilon t + \phi_\epsilon)]. \tag{5.4-12}$$

In the interpretation of this output, which is in itself a single-sideband signal at the carrier frequency ω_ϵ with phase angle ϕ_ϵ, we must conclude that the second term is a distortion term for all values of frequency and phase errors except when $\omega_\epsilon = 0$ and $\phi_\epsilon = 0$ simultaneously. For this case the output is a maximum and is distortion-free:

$$s_d(t) = \frac{1}{2}f(t) \qquad \text{(SSB)}. \tag{5.4-13}$$

In order to better examine the effect of phase errors, let only a constant phase error be present. The distortion term in (5.4.12) is then $\pm\hat{f}(t)\sin(\phi_\epsilon)/2$. It has the effect of phase shifting all frequencies in the recovered signal spectrum by a fixed amount ϕ_ϵ. To illustrate this fact, we apply both (5.2-10) and (5.2-11) to the spectrum $S_d(\omega)$ of the SSB signal of (5.4-12) when $\omega_\epsilon = 0$. The spectrum reduces to

$$S_d(\omega) = \frac{1}{2}[U(\pm\omega)e^{-j\phi_\epsilon} + U(\mp\omega)e^{j\phi_\epsilon}]F(\omega), \tag{5.4-14}$$

where the upper signs correspond to an USB-transmitted SSB signal and lower signs mean LSB transmission. Thus, positive frequencies are shifted by $\mp\phi_\epsilon$, while negative frequencies are shifted by $\pm\phi_\epsilon$. Such phase distortion is not too serious for voice transmissions because the human ear is rather insensitive to phase.

ISBN 0-201-05758-1

Frequency errors, on the other hand, are not easily tolerated by the ear for either speech or music. The primary effect is to destroy the harmonic relationship of spectral components because the frequency error simply shifts all components by an equal amount. With frequency error voice takes on a Donald Duck quality, while music acquires an oriental sound. Experience indicates that a frequency error of less than approximately ±20 Hz is necessary for acceptable voice transmission.

Another form of coherent demodulator for SSB signals is similar to Fig. 5.2-5 used in reverse. Such phase demodulators [4] suffer from all the problems of modulators using the phase method. They are, therefore, not preferred to the simple coherent demodulator having a product device and low-pass filter.

VSB Signal Demodulation

If the input to the demodulator of Fig. 5.4-1 is a VSB signal, it becomes most convenient to carry out an analysis in the frequency domain. By Fourier transforming the product device output $s_{\mathrm{VSB}}(t)\cos(\omega_0 t + \phi)$, we obtain the spectrum $S_p(\omega)$ at that point:*

$$S_p(\omega) = \frac{1}{2}[S_{\mathrm{VSB}}(\omega - \omega_0)e^{j\phi} + S_{\mathrm{VSB}}(\omega + \omega_0)e^{-j\phi}], \tag{5.4-15}$$

where $S_{\mathrm{VSB}}(\omega)$ is the spectrum of the input signal $s_{\mathrm{VSB}}(t)$. On substitution of the VSB spectrum of (5.3-2) we have

$$\begin{aligned} S_p(\omega) = {} & \frac{1}{4}F(\omega)[H_v(\omega - \omega_0)e^{-j(\theta_0 - \phi)} + H_v(\omega + \omega_0)e^{j(\theta_0 - \phi)}] \\ & + \frac{1}{4}[H_v(\omega - \omega_0)F(\omega - 2\omega_0)e^{j(\theta_0 + \phi)} \\ & + H_v(\omega + \omega_0)F(\omega + 2\omega_0)e^{-j(\theta_0 + \phi)}]. \end{aligned} \tag{5.4-16}$$

Terms inside the second set of brackets correspond to a VSB signal spectrum where the carrier frequency is $2\omega_0$ with a phase angle $\theta_0 + \phi$. These terms are removed by the low-pass filter. At the filter output the remaining spectrum is

$$S_d(\omega) = \frac{1}{4}F(\omega)H_d(\omega)[H_v(\omega - \omega_0)e^{-j(\theta_0 - \phi)} + H_v(\omega + \omega_0)e^{j(\theta_0 - \phi)}]. \tag{5.4-17}$$

For a perfectly coherent local carrier, $\theta_0 = \phi$, and a distortion-free output spectrum is seen to result if

$$[H_v(\omega - \omega_0) + H_v(\omega + \omega_0)] = \text{constant} \tag{5.4-18}$$

over the band $|\omega| < W_f$ of the signal spectrum. This condition was used in Section 5.3 to develop the relationship between VSB and SSB modulation types. The undistorted output spectrum becomes $F(\omega)/2$, since the filter gain is assumed unity and the constant in (5.4-18) is chosen as 2 [see (5.3-1)]. Finally, the output waveform is

ISBN 0-201-05758-1

*The following results are strictly valid only for ϕ equal to a constant phase error.

$$s_d(t) = \frac{1}{2} f(t) \qquad \text{(VSB)}. \tag{5.4-19}$$

Because VSB is quite similar to SSB, the effects of frequency and phase errors are similar to those already discussed for the demodulation of SSB signals.

★ Generalization of the Coherent Demodulator

It is not necessary that the local signal in Fig. 5.4-1 be a pure sinusoidal waveform. A more general time function $f_d(t)$ may be used. It is necessary only that $f_d(t)$ be periodic, with a fundamental frequency ω_d greater than twice the spectral extent W_f of the information signal, and have a spectral component which is phase-coherent with the carrier of the incoming AM signal. As a direct consequence of this fact, the local carrier, even though it may be designed to be sinusoidal, may contain distortion without degrading demodulator performance.

To prove the above assertion we shall first find the spectrum $S_d(\omega)$ of the output signal $s_d(t)$ from the demodulator. We begin by writing $f_d(t)$ in terms of its complex Fourier series representation and then obtaining its Fourier transform $F_d(\omega)$. Signal and transform are given by the right sides of (2.7-1) and (2.7-2), respectively, with ω_d replacing ω_T. Now the output signal from the product device is the product of the input AM signal $s(t)$ and $f_d(t)$. From the convolution property (2.4-17) of Fourier transforms we know that the spectrum of this product is $1/2\pi$ times the convolution of the two individual spectra $S(\omega)$ and $F_d(\omega)$. Carrying out these operations and adding the effect of the low-pass filter leads to

$$S_d(\omega) = H_d(\omega) \sum_{n=-\infty}^{\infty} C_n S(\omega - n\omega_d). \tag{5.4-20}$$

The action of the filter, of course, is to pass only those spectral components in the sum which fall in the band $-W_f < \omega < W_f$.

Pausing to study (5.4-20), let us recall that the spectrum of any of the various types of AM signals consists of a component clustered near ω_0 and another component clustered near $-\omega_0$. For positive values of n, each term in the sum is a replica of the AM spectrum shifted higher in frequency by an amount $n\omega_d$. Thus, for increasing n there will be a value, say N, which will move the spectral component originally clustered near $-\omega_0$ to near zero. In this case $N\omega_d = \omega_0$ or $N = \omega_0/\omega_d$. If $\omega_d > 2W_f$, no two replicas will overlap in the band of the low-pass filter. An analogous discussion verifies that, for negative values of n, the AM component clustered around ω_0 is shifted to zero frequency when $N = \omega_0/\omega_d$. The two components about zero pass through the low-pass filter to form an undistorted information signal.

Hence, the output spectrum becomes

$$\begin{aligned} S_d(\omega) &= H_d(\omega)[C_N S(\omega - N\omega_d) + C_{-N} S(\omega + N\omega_d)] \\ &= H_d(\omega)[C_N S(\omega - \omega_0) + C_N^* S(\omega + \omega_0)], \end{aligned} \tag{5.4-21}$$

since the Fourier coefficients satisfy $C_{-n} = C_n^*$. We next illustrate the various specific AM types.

Substituting the standard AM spectrum of (5.1-6) and letting $C_N = |C_N| \exp(j\phi_N)$, we have

ISBN 0-201-05758-1

$$S_d(\omega) = 2\pi |C_N| A_0 \cos(\theta_0 - \phi_N)\delta(\omega) + |C_N| \cos(\theta_0 - \phi_N)F(\omega) \qquad \text{(standard AM)}. \qquad (5.4\text{-}22)$$

For a DSB signal we set $A_0 = 0$ in the above:

$$S_d(\omega) = |C_N| \cos(\theta_0 - \phi_N)F(\omega) \qquad \text{(DSB)}. \qquad (5.4\text{-}23)$$

For SSB we substitute (5.2-10) or (5.2-11) in (5.4-21). Both results may be combined as

$$S_d(\omega) = |C_N| [U(\mp\omega)e^{-j(\theta_0 - \phi_N)} + U(\pm\omega)e^{j(\theta_0 - \phi_N)}] F(\omega) \qquad \text{(SSB)}. \qquad (5.4\text{-}24)$$

Finally, we use (5.3-2) and obtain the demodulator output spectrum for VSB:

$$S_d(\omega) = \frac{|C_N|}{2} F(\omega)H_d(\omega)[H_v(\omega - \omega_0)e^{-j(\theta_0 - \phi_N)} + H_v(\omega + \omega_0)e^{j(\theta_0 - \phi_N)}] \qquad \text{(VSB)}. \qquad (5.4\text{-}25)$$

We may now compare the above four spectra with the previously developed counterparts, which are the Fourier transform of (5.4-4) for standard AM and DSB (with $A_0 = 0$), (5.4-14) for SSB, and (5.4-17) for VSB. In every case the above results equal earlier results multiplied by a factor $2|C_N|$. Hence, the effect of using an arbitrary periodic waveform $f_d(t)$ is only to alter the magnitude of the demodulated waveform by this factor. Since ϕ_N is the phase of the spectral component of $f_d(t)$ at frequency ω_0, all our results verify that the component must be coherent with the transmitted carrier. With $\phi_N = \theta_0$ the above four expressions produce

$$s_d(t) = \begin{cases} |C_N|[A_0 + f(t)] & \text{(standard AM)} \\ |C_N| f(t) & \text{(DSB, SSB, VSB)}. \end{cases} \qquad (5.4\text{-}26)$$

5.5. ENVELOPE DEMODULATION OF STANDARD AM

The most common method of demodulating standard AM employs the *envelope detector* shown in Fig. 5.5-1(*a*). It is the method used in ordinary broadcast AM receivers. Output occurs only during positive half-cycles of the carrier as shown in (*b*). When the carrier is positive the diode conducts and charges the capacitor C until the peak voltage occurs. As the carrier level decreases the diode ceases to conduct because the voltage on C exceeds the carrier voltage, thereby back-biasing the diode. During the negative half-cycle time the charged capacitor discharges through resistor R, causing $s_d(t)$ to decay until the carrier level, during the next positive half-cycle, again exceeds $s_d(t)$, causing the process to repeat.

If the time constant of the network, given by the product RC, is too large, the decay rate between carrier cycle peaks will be too slow to follow the envelope changes caused by the modulation signal $f(t)$. Conversely, if RC is too small the "valleys" between carrier peaks are large and the output contains excessive "ripple" due to the carrier frequency and low output level occurs. A good design compromise is to make

$$W_f << \frac{1}{RC} << \omega_0, \qquad (5.5\text{-}1)$$

ISBN 0-201-05758-1

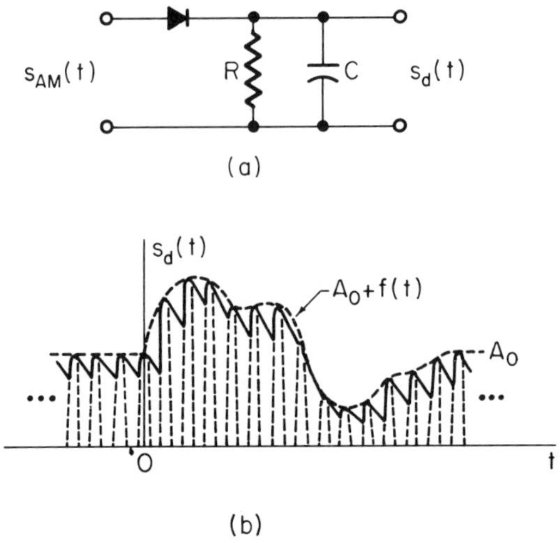

Fig. 5.5-1. Standard AM envelope detector (a) and its response (b).

where W_f is the highest significant radian frequency present in $f(t)$. In this case, the output is approximately

$$s_d(t) = A_0 + f(t). \tag{5.5-2}$$

Even if the capacitor is removed in Fig. 5.5-1 it is still possible to obtain demodulation as long as later circuits filter out the carrier ripple. Such a demodulator is sometimes called a *rectifier detector.* The rectifier detector is less efficient than the envelope detector as evidenced by the fact that its output signal component is $1/\pi$ times that of the envelope detector (problem 5-16).

Both the envelope detector and the rectifier detector may be classed as noncoherent demodulators, since no local coherent carrier is involved. Other noncoherent methods are also possible but are mostly of less importance.

5.6. SUPPRESSED CARRIER DEMODULATION BY CARRIER REINSERTION

Demodulation of suppressed carrier signals using the coherent detector was described in Section 5.4. The method requires a product device and a low-pass filter. It is also possible to obtain demodulation by simply *adding* a local carrier $A_d \cos(\omega_0 t + \phi)$ to the received signal and detecting the sum with an envelope detector. The result is demodulation by *carrier reinsertion.* The receiver block diagram is shown in Fig. 5.6-1. As will be seen, the amplitude A_d

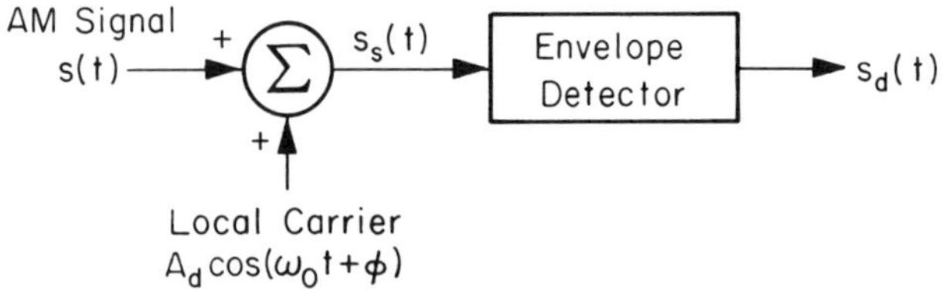

Fig. 5.6-1. Block diagram of the carrier reinsertion method for demodulation of suppressed carrier signals.

ISBN 0-201-05758-1

of the local carrier must be large relative to the maximum received signal magnitude, and its phase must be coherent with the transmitted carrier. Thus, even though an envelope detector is used as part of the demodulation process, we should classify carrier reinsertion as a coherent method.

In following paragraphs the response $s_d(t)$ of Fig. 5.6-1 is found for DSB and SSB modulation.

DSB Signals

For this case the sum signal may be expressed as

$$s_s(t) = A_d \cos(\omega_0 t + \phi) + f(t) \cos(\omega_0 t + \theta_0)$$
$$= A(t) \cos(\omega_0 t + \psi), \tag{5.6-1}$$

where

$$A(t) = [A_d^2 + f^2(t) + 2A_d f(t) \cos(\theta_0 - \phi)]^{1/2}, \tag{5.6-2}$$

$$\psi = \tan^{-1}\left[\frac{A_d \sin(\phi) + f(t) \sin(\theta_0)}{A_d \cos(\phi) + f(t) \cos(\theta_0)}\right]. \tag{5.6-3}$$

Now if A_d is large—that is, if

$$A_d \gg |f(t)|_{max}, \tag{5.6-4}$$

then the envelope $A(t)$, which is the final output, is approximately given by

$$s_d(t) = A(t) = A_d + f(t) \cos(\theta_0 - \phi). \tag{5.6-5}$$

Relative to the term involving the information signal, except for a factor of $1/2$ this result is the same as the output of the coherent demodulator given by (5.4-4). Consequently, all the coherence considerations discussed for the coherent demodulator apply to the carrier reinsertion technique, and neither method is preferred from a coherence or stability standpoint. The main advantage enjoyed by the carrier reinsertion scheme is that circuitry is simpler.

SSB Signals

Here the local carrier is added to the SSB waveform of (5.2-12). The sum becomes

$$s_s(t) = A_d \cos(\omega_0 t + \phi) + f(t) \cos(\omega_0 t + \theta_0) \mp \hat{f}(t) \sin(\omega_0 t + \theta_0)$$
$$= A(t) \cos(\omega_0 t + \psi), \tag{5.6-6}$$

where now

$$A(t) = [A_d^2 + \hat{f}^2(t) + f^2(t) + 2A_d f(t) \cos(\theta_0 - \phi) \mp 2A_d \hat{f}(t) \sin(\theta_0 - \phi)]^{1/2}, \tag{5.6-7}$$

ISBN 0-201-05758-1

$$\psi(t) = \tan^{-1}\left[\frac{A_d \sin(\phi) + f(t)\sin(\theta_0) \pm \hat{f}(t)\cos(\theta_0)}{A_d \cos(\phi) + f(t)\cos(\theta_0) \mp \hat{f}(t)\sin(\theta_0)}\right]. \tag{5.6-8}$$

It can be shown that the magnitude of the SSB signal is

$$|s_{SSB}(t)| = [f^2(t) + \hat{f}^2(t)]^{1/2}. \tag{5.6-9}$$

If $A_d >> |s_{SSB}(t)|_{max}$, the output signal, which is equal to $A(t)$, becomes approximately

$$s_d(t) = A(t) = A_d + f(t)\cos(\theta_0 - \phi) \mp \hat{f}(t)\sin(\theta_0 - \phi). \tag{5.6-10}$$

Except for the dc term and a factor of 1/2, this expression is identical to the output from the coherent demodulator as given by (5.4-11).

In a practical system there is little difficulty in making A_d large enough to make carrier reinsertion a practical demodulation method. The real problem in practice arises from the difficulty in maintaining the local carrier phase coherent with the transmitted carrier. Some methods for achieving coherence are presented in the following section.

5.7. LOCAL CARRIER GENERATION METHODS

All the coherent demodulators for the various AM signals must have a local carrier available which is phase-coherent with that of the transmitter. Such a local carrier allows recovery of the information signal without distortion. If phase or frequency (or both) errors exist, the recovered message may suffer in many respects, varying from simple attenuation all the way to gross distortion or even complete elimination. Many methods exist for locally generating the required carrier. Some of the more important ones are discussed below.

In every case the receiver must develop the carrier from information contained in the incoming modulated signal. If sufficient knowledge is not present, such as with SSB waveforms for example, the transmitted signal may purposely be altered to allow proper local carrier recovery. The recovery problem, which may also be referred to as *receiver synchronization,* is most severe for suppressed carrier modulation, and our discussion is almost entirely limited to this type of signal. Standard AM is rarely demodulated by any method except the noncoherent envelope detector and, consequently, deserves little further attention.

Partial Carrier Method for SC Signals

Figure 5.7-1(*a*) illustrates a method of carrier recovery which is especially simple and applies to DSB, SSB, or VSB modulation. It assumes that the transmitted carrier is not totally suppressed and that a *partial carrier,* sometimes also called a *pilot carrier,* is purposely included to aid in receiver synchronization. A filter, having a narrow bandwidth in relation to the modulated signal bandwidth, is used to select the partial carrier which is then amplified to form the local carrier.

It should be observed that the partial carrier required is typically small in relation to that which would be present without carrier suppression. Thus, the added power required may be negligible in relation to the sideband power and the waveform may still be regarded as a suppressed carrier signal. The required partial carrier is easily obtained by a slight unbalance of the balanced modulator involved in transmitted signal generation.

A possible disadvantage in some cases is the need for the narrowband filter. It may be necessary to implement it in the form of a *phase-locked loop* (PLL). These loops are discussed

ISBN 0-201-05758-1

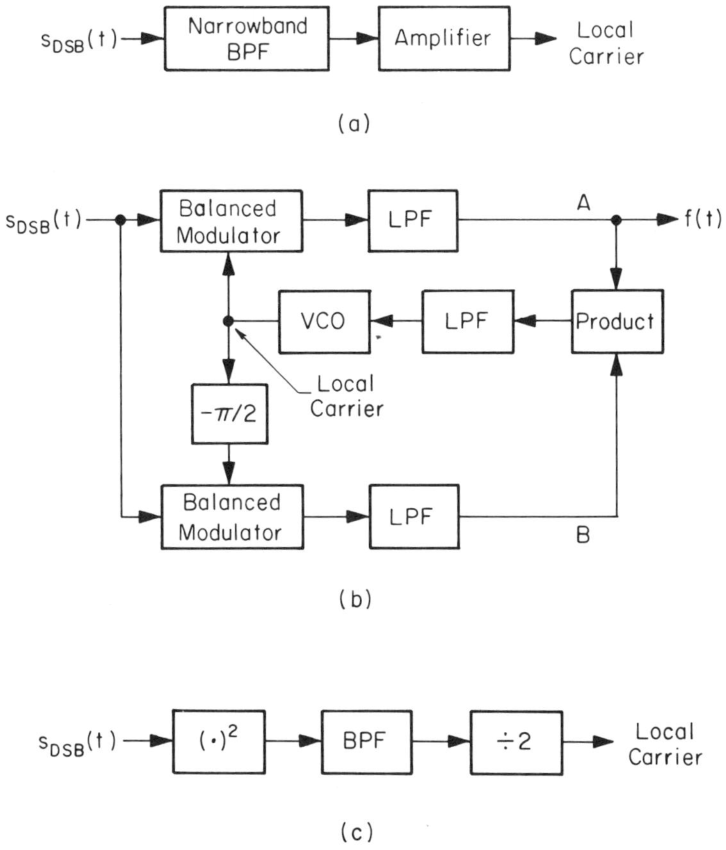

Fig. 5.7-1. Local carrier generation methods for suppressed carrier modulated signals. (a) Partial carrier method for DSB, SSB, or VSB; (b) Costas' method for DSB; and (c) the nonlinear method for DSB.

in Chapter 6. Another disadvantage is that the information signal must not contain significant low-frequency energy. Otherwise, this energy will pass through the narrowband filter and perturb the recovered carrier. For some messages it may be necessary to filter out these low frequencies at the transmitter prior to modulation to guarantee carrier purity.

Costas' Method for DSB Signals

Figure 5.7-1(*b*) illustrates the method of Costas [5]. It is more complicated than the partial carrier technique but has the advantages of operating with a truly suppressed carrier and not requiring the message to be void of low-frequency content. The method uses a closed-loop system so that the local carrier automatically follows the phase of the carrier at the transmitter.

The output of the *voltage-controlled oscillator* (VCO) forms the local carrier. It is an oscillator having output frequency changes (relative to some nominal frequency such as ω_0) which are proportional to a voltage, here derived from a low-pass filter.

To better understand operation, let the input be the DSB signal $f(t)\cos(\omega_0 t + \theta_0)$. If the local carrier is initially presumed to be $\cos(\omega_0 t + \phi)$, the outputs at points A and B are easily found to be $[f(t)/2]\cos(\theta_0 - \phi)$ and $-[f(t)/2]\sin(\theta_0 - \phi)$, respectively, if the low-pass filters following the balanced modulators have bandwidths equal to the message bandwidth W_f. The product of these two outputs, which may actually be implemented with a

ISBN 0-201-05758-1

low-frequency balanced modulator, becomes $-[f^2(t)/8]\sin[2(\theta_0-\phi)]$. If the difference angle is small, the output of the product is approximately given by $-f^2(t)(\theta_0-\phi)/4$. The effect of the low-pass filter to which this signal is applied is to replace $f^2(t)$ by the time average $\overline{f^2(t)}$, a result achieved if the filter bandwidth is very narrow relative to W_f. Finally, the loop "error" signal $-\overline{f^2(t)}(\theta_0-\phi)/4$ acts on the VCO in such a way as to cause the original phase difference $\theta_0-\phi$ to approach zero. Under a stable condition $\phi\approx\theta_0$ and the loop error signal is zero.

Even if carrier phase θ_0 varies slowly with time, Costas' method will maintain synchronization of the local carrier. In other words, even for small *frequency* changes of the transmitted carrier, the loop will follow the changes. For greater frequency variations a modification of Costas' method may be incorporated [6] to maintain synchronization.

Nonlinear Method for DSB Signals

Figure 5.7-1(c) illustrates a third method of carrier recovery with DSB modulation. The input waveform is passed through a squarer to produce $f^2(t)\cos^2(\omega_0 t+\theta_0)=f^2(t)[1+\cos(2\omega_0 t+2\theta_0)]/2$. The band-pass filter removes the baseband term and passes the term $[f^2(t)/2]\cos(2\omega_0 t+2\theta_0)$, which is a DSB waveform where the "message" is $f^2(t)/2$. A limiter may be used prior to the frequency divider to strip off the modulation. An approximately equivalent operation is to make the filter have a very narrow bandwidth. Its output would then become $[\overline{f^2(t)}/2]\cos(2\omega_0 t+2\theta_0)$. The filter could be implemented with a phase-locked loop. Indeed, by proper design, such a loop can provide both the filtering and dividing functions.

Regardless of the implementation, the final output is $\cos(\omega_0 t+\theta_0)$. Although perhaps not obvious, there is a phase ambiguity in the output of π radians. This can be demonstrated by retracing the above discussion assuming a carrier of $\cos(\omega_0 t+\theta_0+\pi)$; the same final output is obtained. The phase ambiguity is not a problem for voice transmission but may be intolerable for other message types.

Two-Frequency Pilot Method for SSB or VSB Signals

When a SSB or VSB waveform is involved, the local carrier cannot be reconstructed from information in the sideband alone as was the case for DSB using the Costas method. Some modification of the suppressed carrier transmitted signal is necessary. As has already been noted, the addition of a partial carrier will allow local carrier generation, but the message is restricted to have little or no low-frequency spectral content. A method [7] which does not have this restriction involves adding two *pilot frequencies* to the transmitted spectrum as illustrated in Fig. 5.7-2. Both SSB and VSB spectra are illustrated, assuming USB transmission. Similar sketches would illustrate the LSB cases. The added frequencies are ω_1 and ω_2, where it is assumed that $\omega_1<\omega_2$ for either USB or LSB situations. The two frequencies are shifted by an amount $\delta\omega$ from either the carrier or the outer information band edge. For VSB the value of $\delta\omega$ is determined by the frequency extent of the attenuated sideband. For SSB it is chosen for convenience of design. Thus, for the various situations, ω_1 and ω_2 will have the following values:

$$\begin{aligned}\omega_1 &= \omega_0-\delta\omega, && \text{USB}\\ &= \omega_0-\delta\omega-W_f, && \text{LSB,} \qquad (5.7\text{-}1)\end{aligned}$$

$$\begin{aligned}\omega_2 &= \omega_0+\delta\omega+W_f, && \text{USB}\\ &= \omega_0+\delta\omega, && \text{LSB.} \qquad (5.7\text{-}2)\end{aligned}$$

ISBN 0-201-05758-1

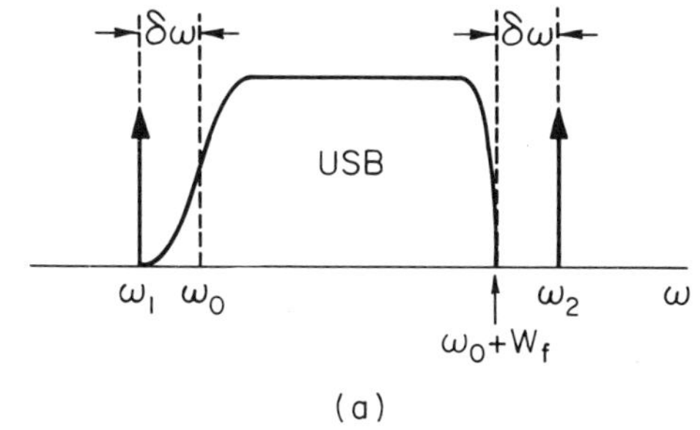

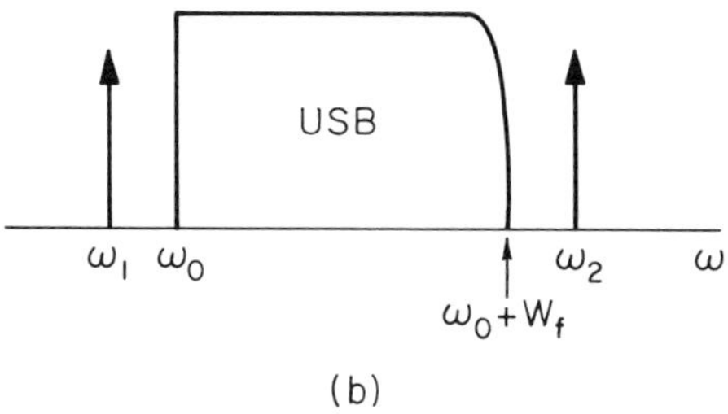

Fig. 5.7-2. A VSB spectrum (a) and a SSB spectrum (b) illustrating pilot frequencies added.

Carrier reconstruction may be accomplished using the network illustrated in Fig. 5.7-3. The spectral components at ω_1 and ω_2 are passed by narrowband band-pass filters to form the inputs to a product device (balanced modulator):

$$e_A(t) = a_1 \cos(\omega_1 t + \theta_0 + \phi_1), \tag{5.7-3}$$

$$e_B(t) = a_2 \cos(\omega_2 t + \theta_0 + \phi_2). \tag{5.7-4}$$

Here a_1 and a_2 are the amplitudes, while ϕ_1 and ϕ_2 are the phases of the pilot components.

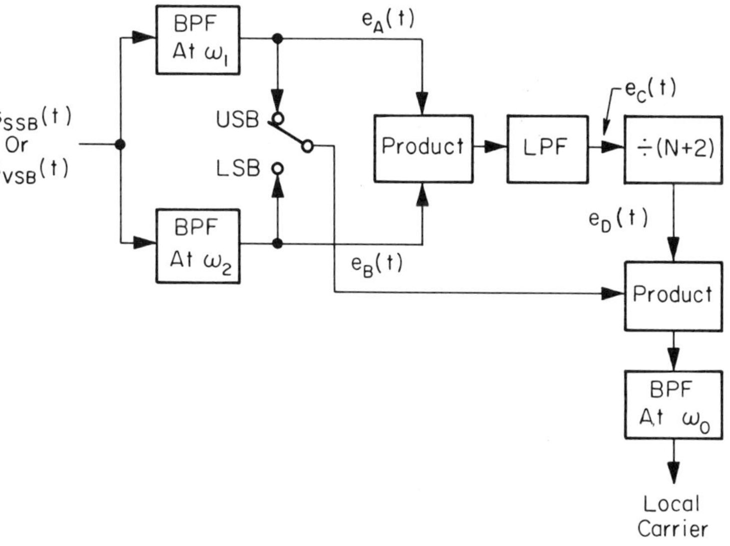

Fig. 5.7-3. Local carrier reconstruction using the two-frequency pilot method.

ISBN 0-201-05758-1

The phase θ_0 is that of the transmitter carrier. The low-pass filtered product device output is readily found to be

$$e_C(t) = \frac{a_1 a_2}{2} \cos\,[(W_f + 2\delta\omega)t + \phi_2 - \phi_1] \qquad (5.7\text{-}5)$$

for either USB or LSB cases. Next, suppose in the selection of $\delta\omega$ the ratio $W_f/\delta\omega$ is made to be an integer N. That is,

$$N = W_f/\delta\omega. \qquad (5.7\text{-}6)$$

The output of the frequency divider will now be

$$e_D(t) = \cos\left[\delta\omega t + \frac{\phi_2 - \phi_1}{N + 2}\right]. \qquad (5.7\text{-}7)$$

Finally, the local carrier output may be found to be proportional to $\cos\,[\omega_0 t + \theta_0]$ if the phases ϕ_1 and ϕ_2 satisfy the relationship

$$\phi_2 = \frac{-\phi_1(N \pm N + 2)}{N \mp N + 2}, \qquad (5.7\text{-}8)$$

where upper and lower signs correspond to USB and LSB signals, respectively. The reader may wish to develop the omitted steps leading to (5.7-8) as an exercise.

5.8. NOISE PERFORMANCE USING A COHERENT DEMODULATOR

The ultimate performance limit in an AM system is determined by the amount of random noise present. The noise may originate from sources external to the receiver or may be generated within the receiver itself. Regardless of the origin of the noise, the total effect on the output of the receiver can be found by modeling all noise sources as a single equivalent source at the receiving system input. Detailed justifications for this approach are considered in Chapter 9. However, at this point it is sufficient to accept the model as valid. Many significant noise sources are additive in nature.* This fact means that the total received signal may be considered to be the sum of the transmitted waveform plus an equivalent noise.

A very important measure of performance of any of the various AM systems is the ratio of average power in the output demodulated signal to the average power in the output noise. Intuition tells us that the larger this output *signal-to-noise power ratio,* (S_o/N_o), the easier it will be to distinguish signal from noise. Thus, (S_o/N_o) will be the most important quantity of interest in performance. Another quantity of interest is the input signal-to-noise ratio for each particular system.

Having established the most important measure of system performance, various systems may be compared by comparing corresponding values of (S_o/N_o). In these comparisons it is helpful to establish a reference point to serve for all systems. A logical reference to choose is the simple, two-wire *baseband communication system.*

*There are also multiplicative noise sources where the signal is modified in some way by the noise as a product. An example is a signal passing through a turbulent channel which results in a randomly time-varying attenuation and/or phase shift. Multiplicative noise effects are not considered in this book.

ISBN 0-201-05758-1

Baseband System and Its Noise Performance

The most elementary baseband system transmits the message $f(t)$ directly to the receiver as depicted in Fig. 5.8-1. The receiver consists of a low-pass filter (assumed ideal) which has a bandwidth just wide enough to pass the largest message spectrum frequency W_f, but no wider, in order to prevent excessive output noise power. Clearly, the input and output signals are the same:

$$s_i(t) = f(t), \tag{5.8-1}$$

$$s_d(t) = f(t), \tag{5.8-2}$$

and carry the same power

$$S_i = \overline{f^2(t)}, \tag{5.8-3}$$

$$S_o = \overline{f^2(t)}. \tag{5.8-4}$$

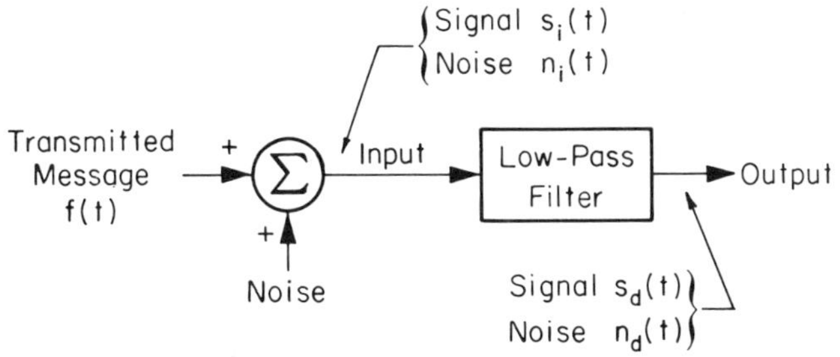

Fig. 5.8-1. Block diagram of a baseband communication system.

To determine the applicable noise powers, let us assume that the input noise $n_i(t)$ is white with power density $\mathcal{N}_0/2$ (watts/hertz). By use of (4.9-10) the output average noise power in the signal band is

$$N_o = \overline{n_d^2(t)} = \frac{\mathcal{N}_0 W_f}{2\pi}. \tag{5.8-5}$$

We shall define the input noise power as that which falls *in the signal band.* Hence,

$$N_i = \frac{\mathcal{N}_0 W_f}{2\pi} = N_o. \tag{5.8-6}$$

Signal-to-noise ratios follow from the above four expressions. Using a subscript B to indicate a baseband system, we have

$$\left(\frac{S_o}{N_o}\right)_B = \left(\frac{S_i}{N_i}\right)_B = \frac{2\pi\overline{f^2(t)}}{\mathcal{N}_0 W_f} = \frac{2\pi S_i}{\mathcal{N}_0 W_f}. \tag{5.8-7}$$

ISBN 0-201-05758-1

Noise in Coherent Demodulation of All AM Signals

Since a coherent demodulator will demodulate any of the various AM signals, let us examine the performance of this type of receiver. The block diagram applicable to all modulation types is shown in Fig. 5.8-2(*a*). Noise is added to the AM signal of interest. The total signal is applied to the receiver as illustrated in (*b*). A practical receiver would consist of various components, possibly a radio-frequency amplifier and filter followed by a mixer and intermediate-frequency amplifier and filter. However, for purposes of analysis, all these components may be replaced by an equivalent band-pass filter. The filter has a bandwidth equal to the modulated signal bandwidth so as to pass this waveform undistorted. It also acts to limit the noise power applied to the demodulator, since its bandwidth is no wider than necessary. Thus, the filter bandwidth is $2W_f$ for standard AM and DSB, is W_f for SSB, and will be approximately W_f for VSB if the sloping region of the VSB spectrum is small. Here, as always, W_f is the maximum frequency extent of the message $f(t)$.

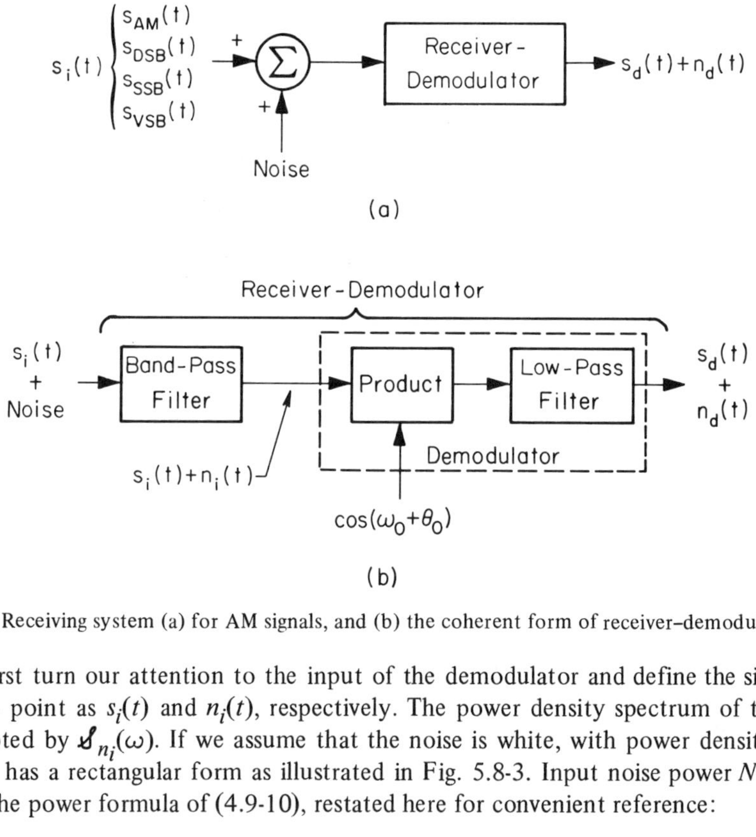

Fig. 5.8-2. Receiving system (a) for AM signals, and (b) the coherent form of receiver–demodulator.

Let us first turn our attention to the input of the demodulator and define the signal and noise at this point as $s_i(t)$ and $n_i(t)$, respectively. The power density spectrum of the noise $n_i(t)$ is denoted by $\mathcal{S}_{n_i}(\omega)$. If we assume that the noise is white, with power density $\mathcal{N}_0/2$, then $\mathcal{S}_{n_i}(\omega)$ has a rectangular form as illustrated in Fig. 5.8-3. Input noise power N_i follows easily from the power formula of (4.9-10), restated here for convenient reference:

$$P = \frac{1}{2\pi}\int_{-\infty}^{\infty} \mathcal{S}(\omega)\, d\omega. \tag{5.8-8}$$

Direct calculation gives the input noise powers:

$$N_i = \mathcal{N}_0 W_f/\pi, \qquad \text{standard AM and DSB,} \tag{5.8-9}$$

ISBN 0-201-05758-1

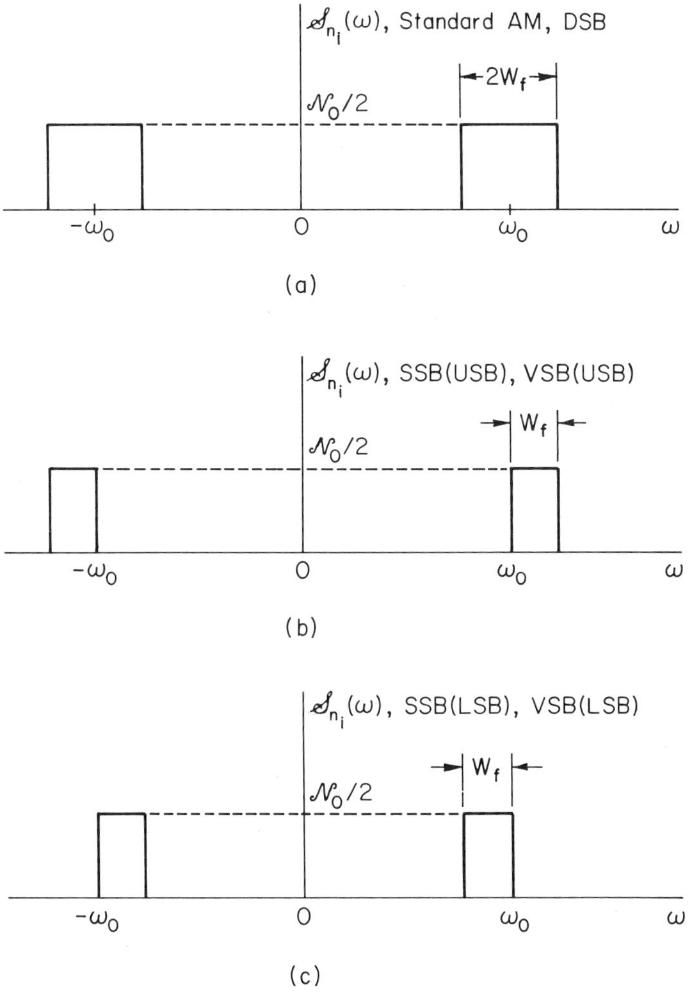

Fig. 5.8-3. Demodulator input noise spectra. (a) For standard AM and DSB signals. (b) For SSB(USB) and approximate spectra for VSB(USB). (c) LSB cases analogous to (b).

$$N_i = \mathcal{N}_0 W_f/(2\pi), \qquad \text{SSB and VSB.} \tag{5.8-10}$$

Signal power S_i at the demodulator input is given by $\overline{s_i^2(t)}$. The various signal expressions involved are summarized from previous work:

$$s_{\text{AM}}(t) = [A_0 + f(t)] \cos(\omega_0 t + \theta_0), \tag{5.8-11}$$

$$s_{\text{DSB}}(t) = f(t) \cos(\omega_0 t + \theta_0), \tag{5.8-12}$$

$$s_{\text{SSB}}(t) = f(t) \cos(\omega_0 t + \theta_0) \mp \hat{f}(t) \sin(\omega_0 t + \theta_0), \tag{5.8-13}$$

$$s_{\text{VSB}}(t) \approx s_{\text{SSB}}(t). \tag{5.8-14}$$

The last result derives from the assumption of a small sloping region in the VSB spectrum. The reader may readily verify that the average input powers are

ISBN 0-201-05758-1

$$S_i = [A_0^2 + \overline{f^2(t)}]/2, \qquad \text{standard AM,} \tag{5.8-15}$$

$$S_i = \overline{f^2(t)}/2, \qquad \text{DSB,} \tag{5.8-16}$$

$$S_i = \overline{f^2(t)}, \qquad \text{SSB,} \tag{5.8-17}$$

$$S_i \approx \overline{f^2(t)}, \qquad \text{VSB.} \tag{5.8-18}$$

The above signal and noise powers may now be used to give the input signal-to-noise power ratios:

$$\left(\frac{S_i}{N_i}\right)_{\text{AM}} = \frac{\pi[A_0^2 + \overline{f^2(t)}]}{2\mathscr{N}_0 W_f}, \tag{5.8-19}$$

$$\left(\frac{S_i}{N_i}\right)_{\text{DSB}} = \frac{\pi\overline{f^2(t)}}{2\mathscr{N}_0 W_f}, \tag{5.8-20}$$

$$\left(\frac{S_i}{N_i}\right)_{\text{SSB or VSB}} = \frac{2\pi\overline{f^2(t)}}{\mathscr{N}_0 W_f}. \tag{5.8-21}$$

Turning our attention next to the demodulator output, some reflection will show that superposition applies. In other words, the components of signal and noise at the output may be determined as the responses of the demodulator acting separately on the input signal and noise components. Previous work has already developed the output signal responses. The portions due to the information signal are all identical. They are

$$s_d(t) = f(t)/2, \qquad \text{standard AM, DSB, SSB, and VSB,} \tag{5.8-22}$$

from (5.4-7), (5.4-8), (5.4-13), and (5.4-19). Output powers due to the message become

$$S_o = \overline{f^2(t)}/4, \qquad \text{standard AM, DSB, SSB, and VSB.} \tag{5.8-23}$$

To find the output average noise power, we turn to a development in Chapter 4. Equation (4.11-6) was found to represent the power density spectrum at the output of a product device in terms of the input power density spectrum. Applying (4.11-6) to the coherent demodulator, and recognizing that the effect of the low-pass filter will be to confine ω to the interval $-W_f < \omega < W_f$, we have

$$\mathscr{S}_{n_d}(\omega) = \frac{1}{4}[\mathscr{S}_{n_i}(\omega - \omega_0) + \mathscr{S}_{n_i}(\omega + \omega_0)], \qquad |\omega| < W_f, \tag{5.8-24}$$

where $\mathscr{S}_{n_d}(\omega)$ is the power density spectrum of the demodulator output noise $n_d(t)$. On substitution of the various AM power spectra, as shown in Fig. 5.8-3, into (5.8-24) and performing the integration suggested by (5.8-8), the output average noise powers are found to be

$$N_o = \mathscr{N}_0 W_f/(4\pi), \qquad \text{standard AM, and DSB,} \tag{5.8-25}$$

ISBN 0-201-05758-1

$$N_o = \mathcal{N}_0 W_f/(8\pi), \qquad \text{SSB, and VSB.} \tag{5.8-26}$$

Equivalently,

$$N_o = N_i/4, \qquad \text{standard AM, DSB, SSB, and VSB.} \tag{5.8-27}$$

From a ratio of (5.8-23) and (5.8-27) the output signal-to-noise power ratio is

$$\left(\frac{S_o}{N_o}\right) = \frac{\overline{f^2(t)}}{N_i}, \qquad \text{all AM types.} \tag{5.8-28}$$

More explicit results follow from the substitutions for N_i:

$$\left(\frac{S_o}{N_o}\right)_{\text{AM}} = \frac{\pi \overline{f^2(t)}}{\mathcal{N}_0 W_f}, \tag{5.8-29}$$

$$\left(\frac{S_o}{N_o}\right)_{\text{DSB}} = \frac{\pi \overline{f^2(t)}}{\mathcal{N}_0 W_f}, \tag{5.8-30}$$

$$\left(\frac{S_o}{N_o}\right)_{\text{SSB or VSB}} = \frac{2\pi \overline{f^2(t)}}{\mathcal{N}_0 W_f}. \tag{5.8-31}$$

Although we shall not prove it, these identical results are achieved when demodulation is accomplished by carrier reinsertion if the local carrier level is large.

On a first glance at the above three expressions one may be tempted to say there is no performance difference in standard AM as compared to DSB. At the same time, the temptation is to presume that SSB and VSB give twice the output signal-to-noise ratio as does DSB. These conclusions may or may not be correct, depending on how the systems are compared. Indeed, the conclusions are correct if the same message (W_f same in all cases), the same message power [$\overline{f^2(t)}$ same in all cases], and the same system noise ($\mathcal{N}_0$ same in all cases) are used. However, this is not really a fair comparison, since the *total power* transmitted by the various systems may be radically different from (5.8-15) through (5.8-18).

A fair comparison of AM systems would allow each system to transmit the same total power. In terms of input signal power, (5.8-29) through (5.8-31) may be altered to read

$$\left(\frac{S_o}{N_o}\right)_{\text{AM}} = \frac{\overline{f^2(t)}}{A_0^2 + \overline{f^2(t)}} \left(\frac{2\pi S_i}{\mathcal{N}_0 W_f}\right), \tag{5.8-32}$$

$$\left(\frac{S_o}{N_o}\right)_{\text{DSB or SSB or VSB}} = \left(\frac{2\pi S_i}{\mathcal{N}_0 W_f}\right). \tag{5.8-33}$$

Thus, for identical message spectral extent W_f, receiver noise density $\mathcal{N}_0/2$, and power S_i, all but the standard AM system give the same performance. To prevent overmodulation in standard AM, $|f(t)| \leqslant A_0$, which prevents the factor $\overline{f^2(t)}/[A_0^2 + \overline{f^2(t)}]$ from exceeding 0.5. It is usually even smaller, leading to the conclusion that standard AM is inferior to other AM modulation methods by more than 3 dB.

ISBN 0-201-05758-1

Sometimes a quantity called *demodulation gain* is defined in the literature. It is the ratio $(S_o/N_o)/(S_i/N_i)$. For the various AM systems the demodulation gains are

$$\frac{(S_o/N_o)}{(S_i/N_i)} \begin{cases} = \dfrac{2\overline{f^2(t)}}{A_0{}^2 + \overline{f^2(t)}}, & \text{standard AM} \\ = 2, & \text{DSB} \\ = 1, & \text{SSB or VSB.} \end{cases} \qquad (5.8\text{-}34)$$

Comparison to Baseband System Performance

From (5.8-7) the factor $2\pi S_i/(\mathcal{N}_0 W_f)$ in both the above expressions, (5.8-32) and (5.8-33), is the same in form as the output signal-to-noise ratio of a baseband system. Hence, if all systems transmit the same average power, use the same message, and have the same value of $\mathcal{N}_0$:

$$\left(\frac{S_o}{N_o}\right)_{\text{AM}} = \frac{\overline{f^2(t)}}{A_0{}^2 + \overline{f^2(t)}} \left(\frac{S_o}{N_o}\right)_{\text{B}}, \qquad (5.8\text{-}35)$$

$$\left(\frac{S_o}{N_o}\right)_{\text{DSB or SSB or VSB}} = \left(\frac{S_o}{N_o}\right)_{\text{B}}. \qquad (5.8\text{-}36)$$

These equations verify that the process of carrier modulation leads to no degradation in performance, as compared to a baseband system, except when standard AM is used. The degradation in standard AM is typically larger than 3 dB.

5.9. NOISE PERFORMANCE USING AN ENVELOPE DETECTOR

In standard AM the most common demodulation method is the envelope detector. The performance of this method with noise may be determined from an analysis of the applicable receiver–demodulator illustrated in Fig. 5.9-1. We shall find, for a large input signal-to-noise ratio, that (5.8-29) again applies. For a small ratio, the message cannot be recovered using the envelope detector.

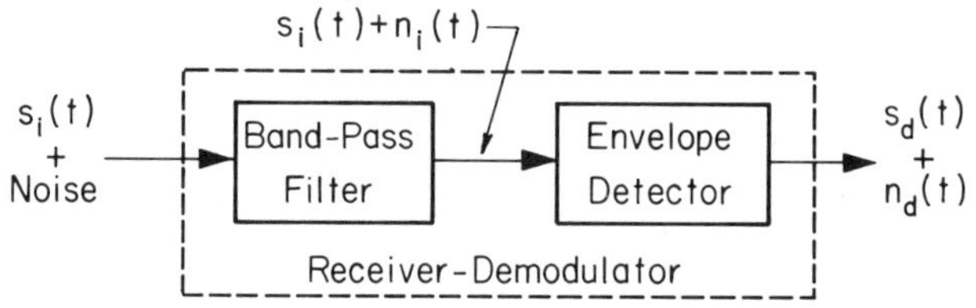

Fig. 5.9-1. Envelope detector form of receiving system for standard AM demodulation.

We may begin the performance analysis by observing that the band-pass filter bandwidth is $2W_f$, which means that the signal power S_i and noise power N_i at the envelope detector input are the same as those at the input to the demodulator of the coherent detector analyzed in the last section [compare Fig. 5.8-2(*b*) with Fig. 5.9-1]:

ISBN 0-201-05758-1

$$S_i = [A_0^2 + \overline{f^2(t)}]/2, \tag{5.9-1}$$

$$N_i = \mathcal{N}_0 W_f/\pi. \tag{5.9-2}$$

Output signal and noise powers must be determined from the response of the envelope detector. This response is most easily found from a time-domain description of the inputs. The signal component is

$$s_i(t) = [A_0 + f(t)] \cos(\omega_0 t + \theta_0), \tag{5.9-3}$$

as usual. The noise, being band-pass, can be expressed as

$$n_i(t) = n_c(t) \cos(\omega_0 t + \theta_0) - n_s(t) \sin(\omega_0 t + \theta_0), \tag{5.9-4}$$

from an application of (4.11-9).* Here $n_c(t)$ is the *in-phase* noise, while $n_s(t)$ is the *quadrature-phase* noise. The sum of signal plus noise at the detector input may now be expressed as

$$\begin{aligned} s_i(t) + n_i(t) &= [A_0 + f(t) + n_c(t)] \cos(\omega_0 t + \theta_0) \\ &\quad - n_s(t) \sin(\omega_0 t + \theta_0) \\ &= A(t) \cos[\omega_0 t + \theta_0 + \psi(t)]. \end{aligned} \tag{5.9-5}$$

Here the envelope $A(t)$ and phase $\psi(t)$ are given by

$$A(t) = \left\{[A_0 + f(t) + n_c(t)]^2 + n_s^2(t)\right\}^{1/2}, \tag{5.9-6}$$

$$\psi(t) = \tan^{-1}\left[\frac{n_s(t)}{A_0 + f(t) + n_c(t)}\right]. \tag{5.9-7}$$

The detector output, assuming a *linear* envelope detector, is equal to (5.9-6). We consider two cases.

Large Signal Case

Equation (5.9-6) may be restated in the form

$$A(t) = [A_0 + f(t)]\left\{1 + \frac{2n_c(t)}{A_0 + f(t)} + \frac{n_c^2(t) + n_s^2(t)}{[A_0 + f(t)]^2}\right\}^{1/2}. \tag{5.9-8}$$

Now if the magnitude of the input signal is nearly always large in relation to the input noise magnitude—that is, if

$$|A_0 + f(t)| >> \sqrt{n_c^2(t) + n_s^2(t)} \tag{5.9-9}$$

*The addition of a phase angle θ_0 only amounts to an origin shift in (4.11-9). The exact correspondences between (5.9-4) and (4.11-9) are: $n_i(t) = N[t + (\theta_0/\omega_0)]$, $n_c(t) = X_c[t + (\theta_0/\omega_0)]$, and $n_s(t) = X_s[t + (\theta_0/\omega_0)]$.

ISBN 0-201-05758-1

most of the time, then (5.9-8) may be approximated by

$$A(t) = A_0 + f(t) + n_c(t). \tag{5.9-10}$$

The often helpful approximation

$$(1 \pm x)^p \approx 1 \pm px, \qquad |x| \ll 1, \tag{5.9-11}$$

has been used, where p is a real number.

The envelope detector output is $A(t)$ as expressed by (5.9-10). The message component $f(t)$ gives an output signal power of

$$S_o = \overline{f^2(t)}. \tag{5.9-12}$$

The noise power is $N_o = \overline{n_c^2(t)}$. However, for band-pass noise we have already shown in (4.11-19) that the power in $n_c(t)$ is the same as the power in $n_i(t)$. Hence,

$$N_o = \overline{n_c^2(t)} = \overline{n_i^2(t)} = \mathcal{N}_0 W_f/\pi. \tag{5.9-13}$$

From the ratio of (5.9-12) and (5.9-13)

$$\left(\frac{S_o}{N_o}\right)_{\text{AM}} = \frac{\pi \overline{f^2(t)}}{\mathcal{N}_0 W_f} = \frac{\overline{f^2(t)}}{A_0^2 + \overline{f^2(t)}} \left(\frac{2\pi S_i}{\mathcal{N}_0 W_f}\right), \tag{5.9-14}$$

or, in terms of a baseband system transmitting the same signal power,

$$\left(\frac{S_o}{N_o}\right)_{\text{AM}} = \frac{\overline{f^2(t)}}{A_0^2 + \overline{f^2(t)}} \left(\frac{S_o}{N_o}\right)_{\text{B}}. \tag{5.9-15}$$

These expressions are the same as those previously found for the coherent detector. Therefore, we conclude that both demodulation methods give the same performance for standard AM when the input signal-to-noise ratio is large.

Small Signal Case

If the magnitude of the input noise is nearly always large relative to the input signal magnitude, the reverse of (5.9-9) holds:

$$|A_0 + f(t)| \ll \sqrt{n_c^2(t) + n_s^2(t)}. \tag{5.9-16}$$

Using this assumption in (5.9-6), the envelope of the input waveform to the envelope detector is approximately

$$A(t) = \sqrt{n_c^2(t) + n_s^2(t)} + \frac{n_c(t)}{\sqrt{n_c^2(t) + n_s^2(t)}} [A_0 + f(t)]. \tag{5.9-17}$$

Since this equation also is an expression of the output, it is obvious that no term is present

ISBN 0-201-05758-1

which is proportional to $f(t)$. The third term is $f(t)$ multiplied by a random waveform and thus actually represents noise.

It may seem inconsistent that, as a function of the signal level, the output can shift from a case where demodulated message is present to a case where the message disappears. It must be kept in mind, however, that we have taken two analysis extremes. A more detailed study would reveal two facts. First, for large input signal-to-noise ratio (S_i/N_i) the output ratio (S_o/N_o) varies proportional to (S_i/N_i). Second, when (S_i/N_i) falls below a transition point where carrier power is approximately equal to input noise power, (S_o/N_o) begins to decrease much more rapidly than does (S_i/N_i). This transition of behavior is usually called the *threshold effect,* and the value of (S_i/N_i) at the transition point is called the *threshold* signal-to-noise ratio.

Threshold effect is not unique to the envelope detector and standard AM. It will also be an important consideration in the study of angle modulation in Chapter 6.

5.10. FREQUENCY-DIVISION MULTIPLEXING

A very useful technique for simultaneously transmitting several information signals involves the assignment of a portion of the final spectrum to each signal. It is known as *frequency-division multiplexing* (FDM), and we encounter it almost daily, often without giving it much thought. Larger cities usually have one, two, or several AM radio stations and may even have several television stations. In addition, firemen, police, taxicabs, amateur operators, and many other radio sources exist within the city. All these sources are frequency multiplexed into the radio spectrum by means of carrier frequency and bandwidth allocations. These are more or less obvious forms of FDM. The main intent of this section is to present some not so obvious methods of multiplexing messages which allow many signals to be transmitted with a single carrier.

The Direct Method

One method of implementation is illustrated by the block diagram of Fig. 5.10-1(a). For simplicity in discussion, we shall assume that N similar baseband signals $f_n(t)$, $n = 1, 2, \ldots, N$, are to be multiplexed. Thus, we assume the message spectral extents are all the same (W_f), such as in voice transmission. At the end of the discussion it should be no problem for the reader to generalize the principles to include information signals of different bandwidths.

The N signals in Fig. 5.10-1(a) are passed through low-pass filters which act to ensure that the messages do not exceed the allowable bandwidth. Each filtered waveform then modulates a *sub-carrier* having a frequency different from all the other sub-carriers allocated to the other signals. The modulated signals may be any of the modulation types already discussed, although we illustrate SSB because it is often selected to conserve bandwidth. These signals are next added together to form the sum signal $f_s(t)$.

The spectrum $F_s(\omega)$ of $f_s(t)$ is illustrated in Fig. 5.10-1(b), where $N = 4$ and USB operation are assumed as an example. The individual spectra are frequency multiplexed, and they may be recovered in the receiver so long as they do not overlap. If overlap did occur to a significant extent, unintelligble *crosstalk* would exist between the message channels. To aid in preventing crosstalk, and to help in the receiver demodulation process, the sub-carrier frequencies are separated by an amount W_s which is larger than the minimum value W_f by an amount W_g called the *guard band.* Thus,

$$W_s = W_f + W_g. \tag{5.10-1}$$

ISBN 0-201-05758-1

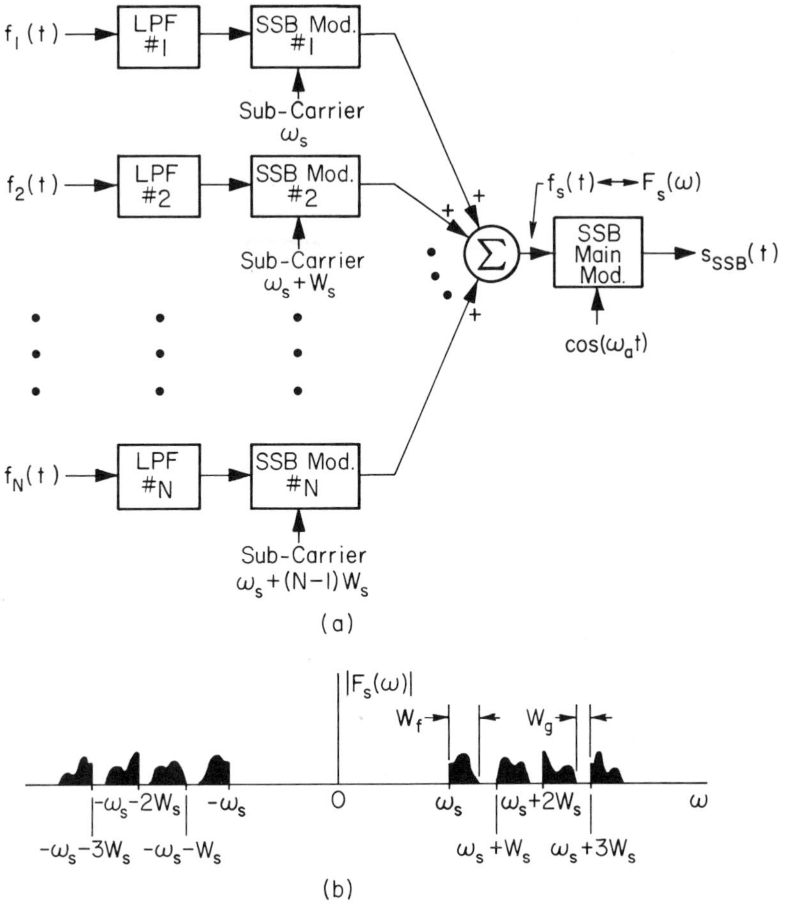

Fig. 5.10-1. Frequency-division multiplexing of signals. (a) Block diagram of multiplexer and (b) the spectrum magnitude of the sum signal $f_s(t)$.

In some applications the sum signal $f_s(t)$ may be transmitted without further processing. For this situation, the minimum required bandwidth is

$$W_{SSB} = NW_f + (N-1)W_g = W_f + (N-1)W_s. \tag{5.10-2}$$

In other applications, such as microwave FDM links, $f_s(t)$ must modulate a carrier for transmission. As with the sub-carriers, the main carrier modulator may have any form. Although it is often a frequency modulator (see Chapter 6), Fig. 5.10-1(a) illustrates SSB, since we have yet to cover frequency modulation. For USB or LSB operation the effective carrier frequency ω_0 becomes $\omega_0 = \omega_a + \omega_s$ and $\omega_0 = \omega_a - \omega_s$, respectively. Bandwidth is given by (5.10-2).

Demodulation of FDM signals is essentially the reverse operation to modulation. Figure 5.10-2 shows a demodulator block diagram. If a main carrier is involved, a band-pass filter preceding the main carrier demodulator is used to remove any out-of-band noise at the input. The main demodulator recovers the transmitted sum signal $f_s(t)$ which is applied to a bank of band-pass filters. These filters actually form the heart of the demodulator and are used to select the portion of the sum signal spectrum which corresponds to a given message. A given filter, say number n, must have a flat passband in the interval $\omega_s + (n-1)W_s$ to $\omega_s + (n -$

ISBN 0-201-05758-1

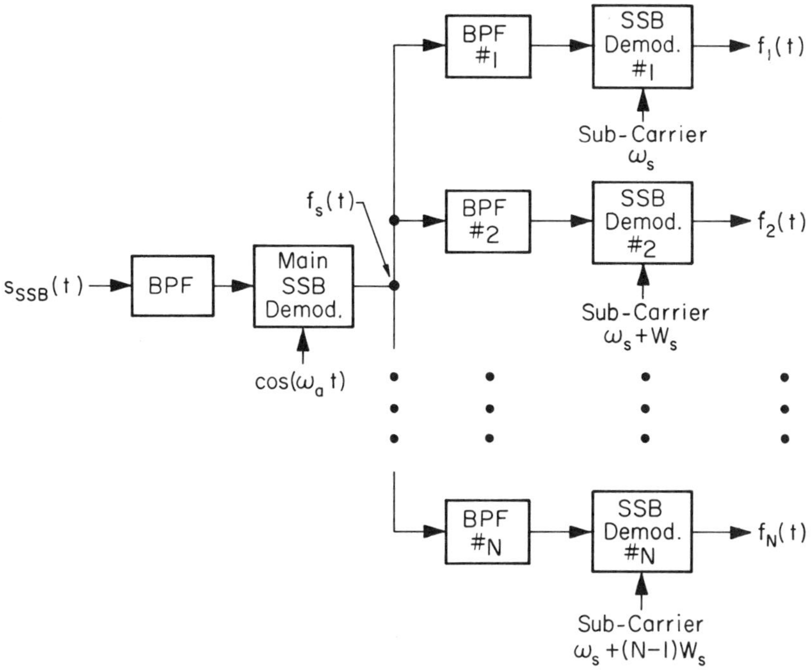

Fig. 5.10-2. Demodulator for the frequency-division multiplexed signal generated according to Fig. 5.10-1.

$1)W_s + W_f$ and then roll off to a negligible level over the guard bands adjacent to this interval. The filter outputs are finally used to recover the various messages via the sub-carrier demodulators.

In principle, the number of messages that may be frequency multiplexed is limited only by the allowable bandwidth and cost. Although some techniques exist to reduce cost, as we shall discuss below, it is always a factor in design. In fact, expense and size of the band-pass filters is one of the disadvantages of FDM. Another disadvantage is vulnerability to crosstalk. When system nonlinearities exist, such as in the main channel or in the amplifier which processes the sum signal, it is possible for one information signal to partially modulate another's sub-carrier. During demodulation intelligible crosstalk can then occur.

Of course, the main advantage of FDM is its ability to convey many simultaneous messages. However, there may also be a tremendous power savings, at least for the case of voice communications. For N voice messages we would expect that the total required power would be N times that for one message (recall that the power in the sum of independent random signals equals the sum of the individual powers, as developed in Chapter 4). Indeed, this would be the case if it were not for the fact that users do not talk continuously. Measurements [8] for conversational speech indicate that the required powers for 10, 100, 500, and 1000 messages are, respectively, 6, 9, 13, and 16 dB larger than for a single signal. These values correspond to overload of the system only 1% of the time and represent large savings. They indicate that required power increases approximately as $\sqrt{N}$ rather than N. An example further demonstrates the possible power savings.

Example 5.10-1

A FDM signal contains 1860 voice channels. The practical power required is approximately

ISBN 0-201-05758-1

$\sqrt{1860} = 43.1$ times that for one channel. Compared to a calculation based on full-time simultaneous use of all channels, where required power is 1860 times that of one channel, a savings of 43.1 or 16.3 dB results.

Multiple-Stage Method

When N is large it may be more convenient to implement the FDM system in a different manner from that shown in Fig. 5.10-1. Hardware savings are possible by creating the composite signal $f_s(t)$ using several stages of modulation. For example, consider two stages. In the first stage we divide the N signals into N_2 groups of N_1 signals per group where $N_1N_2 \geqslant N$. We assign N_1 *sub-sub-carrier* frequencies to the first stage and each group of N_1 messages modulates the same sub-sub-carriers. The N_1 modulated sub-sub-carrier signals in each group are then added to form N_2 signals which then modulate N_2 sub-carriers in the second stage. The modulated sub-carrier waveforms become $f_s(t)$ when added. Signal paths would resemble a tree with the tree trunk carrying $f_s(t)$. Neglecting the final carrier modulation stage, the minimum numbers of components required are

$$N_1N_2 + N_2 - 1 \text{ modulators and} \tag{5.10-3}$$

$$N_1 + N_2 - 1 \text{ sub-sub and sub-carriers.} \tag{5.10-4}$$

Some reflection will show (see problems 5-27 and 5-28) that, at the expense of a small number of modulators, the number of sub-sub- and sub-carrier sources is significantly reduced. An example will illustrate this point.

Example 5.10-2

A two-stage modulator uses $N_1 = 32$ and $N_2 = 30$ for $N_1N_2 = 960$ voice channels. We find the hardware savings. From (5.10-3) there are $32(30) + 30 - 1 = 989$ modulators needed and $32 + 30 - 1 = 61$ carrier sources required from (5.10-4). In the direct method we would need 959 modulators and 959 carrier sources. Thus, at the expense of 30 modulators we may save 898 carrier sources.

PROBLEMS

5-1. A carrier with peak amplitude A_0 is modulated using standard AM by a square wave (without dc) with peak-to-peak amplitude $2A_m$. (*a*) Sketch the time waveform. (*b*) Find and sketch the spectrum of the modulated signal. (*c*) What value of A_m corresponds to the onset of overmodulation?

5-2. (*a*) Find the total sideband power in the modulated signal of problem 5-1 for several values of A_m from zero to $A_m = A_0$. (*b*) Find and sketch a curve of efficiency versus A_m/A_0.

5-3. Explain why the tube in a plate-modulated class C amplifier must be capable of handling a total plate voltage of at least 4 times the plate supply voltage E_{bb}.

★ 5-4. (*a*) If $f(t)$ is a random signal and has no average value, prove that the crosspower term in (5.1-8) is zero. (*b*) Prove that carrier and signal powers are given by (5.1-9) and (5.1-10).

ISBN 0-201-05758-1

★ 5-5. Show that (5.1-13) follows from (5.1-12) if $f(t)$ is a random information signal and has no average value.

5-6. An information signal $f(t)$ has an autocorrelation function $R_f(\tau) = P_0 \exp(-|\tau|)$. (*a*) Find $\mathscr{S}_f(\omega)$. (*b*) Find P_0 such that standard AM modulation efficiency is 25% (neglect possibility of overmodulation).

5-7. (*a*) Find and sketch the efficiency of sinusoidal standard AM as given by (5.1-24) versus A_m/A_0, where A_m is the peak value of the modulating sinusoid. Compare the result to that of a square wave obtained in problem 5-2(*b*). (*b*) Compute efficiency by use of (5.1-17) and $\mathscr{S}_f(\omega)$ from example 5.1-2, and verify the work in part (*a*).

5-8. Obtain and sketch the spectrum magnitude of a standard AM signal where $f(t)$ is the periodic triangular waveform shown in Fig. P5-8 and $A_m \leqslant A_0$.

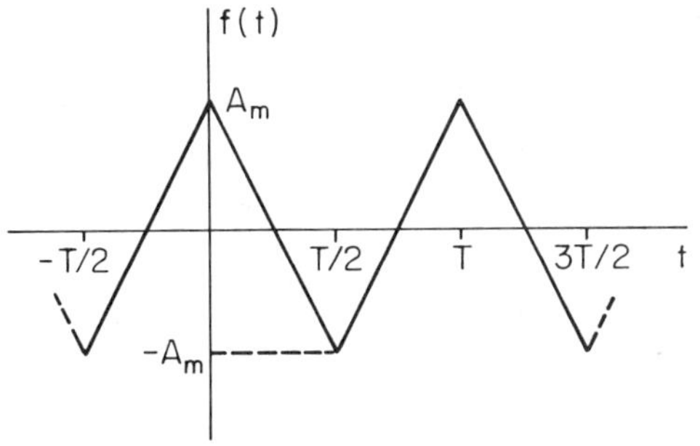

Fig. P5-8

5-9. The sum $x(t)$ of a carrier $A_0 \cos(\omega_0 t)$ and signal $f(t)$ is applied to a square-law device where the output is $y(t) = Kx^2(t)$ with K a constant. (*a*) Find the output signal. What type of AM is present in the output? (*b*) How must ω_0 be related to signal bandwidth to obtain a distortion-free AM signal?

5-10. (*a*) Sketch the spectrum of the signal

$$x(t) = f(t) \cos^4 (\omega_0 t).$$

(*b*) Where should a bandpass filter be placed to obtain a DSB signal, and what bandwidth should it have?

5-11. If the forward and reverse voltage impedances of the diodes of Fig. 5.2-2(*b*) are R_{fi} and R_{bi}, respectively, for $i = 1, \ldots, 4$, show that the main effect relative to the ideal output (when $R_{fi} = 0$ and $R_{bi} = \infty$) is additional attenuation if Q_1 is identical to Q_2 and Q_3 is identical to Q_4.

5-12. (*a*) Draw an equivalent circuit for the network of Fig. 5.2-3(*a*) when the diodes are caused to conduct by the carrier voltage. Assume the output reflects a resistive impedance (R_L) into the primary of T at all frequencies, the impedance of the carrier source as seen across the diodes is resistive (R_s), and diodes are unequal but resistive with values R_{f1} and R_{f2}. Explain how R may be adjusted to give a null in the carrier component of the output. (*b*) Will the same adjustment prevent signal current from flowing in the carrier source? Explain.

5-13. Show that, if an arbitrary phase $\theta(\omega)$ is present in the two signal paths of Fig. 5.2-5, the resulting phase method is an ideal SSB generator for the signal having the spectrum $F(\omega) \exp[j\theta(\omega)]$, where $F(\omega)$ is the spectrum of $f(t)$.

ISBN 0-201-05758-1

★ 5-14. (*a*) If the signal $f(t)$ and frequency ω_b are the same as in Weaver's SSB method, sketch the various spectra for the circuit of Fig. P5-14 if it is to operate similar to Weaver's method using a high-pass filter. (*b*) What are the required band-pass filter characteristics to produce USB and LSB operations? What are the corresponding carrier frequencies?

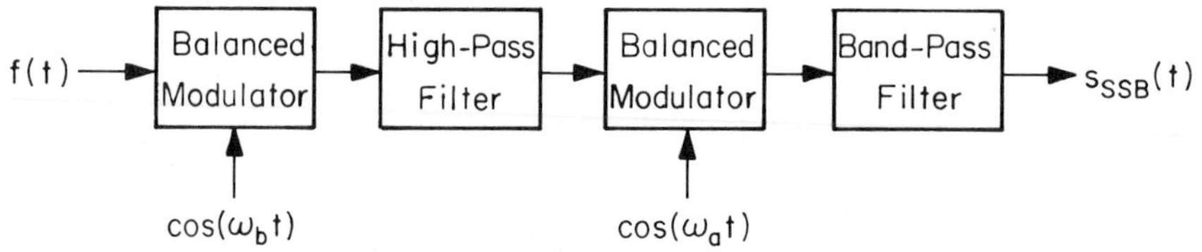

Fig. P5-14

★ 5-15. For the network of Fig. P5-14 discuss the effect of unbalance in (*a*) the first balanced modulator, and (*b*) the second modulator, when the network operates as described in problem 5-14.

5-16. Prove that the output of the rectifier detector for demodulation of standard AM is $f(t)/\pi$.

5-17. What would be a reasonable value for RC in Fig. 5.5-1 if the waveform of problem 5-8 is to be demodulated?

5-18. Show, for demodulation by carrier reinsertion of a SSB signal containing upper sidebands, that the spectrum of the demodulated signal (5.6-10) is, on use of (5.4-5),

$$S_d(\omega) = 2\pi A_d \delta(\omega) + F(\omega)\left\{\cos(\phi_\epsilon) - j[2U(\omega) - 1]\sin(\phi_\epsilon)\right\}$$

if $\omega_\epsilon = 0$. What is $S_d(\omega)$ when $\omega_\epsilon \neq 0$?

5-19. In a certain standard AM system the root-mean-squared message level is 0.25 times the carrier peak amplitude of 10 volts. If the receiver input noise is white with power density $\mathcal{N}_0/2 = 1.33(10^{-4})$ watt/hertz, what is the maximum allowable message spectral extent (in hertz) that will allow at least a 10 dB input signal-to-noise power ratio?

5-20. In an AM receiver the input noise power density spectrum can be approximated by

$$\mathcal{S}_{n_i}(\omega) = P_0\left\{\text{rect}\left(\frac{\omega - \omega_0}{W_{IF}}\right)\cos\left[\frac{\pi(\omega - \omega_0)}{W_{IF}}\right] + \text{rect}\left(\frac{\omega + \omega_0}{W_{IF}}\right)\cos\left[\frac{\pi(\omega + \omega_0)}{W_{IF}}\right]\right\},$$

where P_0 and W_{IF} are constants. If $W_{IF} \geqslant W_f$, what is the average noise power emerging from a coherent demodulator having an ideal low-pass filter with bandwidth W_f?

5-21. Plot the ratio of output to input signal-to-noise ratios for standard AM as a function of the ratio x of sideband (information) power to carrier power. To what does the result approach as x approaches ∞?

5-22. For a certain message it is known that $\overline{f^2(t)}/|f(t)|^2_{max} = 0.09$. If the message can be transmitted over either a DSB system or a standard AM system (no overmodulation), how will the output signal-to-noise ratios of the two systems compare assuming both have the same total transmitted average power and both have the same input noise power density?

ISBN 0-201-05758-1

5-23. A receiver produces an output signal-to-noise power ratio of 20 dB when the output noise power is 10^{-9} watt. A total loss of 100 dB exists over the path from transmitter to receiver input. (*a*) What power level must a DSB transmitter have to produce the output? (*b*) Repeat (*a*) for a SSB transmitter.

★ 5-24. In a practical communication system a power loss L (number not less than unity) will exist between the transmitter and receiver. For the practical input noise power density spectrum of problem 5-20 find input and output signal-to-noise power ratio expressions for (*a*) standard AM and (*b*) DSB systems. Compare the results with expressions of the text by letting $P_0 = \mathcal{N}_0/2$.

5-25. Voice signals are SSB (USB) frequency multiplexed after being passed through Butterworth low-pass filters with cutoff frequencies of 2.7 kHz. If a guard band of 0.6 kHz is allowed, what filter complexity—i.e., how many poles—is required to guarantee that spectral overlap is down by 20 dB at band edges?

5-26. Explain how (5.10-3) and (5.10-4) result?

5-27. For a two-stage SSB frequency multiplexer verify that the total number of carriers and modulators is minimum when $N_1 = \sqrt{2N}$ and $N_2 = \sqrt{N/2}$. (Assume that $N_1 N_2 = N$ where N is fixed.)

5-28. In problem 5-27 show that if the number of carriers alone is minimum then $N_1 = N_2 = \sqrt{N}$.

REFERENCES

[1] Kennedy, G., *Electronic Communication Systems,* McGraw-Hill Book Co., New York, 1970.

[2] Norgaard, D. E., The Phase Shift Method of Single-Sideband Signal Generation, *Proceedings of the IRE,* December, 1956, p. 1718.

[3] Weaver, D. K., A Third Method of Generation and Detection of SSB Signals, *Proceedings of the IRE,* December, 1956, pp. 1703–1705.

[4] Simpson, R. S., and Houts, R. C., *Fundamentals of Analog and Digital Communication Systems,* Allyn and Bacon, Boston, Massachusetts, 1971.

[5] Costas, J. P., Synchronous Communication, *Proceedings of the IRE,* December, 1956, pp. 1713–1718.

[6] Lathi, B. P., *Communication Systems,* John Wiley & Sons, New York, 1968.

[7] Lucky, R. W., Salz, J., and Weldon, E. J., Jr., *Principles of Data Communication,* McGraw-Hill Book Co., New York, 1968.

[8] Holbrook, B. D., and Dixon, J. T., Load Rating Theory for Multichannel Amplifiers, *Bell System Technical Journal,* Vol. 18, October, 1939, pp. 624–644.

ISBN 0-201-05758-1

Chapter 6

Angle Modulation

6.0. INTRODUCTION

In the preceding chapter ways were discussed for modulating the amplitude A of a carrier signal of the form $A \cos(\omega_0 t + \theta_0)$, where the angle of the carrier was not changed; that is, ω_0 and θ_0 were constants independent of the information signal $f(t)$. In this chapter we shall be concerned with the opposite situation. Here A will be maintained constant, while the angle of the carrier will vary in some manner according to the message waveform.

Angle can be varied by phase or frequency changes. If phase is made to vary linearly as a function of the message $f(t)$, the carrier is said to have *phase modulation* (PM) imparted upon it. *Frequency modulation* (FM) corresponds to linearly* varying instantaneous frequency with $f(t)$. FM and PM analyses are intimately related. Because of this fact, some detailed discussions will be limited to FM, and the analogous PM developments will only be outlined.

The most obvious example of a frequency-modulated system is ordinary broadcast FM radio where the carrier frequency falls in the band 88 MHz to 108 MHz. Still other uses of FM are in telemetry systems and broadcast television. The latter uses FM to impart audio information onto the audio sub-carrier. One of the chief applications of PM is in digital systems using phase shift keying (PSK). PSK is discussed in Chapter 8.

A valid question is: Why angle modulation? Stated very simply, angle modulation produces some distinct advantages that AM does not. Also important is the fact that the advantages are realized with little or no increase in equipment complexity. Perhaps the greatest single advantage is a reduction in effects of noise and other interference. The reader has no doubt experienced difficulty with AM reception in an automobile while driving in, or near, thunderstorms. By switching to FM, on radios so equipped, reception is greatly improved and effects of interference due to radiation from lightning may no longer be heard. The improvement results from the fact that an angle-modulated signal may have a bandwidth many times that of the information signal. It may be recalled that this is not possible with AM waveforms.

As noted in Chapter 5, it is sometimes convenient to interpret the information signal $f(t)$ as a deterministic waveform. At other times, it is assumed to be a random quantity. We again use both interpretations during subsequent discussions about FM and PM. The reader should refer again to the comments in the introduction to Chapter 5 which define in detail the applicable assumptions.

6.1. ANGLE MODULATION, PHASE AND FREQUENCY

We may define a sinusoidal signal having a constant peak amplitude A and *instantaneous* or *generalized angle* $\theta(t)$ as

ISBN 0-201-05758-1

References start on p. 285.

*Although the parameter, phase or frequency, being varied does so as a linear function of $f(t)$, this does not mean that angle modulation is linear modulation as in AM. On the contrary, angle modulation is a nonlinear operation, since the principle of superposition does not hold.

$$s(t) = A \cos [\theta(t)], \tag{6.1-1}$$

where $\theta(t)$ will be a function of time. The *instantaneous* or *generalized frequency* of (6.1-1) is the time derivative of instantaneous phase:

$$\omega(t) = d\theta(t)/dt. \tag{6.1-2}$$

Angle and frequency of ordinary sinusoids are special cases of these definitions, as the following simple example shows.

Example 6.1-1

For an ordinary sinusoid $A \cos (\omega_0 t + \theta_0)$, where ω_0 and θ_0 are constants, the angle is $\theta(t) = \omega_0 t + \theta_0$. The instantaneous frequency is $\omega(t) = d\theta(t)/dt = \omega_0$, a constant, as it should be.

The PM Waveform

A sinusoid is said to be *phase-modulated* if instantaneous angle is a linear function of the information signal $f(t)$. Thus,

$$\theta_{PM}(t) = \omega_0 t + \theta_0 + k_{PM} f(t), \tag{6.1-3}$$

where ω_0 and k_{PM} are assumed to be positive constants, while θ_0 is an arbitrary phase angle. Using (6.1-2), the instantaneous frequency of the PM waveform is

$$\omega_{PM}(t) = \omega_0 + k_{PM} \frac{df(t)}{dt}, \tag{6.1-4}$$

while the PM waveform itself is

$$s_{PM}(t) = A \cos [\omega_0 t + \theta_0 + k_{PM} f(t)]. \tag{6.1-5}$$

The FM Waveform

A sinusoid is said to be *frequency-modulated* if instantaneous frequency is a linear function of the information signal $f(t)$. Thus,

$$\omega_{FM}(t) = \omega_0 + k_{FM} f(t), \tag{6.1-6}$$

where ω_0 and k_{FM} are assumed to be positive constants. The FM signal's instantaneous angle becomes

$$\theta_{FM}(t) = \omega_0 t + \theta_0 + k_{FM} \int f(t)\, dt, \tag{6.1-7}$$

while the FM signal itself is

$$s_{FM}(t) = A \cos [\omega_0 t + \theta_0 + k_{FM} \int f(t)\, dt]. \tag{6.1-8}$$

ISBN 0-201-05758-1

Again θ_0 is an arbitrary phase angle.

Relationship between PM and FM

A study of the FM waveform as given by (6.1-8) reveals that it may be viewed as a PM waveform if the modulating signal is taken to be $\int f(t)\,dt$. Similarly, the PM signal of (6.1-5) is equivalent to an FM waveform if the modulating signal is taken to be $df(t)/dt$. These relationships may be further demonstrated by considering the waveforms of Fig. 6.1-1, which assume $f(t)$ is the triangular message of (*a*). In the FM case, instantaneous frequency varies linearly with $f(t)$, as shown in (*b*). The FM waveform of (*c*) depicts the frequency increasing to a time $T/2$ and then decreasing to its minimum value at time T. The peak frequency deviation $\Delta\omega$ from the carrier ω_0 is

$$\Delta\omega = k_{FM}\,|f(t)|_{max}. \tag{6.1-9}$$

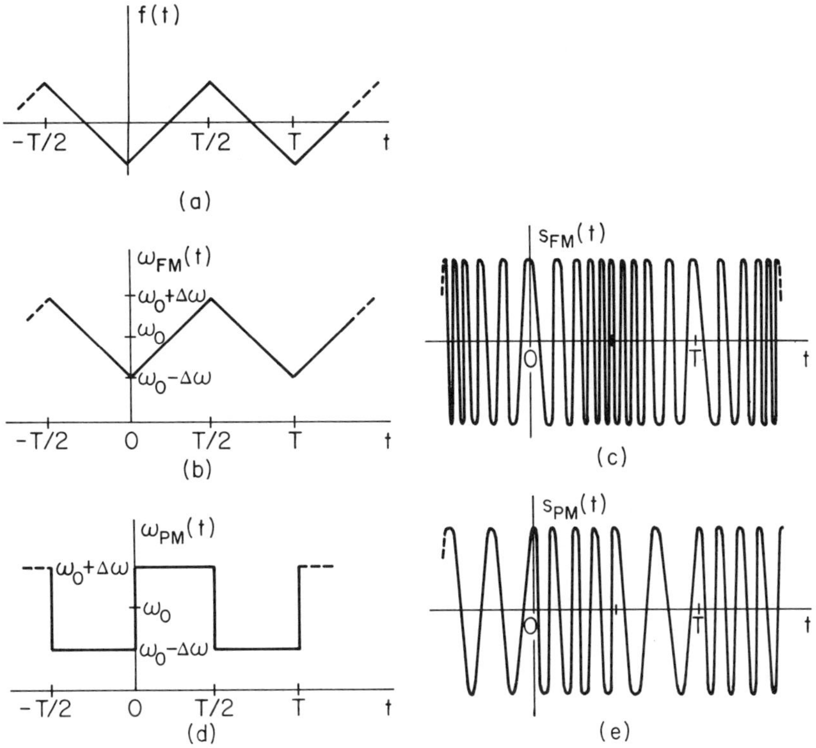

Fig. 6.1-1. FM and PM waveforms. (a) Information signal, (b) instantaneous frequency of FM signal, (c) FM signal, (d) instantaneous frequency of PM signal, and (e) PM signal.

If the triangular signal is used to produce PM, the instantaneous frequency, obtained from (6.1-4), is a rectangular function as sketched in (*d*). The corresponding PM waveform is shown in (*e*). The peak frequency deviation of the PM wave is

$$\Delta\omega = k_{PM}\left|\frac{df(t)}{dt}\right|_{max}. \tag{6.1-10}$$

ISBN 0-201-05758-1

The most obvious ways of generating FM or PM waveforms are to apply $f(t)$ to frequency or phase modulators as shown in Fig. 6.1-2(*a*) and (*b*). However, in FM phase change due to modulation is proportional to the integral of $f(t)$ from (6.1-7), so that a phase modulator can also be used to produce FM using the system of (*c*) if $f(t)$ is first integrated. Similarly, PM can be generated from a frequency modulator using a differentiator as illustrated in (*d*). We shall later find that the methods of (*a*) and (*d*) may be used for generating any FM or PM signal, while the methods of (*b*) and (*c*) are mainly useful in generating narrowband angle-modulated waveforms. We consider the narrowband problem in the next section.

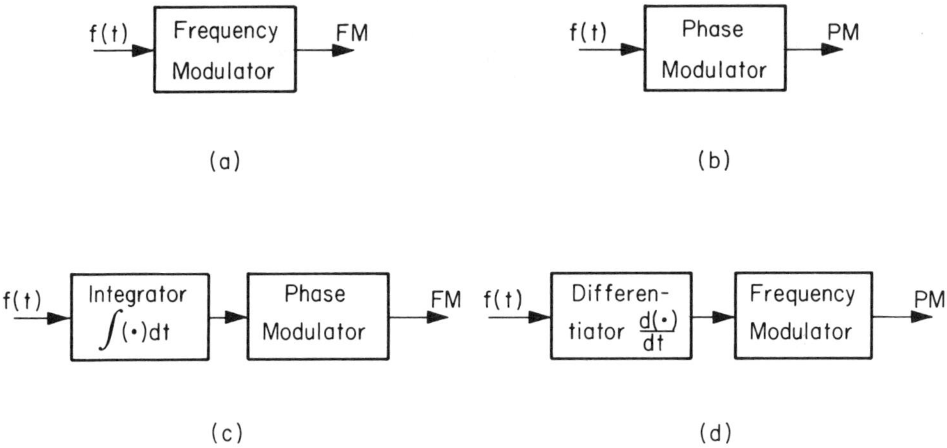

Fig. 6.1-2. FM and PM generators. (a) and (c) Generators for FM. (b) and (d) Generators for PM.

6.2. NARROWBAND ANGLE MODULATION

Analysis of an FM or PM signal to obtain a general expression for its Fourier spectrum is a difficult task. However, in the simplified problem where the maximum phase deviation due to modulation is kept small, the spectrum of the modulated signal may be found for an arbitrary information signal $f(t)$. The effect of small phase deviation is to restrict bandwidth to a small value as compared with a large-deviation case. The restricted waveforms are termed *narrowband FM* (NBFM) and *narrowband PM* (NBPM).

Narrowband Frequency Modulation

Assume in (6.1-8) that the maximum phase deviation due to modulation is small. That is, assume

$$k_{\mathrm{FM}} \left| \int f(t)\, dt \right|_{\max} \ll \frac{\pi}{6}. \tag{6.2-1}$$

With this restriction and using the trigonometric expansion of cos $(A + B)$,

$$s_{\mathrm{NBFM}}(t) = A \cos(\omega_0 t + \theta_0) \cos \left[k_{\mathrm{FM}} \int f(t)\, dt\right]$$

ISBN 0-201-05758-1

$$- A \sin(\omega_0 t + \theta_0) \sin [k_{FM} \int f(t)\, dt]$$

$$s_{NBFm}(t) \approx A \cos(\omega_0 t + \theta_0) - Ak_{FM} \int f(t)\, dt \sin(\omega_0 t + \theta_0) \qquad (6.2\text{-}2)$$

is the narrowband FM waveform. Here we have used the approximations $\cos(x) \approx 1$ and $\sin(x) \approx x$ if x is small.

The spectrum of $s_{NBFM}(t)$ is easily found. If $f(t)$ has the spectrum $F(\omega)$,

$$f(t) \leftrightarrow F(\omega), \qquad (6.2\text{-}3)$$

then

$$\int f(t)\, dt \leftrightarrow F(\omega)/j\omega, \qquad (6.2\text{-}4)$$

where $F(0)$ is assumed to be zero. Using this result with Fourier transform pairs of Chapter 2, we obtain the spectrum of $s_{NBFM}(t)$:

$$S_{NBFM}(\omega) = \pi A [\delta(\omega - \omega_0)e^{j\theta_0} + \delta(\omega + \omega_0)e^{-j\theta_0}] + \frac{Ak_{FM}}{2}\left[\frac{F(\omega - \omega_0)}{(\omega - \omega_0)} e^{j\theta_0} - \frac{F(\omega + \omega_0)}{(\omega + \omega_0)} e^{-j\theta_0}\right]. \qquad (6.2\text{-}5)$$

There is considerable similarity between the narrowband FM signal spectrum and that of standard AM [compare (6.2-5) with (5.1-6)]. Both spectra have identical "carrier" terms, corresponding to the impulses, and both have positive- and negative-frequency spectral components, due to modulation, that are centered about frequencies of ω_0 and $-\omega_0$, respectively. If $f(t)$ is bandlimited to have a maximum spectral extent W_f, then these components in both standard AM and NBFM are constrained to a band $2W_f$. However, the NBFM components are seen to differ in form owing to the factors $1/(\omega - \omega_0)$ and $1/(\omega + \omega_0)$. Another difference is reflected in the 180-degree phase reversal of the NBFM negative-frequency component not present in the AM spectrum.

Narrowband FM finds applications in amateur or "ham" radio and in *mobile communication* systems.

Narrowband Phase Modulation

Assuming maximum phase deviation to be small, that is,

$$k_{PM} |f(t)|_{max} << \pi/6, \qquad (6.2\text{-}6)$$

will allow (6.1-5) to be written as

$$s_{NBPM}(t) \approx A \cos(\omega_0 t + \theta_0) - Ak_{PM} f(t) \sin(\omega_0 t + \theta_0). \qquad (6.2\text{-}7)$$

ISBN 0-201-05758-1

Fourier transformation gives the NBPM spectrum:

$$S_{\text{NBPM}}(\omega) = \pi A\left[\delta(\omega - \omega_0)e^{j\theta_0} + \delta(\omega + \omega_0)e^{-j\theta_0}\right] + \frac{jAk_{\text{PM}}}{2}\left[F(\omega - \omega_0)e^{j\theta_0} - F(\omega + \omega_0)e^{-j\theta_0}\right]. \tag{6.2-8}$$

As in the NBFM case earlier, this spectrum also resembles that of standard AM. A comparison with (5.1-6) reveals that the form and extent of the positive- and negative-frequency spectral components centered at ω_0 and $-\omega_0$ are identical. However, these components are phase-shifted by $\pi/2$ and $-\pi/2$, respectively, as compared to standard AM.

Both NBPM and NBFM are the result of linear modulation. The reader is encouraged to verify this fact. The procedure is to assume that the modulating signal is a sum of several components and rederive the appropriate spectrum under the assumption of a small maximum total phase deviation. It will be found that the modulated signal spectrum is the sum of components which would have been produced by individual messages taken singly. Thus, superposition applies and modulation is linear.

NB Modulation Methods

The similarity of the NBPM and NBFM waveforms to standard AM suggests that modulation methods similar to those of Fig. 5.1-4 are possible. Indeed, such realizations are illustrated in Fig. 6.2-1.

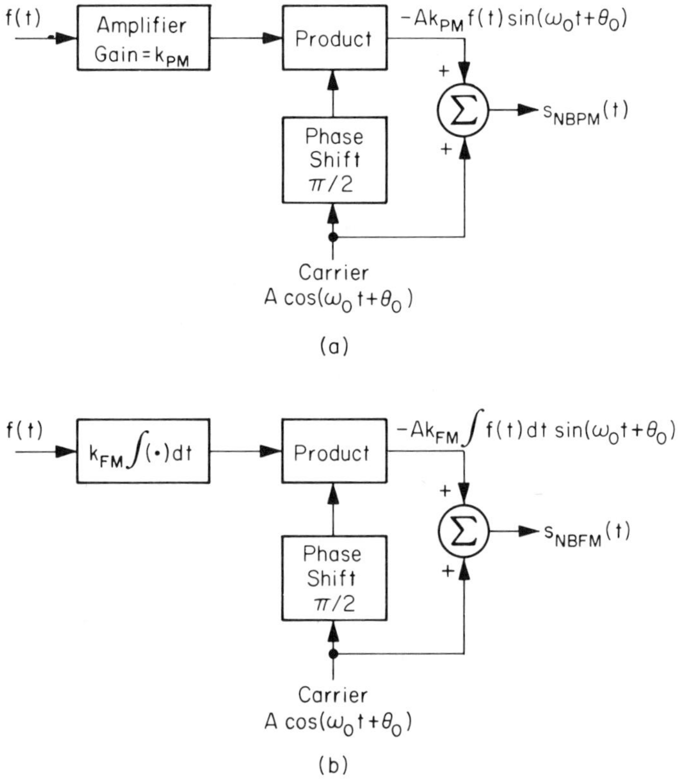

Fig. 6.2-1. Narrowband PM and FM modulation methods. (a) PM and (b) FM.

ISBN 0-201-05758-1

NB Demodulation Methods

Recovery of $f(t)$ from the angle-modulated wave may be accomplished by using a product device. Block diagrams of demodulators for both PM and FM are illustrated in Fig. 6.2-2. The purpose of the band-pass filter (assumed ideal) is to pass the modulated signal while rejecting undesired noise. The bandwidth of the filter will therefore be equal to the spectral extent ($2W_f$) of the modulated signal. The bandwidth of the low-pass filter (assumed ideal) is W_f. Its purpose is to pass only spectral components within the message band. It thereby removes undesired spectral components which are produced by the product device.

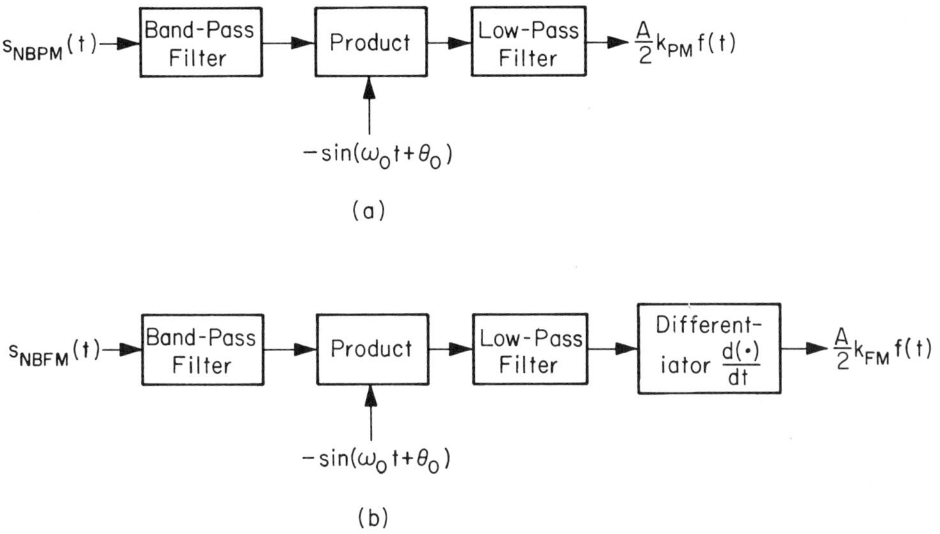

Fig. 6.2-2. Demodulators. (a) For narrowband phase-modulated signal, and (b) for narrowband frequency-modulated signal.

6.3. WIDEBAND FM WITH A SINUSOIDAL SIGNAL

As already stated, the general analysis of FM is quite difficult, and one tends to lose sight of the physical picture. In Section 6.4 we shall give some consideration to FM with arbitrary waveforms. However, to gain insight, let us first discuss in some detail the simplest possible wideband case, that of a single sinusoidal waveform. Thus,

$$f(t) = A_f \cos(\omega_f t) \tag{6.3-1}$$

is the assumed information signal where A_f and ω_f are constants.

Waveform and Modulation Index

From (6.1-8) the FM waveform is*

$$s_{\text{FM}}(t) = A \cos[\omega_0 t + \beta_{\text{FM}} \sin(\omega_f t)], \tag{6.3-2}$$

*We choose to drop the arbitrary carrier phase angle θ_0 for convenience. Results for $\theta_0 \neq 0$ are easy extensions of the procedure outlined (see problem 6-8).

ISBN 0-201-05758-1

where (6.3-1) has been integrated, and we define a quantity

$$\beta_{\mathrm{FM}} = A_f k_{\mathrm{FM}}/\omega_f = \Delta\omega/\omega_f \tag{6.3-3}$$

called the FM *modulation index.* It is the ratio of the maximum frequency deviation

$$\Delta\omega = k_{\mathrm{FM}}|f(t)|_{\max} = A_f k_{\mathrm{FM}} \tag{6.3-4}$$

to the signal frequency ω_f. The modulation index is nothing more than a measure of how intense the modulation is relative to the rapidity of modulation.

Using the well-known Jacobi equations

$$\cos[\beta \sin(x)] = J_0(\beta) + 2\sum_{k=1}^{\infty} J_{2k}(\beta)\cos(2kx), \tag{6.3-5}$$

$$\sin[\beta \sin(x)] = 2\sum_{k=1}^{\infty} J_{2k-1}(\beta)\sin[(2k-1)x], \tag{6.3-6}$$

we may write (6.3-2) in the form

$$s_{\mathrm{FM}}(t) = A\sum_{n=-\infty}^{\infty} J_n(\beta_{\mathrm{FM}})\cos[(\omega_0 + n\omega_f)t]. \tag{6.3-7}$$

The quantities $J_n(\beta)$ are coefficients called *Bessel functions* of the first kind of order n. They are constants in time but are functions of β. A property of Bessel functions used in obtaining (6.3-7) is

$$J_{-n}(\beta) = (-1)^n J_n(\beta). \tag{6.3-8}$$

Examination of (6.3-7) shows that $s_{\mathrm{FM}}(t)$ contains an infinite number of frequencies. Thus, many frequencies are generated with FM which are not present in the information signal. Because of this fact the FM waveform may have a much larger bandwidth than the signal. The exact form of the spectrum, and therefore its bandwidth, derives easily from (6.3-7). These quantities will shortly be found. However, we must first discuss some of the properties of the Bessel functions, since the spectrum behavior depends on the Bessel coefficients and their behavior.

Properties of Bessel Functions

Figure 6.3-1 shows the behavior of the first six Bessel functions. Data for other values of n and β may be found tabulated in Appendix D and in [1]. Each function exhibits a cyclic or lobing behavior with decreasing peak values as β increases. As a function of n with β fixed,*

$$J_n(\beta) \approx 0, \qquad n > \beta + 1. \tag{6.3-9}$$

*The approximation of (6.3-9) becomes more accurate as β becomes large relative to unity.

ISBN 0-201-05758-1

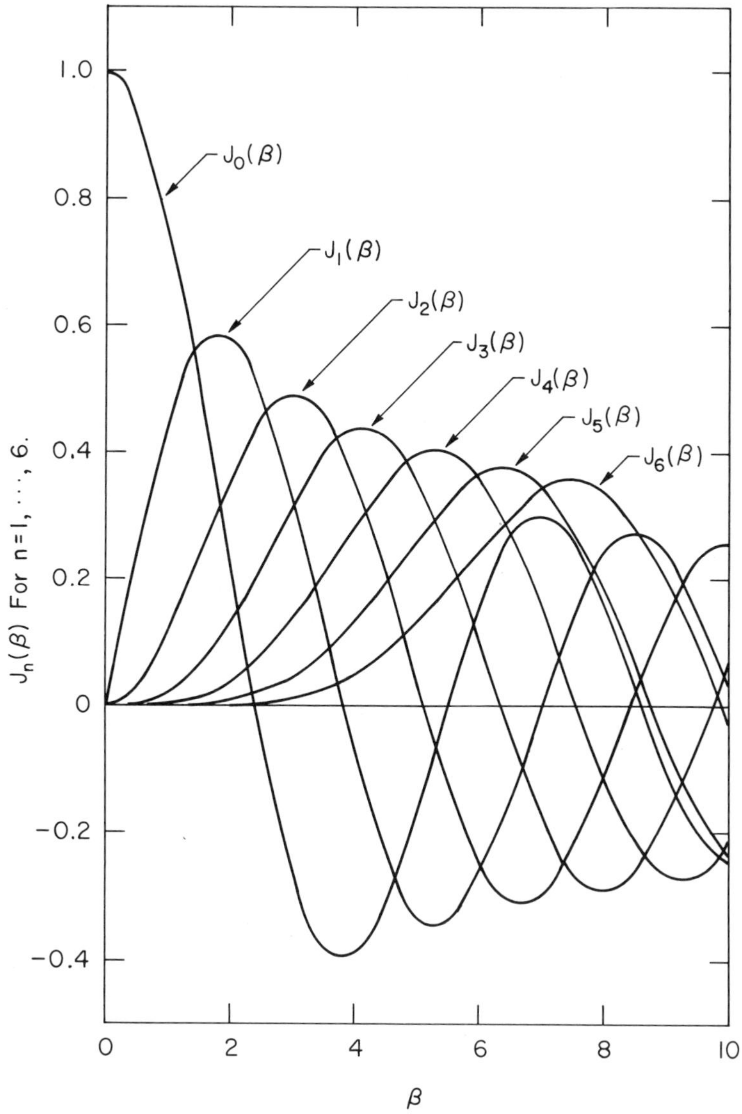

Fig. 6.3-1. Plots of the first six Bessel functions.

Two other useful properties involve the sum of squares:

$$\sum_{n=-\infty}^{\infty} J_n^{\,2}(\beta) = 1, \qquad \text{all } \beta, \tag{6.3-10}$$

and the symmetry condition:

$$J_n(-\beta) = (-1)^n J_n(\beta). \tag{6.3-11}$$

Bessel functions in general are defined for fractional as well as integral orders. All our work will involve only integral orders.

ISBN 0-201-05758-1

FM Spectrum

We are now in a position to study the behavior of the FM spectrum as a function of modulation index β_{FM} and frequency ω_f. The spectrum of $s_{FM}(t)$ is the Fourier transform of (6.3-7). Standard transforms yield

$$S_{FM}(\omega) = \pi A \sum_{n=-\infty}^{\infty} J_n(\beta_{FM})[\delta(\omega - \omega_0 - n\omega_f) + \delta(\omega + \omega_0 + n\omega_f)]. \tag{6.3-12}$$

This spectrum has two main parts. They are centered about frequencies of $+\omega_0$ and $-\omega_0$. The magnitudes of the parts are identical. The carrier level is proportional to $J_0(\beta_{FM})$, while the levels of the first sideband frequencies, at $\pm\omega_f$ relative to the carrier, are proportional to $|J_1(\beta_{FM})|$. Similarly, the level of the nth sideband frequency is proportional to $|J_n(\beta_{FM})|$. These facts may be illustrated by taking a numerical example.

Example 6.3-1

We find the spectrum component amplitudes when $\omega_f = 2\pi(10^4)$ and $\Delta\omega = 6\pi(10^4)$. Here $\beta_{FM} = 3$. Using tables (Appendix D) or Fig. 6.3-1 we obtain

$J_0(3) = -0.260$	$J_3(3) = 0.309$	$J_6(3) = 0.011$
$J_1(3) = 0.339$	$J_4(3) = 0.132$	
$J_2(3) = 0.486$	$J_5(3) = 0.043$	

The component amplitudes are then $\pi A J_n(3)$.

A sketch of the spectrum of example 6.3-1 is shown in Fig. 6.3-2. The significant bandwidth of this spectrum is $W_{FM} = 8\omega_f$ as shown.

Bandwidth

In terms of β_{FM} we might write the bandwidth of example 6.3-1 as

$$W_{FM} = 2(\beta_{FM} + 1)\omega_f. \tag{6.3-13}$$

This result is a good rule-of-thumb for bandwidth in general for modulation by sinusoids. It can be justified by use of (6.3-9). Since the magnitude of the nth sideband frequency is proportional to $|J_n(\beta_{FM})|$, the frequencies for $|n| > \beta_{FM} + 1$ are negligible,* and only $2n = 2(\beta_{FM} + 1)$ sideband frequencies are important. Accounting for the fact that frequency separation is ω_f, we arrive at (6.3-13) as the bandwidth.

When $\beta_{FM} << 1$ the bandwidth approaches $2\omega_f$ as we would expect (the narrowband case). If $\beta_{FM} >> 1$, which is the very broadband case,

$$W_{FM} = 2\beta_{FM}\omega_f = 2\Delta\omega. \tag{6.3-14}$$

In words, the FM bandwidth for large modulation index equals twice the peak frequency deviation of the frequency excursions and is independent of the modulation frequency.

*Amplitudes of frequencies for $|n| > \beta_{FM} + 1$ are less than approximately 15% of the unmodulated carrier level. They become less than 10% for $n > \beta_{FM} + 2$.

ISBN 0-201-05758-1

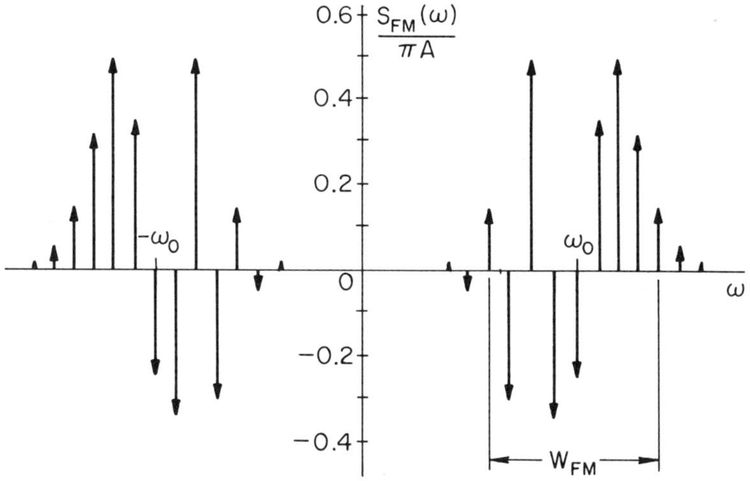

Fig. 6.3-2. FM spectrum for sinusoidal modulation with $\beta_{FM} = 3$.

To illustrate bandwidth dependence on $\Delta\omega$, assume ω_f fixed and let $\Delta\omega$ vary so that $\beta_{FM} = 0.5$, 2, and 8. Resulting spectrum magnitudes are shown in Fig. 6.3-3 along with values of bandwidth W_{FM} obtained from (6.3-13). The opposite case is obtained by letting $\Delta\omega$ be fixed and allowing ω_f to vary. Results are given in Fig. 6.3-4 for several values of β_{FM}. These figures demonstrate that bandwidth is mainly controlled by peak frequency deviation $\Delta\omega$, while ω_f determines how "filled" the significant band is for a given value of $\Delta\omega$.

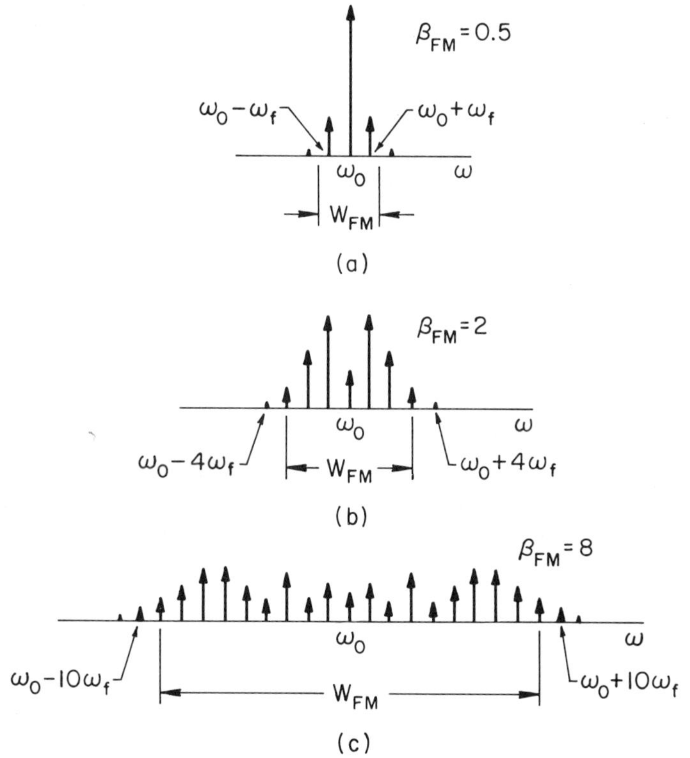

Fig. 6.3-3. FM magnitude spectra for ω_f fixed and $\Delta\omega$ variable in order to vary β_{FM}. (a) $\Delta\omega/\omega_f = 0.5$, (b) $\Delta\omega/\omega_f = 2$, and (c) $\Delta\omega/\omega_f = 8$.

ISBN 0-201-05758-1

In commercial broadcast FM the value of $\Delta\omega/2\pi$ is limited to 75 kHz by the Federal Communications Commission. By noting that (6.3-13) can be put in the form $W_{FM} = 2(\Delta\omega + \omega_f)$, we find that the highest frequency to be broadcast determines the required FM bandwidth. Assuming a maximum of $\omega_f/2\pi = 15$ kHz leads to a transmission bandwidth of approximately $W_{FM}/2\pi = 180$ kHz. Actually, (6.3-13) tends to somewhat underestimate the required bandwidth and a value of around 210 kHz becomes more realistic [2].

In broadcast television audio is included via FM of a sub-carrier. Here the value of $\Delta\omega/2\pi$ is limited to 25 kHz. Again assuming a 15-kHz modulating signal, the rule-of-thumb bandwidth calculates to $W_{FM}/2\pi = 80$ kHz.

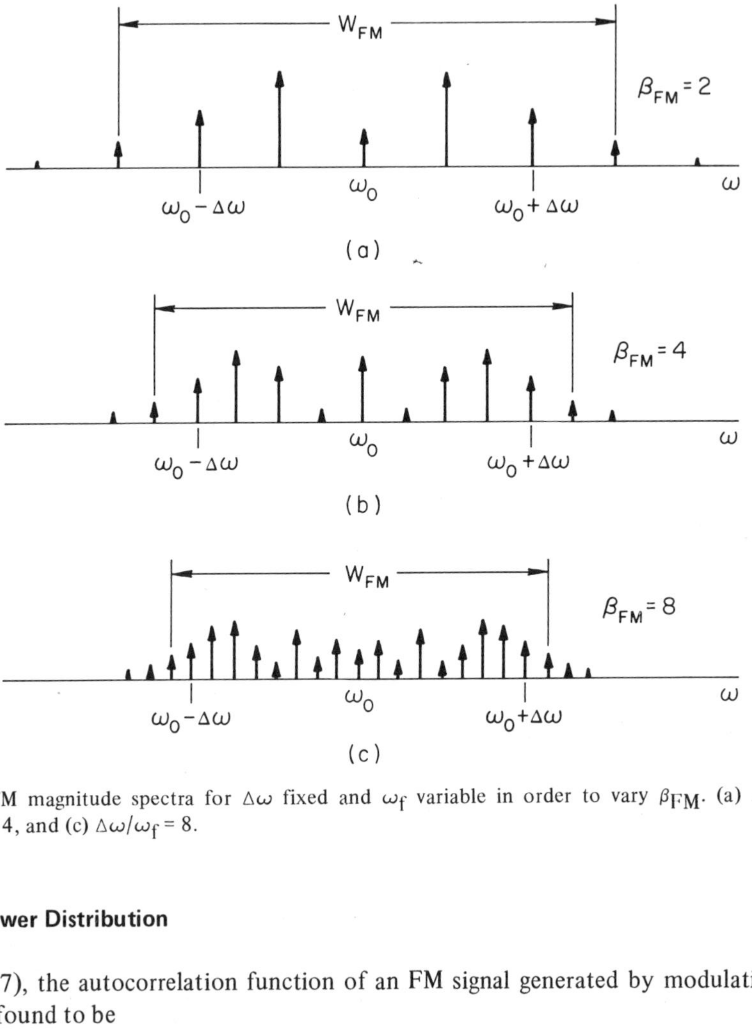

Fig. 6.3-4. FM magnitude spectra for $\Delta\omega$ fixed and ω_f variable in order to vary β_{FM}. (a) $\Delta\omega/\omega_f = 2$, (b) $\Delta\omega/\omega_f = 4$, and (c) $\Delta\omega/\omega_f = 8$.

Spectral Power Distribution

Using (6.3-7), the autocorrelation function of an FM signal generated by modulation with a sinusoid is found to be

$$R_{FM}(\tau) = \frac{A^2}{2} \sum_{n=-\infty}^{\infty} J_n^2(\beta_{FM}) \cos\left[(\omega_0 + n\omega_f)\tau\right]. \qquad (6.3\text{-}15)$$

Fourier transformation leads to the power density spectrum:

ISBN 0-201-05758-1

$$\mathscr{S}_{\text{FM}}(\omega) = \frac{\pi A^2}{2} \sum_{n=-\infty}^{\infty} J_n^2(\beta_{\text{FM}})[\delta(\omega - \omega_0 - n\omega_f) + \delta(\omega + \omega_0 + n\omega_f)]. \tag{6.3-16}$$

By integrating this result according to (2.8-6) we obtain the total power P_{FM} in the FM signal:

$$P_{\text{FM}} = \frac{1}{2\pi} \int_{-\infty}^{\infty} \mathscr{S}_{\text{FM}}(\omega)\, d\omega = \frac{A^2}{2} \sum_{n=-\infty}^{\infty} J_n^2(\beta_{\text{FM}}). \tag{6.3-17}$$

Since the sum is unity from (6.3-10) for all values of β_{FM}, total power is $A^2/2$ which is independent of the FM modulation process. This result is not too surprising, since we might suspect that power would be related to signal amplitude, which is constant for FM, and not necessarily dependent on the signal's phase. An example will illustrate power distribution.

Example 6.3-2

We find the percentage of total power carried by frequencies within the bandwidth of the signal of example 6.3-1. For the carrier power we have [from (5.8-8)]:

$$\text{carrier power} = \frac{2}{2\pi}\left[\frac{\pi A^2}{2} J_0^2(3)\right] = \frac{A^2}{2} J_0^2(3) = \frac{A^2}{2}(0.0676).$$

For the four significant side frequencies:

$$\text{sideband power} = \frac{4}{2\pi} \cdot \frac{\pi A^2}{2}\left[J_1^2(3) + J_2^2(3) + J_3^2(3) + J_4^2(3)\right]$$

$$= \frac{A^2}{2}[0.2298 + 0.4724 + 0.1910 + 0.0348]$$

$$= \frac{A^2}{2}[0.9280].$$

Thus, approximately 6.8% of total power is in the carrier, 92.8% is in the sidebands, and 99.6% is within the bandwidth of the signal.

6.4. WIDEBAND FM WITH ARBITRARY SIGNALS

We now turn our attention to the problem of finding the spectrum and bandwidth of an FM signal when the modulating waveform is arbitrary within a class. No attempt will be made to solve the general problem, as this is well beyond the scope of this book. Rather, we limit discussion to two classes of signals, periodic and random. First we shall extend the earlier single sinusoid results to include an arbitrary periodic signal. Although the solution will be obtained in an explicit form, the numerical evaluation of certain coefficients may remain a real task. In any particular example, coefficient evaluation may have to be done by numerical methods.

ISBN 0-201-05758-1

Most information signals may be treated as random. Validity of such a model derives from the fact that we (the receiver) never quite know what the signal (from the sender) will be. If we did, the message would not convey any real information. To be more specific, many typical information signals such as speech, music, telemetry, or some other binary signals may be treated using the random model. Our work, assuming a random signal, will lead to an approximation to the FM signal power density spectrum which becomes more accurate as the modulated signal bandwidth becomes larger.

Periodic Signal

We seek the spectrum of the general FM signal (6.1-8), which may be written in the form

$$s_{\mathrm{FM}}(t) = \frac{A}{2}\left[q(t)e^{j\omega_0 t+j\theta_0} + q^*(t)e^{-j\omega_0 t-j\theta_0}\right]. \tag{6.4-1}$$

Here * represents the complex conjugate and $q(t)$ is defined as

$$q(t) = e^{jk_{\mathrm{FM}}\int f(t)\,dt}. \tag{6.4-2}$$

Since $f(t)$ is periodic by assumption, then so is $q(t)$. As a result, it can be represented by its complex Fourier series:

$$q(t) = \sum_{n=-\infty}^{\infty} \gamma_n e^{jn\omega_f t}, \tag{6.4-3}$$

where ω_f is the fundamental frequency of $f(t)$ and the Fourier coefficients γ_n are (see Chapter 2)

$$\gamma_n = \frac{1}{T}\int_{-T/2}^{T/2} q(t)e^{-jn\omega_f t}\,dt, \qquad \text{all } n, \tag{6.4-4}$$

with the period $T = 2\pi/\omega_f$.

The FM signal, by substitution of (6.4-3) in (6.4-1), becomes

$$s_{\mathrm{FM}}(t) = \frac{A}{2}\sum_{n=-\infty}^{\infty}\left[\gamma_n e^{j\theta_0}e^{j(\omega_0+n\omega_f)t} + \gamma_n^* e^{-j\theta_0}e^{-j(\omega_0+n\omega_f)t}\right]. \tag{6.4-5}$$

The spectrum follows from the Fourier transform of this expression:

$$S_{\mathrm{FM}}(\omega) = \pi A\sum_{n=-\infty}^{\infty}\left[\gamma_n e^{j\theta_0}\delta(\omega - \omega_0 - n\omega_f) + \gamma_n^* e^{-j\theta_0}\delta(\omega + \omega_0 + n\omega_f)\right]. \tag{6.4-6}$$

Hence, the FM spectrum corresponding to a periodic modulating signal is given by (6.4-6). The only difficulty with obtaining numerical results for a given signal is with solving (6.4-4).

ISBN 0-201-05758-1

Solutions will depend on the complexity of $f(t)$. For some simple forms γ_n may result readily, but in general the integral may be difficult to solve except by numerical methods.

Example 6.4-1

We find the coefficients γ_n for a square-wave modulation signal defined in one period by

$$f(t) = \frac{\Delta\omega}{k_{\mathrm{FM}}}, \qquad 0 < t < T/2$$

$$= -\frac{\Delta\omega}{k_{\mathrm{FM}}}, \qquad -T/2 < t < 0.$$

Using (6.4-4) and (6.4-2)

$$\gamma_n = \frac{1}{T}\int_0^{T/2} e^{j(\Delta\omega - n\omega_f)t}\, dt + \frac{1}{T}\int_{-T/2}^{0} e^{-j(\Delta\omega + n\omega_f)t}\, dt$$

$$= \frac{e^{j(\beta_{\mathrm{FM}}-n)\pi/2}}{2}\,\frac{\sin[(\beta_{\mathrm{FM}} - n)\pi/2]}{(\beta_{\mathrm{FM}} - n)\pi/2} + \frac{e^{j(\beta_{\mathrm{FM}}+n)\pi/2}}{2}\,\frac{\sin[(\beta_{\mathrm{FM}} + n)\pi/2]}{(\beta_{\mathrm{FM}} + n)\pi/2},$$

where we define $\beta_{\mathrm{FM}} = \Delta\omega/\omega_f$.

Random Signal

As we have already had occasion to observe, information signals may often be treated as random. For such waveforms we may state an approximate result for the modulated signal power density spectrum.* In doing so, it will be necessary to make use of the probability density function, $p_f(f)$, of a random signal $f(t)$.

Figure 6.4-1(a) depicts a random voltage $f(t)$ having an average or dc component $\overline{f} = \overline{f(t)}$. A possible probability density function for the signal is depicted in (b). The function $p_f(f)$

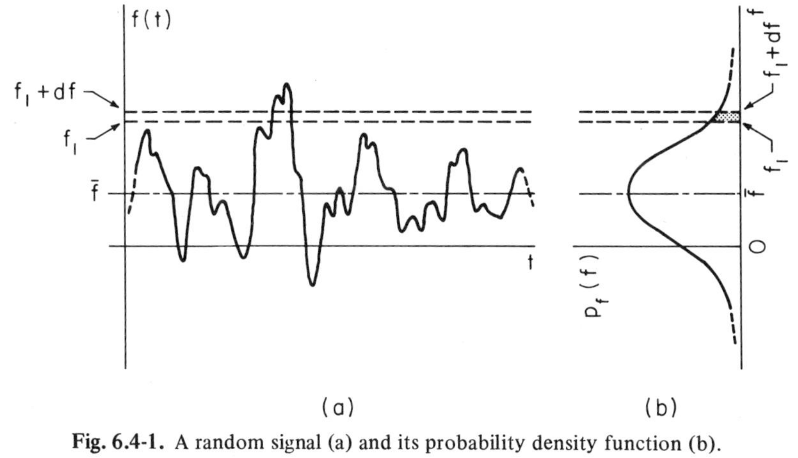

Fig. 6.4-1. A random signal (a) and its probability density function (b).

*Recall that the spectrum (Fourier transform) of a random signal has little meaning, while the power density spectrum represents a meaningful physical quantity that can be measured.

ISBN 0-201-05758-1

describes the likelihood or probability that the voltage f will have certain values. For example, consider the probability that $f(t)$ will have a value above some voltage f_1 but not greater than the level $f_1 + df$. Such a probability, also equal to the average fraction of time that $f(t)$ lingers between the two levels, is given by $p_f(f_1)\, df$, the shaded area in (*b*).

For physically realizable information signals it is unlikely that $f(t)$ will have very large positive ($f \to +\infty$) or very large negative ($f \to -\infty$) values for very long. Thus, the function $p_f(f)$ approaches zero at these extremes.

On the other hand, we would expect that the voltage would linger near its average value quite often so that $p_f(f)$ tends to peak near its average (mean) value $\overline{f(t)}$ as shown.* For many random signals $\overline{f(t)}$ is zero. Those that do have nonzero mean values may be represented as the sum of a dc component plus a zero-mean information signal. As a result, if the carrier frequency is ω_1 initially, the effect of a dc component is to cause the effective carrier frequency to be $\omega_0 = \omega_1 + k_{FM}\overline{f(t)}$. Thus, by appropriate interpretation of the carrier frequency, we only need consider random signals without average values.

Returning to the problem at hand, it is the power density spectrum $\mathscr{S}_{FM}(\omega)$ of the FM waveform which best describes the effect of random signal modulation. If the rms frequency deviation is large in relation to W_f, the maximum frequency extent of the information signal $f(t)$, we shall subsequently show that a technique called the *quasi-static approximation method* leads to the FM signal power spectrum:

$$\mathscr{S}_{FM}(\omega) = \frac{A^2\pi}{2k_{FM}}\left[p_f\left(\frac{\omega - \omega_0}{k_{FM}}\right) + p_f\left(\frac{-\omega - \omega_0}{k_{FM}}\right)\right], \tag{6.4-7}$$

where again $p_f(f)$ is the probability density function of $f(t)$. Thus, the power density spectrum of the FM waveform has the same form as the probability density function of the information signal. The condition of large rms frequency deviation limits use of (6.4-7) to wideband FM, which, of course, is the only case of interest, since the narrowband problem was solved previously in general. For the signal of Fig. 6.4-1, $\mathscr{S}_{FM}(\omega)$ appears as illustrated in Fig. 6.4-2.

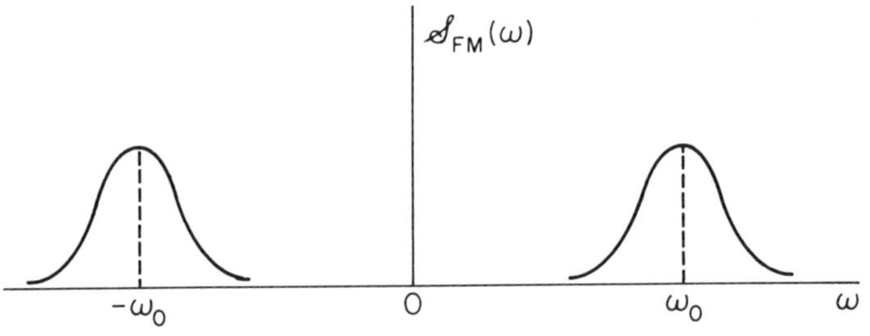

Fig. 6.4-2. Power density spectrum of the FM waveform produced by the random signal of Fig. 6.4-1.

Each term of (6.4-7) contributes half the total power in the FM signal. The first term usually has significant values only near ω_0 and may be considered as the contribution from positive frequencies. Similarly the second term is primarily the result of negative frequencies. An intuitive argument [3, 4] may be used to justify (6.4-7). We present the argument only in terms of the first term for positive frequencies; an analogous argument justifies the second term.

*This type of simple behavior is described because it is helpful in gaining a physical picture of the problem. There are, of course, exceptions to this behavior. Examples include densities with multiple peaks and those with little significant area near the mean value.

ISBN 0-201-05758-1

Imagine that the FM signal is applied to an ideal filter of narrow bandwidth $d\omega$ centered at a (positive) frequency ω. By considering the power emerging from the filter from both a time- and a frequency-domain approach, two expressions may be found, which, when equated, will yield the power density spectrum of the FM signal. Working first in the frequency domain, we apply (4.9-21) and (4.10-11) to achieve the first expression:

$$dP = \frac{1}{2\pi} \mathscr{S}_{\text{FM}}(\omega)\, d\omega. \tag{6.4-8}$$

Turning to the time domain, we assume that the rms frequency deviation is large relative to the highest frequency (W_f) contained in the modulating signal [3, 4]. The filter output average power will now be that portion of the total FM signal power $A^2/2$ determined by the fraction of time $p_f[f(t)]\ df(t)$ that the signal is at a level to cause the instantaneous frequency to fall in the filter passband. Thus, using a factor of ½ to account for just the positive frequency power contribution, the filter output power is

$$dP = \frac{A^2}{4} \cdot p_f(f)\, df. \tag{6.4-9}$$

Next, since instantaneous frequency and $f(t)$ are related by

$$\omega = \omega_0 + k_{\text{FM}} f(t), \tag{6.4-10}$$

then

$$f(t) = (\omega - \omega_0)/k_{\text{FM}}, \tag{6.4-11}$$

$$df(t) = d\omega/k_{\text{FM}}, \tag{6.4-12}$$

and (6.4-9) becomes

$$dP = \frac{A^2}{4k_{\text{FM}}}\, p_f\left(\frac{\omega - \omega_0}{k_{\text{FM}}}\right) d\omega. \tag{6.4-13}$$

On equating (6.4-8) and (6.4-13) we obtain the first term of (6.4-7). Repeating the analysis for negative frequencies leads to the second term. An example will demonstrate the use of (6.4-7).

Example 6.4-2

Suppose $f(t)$ is equally likely to have values from $-F$ to $+F$. The probability density function of $f(t)$ is uniform and is given by

$$p_f[f(t)] = \frac{1}{2F}, \qquad -F < f < F$$

$$= 0, \qquad \text{elsewhere.}$$

The power density spectrum of the FM wave is, from (6.4-7),

ISBN 0-201-05758-1

$$\mathscr{S}_{FM}(\omega) = \frac{A^2\pi}{4k_{FM}F}, \qquad |\omega - \omega_0| < k_{FM}F$$

$$= \frac{A^2\pi}{4k_{FM}F}, \qquad |-\omega - \omega_0| < k_{FM}F$$

$$= 0, \qquad \text{elsewhere.}$$

The power density spectrum of (6.4-7) has been developed and described as it applies to random signals. It can also be applied to nonrandom signals for which $p_f(f)df$ is meaningful in the sense of average fractional amount of time that the waveform has amplitudes from f to $f + df$. This includes periodic (power-type) signals but excludes finite duration (energy-type) waveforms [2]. However, when the quasi-static approximation method is used to approximate the power density spectrum of a periodic waveform, the fine-grain (line) structure of the power spectrum is not preserved. Only the general shape or envelope is obtained. The quasi-static method is also known as the *quasi-stationary* method [5], *adiabatic theorem* [3], and *Woodward's theorem* [4].

The above proof of (6.4-7) is adequate for those who are most inclined to the practical side of communications. However, it may not satisfy the desire of some readers for additional rigor. In response to this need, a more complete proof of the important quasi-static approximation is developed in Section 6.6.

Bandwidth

Recall that for sinusoidal FM the bandwidth of the modulated signal was $2(\beta_{FM} + 1)\omega_f$, where ω_f was the frequency of the sinusoid and β_{FM} was the ratio of peak frequency deviation $\Delta\omega$ and ω_f. A similar bandwidth expression for modulation by an arbitrary information signal $f(t)$ derives from extrapolating the sinusoidal FM result [2].

Let us define a *frequency deviation ratio* as

$$D_{FM} = \Delta\omega/W_f. \tag{6.4-14}$$

It is the peak frequency deviation

$$\Delta\omega = k_{FM}|f(t)|_{max} \tag{6.4-15}$$

divided by the maximum frequency extent W_f of $f(t)$. In terms of deviation ratio the FM waveform bandwidth W_{FM} is

$$W_{FM} = 2(D_{FM} + 1)W_f, \tag{6.4-16}$$

which is known as *Carson's rule* [6]. If $f(t)$ is not suitably bounded, such that $\Delta\omega$ is bounded, then W_{FM} may be determined by the frequency interval which is exceeded during modulation only for a time having acceptably small probability. For example, with a Gaussian signal (wideband) 98% of the signal power is contained in the band $W_{FM} = 4.66k_{FM}\sqrt{\overline{f^2(t)}}$. For practical applications Carson's rule underestimates bandwidth. For $D > 2$ a better result for equipment design is [2]

$$W_{FM} = 2(D_{FM} + 2)W_f. \tag{6.4-17}$$

ISBN 0-201-05758-1

6.5. ANALOGOUS RESULTS FOR WIDEBAND PM

In this section we shall develop results for wideband PM which are analogous to those for FM found in the last two sections. That is, we explore PM with a single sinusoidal signal, and with both periodic and random waveforms. However, because the procedures are quite similar to those used for FM, we shall only outline the main points and conclusions.

Wideband PM with a Sinusoid

For the information signal

$$f(t) = A_f \cos(\omega_f t) \tag{6.5-1}$$

the PM waveform is

$$s_{PM}(t) = A \cos[\omega_0 t + \beta_{PM} \cos(\omega_f t)], \tag{6.5-2}$$

where we define the PM *modulation index* β_{PM} by

$$\beta_{PM} = k_{PM} A_f = \Delta\theta. \tag{6.5-3}$$

Here $\Delta\theta$ is the peak phase deviation. The peak frequency deviation is

$$\Delta\omega = \beta_{PM}\omega_f. \tag{6.5-4}$$

In terms of a Bessel coefficient expansion, the PM signal may be written in the form

$$s_{PM}(t) = A \sum_{n=-\infty}^{\infty} J_n(\beta_{PM}) \cos[(\omega_0 + n\omega_f)t + n(\pi/2)]. \tag{6.5-5}$$

The PM signal spectrum is the Fourier transform of (6.5-5):

$$S_{PM}(\omega) = \pi A \sum_{n=-\infty}^{\infty} J_n(\beta_{PM}) \Big[e^{jn(\pi/2)}\, \delta(\omega - \omega_0 - n\omega_f) + e^{-jn(\pi/2)}\, \delta(\omega + \omega_0 + n\omega_f) \Big]. \tag{6.5-6}$$

As in FM we see that the amplitudes of the spectral components are proportional to Bessel functions. Indeed, the only difference between the form of the PM signal spectrum and that of the FM spectrum is the progressive phase shift of spectral lines by $\pi/2$, as can be seen by comparing (6.5-6) with (6.3-12). Since (6.3-9) again applies, only $2(\beta_{PM} + 1)$ side frequencies about $+\omega_0$ and $-\omega_0$ are significant. Thus, the bandwidth is again given by the right side of (6.3-13) with β_{FM} replaced by β_{PM}.

Although both FM and PM bandwidth expressions appear to be identical, there are some important differences. They may be compared by writing them as

$$W_{FM} = 2(\Delta\omega + \omega_f) \rightarrow 2\Delta\omega, \qquad \beta_{FM} \gg 1, \tag{6.5-7}$$

ISBN 0-201-05758-1

$$W_{PM} = 2(\Delta\theta + 1)\omega_f \to 2\Delta\theta\omega_f, \quad \beta_{PM} >> 1. \tag{6.5-8}$$

For fixed $\Delta\omega$ in wideband FM, bandwidth is approximately constant at $2\Delta\omega$ regardless of how ω_f varies (see Fig. 6.3-4). In PM with $\Delta\theta$ fixed, the bandwidth increases with increasing ω_f. On the other hand, if ω_f is fixed and modulation index is allowed to vary by changing $\Delta\omega$ or $\Delta\theta$, the behavior of both FM and PM spectra is similar (such as in Fig. 6.3-3). Both these effects result from the modulation index's being independent of ω_f in PM but not in FM. As in FM, when the PM modulation index is small, $W_{PM} \approx 2\omega_f$.

The PM power density spectrum is identical to that of FM in form and is given by (6.3-16) with β_{FM} replaced by β_{PM}.

Arbitrary Modulating Signals

For an arbitrary periodic signal $f(t)$, if we define

$$q(t) = e^{jk_{PM}f(t)}, \tag{6.5-9}$$

then the forms of all previous FM results hold. The PM signal is given by the right side of (6.4-5), while its spectrum is given by the right side of (6.4-6). However, the coefficients γ_n are now given by (6.4-4) using (6.5-9).

For random information signals the instantaneous frequency ω is $\omega_0 + k_{PM}[df(t)/dt]$. Defining

$$\dot{f}(t) = df(t)/dt, \tag{6.5-10}$$

we have

$$\omega = \omega_0 + k_{PM}\dot{f}(t). \tag{6.5-11}$$

Using this relationship, and assuming $\dot{f}(t)$ has the probability density function $p_{\dot{f}}(\dot{f})$, the power density spectrum of the PM waveform is given by

$$\mathscr{S}_{PM}(\omega) = \frac{A^2\pi}{2k_{PM}}\left[p_{\dot{f}}\left(\frac{\omega - \omega_0}{k_{PM}}\right) + p_{\dot{f}}\left(\frac{-\omega - \omega_0}{k_{PM}}\right)\right]. \tag{6.5-12}$$

It may not always be easy to specify $p_{\dot{f}}(\dot{f})$. For the important class of Gaussian signals, defined by the density function

$$p_f(f) = \frac{1}{\sqrt{2\pi}\,f_{rms}} \exp\left[-(f - \bar{f})^2/(2f_{rms}{}^2)\right], \tag{6.5-13}$$

where

$$\bar{f} = \overline{f(t)}, \tag{6.5-14}$$

$$f_{rms}{}^2 = \overline{[f(t) - \bar{f}]^2} \tag{6.5-15}$$

are the mean value and variance, respectively, of $f(t)$, it is known that $\dot{f}(t)$ is also Gaussian. The mean and variance of $\dot{f}$ will, in general, be different from those for f, but both f and $\dot{f}$ will

ISBN 0-201-05758-1

have the density function *form* of (6.5-13). It can be shown that $\overline{\dot{f}} = 0$ and $\dot{f}_{\text{rms}}{}^2 = \overline{\dot{f}^2} = W_{\text{rms}}{}^2 f_{\text{rms}}{}^2$, where W_{rms} is the rms bandwidth of the power density spectrum of $f(t)$.

Bandwidth for Arbitrary Modulation

Peak frequency deviation in a PM signal is given by

$$\Delta\omega = k_{\text{PM}} \left| \frac{df(t)}{dt} \right|_{\text{max}}. \qquad (6.5\text{-}16)$$

Since the PM signal is equivalent to an FM signal modulated by $df(t)/dt$, and since the bandwidth of the FM signal was given by Carson's rule (6.4-16), we conclude that the bandwidth of the PM signal is

$$W_{\text{PM}} = 2(D_{\text{PM}} + 1)W_f \qquad (6.5\text{-}17)$$

if we are careful to define D_{PM} by

$$D_{\text{PM}} = \Delta\omega / W_f \qquad (6.5\text{-}18)$$

with $\Delta\omega$ given by (6.5-16).

An alternative argument [2] is to observe that D_{FM} is the maximum phase deviation of an FM wave under worst-case bandwidth conditions. Thus, D_{FM} and peak phase deviation $\Delta\theta$ for the PM wave are equivalent parameters and PM bandwidth must also be given by

$$W_{\text{PM}} = 2(\Delta\theta + 1)W_f, \qquad (6.5\text{-}19)$$

where

$$\Delta\theta = k_{\text{PM}} |f(t)|_{\text{max}}. \qquad (6.5\text{-}20)$$

A result of the above two bandwidth arguments is a relation linking the two that will subsequently be useful in comparing FM and PM performance with noise present. By equating D_{PM} to $\Delta\theta$ and using (6.5-16), we have

$$\left| \frac{df(t)}{dt} \right|_{\text{max}} = W_f |f(t)|_{\text{max}} \qquad (6.5\text{-}21)$$

under worst-case bandwidth conditions.

★ 6.6. DERIVATION OF QUASI-STATIC APPROXIMATION

In the last two sections, approximate expressions for the power density spectra of FM and PM signals were stated. They were the result of applying the quasi-static approximation method when the modulating signal was random. An intuitive argument was used to lend support to the FM expression, although a rigorous proof was not given. In this section such a proof is presented. Readers uninterested in the more theoretical aspects of the method may omit this section without great loss. However, for those with the desire to pursue the topic, the material should prove to be both interesting and rewarding.

ISBN 0-201-05758-1

Our developments will, in addition to providing the desired proof, place some light on the accuracy of the quasi-static method. The procedure used will follow Blachman [3], with some minor variations. We begin by deriving the autocorrelation function of a general angle-modulated waveform.

Angle-Modulated Signal Autocorrelation Function and Spectrum

For either FM or PM the modulated signal may be expressed as

$$s_A(t) = A \cos [\omega_0 t + \theta_0 + km(t)], \tag{6.6-1}$$

where A is peak amplitude and ω_0 is the carrier frequency. The phase angle θ_0 is assumed to be a random variable uniformly distributed over $-\pi$ to π. The phase variation due to modulation is $km(t)$, where we identify

$$km(t) = k_{PM} f(t) \qquad \text{for PM,} \tag{6.6-2}$$

$$km(t) = k_{FM} \int f(t)\, dt \qquad \text{for FM.} \tag{6.6-3}$$

We assume the "message" $m(t)$ is a sample function of a random process which is independent of θ_0 and k is a positive constant.

The power density spectrum of $s_A(t)$ is the Fourier transform of the autocorrelation function, which we first find. By direct substitution,

$$\begin{aligned} R_s(t, t + \tau) &= E\left\{s_A(t)s_A(t + \tau)\right\} \\ &= \frac{A^2}{2} E\Big\{\cos [2\omega_0 t + \omega_0 \tau + 2\theta_0 + km(t + \tau) + km(t)] \\ &\qquad + \cos [\omega_0 \tau + km(t + \tau) - km(t)]\Big\}, \end{aligned} \tag{6.6-4}$$

where the expansion $\cos (A) \cos (B) = [\cos (A + B) + \cos (A - B)]/2$ has been used. By initially taking the expected value with respect to θ_0, the first right-side term is zero. Thus,

$$R_s(t, t + \tau) = \frac{A^2}{2} \operatorname{Re} \left[e^{j\omega_0 \tau} E\left\{e^{jkm(t+\tau) - jkm(t)}\right\}\right], \tag{6.6-5}$$

where Re $[\cdot]$ represents the real part. Now if we define the complex random process

$$q(t) = e^{jkm(t)}, \tag{6.6-6}$$

we recognize the factor in (6.6-5) involving the expected value as the complex conjugate of the autocorrelation function of $q(t)$. It is known, [7, p. 109], that $q(t)$ is wide-sense stationary if $km(t)$ is strictly stationary. In fact, $q(t)$ remains wide-sense stationary even if $km(t)$ is not strictly stationary, so long as $k\dot{m}(t) = k\, dm(t)/dt$ is strictly stationary [7, p. 112]. We recognize $k\dot{m}(t)$ as the instantaneous frequency deviation about the carrier. Thus, if $k\dot{m}(t)$ is assumed strictly stationary, $q(t)$ is wide-sense stationary and its autocorrelation function does

ISBN 0-201-05758-1

not depend on absolute time. It is then given by

$$R_q^*(\tau) = E\left\{q^*(t)q(t + \tau)\right\} = E\left\{e^{jk[m(t+\tau)-m(t)]}\right\}. \tag{6.6-7}$$

By substituting this expression into (6.6-5) we find that the autocorrelation function $R_s(t, t + \tau)$ depends only on τ. Hence, the angle-modulated signal is wide-sense stationary, and its autocorrelation function is

$$R_s(\tau) = \frac{A^2}{2} \operatorname{Re}\,[R_q^*(\tau)e^{j\omega_0\tau}]. \tag{6.6-8}$$

By using the relationship [see (4.8-53)]

$$R_q^*(\tau) = R_q(-\tau), \tag{6.6-9}$$

we obtain the more convenient form

$$R_s(\tau) = \frac{A^2}{4}\left[R_q(-\tau)e^{j\omega_0\tau} + R_q(\tau)e^{-j\omega_0\tau}\right]. \tag{6.6-10}$$

By straightforward Fourier transformation of (6.6-10), the angle-modulated signal power density spectrum becomes

$$\mathscr{S}_s(\omega) = \frac{A^2}{4}\left\{\mathscr{S}_q(-\omega + \omega_0) + \mathscr{S}_q(\omega + \omega_0)\right\}, \tag{6.6-11}$$

where $\mathscr{S}_q(\omega)$ is defined as the Fourier transform of $R_q(\tau)$ given by (6.6-7),

$$R_q(\tau) \leftrightarrow \mathscr{S}_q(\omega). \tag{6.6-12}$$

$R_s(\tau)$ and $\mathscr{S}_s(\omega)$ are our principal results. They apply to any general angle-modulated wave, having a uniformly distributed random phase angle, for which the derivative $k\dot{m}(t)$ of the phase deviation in the signal is strictly stationary. Within these constraints, the expressions (6.6-10) and (6.6-11) are exact and apply to either PM or FM, with (6.6-2) or (6.6-3) applying, respectively. From the form of (6.6-11), the power spectrum clearly comprises two parts, centered at ω_0 and $-\omega_0$. Both are related to $q(t)$, and we only need determine the power spectrum $\mathscr{S}_q(\omega)$ of $q(t)$ for a complete solution. The quasi-static approximation results from efforts to evaluate $\mathscr{S}_q(\omega)$.

Quasi-Static Approximation

If an approximate evaluation of $R_q(\tau)$ is found, $\mathscr{S}_q(\omega)$ results from Fourier transformation. Equation (6.6-7) may be written as

$$R_q^*(\tau) = E\left\{e^{jk\int_t^{t+\tau} \dot{m}(\xi)d\xi}\right\}. \tag{6.6-13}$$

Generally, we are interested in the broadband case. From the form of (6.6-11), this means that $\mathscr{S}_q(\omega)$ is broadband. Consequently, only small values of τ [on the order of the period of

ISBN 0-201-05758-1

the worst-case or highest frequency present in $\dot{m}(t)$] determine the principal behavior of $R_q^*(\tau)$. Thus, over the small range of integration in (6.6-13), $\dot{m}(\xi)$ is approximately constant and is well represented by the first two terms of its Taylor series expansion about the point $\xi = t$. We then have

$$\int_t^{t+\tau} \dot{m}(\xi)\, d\xi \approx \int_t^{t+\tau} [\dot{m}(t) + \ddot{m}(t)(\xi - t)]\, d\xi$$

$$= \dot{m}(t)\tau + \ddot{m}(t)\frac{\tau^2}{2}. \tag{6.6-14}$$

Combining this expression with (6.6-13), we obtain

$$R_q^*(\tau) \approx E\left\{e^{jk\dot{m}(t)\tau}\, e^{jk\ddot{m}(t)(\tau^2/2)}\right\}. \tag{6.6-15}$$

The expectation is now taken with respect to the two random processes $\dot{m}(t)$ and $\ddot{m}(t)$.

If $|k\ddot{m}(t)\tau^2/2| << 1$ most of the time in the second exponential factor of (6.6-15), we may use the approximation

$$e^{jB} \approx 1 - \frac{B^2}{2} + jB, \qquad |B| << 1, \tag{6.6-16}$$

and write

$$R_q^*(\tau) \approx E\left\{e^{jk\dot{m}(t)\tau}\left[1 - \frac{k^2\ddot{m}^2(t)\tau^4}{8} + jk\ddot{m}(t)\frac{\tau^2}{2}\right]\right\}. \tag{6.6-17}$$

The condition $|k\ddot{m}(t)\tau^2/2| << 1$ may be interpreted in a statistical sense giving the corresponding requirement $E\left\{k^2\ddot{m}^2(t)\tau^4\right\} << 4$. Now over a small time interval τ instantaneous frequency deviation has a nominal value $k\dot{m}(t)$ and changes by an amount $k\ddot{m}(t)\tau$. Thus, in words, the condition means that the largest significant value of τ, which is on the order of the reciprocal of the rms frequency deviation $[E\left\{k^2\dot{m}^2(t)\right\}]^{1/2}$, must be small enough so that the rms frequency change $[E\left\{k^2\ddot{m}^2(t)\tau^2\right\}]^{1/2}$ is much less than $2/\tau$. Hence a requirement for (6.6-17) to be valid is

$$E\left\{k^2\ddot{m}^2(t)\tau^2\right\} << 4E\left\{k^2\dot{m}^2(t)\right\}. \tag{6.6-18}$$

We subsequently consider this condition further and find that it is satisfied for waveforms having large frequency deviation.

Next, if the terms in $\ddot{m}$ are not negligible, we assume that $\ddot{m}$ is statistically independent of $\dot{m}$ in order that we may determine at least the order of magnitude of the effect of including the $\ddot{m}$ terms without introducing undue complexity into the analysis. If $\dot{m}(t)$ is a Gaussian signal, then $\dot{m}$ and $\ddot{m}$ *are* statistically independent [3]. Equation (6.6-17) now becomes

$$R_q^*(\tau) \approx E\left\{e^{jk\dot{m}(t)\tau}\right\}\left[1 - \frac{k^2\tau^4}{8}E\left\{\ddot{m}^2(t)\right\} + j\frac{k\tau^2}{2}E\left\{\ddot{m}(t)\right\}\right]. \tag{6.6-19}$$

ISBN 0-201-05758-1

It is readily shown that $E\left\{\ddot{m}(t)\right\} = 0$, since $\dot{m}(t)$ is assumed strictly stationary, so

$$R_q^*(\tau) \approx E\left\{e^{jk\dot{m}(t)\tau}\right\}\left[1 - \frac{k^2\tau^4}{8}E\left\{\ddot{m}^2(t)\right\}\right]. \tag{6.6-20}$$

Consider first the leading term. It must be an autocorrelation function [to show this, retrace the analysis assuming only a one-term Taylor series expansion of the integrand of (6.6-13)]. Call this function $R_{q0}^*(\tau)$; it is given by

$$R_{q0}^*(\tau) = E\left\{e^{jk\dot{m}(t)\tau}\right\} = \frac{1}{2\pi}\int_{-\infty}^{\infty} 2\pi p_{\dot{m}}(\dot{m})e^{jk\dot{m}\tau}\,d\dot{m}, \tag{6.6-21}$$

where $p_{\dot{m}}(\dot{m})$ is the probability density function of $\dot{m}$. Using the variable change $\omega = k\dot{m}$, we obtain

$$R_{q0}^*(\tau) = \frac{1}{2\pi}\int_{-\infty}^{\infty}\left[\frac{2\pi}{k}p_{\dot{m}}\left(\frac{\omega}{k}\right)\right]e^{j\omega\tau}\,d\omega. \tag{6.6-22}$$

The term $(2\pi/k)p_{\dot{m}}(\omega/k)$ is recognized as the power density spectrum associated with $R_{q0}^*(\tau)$. From simple Fourier transform properties, that associated with $R_{q0}(\tau)$ is

$$R_{q0}(\tau) \leftrightarrow \mathscr{S}_{q0}(\omega) = \frac{2\pi}{k}p_{\dot{m}}\left(\frac{-\omega}{k}\right). \tag{6.6-23}$$

Finally, from (6.6-11), the first-order approximation to the power spectrum of an angle-modulated signal is

$$\mathscr{S}_s(\omega) = \frac{A^2\pi}{2k}\left\{p_{\dot{m}}\left(\frac{\omega - \omega_0}{k}\right) + p_{\dot{m}}\left(\frac{-\omega - \omega_0}{k}\right)\right\}, \tag{6.6-24}$$

which is a statement of the quasi-static approximation. For FM, $k = k_{FM}$, $\dot{m}(t) = f(t)$, and our expression becomes identical to (6.4-7). For PM, $k = k_{PM}$, $\dot{m}(t) = \dot{f}(t)$, and we obtain (6.5-12), as we should.

In summary, the quasi-static approximation (6.6-24) to the power density spectrum of a waveform having angle modulation $km(t)$ has been derived for messages where $k\dot{m}(t)$ is strictly stationary. An additional assumption requires that $\dot{m}(\xi)$ in (6.6-13) be approximately constant over any time interval t to $t + \tau$. This assumption means that the second term in the Taylor series expansion of $\dot{m}(\xi)$ should be negligible. If the second term is not negligible but is small in relation to the first term, that is, if

$$E\left\{\ddot{m}^2(t)\tau^2\right\} << 4E\left\{\dot{m}^2(t)\right\}, \tag{6.6-25}$$

then a two-term expansion of $\dot{m}(\xi)$ leads to (6.6-20), which has a "correction" term involving $\ddot{m}^2(t)$. The derivation of (6.6-20) assumed $\dot{m}(t)$ and $\ddot{m}(t)$ statistically independent. The correction term alters the power spectrum $\mathscr{S}_s(\omega)$, which we next find.

ISBN 0-201-05758-1

Power Spectrum with Correction Term

Using (6.6-21) with the correction term in (6.6-20) we have

$$-\frac{k^2 E\left\{\ddot{m}^2(t)\right\}}{8}\tau^4 E\left\{e^{jk\dot{m}(t)\tau}\right\} = -\frac{k^2 E\left\{\ddot{m}^2(t)\right\}}{8}\tau^4 R_{q0}^*(\tau). \tag{6.6-26}$$

Since

$$R_{q0}(\tau) \leftrightarrow \mathcal{S}_{q0}(\omega), \tag{6.6-27}$$

we apply the time differentiation property of Fourier transforms to obtain

$$\tau^4 R_{q0}(\tau) \leftrightarrow \frac{d^4 \mathcal{S}_{q0}(\omega)}{d\omega^4}. \tag{6.6-28}$$

Using this result allows the Fourier transform of (6.6-20) to be written

$$\mathcal{S}_q(\omega) = \frac{2\pi}{k}\left[p_{\dot{m}}\left(\frac{-\omega}{k}\right) - \frac{k^2 E\left\{\ddot{m}^2(t)\right\}}{8}\frac{d^4 p_{\dot{m}}(-\omega/k)}{d\omega^4}\right]. \tag{6.6-29}$$

The modulated signal power spectrum with correction term added becomes

$$\mathcal{S}_s(\omega) = \frac{A^2\pi}{2k}\left\{p_{\dot{m}}\left(\frac{\omega-\omega_0}{k}\right) + p_{\dot{m}}\left(\frac{-\omega-\omega_0}{k}\right)\right.$$

$$\left. - \frac{k^2 E\left\{\ddot{m}^2(t)\right\}}{8}\left[\frac{d^4 p_{\dot{m}}\left(\frac{\omega-\omega_0}{k}\right)}{d\omega^4} + \frac{d^4 p_{\dot{m}}\left(\frac{-\omega-\omega_0}{k}\right)}{d\omega^4}\right]\right\}. \tag{6.6-30}$$

The validity of this expression is related to the required condition (6.6-18). We next examine this condition to see when (6.6-30) is applicable and to determine the approximate magnitude of the spectral correction term.

Correction Term Validity and Magnitude

Clearly, (6.6-18) may be stated in the equivalent form:

$$E\left\{\ddot{m}^2(t)\right\}\tau^2/4 << E\left\{\dot{m}^2(t)\right\}. \tag{6.6-31}$$

Now if $\mathcal{S}_{\dot{m}}(\omega)$ is the power density spectrum of $\dot{m}(t)$, the power spectrum of $\ddot{m}(t)$ is given by (4.9-23). Thus,

$$E\left\{\ddot{m}^2\right\} = \frac{1}{2\pi}\int_{-\infty}^{\infty} \omega^2 \mathcal{S}_{\dot{m}}(\omega)\, d\omega$$

ISBN 0-201-05758-1

$$= \frac{\int_{-\infty}^{\infty} \omega^2 \mathscr{S}_{\dot{m}}(\omega)\, d\omega}{\int_{-\infty}^{\infty} \mathscr{S}_{\dot{m}}(\omega)\, d\omega} \cdot \frac{1}{2\pi} \int_{-\infty}^{\infty} \mathscr{S}_{\dot{m}}(\omega)\, d\omega = W_{\dot{m},\mathrm{rms}}{}^2 E\{\dot{m}^2\}, \tag{6.6-32}$$

from (4.9-35). Here $W_{\dot{m},\mathrm{rms}}$ is the rms bandwidth of the signal $\dot{m}(t)$. From (6.6-31), we now require

$$W_{\dot{m},\mathrm{rms}}\tau << 2. \tag{6.6-33}$$

Since the largest significant value of τ is on the order of $1/\Delta\omega_{\mathrm{rms}}$, where $\Delta\omega_{\mathrm{rms}}$ is the rms frequency deviation in the angle-modulated waveform, we use this fact with (6.6-33) to write

$$\beta_{\mathrm{rms}} = \Delta\omega_{\mathrm{rms}}/W_{\dot{m},\mathrm{rms}} >> \frac{1}{2}. \tag{6.6-34}$$

Here we define β_{rms} as the *rms modulation index* (or *rms deviation ratio*) of the angle-modulated wave. It is a reasonable definition for *both* PM and FM, because in either case $k\dot{m}(t)$ is the instantaneous frequency deviation of the angle-modulated waveform from the carrier. Hence it is the ratio of the rms frequency deviation to the rms frequency (bandwidth) of the "frequency"-modulating "message" $\dot{m}(t)$.

The final result (6.6-34) simply indicates that the corrected quasi-static power spectrum (6.6-30) is valid if, under already stated assumptions, the rms modulation index is large with respect to $1/2$.

To examine the magnitude of the correction term in (6.6-30) we consider (6.6-29) to obtain the same result more simply. We define the magnitude of the correction term on a percentage basis as 100 times the ratio of the maximum magnitude of the second term to the maximum magnitude of the first term

$$\frac{\epsilon_{\%}}{100} = \frac{1}{8} W_{\dot{m},\mathrm{rms}}{}^2 \Delta\omega_{\mathrm{rms}}{}^2 \frac{\left|\dfrac{d^4 p_{\dot{m}}(-\omega/k)}{d\omega^4}\right|_{\max}}{|p_{\dot{m}}(-\omega/k)|_{\max}}. \tag{6.6-35}$$

Here we have used (6.6-32) and recognized that $\Delta\omega_{\mathrm{rms}}{}^2 = k^2 E\{\dot{m}^2\}$. Numerical evaluation of this expression requires that $p_{\dot{m}}(\dot{m})$ be specified. To obtain example results, let us assume $\dot{m}(t)$ is Gaussian. Then $p_{\dot{m}}(\dot{m}) = [2\pi E\{\dot{m}^2\}]^{-1/2} \exp[-\dot{m}^2/2E\{\dot{m}^2\}]$. By substitution into (6.6-35) we find [3] that the maximum of the fourth derivative of (6.6-35) is $3/[\sqrt{2\pi}\, k^4(E\{\dot{m}^2\})^{5/2}]$, while the denominator factor maximum is $[2\pi E\{\dot{m}^2\}]^{-1/2}$. Thus, (6.6-35) finally becomes

$$\epsilon_{\%} = 300/8\beta_{\mathrm{rms}}{}^2. \tag{6.6-36}$$

This percentage correction term is plotted in Fig. 6.6-1 along with two results given by Blachman and McAlpine [4]. Our result differs from that of Blachman [3] by a factor of 3. The difference arises from the details of analysis (Blachman used a three-term Taylor series

ISBN 0-201-05758-1

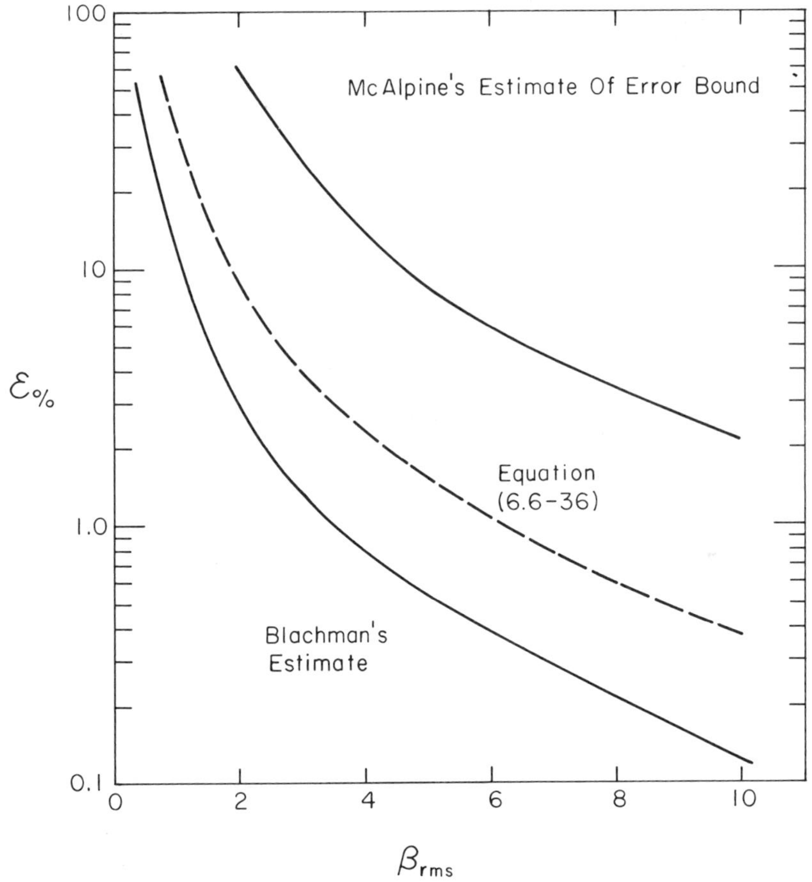

Fig. 6.6-1. Approximate magnitude of the correction (error) term applicable to the quasi-static power spectrum approximation. Error is shown as a percentage of the approximation. Curves due to Blachman and McAlpine have been adapted from reference [4] with permission.

approximation, where we have used only two). For an rms modulation index larger than 5, the correction term maximum magnitude is expected to be about 1.5% of the maximum magnitude of the quasi-stationary (uncorrected) power spectrum. It could, however, be larger due to neglected higher-order terms. In any case, it should not exceed about 8.5% for Gaussian modulation from McAlpine's curve. The correction term becomes rapidly unimportant as β_{rms} increases above 5.

6.7. WIDEBAND MODULATION METHODS

There are two fundamental methods for generating wideband FM and PM signals. These may be called *direct* and *indirect methods.* The direct method depends on varying the frequency of an oscillator linearly with $f(t)$ for FM or with $df(t)/dt$ for PM. The direct method is capable of generating FM or PM signals with large deviation ratio (wideband FM or PM).

The indirect method depends on first generating a narrowband FM or PM signal and then using a multiplication technique whereby the deviation ratio can be raised to a large value.

ISBN 0-201-05758-1

The Direct Method, Voltage-Controlled Oscillator

Direct wideband FM or PM can be generated by using the techniques of Fig. 6.7-1. In (*a*), $f(t)$ is applied to a *voltage-controlled oscillator* (VCO) to produce FM. The VCO frequency is made to be a linear function of $f(t)$ and nominally equals ω_0 when $f(t) = 0$. The VCO is used in (*b*) to produce PM where phase modulation is a linear function of $f(t)$.

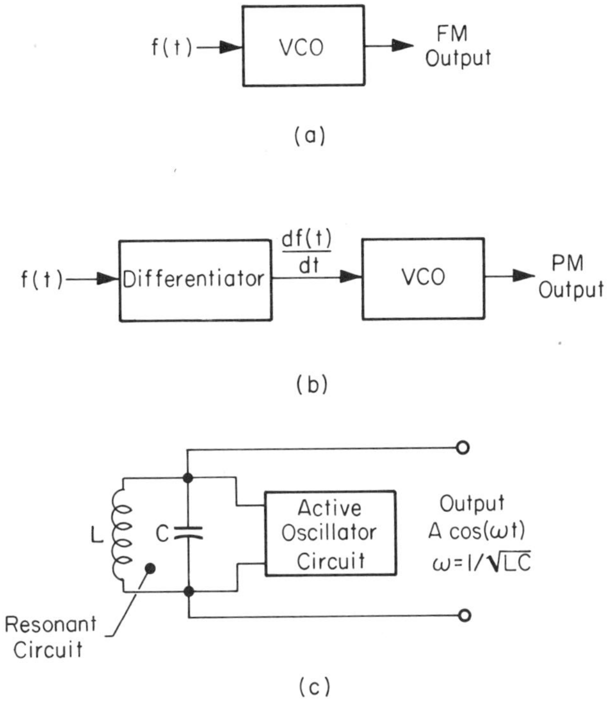

Fig. 6.7-1. Direct methods of angle modulation which use a voltage-controlled oscillator. (a) FM, (b) PM, and (c) diagram of an oscillator illustrating the VCO principle.

A common form of VCO construction depends on varying the resonant frequency of a tuned circuit, as shown in (*c*), which controls the frequency of an oscillator. Clearly, by varying either L or C a small amount about a nominal value, the desired frequency variations are achieved. Thus, for this form of VCO the problem reduces to one of achieving either an inductive or capacitive voltage-variable reactance.

Variable Reactance Techniques

Several methods are possible for obtaining a variable reactance. One, called the saturable reactor, is a variable inductance. It depends on the fact that some core materials for coils, such as ferrite, have a small-signal permeability which is a function of the magnetic flux within the core. Thus, by causing the core flux to vary as a function of $f(t)$, such as with a special winding to which $f(t)$ is applied, permeability and therefore inductance becomes a function of $f(t)$.

One especially simple way of varying capacitance uses a back-biased semiconductor junction diode. A characteristic of such diodes is that the small-signal junction capacitance is a

ISBN 0-201-05758-1

function of the bias voltage. By letting total bias be composed of a dc voltage to establish operating point and the voltage $f(t)$, capacitance, and therefore oscillator frequency, may be made to vary as a function of $f(t)$. The diodes are usually called either *varicaps* or *varactors.* Varactor-controlled oscillators may be made to have a quite linear frequency-voltage characteristic but often have only small frequency deviation as compared with other means of achieving reactance variation.

Other forms of voltage-controlled oscillators are multivibrators and klystrons. In the latter case the repeller voltage of a reflex klystron can be varied according to $f(t)$ to generate FM.

A popular method for obtaining either a variable inductance or capacitance for oscillator control uses the *reactance modulator.* A typical circuit is shown in Fig. 6.7-2(a), where a *field effect transistor* (FET) is illustrated. It is attractive for use in the reactance modulator because its gate presents only a small loading impedance in parallel with the impedance Z_2. Small circuit loading due to the active device is desirable for best performance. Because vacuum tubes also possess these properties, they may also be used as the active device. Common tube implementation uses either a tetrode or pentode.

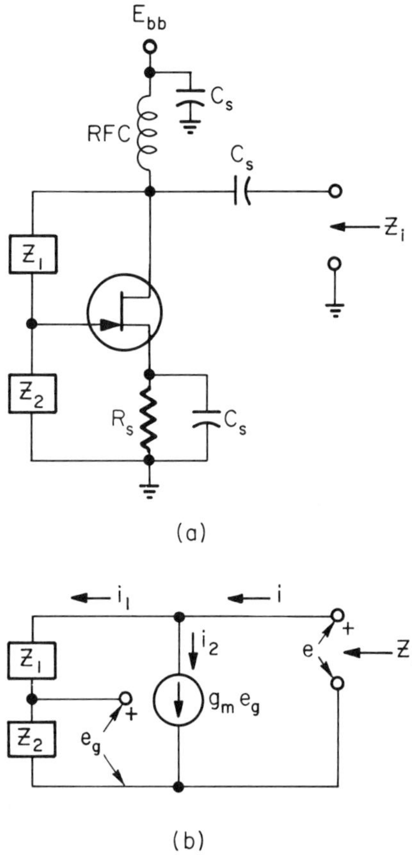

Fig. 6.7-2. Reactance modulator circuit (a) and ac equivalent circuit (b).

Analysis of the Reactance Modulator

We shall give a simplified analysis of the circuit of Fig. 6.7-2(a) which holds for use of either a vacuum tube or FET. All capacitors C_s are designed to have very low impedance for frequencies of interest (near the carrier) and may be replaced by short circuits in the small-signal ac

ISBN 0-201-05758-1

equivalent circuit of (*b*). R_s is similarly used only for dc biasing and does not appear in (*b*). The active device is replaced by its equivalent circuit, a constant current generator, where g_m is its transconductance. Finally, the *radio-frequency choke* (RFC) will have a high impedance at frequencies of interest and becomes an open circuit in (*b*).

Two simplifying assumptions will lead to the input impedance Z_i being approximately a pure reactance. First, assume $|Z_1 + Z_2|$ is large enough so that $|i_1| << |i_2|$. Then

$$Z_i \approx \frac{e}{i_2} = \frac{e}{g_m e_g} = \frac{Z_2 + Z_1}{g_m Z_2} = \frac{1}{g_m} + \frac{Z_1}{g_m Z_2}. \tag{6.7-1}$$

The term $1/g_m$ is a resistive component while the second term is purely reactive if either Z_1 or Z_2 is reactive while the other is resistive. Next, if we make the second assumption that

$$|Z_1| >> |Z_2|, \tag{6.7-2}$$

then (6.7-1) becomes approximately

$$Z_i \approx Z_1/g_m Z_2. \tag{6.7-3}$$

Four possibilities exist for making Z_i reactive. They are tabulated in Table 6.7-1. Two result in an equivalent inductance and two in a capacitance.

TABLE 6.7-1. Impedance Combinations for the Reactance Modulator

Case	Z_1	Z_2	Equivalent Element for Z_i
1	$\frac{1}{j\omega C}$	R	Capacitance = $g_m RC$
2	R	$j\omega L$	Capacitance = $\frac{g_m L}{R}$
3	$j\omega L$	R	Inductance = $\frac{L}{g_m R}$
4	R	$\frac{1}{j\omega C}$	Inductance = $\frac{RC}{g_m}$

Since transconductance g_m is a function of gate bias voltage in general, Z_i is a function of bias voltage. If the signal voltage $f(t)$ is added to a dc bias (establishes the operating point) to establish the total gate bias voltage, Z_i is readily made to vary as a function of $f(t)$.

Stabilization of VCO's

All the oscillators used in the direct method are subject to carrier-frequency drift, and it may be necessary in some systems to stabilize the carrier frequency.

Figure 6.7-3 illustrates a method whereby the carrier frequency ω_0 is forced to a stable value. The modulating signal is passed through a high-pass filter which only assures that $f(t)$,

ISBN 0-201-05758-1

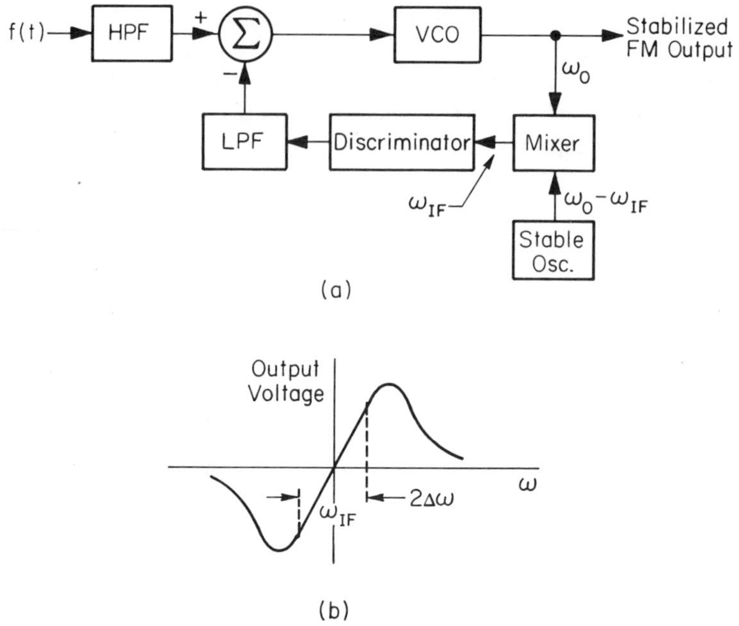

Fig. 6.7-3. Discriminator stabilization of a VCO. (a) Block diagram of stabilization loop and (b) discriminator characteristic.

as applied to the VCO, has no dc or very low-frequency content. The FM output, which may be subject to drift in the carrier frequency ω_0, is applied to a *mixer.* A mixer is a device which simply translates the FM signal to a new "carrier" frequency ω_{IF}. The new frequency is the difference between ω_0 and a "local oscillator" frequency, $\omega_0 - \omega_{IF}$. The mixer output then goes into a *frequency discriminator* which is a device having a transfer characteristic* as shown in (*b*) of Fig. 6.7-3 (see Section 6.8). The output is a voltage proportional to instantaneous frequency changes about ω_{IF}.

The discriminator voltage is next averaged by a very narrowband low-pass filter. Its effect is to remove the voltage fluctuations caused by $f(t)$ during modulation. If ω_0 drifts by a small amount $\delta\omega_0$, the filter output will be a dc voltage proportional to $\delta\omega_0$ which can be used to correct the VCO frequency, thereby reducing $\delta\omega_0$ to nearly zero. Because of the high-pass filter, $f(t)$ has no very low-frequency content and, since the low-pass filter only passes very low frequencies (corresponding to slow drifts in ω_0), the stabilization loop† has little effect on the FM caused by $f(t)$. Similarly, the modulation has little effect on the stabilization process. Discriminator stabilization can hold frequency within a few hertz of the desired frequency. Accuracy is, to a large extent, a function of the stability of the discriminator *"crossover" frequency.*

Indirect or Multiplication Method

As noted earlier, the indirect method involves generating narrowband angle modulation first and multiplying the result to obtain wideband modulation. Figure 6.7-4(*a*) illustrates the steps involved. The narrowband modulator may be any one of the methods illustrated in Fig.

*The reason the mixer is used is to allow the discriminator to be designed at a low frequency so that its "crossover" frequency ω_{IF} can be made stable.

†Loops of this type are called *automatic frequency control* (AFC) loops.

ISBN 0-201-05758-1

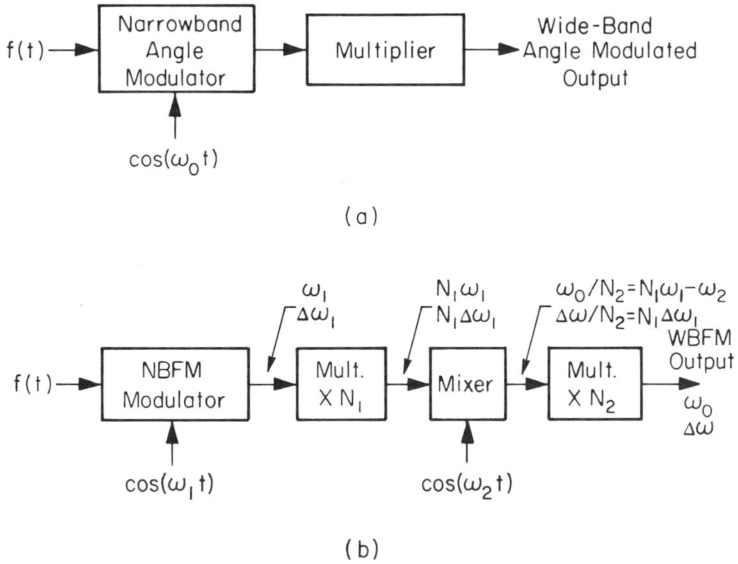

Fig. 6.7-4. Indirect method of angle modulation. (a) General method and (b) practical system for FM generation.

6.2-1. The *multiplier* is a device which multiplies the instantaneous frequency of its input waveform by a factor N. If the input waveform is frequency-modulated with instantaneous frequency $\omega_{0i} + k_{\mathrm{FM}i}f(t)$, the output frequency will be

$$\omega_0 + k_{\mathrm{FM}}f(t) = N\omega_{0i} + Nk_{\mathrm{FM}i}f(t). \tag{6.7-4}$$

Observe that the output carrier frequency ω_0 is N times as large as the input. Similarly, the output peak frequency deviation $\Delta\omega$ is N times as large as the input peak deviation $\Delta\omega_i$:

$$\Delta\omega = Nk_{\mathrm{FM}i}|f(t)|_{\max} = N\,\Delta\omega_i. \tag{6.7-5}$$

The *rate* of modulation is not affected. Thus, a multiplier increases the carrier frequency and bandwidth by the multiplication factor but does not alter the rate of modulation. The same conclusions hold for PM.

In a typical FM broadcast system the required multiplication factor may be on the order of 3000 which could result in an extremely large output carrier frequency if the system of Fig. 6.7-4(a) were used directly. A more practical circuit is shown in (b), where the multiplier is separated into two portions by a mixer which serves to lower the "carrier" frequency at its output by an amount ω_2. The desired final carrier frequency ω_0 is related to the selectable parameters ω_1, ω_2, N_1, and N_2 by

$$\omega_0 = N_2(N_1\omega_1 - \omega_2), \tag{6.7-6}$$

while the output peak frequency deviation is

$$\Delta\omega = N_1N_2\,\Delta\omega_1. \tag{6.7-7}$$

The practical circuit of (b) is called the *Armstrong indirect method* after E. H. Armstrong who did much to advance the cause of FM in the middle 1930's.

ISBN 0-201-05758-1

The main disadvantage of the narrowband modulators of Fig. 6.2-1 is poor linearity. A more complex but highly linear device is the *serrasoid modulator* [2]. It produces a train of narrow pulses at nominal pulse rate ω_1. The effect of modulation is to vary the position of the pulses linearly with $f(t)$. By filtering the pulse train by a narrowband filter centered at ω_1 a highly linear narrowband angle-modulated wave results at a carrier frequency ω_1. However, because the pulse train is rich in harmonics it is possible to tune the filter to a harmonic $N\omega_1$ so that advantage can be taken of the multiplication possibilities of the circuit. While narrowband phase modulators used in Fig. 6.2-1 have a linear phase deviation capability of only about $\pm\pi/6$ rad or less, serrasoid modulators are capable of as much as $\pm 5\pi/6$ rad of deviation with excellent linearity.

6.8. WIDEBAND DEMODULATION METHODS

In demodulating angle-modulated waves we require a device that will give output voltage as a linear function of the modulated quantity. For FM this means we need a voltage linear with frequency, and for PM a voltage linear with phase. We may broadly let all demodulators fall into two classes, *discriminator* and *locked-loop demodulators*. The former class is more commonly used for broadcast and some other systems owing to simplicity and low implementation cost. The latter class gives better performance, however, when signal levels are low and noise is an important consideration, such as in space applications.

Discriminator Demodulation of FM

One of the simplest discriminators possible is illustrated in Fig. 6.8-1(*a*). The envelope detector produces a voltage proportional to the magnitude of the transfer function of the high-pass CR filter as depicted in (*b*). R and C are chosen to center the linear part of the sloping region of $|H(\omega)|$ at the carrier frequency ω_0. As instantaneous frequency variations proportional to

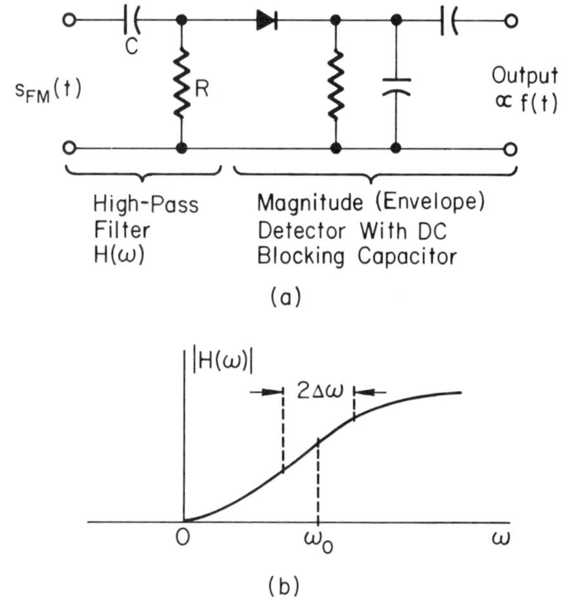

Fig. 6.8-1. Simple FM discriminator. (a) Circuit and (b) discriminator characteristic.

ISBN 0-201-05758-1

$f(t)$ take place within an interval $2\Delta\omega$ centered on ω_0, the detected voltage will vary proportional to $f(t)$ about a dc level. A blocking capacitor removes the dc component. Although a high-pass filter is shown, either a low-pass or a band-pass filter could also be used. These simple discriminators have been called *slope detectors* [8].

An improved discriminator using two tuned circuits is illustrated in Fig. 6.8-2(*a*). Its response characteristic is shown in (*b*). The circuit has been called a Travis detector [8], after its inventor. Its major disadvantage is in alignment, since two tuned circuits must be adjusted to resonance at appropriate frequencies ω_1 and ω_2 by adjustment of C_1 and C_2, respectively. Resistors R_1 and R_2 control the bandwidth of each circuit and therefore the discriminator bandwidth.

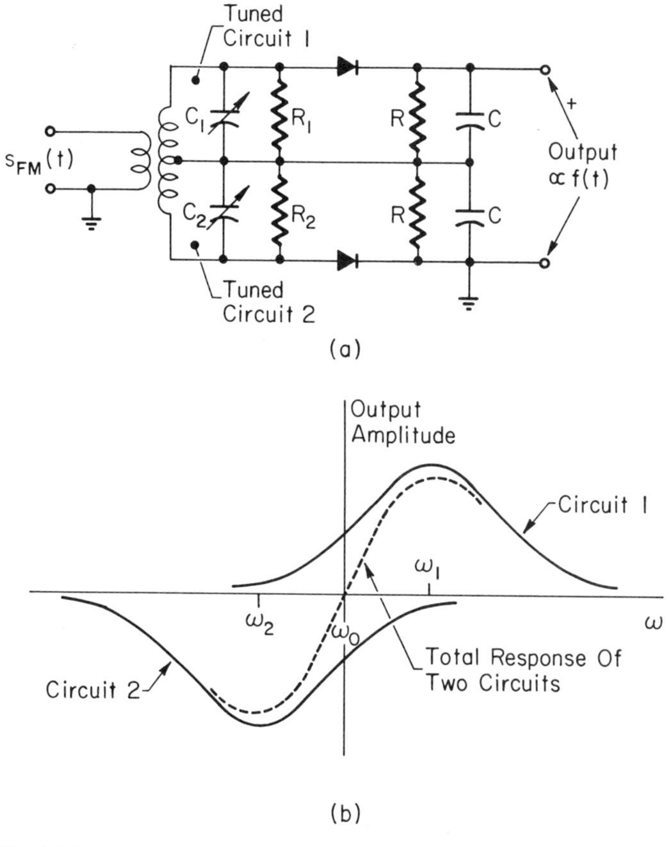

Fig. 6.8-2. Travis discriminator. (a) Circuit and (b) characteristic response.

The Foster-Seeley discriminator (after the inventors) shown in Fig. 6.8-3(*a*) is an improvement of the Travis discriminator. All tuned circuits are resonant at the "crossover" or carrier frequency ω_0 of the FM wave which makes alignment simple. Operation is explained with the aid of the vector plots of (*b*), (*c*), and (*d*). For frequencies near ω_0 the capacitor C_0 is a coupling capacitor (short) while the choke is an effective open circuit. Hence, the voltages applied to diodes Q_1 and Q_2 are $E_A = E_1 + E_2$ and $E_B = E_1 - E_2$, respectively. At resonance where $\omega = \omega_0$, E_2 will be 90 degrees out of phase with E_1 owing to transformer action, and the resultant vector diagram of (*b*) applies. The output due to Q_1 is proportional to $|E_A|$ and that of Q_2 to $-|E_B|$. The total output is $|E_A| - |E_B|$, which is zero for $\omega = \omega_0$. When $\omega > \omega_0$, E_2 shifts in phase by an additional amount $\delta\theta$ relative to E_1, and the diagram of (*c*)

ISBN 0-201-05758-1

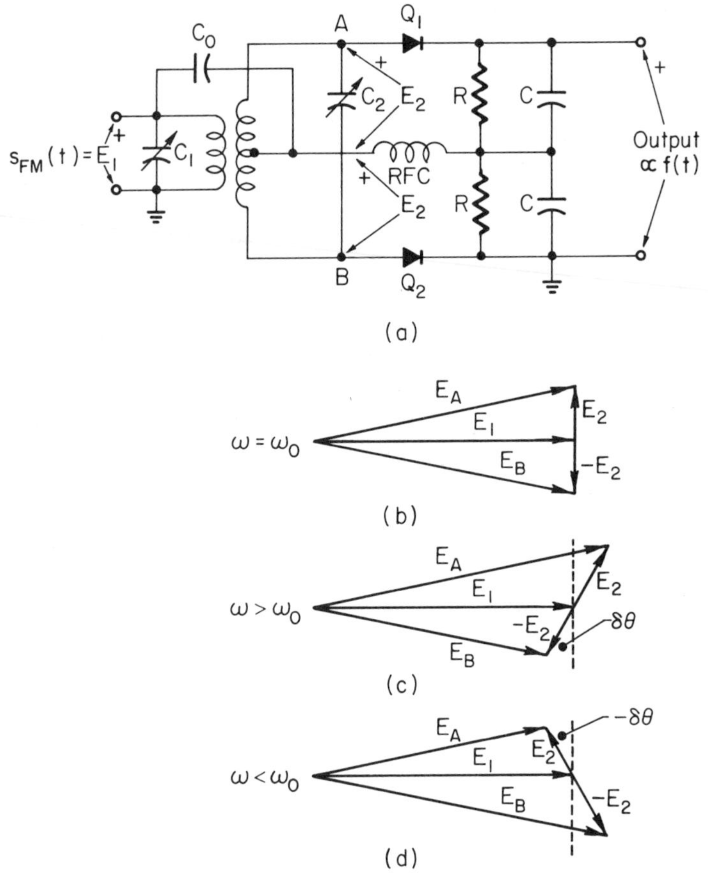

Fig. 6.8-3. Foster-Seeley discriminator. (a) Circuit and (b) – (d) applicable phasor diagrams.

applies. Now the output becomes positive, since $|E_A| - |E_B| > 0$. When $\omega < \omega_0$, a phase shift of $-\delta\theta$ causes $|E_A| - |E_B| < 0$ to give a negative output. The overall response of the Foster-Seeley discriminator is similar to Fig. 6.8-2(*b*) but has better linearity than the Travis device. The useful linear range of frequencies corresponds approximately to the bandwidth of the transformer.

All the above discriminators are sensitive to input level changes, and it is customary to precede the circuit by a *"hard" amplitude limiter* to keep the drive level constant.* A "hard" limiter is one that removes both slow and fast amplitude changes.

The Foster-Seeley circuit can be modified as illustrated in Fig. 6.8-4 so as to produce limiting of fast amplitude changes. It is known as a *ratio detector*. For constant amplitude but varying frequency it is easy to establish that the output is now $(|E_A| - |E_B|)/2$ and normal discriminator action results. However, for fixed frequency but amplitude varying, a large capacitor C_h is present which effectively absorbs the amplitude changes as long as they are not very slow changes. Thus, the addition of C_h causes the circuit to have self-limiting behavior for all but slow amplitude variations. A good discussion of the action of C_h on limiting is given by Kennedy [8].

*A constant level is desirable to keep amplitude variations caused by noise, interference, propagation path fading, and other effects from appearing in the output. That is, we simply want the output to be related to the frequency and not the amplitude of the input signal.

ISBN 0-201-05758-1

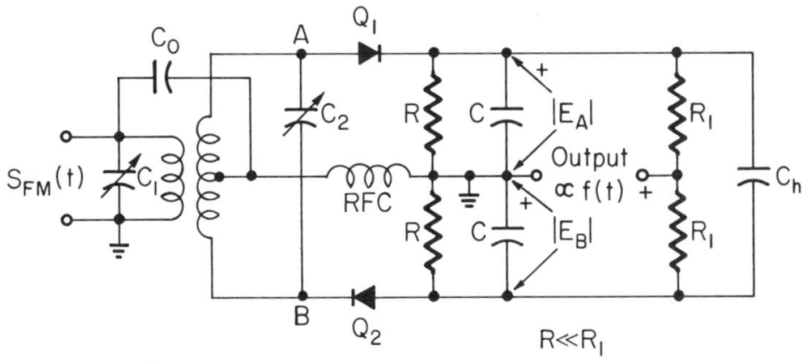

Fig. 6.8-4. Ratio detector circuit.

The ratio detector may still require a limiter preceding it to remove slow amplitude variations. A common means of accomplishing the slow limiting is to use *automatic gain control* (AGC) where the gain of an amplifier is regulated by the long-term average signal level.

The Discriminator Viewed as a Differentiator

A discriminator may be viewed as a differentiator followed by an envelope detector as depicted in Fig. 6.8-5. To show this fact we differentiate the general FM waveform (6.1-8) to obtain

$$\frac{ds_{FM}(t)}{dt} = \frac{dA}{dt} \cos\left[\omega_0 t + \theta_0 + k_{FM}\int f(t)\,dt\right]$$

$$-A\left[\omega_0 + k_{FM}f(t)\right] \sin\left[\omega_0 t + \theta_0 + k_{FM}\int f(t)\,dt\right]. \tag{6.8-1}$$

If amplitude A is constant, as it would be when a limiter precedes the discriminator, we have

$$\frac{ds_{FM}(t)}{dt} = -A\left[\omega_0 + k_{FM}f(t)\right] \sin\left[\omega_0 t + \theta_0 + k_{FM}\int f(t)\,dt\right], \tag{6.8-2}$$

which is similar to a standard AM signal with small deviation ratio. The deviation ratio is small, since usually $k_{FM}|f(t)|_{max} \ll \omega_0$. The response of an envelope detector becomes

$$\text{detector response} = A\left[\omega_0 + k_{FM}f(t)\right]. \tag{6.8-3}$$

The dc term $A\omega_0$ can be removed by a blocking capacitor leaving an output:

$$\text{output} = Ak_{FM}f(t). \tag{6.8-4}$$

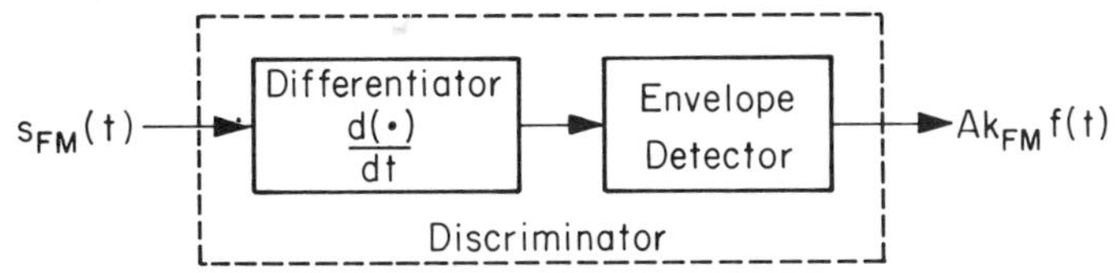

Fig. 6.8-5. Differentiator and envelope detector equivalent of a discriminator.

ISBN 0-201-05758-1

Discriminator for Phase Demodulation

A wideband direct phase demodulator is not possible, and demodulation must be accomplished using a frequency discriminator. The output of the discriminator will be proportional to $df(t)/dt$ for an input PM signal. Hence, an integrator must be provided following the discriminator. Figure 6.8-6 illustrates the general wideband phase demodulator.

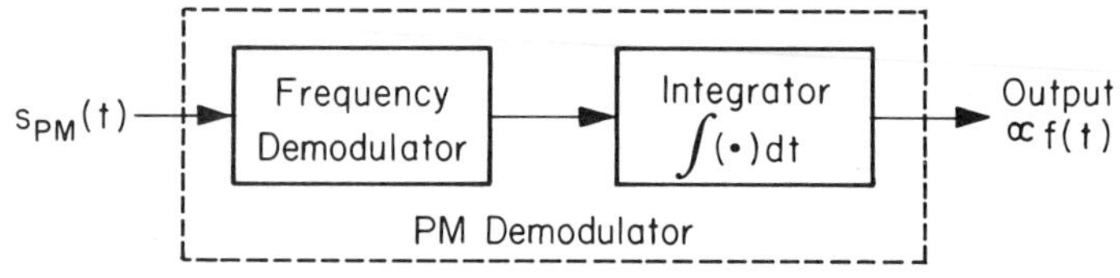

Fig. 6.8-6. General wideband PM demodulator.

Locked-Loop Demodulators

When input signal level is small and noise is an important consideration, FM demodulation performance can be improved, as compared to using a discriminator, by implementing the demodulator in the form of a circuit having a feedback loop. Such a circuit is called a *locked-loop demodulator,* and we describe two forms. In the *phase-locked loop* (PLL) demodulator, illustrated in Fig. 6.8-7, the phase of a VCO output signal is forced to follow (lock to) the phase of the input FM waveform with small error. A second form of loop forces the frequency of a VCO to follow the instantaneous frequency of the FM signal with small error. We shall give only very simplified discussions of these two types of loops. For additional detail the reader is referred to the literature [9]. We begin with the phase-locked loop.

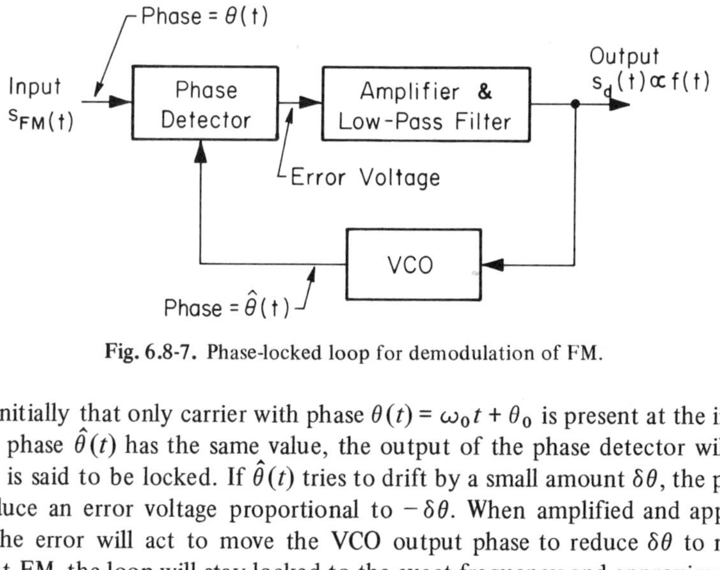

Fig. 6.8-7. Phase-locked loop for demodulation of FM.

Assume initially that only carrier with phase $\theta(t) = \omega_0 t + \theta_0$ is present at the input. If the VCO output phase $\hat{\theta}(t)$ has the same value, the output of the phase detector will be zero,* and the loop is said to be locked. If $\hat{\theta}(t)$ tries to drift by a small amount $\delta\theta$, the phase detector will produce an error voltage proportional to $-\delta\theta$. When amplified and applied to the VCO input the error will act to move the VCO output phase to reduce $\delta\theta$ to nearly zero. Thus, without FM, the loop will stay locked to the exact frequency and approximate phase of the carrier signal.

*A phase detector produces a voltage proportional to the sine of the difference in phase between the two inputs.

ISBN 0-201-05758-1

When modulation is present the loop's action is similar. The VCO phase $\hat{\theta}(t)$ will now follow the *instantaneous* phase $\theta(t)$ of the input signal with only a small phase error. The amplified error voltage appearing at the VCO input will be proportional to $f(t)$, since the VCO acts as an integrator. This fact may also be reasoned as follows. Since FM phase variations in $\theta(t)$ are proportional to $\int f(t)\,dt$ and $\hat{\theta}(t) \approx \theta(t)$, then $\hat{\theta}(t)$ will have phase variations approximately proportional to $\int f(t)\,dt$. However, for a VCO, output *frequency* variations are proportional to input voltage so that, to produce the required *phase* variations in $\hat{\theta}(t)$, the VCO input voltage must be proportional to $f(t)$, since phase is the integral of frequency.

The filter is selected to give the desired closed-loop bandwidth, which is often chosen as narrow as possible to reject noise but broad enough to give demodulation with minor distortion of $f(t)$.

A phase-locked loop may be used to demodulate a PM signal as described earlier. Since it behaves as an FM discriminator, all that is needed is to insert an integrator at the PLL output.

Finally, we conclude this section with a brief discussion of a *frequency-locked loop* demodulator which uses a discriminator. It is shown in Fig. 6.8-8 and is often called an *FM demodulator with feedback* (FMFB) [10]. The key to loop operation is to note that the VCO output frequency is a linear function of the discriminator output voltage $s_d(t)$. If the nominal VCO output frequency is $(\omega_0 - \omega_{IF})$—that is, lower than the input carrier frequency by an amount ω_{IF}—then the output frequency is $\omega_{VCO} = (\omega_0 - \omega_{IF}) + K_{VCO}s_d(t)$, where K_{VCO} is a constant called *VCO gain.* The unit of K_{VCO} is radians per second per volt. Integration gives output phase. The output signal is then

$$s_{VCO}(t) = A_{VCO} \cos \left[(\omega_0 - \omega_{IF})t + \theta_{VCO} + K_{VCO}\int s_d(t)\,dt\right], \qquad (6.8\text{-}5)$$

where A_{VCO} and θ_{VCO} are constants. This signal and the input

$$s_{FM}(t) = A \cos \left[\omega_0 t + \theta_0 + k_{FM}\int f(t)\,dt\right] \qquad (6.8\text{-}6)$$

are applied to the mixer which is basically a product device.

If the *mixer "gain"* is K_M and that of the band-pass filter centered at frequency ω_{IF} is K_F, the *filtered* output to the limiter input is

$$s_L(t) = \frac{AA_{VCO}K_MK_F}{2} \cos\left[\omega_{IF}t + (\theta_0 - \theta_{VCO}) + k_{FM}\int f(t)\,dt - K_{VCO}\int s_d(t)\,dt\right]. \qquad (6.8\text{-}7)$$

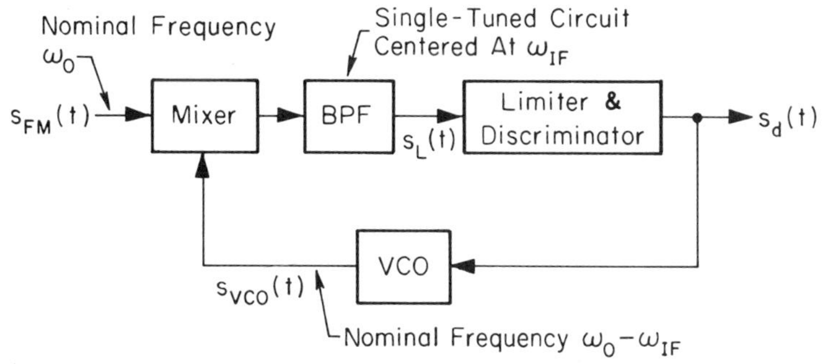

Fig. 6.8-8. FM demodulator with feedback.

ISBN 0-201-05758-1

Finally, the limiter-discriminator output will be proportional to the instantaneous frequency deviations about $\omega_{\rm IF}$ giving

$$s_d(t) = [k_{\rm FM} f(t) - K_{\rm VCO} s_d(t)] K_{\rm D}, \tag{6.8-8}$$

where $K_{\rm D}$ is the *discriminator gain* (volts per radian per second). Solving for the demodulated output $s_d(t)$ we obtain

$$s_d(t) = \frac{K_{\rm D} k_{\rm FM}}{1 + K_{\rm D} K_{\rm VCO}} f(t), \tag{6.8-9}$$

which shows that the control voltage applied to the VCO is proportional to $f(t)$ and the whole circuit is an FM demodulator.

The numerator of (6.8-9) is the output which would occur using a limiter-discriminator alone. Because $K_{\rm D} K_{\rm VCO} > 0$, the output of the FMFB demodulator is less than that of a discriminator alone.

The above brief analysis has neglected the detailed characteristics of the band-pass filter at the output of the mixer. By using (6.8-9) with (6.8-7) we may show that the instantaneous frequency at the mixer output is $\omega_{\rm IF} + [k_{\rm FM} f(t)/(1 + K_{\rm D} K_{\rm VCO})]$. This result demonstrates that the instantaneous frequency deviations in the FM signal are reduced by a factor $1/(1 + K_{\rm D} K_{\rm VCO})$. Thus, the band-pass filter may have a bandwidth smaller than that required at the carrier frequency by about the same factor. The filter is shown as a single-tuned filter because this simple filter gives a stable loop whereas more complicated filters may not [10].

6.9. NOISE PERFORMANCE OF NARROWBAND ANGLE MODULATION

Since narrowband angle modulation finds limited use, we only briefly look at this topic. One of its applications is in amateur or "ham" radio.

NBPM Performance

The NBPM receiver was shown in Fig. 6.2.2(*a*). As with any PM or FM waveform, the power at the receiver input is

$$S_i = A^2/2. \tag{6.9-1}$$

Input noise is band-pass and has spectral components centered about frequencies $+\omega_0$ and $-\omega_0$. If its power density spectrum is defined as $\mathcal{S}_{n_i}(\omega)$, we easily find the input noise power:

$$N_i = \frac{1}{\pi} \int_{\omega_0 - (W_{\rm PM}/2)}^{\omega_0 + (W_{\rm PM}/2)} \mathcal{S}_{n_i}(\omega)\, d\omega. \tag{6.9-2}$$

As usual, $W_{\rm PM}$ is the signal bandwidth, and limits on the integral result from the action of the (ideal) band-pass filter. For white noise with power density $\mathcal{N}_0/2$ we have

$$N_i = \frac{\mathcal{N}_0 W_{\rm PM}}{2\pi} = \frac{\mathcal{N}_0 W_f}{\pi}, \tag{6.9-3}$$

ISBN 0-201-05758-1

since $W_{PM} = 2W_f$ for NBPM. The input signal-to-noise ratio easily follows:

$$\left(\frac{S_i}{N_i}\right)_{NBPM} = \frac{A^2 \pi}{2\mathcal{N}_0 W_f}. \tag{6.9-4}$$

The output signal from Fig. 6.2-2(a) is $Ak_{PM}f(t)/2$. Its average power is

$$S_o = A^2 k_{PM}{}^2 \overline{f^2(t)}/4. \tag{6.9-5}$$

To obtain the average output noise power we draw on a result from Chapter 4 which gives the low-frequency noise power density spectrum at the output of a product device, such as that illustrated in Fig. 6.2-2(a), as

$$\frac{1}{4}[\mathcal{S}_{n_i}(\omega + \omega_0) + \mathcal{S}_{n_i}(\omega - \omega_0)], \qquad |\omega| < W_{PM}/2. \tag{6.9-6}$$

The low-pass filter bandwidth will be W_f. Using this fact, and integrating (6.9-6) for the case of white noise, results in

$$N_o = \mathcal{N}_0 W_f/(4\pi). \tag{6.9-7}$$

Performance is described by the output signal-to-noise power ratio:

$$\left(\frac{S_o}{N_o}\right)_{NBPM} = \frac{A^2 \pi k_{PM}{}^2 \overline{f^2(t)}}{\mathcal{N}_0 W_f} = 2(\Delta\theta)^2 \frac{\overline{f^2(t)}}{|f(t)|_{max}{}^2} \left(\frac{S_i}{N_i}\right)_{NBPM}, \tag{6.9-8}$$

where $\Delta\theta = k_{PM}|f(t)|_{max}$ is the peak phase deviation. We may put this expression in various forms. A useful one is to substitute (5.8-7) and develop a comparison to baseband transmission on the basis of the same transmitted signal power in each case.

$$\left(\frac{S_o}{N_o}\right)_{NBPM} = k_{PM}{}^2 \overline{f^2(t)} \left(\frac{S_o}{N_o}\right)_B. \tag{6.9-9}$$

Next, we recognize that $k_{PM}{}^2\overline{f^2(t)}$ is the mean-squared phase deviation, which we may define as $(\Delta\theta_{rms})^2$. The final form is

$$\left(\frac{S_o}{N_o}\right)_{NBPM} = (\Delta\theta_{rms})^2 \left(\frac{S_o}{N_o}\right)_B \ll 0.27 \left(\frac{S_o}{N_o}\right)_B \tag{6.9-10}$$

on using (6.2-6). This last expression demonstrates that NBPM is much poorer than simple baseband transmission because of its lower signal-to-noise ratio.

NBFM Performance

Most of the analysis of Fig. 6.2-2(b) parallels that for NBPM. In fact, for white noise, input signal and noise powers are again given by (6.9-1) and (6.9-3), respectively. Input signal-to-noise ratio is then the same as for NBPM:

ISBN 0-201-05758-1

$$\left(\frac{S_i}{N_i}\right)_{\text{NBFM}} = \frac{A^2\pi}{2\mathcal{N}_0 W_f}. \tag{6.9-11}$$

Output signal power is easily found to be

$$S_o = A^2 k_{\text{FM}}{}^2 \overline{f^2(t)}/4. \tag{6.9-12}$$

Analysis to find output noise power is identical to that for NBPM up to the output of the product device, which is still given by (6.9-6). However, we now have to consider the differentiation. For it we have $|H(\omega)|^2 = \omega^2$, and by applying (4.10-11) the output noise power density spectrum is

$$\frac{1}{4}\left[\mathcal{S}_{n_i}(\omega + \omega_0) + \mathcal{S}_{n_i}(\omega - \omega_0)\right]\omega^2, \qquad |\omega| < W_f. \tag{6.9-13}$$

On assuming white input noise, integration of this expression gives the output noise power:

$$N_o = \mathcal{N}_0 W_f{}^3/(12\pi). \tag{6.9-14}$$

Output signal-to-noise power ratio is found from (6.9-12) and (6.9-14). It is

$$\left(\frac{S_o}{N_o}\right)_{\text{NBFM}} = \frac{3A^2\pi k_{\text{FM}}{}^2\overline{f^2(t)}}{\mathcal{N}_0 W_f{}^3} = 6\left(\frac{\Delta\omega}{W_f}\right)^2 \frac{\overline{f^2(t)}}{|f(t)|_{\max}{}^2}\left(\frac{S_i}{N_i}\right)_{\text{NBFM}}, \tag{6.9-15}$$

or, on comparing with the output of a baseband system having the same transmitted signal power,

$$\left(\frac{S_o}{N_o}\right)_{\text{NBFM}} = \frac{3k_{\text{FM}}{}^2\overline{f^2(t)}}{W_f{}^2}\left(\frac{S_o}{N_o}\right)_{\text{B}}. \tag{6.9-16}$$

Now recognizing that $k_{\text{FM}}{}^2\,\overline{f^2(t)}$ is the mean-squared frequency deviation, which we define as $(\Delta\omega_{\text{rms}})^2$, we have

$$\left(\frac{S_o}{N_o}\right)_{\text{NBFM}} = 3\left(\frac{\Delta\omega_{\text{rms}}}{W_f}\right)^2\left(\frac{S_o}{N_o}\right)_{\text{B}} < 3\left(\frac{S_o}{N_o}\right)_{\text{B}}, \tag{6.9-17}$$

since $\Delta\omega_{\text{rms}} < W_f$.

Our analysis indicates that NBFM is capable of giving only slightly better performance than baseband, if any, and never better than 3 or 4.8 dB. For many signals $\Delta\omega_{\text{rms}}/W_f < 1/\sqrt{3}$, and no improvement results.

A comparison of NBFM with NBPM through the above expressions for output signal-to-noise ratio indicates the superiority of FM.

ISBN 0-201-05758-1

6.10. NOISE PERFORMANCE OF WIDEBAND FM

Noise performance of narrowband angle-modulated systems was briefly introduced in the last section. We now take up the more interesting case of wideband FM (WBFM). The ratio of output average signal power to output average noise power will continue to be the measure of performance.

The applicable receiver block diagram is given in Fig. 6.10-1. A discriminator is assumed for the demodulator. However, this assumption is for simplicity of discussion. Locked-loop demodulators will give the same performance for a large input signal-to-noise ratio, which is the only case of interest here. When input signal-to-noise ratio becomes small, a threshold effect occurs; we reserve discussion of this phenomenon until Section 6.12.

The band-pass filter in Fig. 6.10-1 is assumed to be ideal with a bandwidth W_{FM} centered at the carrier frequency ω_0. Of course, for WBFM, $W_{\text{FM}} \approx 2\Delta\omega$, approximately twice the peak frequency deviation. The low-pass filter at the discriminator output is assumed to be ideal and has a bandwidth equal to the spectral extent W_f of the modulating signal $f(t)$. Both filters are then chosen just wide enough to pass the signals of interest but not so wide as to pass excessive noise.

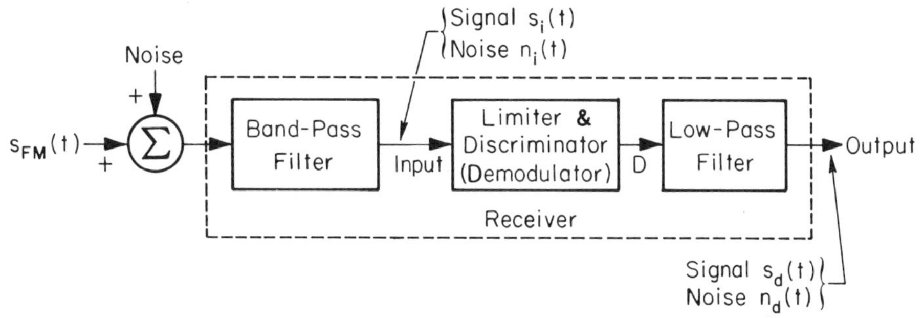

Fig. 6.10-1. Block diagram of an FM receiver which uses a discriminator.

Signal and Noise Powers

Consider the input to the limiter-discriminator first. The input signal power is

$$S_i = A^2/2. \tag{6.10-1}$$

If $\mathscr{S}_{n_i}(\omega)$ is defined as the power density spectrum of the input noise $n_i(t)$, the input average noise power is

$$N_i = \frac{1}{\pi} \int_{\omega_0 - \Delta\omega}^{\omega_0 + \Delta\omega} \mathscr{S}_{n_i}(\omega)\, d\omega. \tag{6.10-2}$$

For white noise at the input to the band-pass filter having power density $\mathscr{N}_0/2$, this expression reduces to

$$N_i = \mathscr{N}_0\, \Delta\omega/\pi. \tag{6.10-3}$$

ISBN 0-201-05758-1

Care must be exercised in solving for the output signal and noise powers because, in general, the process of FM is nonlinear and output noise power is not independent of the presence of the FM signal. Fortunately, reasonable arguments indicate [11, 12] that, for wideband FM with large input signal-to-noise ratio, the output powers may be found independently. The procedure to find output average signal power is to assume noise is zero while the procedure for determining output average noise power is to assume the modulating signal $f(t)$ is zero. The latter procedure retains a nonzero carrier, however.

First assume input noise is zero. The discriminator output voltage will be proportional to instantaneous frequency deviations from the carrier value. If the discriminator constant of proportionality (sometimes called "gain") is K_D, the output signal is $s_d(t) = K_D k_{FM} f(t)$, and the output signal power is

$$S_o = \overline{s_d^{\,2}(t)} = K_D^{\,2} k_{FM}^{\,2}\, \overline{f^2(t)}. \tag{6.10-4}$$

Turning to the more significant task of finding output noise power, we begin by letting $f(t) = 0$. The signal at the demodulator input is simply $A \cos(\omega_0 t + \theta_0)$, while the noise is band-pass. Recall from Chapter 5 that band-pass noise may be expressed as

$$n_i(t) = n_c(t) \cos(\omega_0 t + \theta_0) - n_s(t) \sin(\omega_0 t + \theta_0), \tag{6.10-5}$$

where $n_c(t)$ and $n_s(t)$ are low-pass noises having identical power density spectra and therefore equal powers. Their powers were also equal to that of $n_i(t)$—that is, $\overline{n_i^{\,2}(t)} = \overline{n_c^{\,2}(t)} = \overline{n_s^{\,2}(t)}$. The total demodulator input waveform becomes

$$\begin{aligned} A \cos(\omega_0 t + \theta_0) + n_i(t) &= [A + n_c(t)] \cos(\omega_0 t + \theta_0) - n_s(t) \sin(\omega_0 t + \theta_0) \\ &= A(t) \cos[\omega_0 t + \theta_0 + \psi(t)], \end{aligned} \tag{6.10-6}$$

where

$$A(t) = \left\{ [A + n_c(t)]^2 + n_s^{\,2}(t) \right\}^{1/2}, \tag{6.10-7}$$

$$\psi(t) = \tan^{-1}\left(\frac{n_s(t)}{A + n_c(t)}\right). \tag{6.10-8}$$

Since we must necessarily assume a large input signal-to-noise ratio, we may make the assumption that $|n_c(t)|$ and $|n_s(t)|$ are nearly always small in relation to A. The approximation $\tan^{-1}(x) \approx x$ for small x may be used in (6.10-8). The phase deviation becomes approximately

$$\psi(t) = n_s(t)/A. \tag{6.10-9}$$

Because the discriminator responds to frequency departures from ω_0 rather than phase, its output will be proportional to the time derivative of $\psi(t)$ rather than to $\psi(t)$ directly. Thus, the output noise of the discriminator is

$$\frac{K_D}{A} \frac{dn_s(t)}{dt}. \tag{6.10-10}$$

To find output average noise power we seek the power density spectrum of this noise waveform. This will be $(K_D/A)^2$ times the power density spectrum of $dn_s(t)/dt$. To obtain the

ISBN 0-201-05758-1

latter it is easiest to think of $n_s(t)$ as being applied to a differentiator having a transfer function $H(\omega) = j\omega$. If $\mathscr{S}_{n_s}(\omega)$ is the power density spectrum of $n_s(t)$, that of the overall output noise $n_d(t)$ becomes

$$\mathscr{S}_{n_d}(\omega) = \left(\frac{K_D}{A}\right)^2 \omega^2 \mathscr{S}_{n_s}(\omega), \qquad |\omega| < W_f \qquad (6.10\text{-}11)$$
$$= 0, \qquad |\omega| > W_f.$$

The restriction $|\omega| < W_f$ arises from the action of the low-pass filter in Fig. 6.10-1.

Finally, we recall from (4.11-15) that

$$\mathscr{S}_{n_s}(\omega) = \mathscr{S}_{n_i}(\omega + \omega_0) + \mathscr{S}_{n_i}(\omega - \omega_0), \qquad |\omega| < \Delta\omega$$
$$= 0, \qquad |\omega| > \Delta\omega \qquad (6.10\text{-}12)$$

in present notation. Figure 6.10-2 is helpful in visualizing this power density spectrum for the case of white noise at the input. The corresponding output spectrum (6.10-11) is sketched in Fig. 6.10-3. The parabolic shape of the output power spectrum is an important characteristic of FM which is exploited later to improve performance. Output average noise power results from integration of (6.10-11), using (6.10-12), over limits $-W_f$ to W_f which are the result of action of the low-pass filter. Thus,

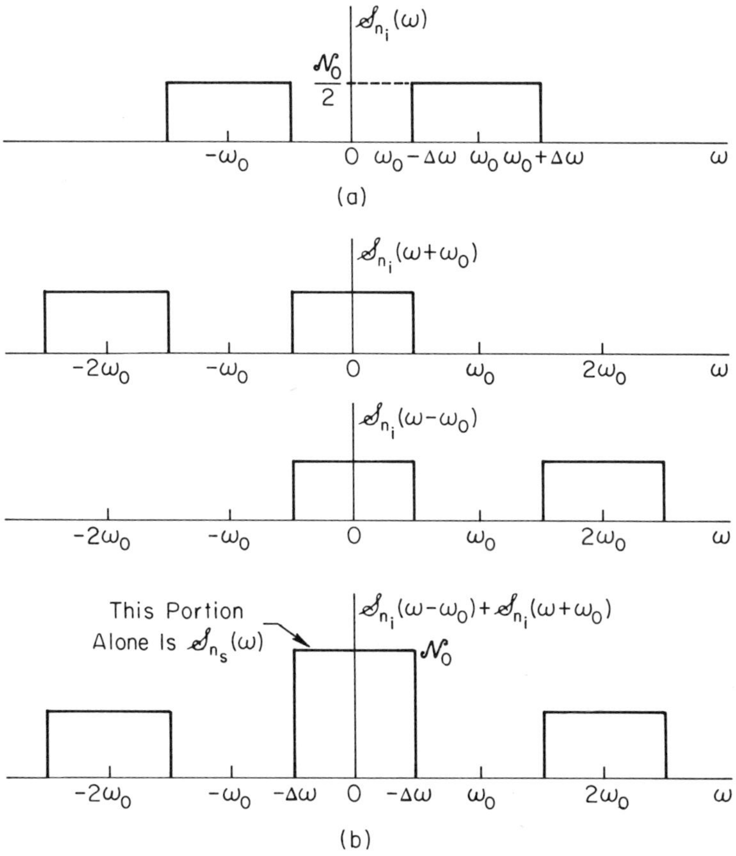

Fig. 6.10-2. Bandlimited white noise spectra. (a) Band-pass noise power density spectrum and (b) power density spectra associated with $\mathscr{S}_{n_s}(\omega)$.

ISBN 0-201-05758-1

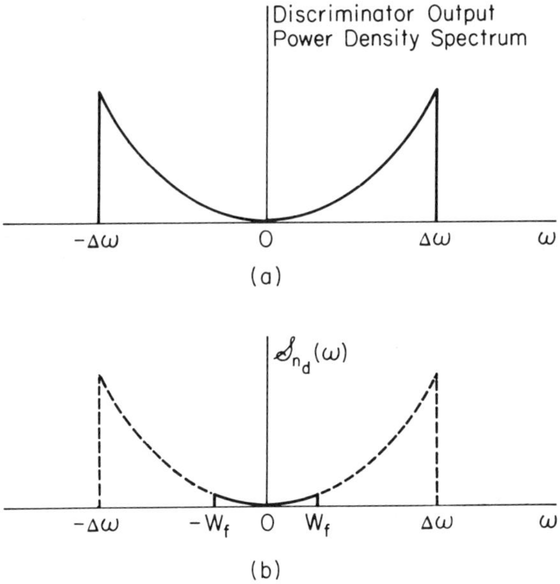

Fig. 6.10-3. FM power density spectra associated with white receiver input noise. (a) At the discriminator output, and (b) at the low-pass filter output.

$$N_o = \overline{n_d{}^2(t)} = \left(\frac{K_D}{A}\right)^2 \frac{1}{2\pi} \int_{-W_f}^{W_f} \omega^2 [\mathscr{S}_{n_i}(\omega + \omega_0) + \mathscr{S}_{n_i}(\omega - \omega_0)]\, d\omega. \tag{6.10-13}$$

For white input noise this expression reduces to

$$N_o = \frac{K_D{}^2 \mathscr{N}_0 W_f{}^3}{3\pi A^2}. \tag{6.10-14}$$

S/N Ratio–Performance

The analysis is now almost completed. Important results are (6.10-1), (6.10-2), (6.10-4), and (6.10-13). From these expressions signal-to-noise power ratios may be written for input $(S_i/N_i)_{FM}$ and output $(S_o/N_o)_{FM}$ in terms of an arbitrary information signal $f(t)$ and arbitrary noise power density spectrum. However, we shall state only ratios for the important white noise case. They are

$$\left(\frac{S_i}{N_i}\right)_{FM} = \frac{A^2\pi}{2\mathscr{N}_0\, \Delta\omega} = \frac{A^2\pi}{2\mathscr{N}_0\, W_f}\left(\frac{W_f}{\Delta\omega}\right), \tag{6.10-15}$$

$$\left(\frac{S_o}{N_o}\right)_{FM} = \frac{3A^2\pi k_{FM}{}^2\, \overline{f^2(t)}}{\mathscr{N}_0 W_f{}^3} = \frac{3A^2\pi}{\mathscr{N}_0 W_f}\left(\frac{\Delta\omega}{W_f}\right)^2 \frac{\overline{f^2(t)}}{|f(t)|_{max}{}^2}. \tag{6.10-16}$$

Other forms of (6.10-16) are useful. In terms of a baseband system having the same transmitted signal power as the FM system,

ISBN 0-201-05758-1

$$\left(\frac{S_o}{N_o}\right)_{\text{FM}} = 3\left(\frac{\Delta\omega}{W_f}\right)^2 \frac{\overline{f^2(t)}}{|f(t)|_{\max}^2} \left(\frac{S_o}{N_o}\right)_{\text{B}} . \tag{6.10-17}$$

We may also obtain

$$\left(\frac{S_o}{N_o}\right)_{\text{FM}} = 6\left(\frac{\Delta\omega}{W_f}\right)^3 \frac{\overline{f^2(t)}}{|f(t)|_{\max}^2} \left(\frac{S_i}{N_i}\right)_{\text{FM}} . \tag{6.10-18}$$

Let us pause and examine our results. Several conclusions may be drawn from these white noise expressions which are not as obvious from more generalized results. We have already defined $D_{\text{FM}} = \Delta\omega/W_f$ as the deviation ratio (it equals modulation index in sinusoidal FM). From (6.10-15) and (6.10-16) we see that, if D_{FM} increases by increasing $\Delta\omega$, the output signal-to-noise ratio increases as D_{FM}^2, while the input *decreases* as $1/D_{\text{FM}}$. In other words, the output increases faster than the input decreases, and a net improvement is possible by increasing bandwidth through D_{FM}. There are limitations to the possible net improvement, however, since a point will eventually be reached where the input signal-to-noise ratio is no longer large as was necessary for the foregoing analysis to be valid. The point is called the FM *threshold*. It is discussed further in Section 6.12.

Another important conclusion derives from (6.10-17). Since $\Delta\omega/W_f >> 1$ is easily possible, WBFM is capable of a much larger output signal-to-noise ratio than in simple baseband transmission. For example, with sinusoidal modulation (6.10-17) becomes

$$\left(\frac{S_o}{N_o}\right)_{\text{FM}} = \frac{3}{2}\,\beta_{\text{FM}}^2 \left(\frac{S_o}{N_o}\right)_{\text{B}} , \tag{6.10-19}$$

which is plotted in Fig. 6.10-4 for the range of values of $(S_o/N_o)_{\text{B}}$ which lie above the threshold points. Curves for several values of β_{FM} are given.

The ratio of output to input signal-to-noise ratios is called *demodulator gain*. From (6.10-18), gain is seen to increase proportional to the cube of $\Delta\omega/W_f$. For sinusoidal modulation the gain is

$$\frac{(S_o/N_o)_{\text{FM}}}{(S_i/N_i)_{\text{FM}}} = 3\beta_{\text{FM}}^3 . \tag{6.10-20}$$

Other conclusions will result from comparing FM output signal-to-noise ratio to that of PM in the next section.

Example 6.10-1

An FM receiver has white noise at its input. If the information signal is known to have a maximum magnitude of four times its root-mean-squared (rms) value, what deviation ratio is required to produce a demodulator gain of 24?

Here $\overline{f^2(t)}\,/\,|f(t)|_{\max}^2 = 1/16$. Using (6.10-18), the deviation ratio required is

$$D_{\text{FM}} = \Delta\omega/W_f = \sqrt[3]{24(16)/6} = 4.$$

ISBN 0-201-05758-1

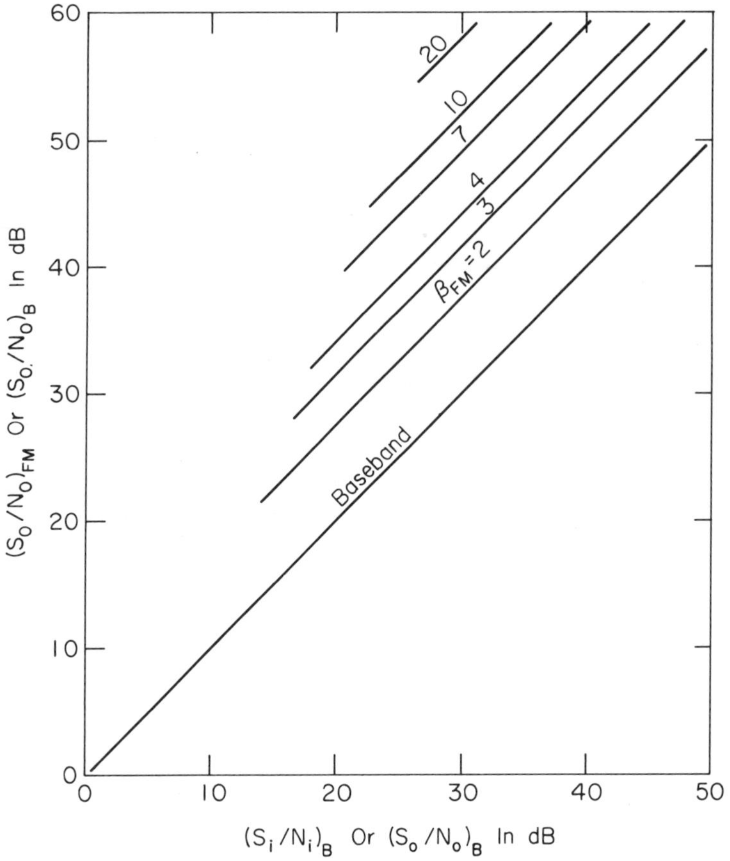

Fig. 6.10-4. FM and baseband system output signal-to-noise ratios as a function of baseband signal-to-noise ratios with FM modulation index as a parameter.

Performance Comparison with AM

We have already developed a comparison of FM performance with baseband transmission through (6.10-17) and shown FM to be far superior. In Chapter 5 similar comparisons of the various AM methods with baseband transmission, such as (5.8-35) and (5.8-36), showed that output signal-to-noise ratios for all AM systems never exceed those of baseband. Thus, FM performance may be considerably better than that of any AM system.

Performance Improvement by Emphasis

In the previous analysis of the noise response of the FM demodulator the output noise power density spectrum (6.10-11) was found to contain a factor ω^2 which was not present in the output signal power density spectrum. This factor places an emphasis on high-frequency output noise. It would seem that a *deemphasis filter* could be placed in the output having a high-frequency roll-off of $1/\omega^2$ to reduce this effect and to reduce the total output noise power.* Indeed, this is the case. However, we note that, if nothing else were done, the deemphasis filter would also distort the demodulated message. To remedy such distortion, we may purposely distort the modulating signal back at the transmitter to counteract the effect of the deemphasis filter on the signal.

*The filter is not restricted to a $1/\omega^2$ characteristic in general.

ISBN 0-201-05758-1

The transmitter filter, called a ***preemphasis filter,*** will necessarily have a transfer function $H_p(\omega)$ which is the reciprocal of the receiver deemphasis filter transfer function $H_d(\omega)$. That is,

$$H_p(\omega) = 1/H_d(\omega), \tag{6.10-21}$$

so that there is no overall message distortion.

Figure 6.10-5(*a*) illustrates the placement of pre- and deemphasis networks in the FM system. The gain constant K is unimportant to the present discussion but enters into subsequent

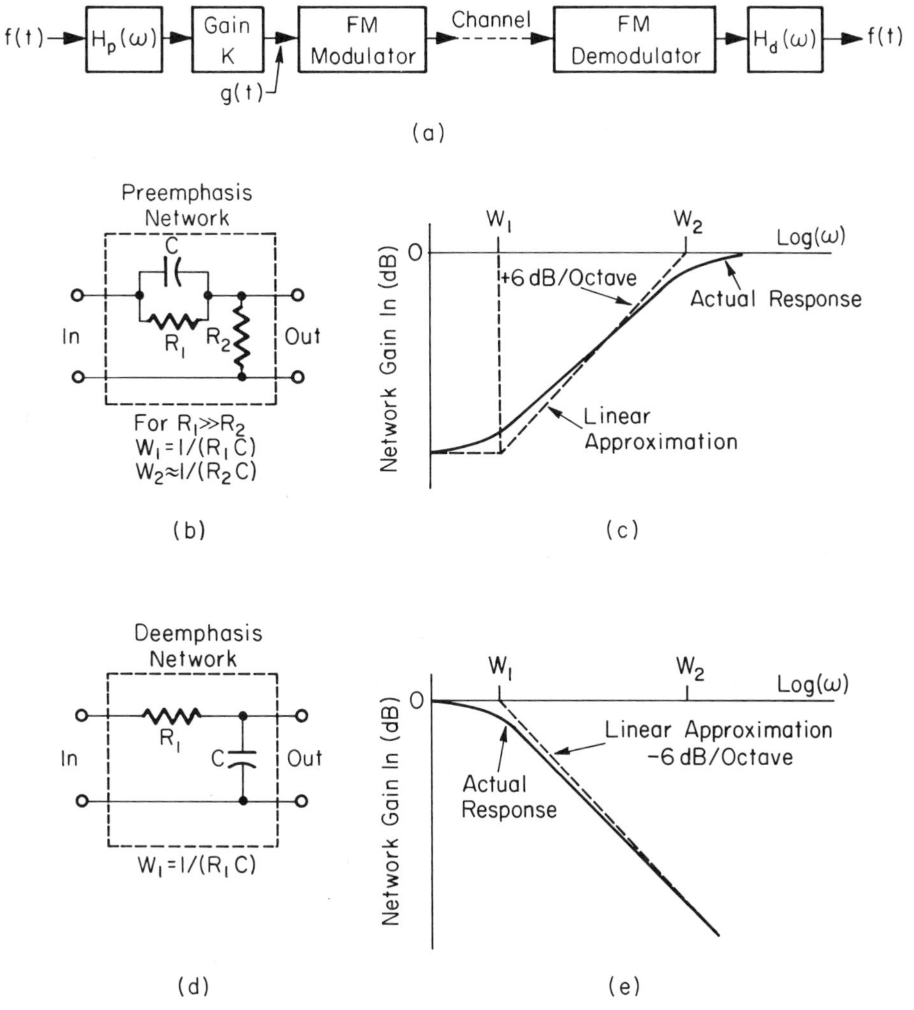

Fig. 6.10-5. (a) Placement of emphasis filters in an FM system. (b) Typical preemphasis network and its response (c) for broadcast FM. (d) Corresponding deemphasis network and its response (e) for broadcast FM.

ISBN 0-201-05758-1

work. If the output signal is unaffected by the emphasis filters, the gain in output signal-to-noise ratio will simply be the ratio of noise power without emphasis to that with emphasis. Applying (6.10-11), this ratio, denoted R_{FM}, becomes

$$R_{\text{FM}} = \frac{\displaystyle\int_{-W_f}^{W_f} \omega^2 \mathcal{S}_{n_s}(\omega)\, d\omega}{\displaystyle\int_{-W_f}^{W_f} \omega^2 \mathcal{S}_{n_s}(\omega)\, |H_d(\omega)|^2\, d\omega}. \qquad (6.10\text{-}22)$$

For white noise $\mathcal{S}_{n_s}(\omega) = \mathcal{N}_0$ within the region $|\omega| < W_f$, and R_{FM} reduces to

$$R_{\text{FM}} = \frac{2W_f^{\,3}}{3\displaystyle\int_{-W_f}^{W_f} \omega^2 |H_d(\omega)|^2\, d\omega}. \qquad (6.10\text{-}23)$$

Numerical evaluation of R_{FM} follows from these last two expressions only for specific noise and filter specifications. The simplest case assumes white noise and allows $H_d(\omega)$ to correspond to a simple low-pass filter, such as that shown in Fig. 6.10-5(*d*). If the filter has a 3-dB bandwidth W_1 its response is

$$|H_d(\omega)|^2 = \frac{1}{1 + (\omega/W_1)^2}, \qquad (6.10\text{-}24)$$

which is sketched in (*e*). Under these assumptions R_{FM} evaluates to

$$R_{\text{FM}} = \frac{(W_f/W_1)^3}{3\left[(W_f/W_1) - \tan^{-1}(W_f/W_1)\right]}. \qquad (6.10\text{-}25)$$

R_{FM} is plotted in Fig. 6.10-6, curve *A*. We consider a practical example.

Example 6.10-2

Consider some parameter values typical of a commercial FM system: $W_f = 30\pi(10^3)$ rad/s and $W_1 = 4.2\pi(10^3)$. Thus $W_f/W_1 = 7.14$ and, from either (6.10-25) for white noise or from Fig. 6.10-6, curve *A*, we have $R_{\text{FM}} = 21.3$ or 13.3 dB.

As this example illustrates, considerable improvement is possible by use of emphasis techniques. Only a portion of the improvement is directly from the use of emphasis, however. Part of the improvement is actually due to an increase in transmitted signal bandwidth. To show this fact, assume a random message $f(t)$ and let the bandwidth without emphasis be $2\Delta\omega$, while bandwidth with preemphasis is defined as $2\Delta\omega_p$.†

Since instantaneous frequency deviations from the carrier without emphasis are proportional to $f(t)$, bandwidth is proportional to the rms value of $f(t)$ (see Section 6.4)

$$(2\Delta\omega)^2 \propto \overline{f^2(t)} = \frac{1}{2\pi}\int_{-W_f}^{W_f} \mathcal{S}_f(\omega)\, d\omega. \qquad (6.10\text{-}26)$$

As usual, $\mathcal{S}_f(\omega)$ is the power density spectrum of $f(t)$. In a similar manner, $2\Delta\omega_p$ will be proportional (with the same proportionality constant as before) to the rms value of the pre-

†In the remainder of Section 6.10 we assume $f(t)$ and the emphasized signal both have the same form of probability density function, such as for a Gaussian message. If this is not true, the following developments are valid if $\Delta\omega$ and $\Delta\omega_p$ are replaced by rms frequency deviations.

ISBN 0-201-05758-1

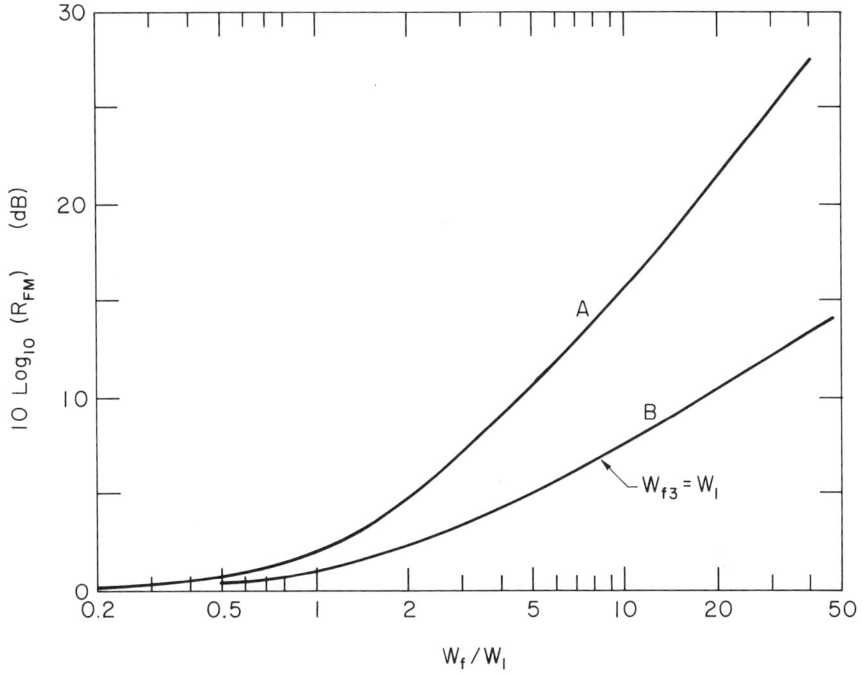

Fig. 6.10-6. FM output signal-to-noise ratio increase due to emphasis. Increase without (curve A) and with (curve B) bandwidth constraint.

emphasized signal $g(t)$ [see Fig. 6.10-5(a), where the constant K is now included in the analysis] :

$$(2\Delta\omega_p)^2 \propto \overline{g^2(t)} = \frac{1}{2\pi}\int_{-W_f}^{W_f} K^2 |H_p(\omega)|^2 \mathcal{S}_f(\omega)\, d\omega. \tag{6.10-27}$$

The ratio of the above two expressions gives

$$\left(\frac{\Delta\omega_p}{\Delta\omega}\right)^2 = \frac{K^2 \displaystyle\int_{-W_f}^{W_f} |H_p(\omega)|^2 \mathcal{S}_f(\omega)\, d\omega}{\displaystyle\int_{-W_f}^{W_f} \mathcal{S}_f(\omega)\, d\omega}. \tag{6.10-28}$$

For the simplest deemphasis filter of (6.10-24) the corresponding preemphasis filter power transfer function is $|H_p(\omega)|^2 = 1 + (\omega/W_1)^2$. Bandwidth ratio reduces to

$$\left(\frac{\Delta\omega_p}{\Delta\omega}\right)^2 = K^2 \left\{ 1 + \frac{\displaystyle\int_{-W_f}^{W_f} \omega^2 \mathcal{S}_f(\omega)\, d\omega}{W_1{}^2 \displaystyle\int_{-W_f}^{W_f} \mathcal{S}_f(\omega)\, d\omega} \right\}. \tag{6.10-29}$$

ISBN 0-201-05758-1

Since the information signal is assumed to be random, the second term is recognized as $(W_{rms}/W_1)^2$, where W_{rms} is the rms bandwidth of $f(t)$ (see Section 4.9). Thus,

$$\left(\frac{\Delta\omega_p}{\Delta\omega}\right)^2 = K^2[1 + (W_{rms}/W_1)^2]. \tag{6.10-30}$$

In deriving (6.10-28), and either (6.10-29) or (6.10-30) for the simple deemphasis filter, we have actually been a little more general than necessary. The generality involves carrying along the constant K and is useful in a subsequent discussion. For the present discussion the improvement ratio R_{FM} of (6.10-25) corresponds to $K = 1$ in (6.10-30). The reader should verify this fact for himself as an exercise. As a consequence of our work we find that $\Delta\omega_p > \Delta\omega$, and part of the improvement in output signal-to-noise ratio is due to this fact.

Emphasis Used in Bandwidth-Limited Applications

From the foregoing discussion, the output signal-to-noise ratio improvement of (6.10-25) is achieved only when the system has no bandwidth constraints. For important situations, such as broadcast FM, there are constraints imposed by law, and $\Delta\omega/2\pi \leqslant 75$ kHz is required. For such a system, we must not increase bandwidth when emphasis is used. This amounts to requiring $\Delta\omega_p = \Delta\omega$ in (6.10-30), which may be satisfied if K is selected such that

$$K^2 = 1/[1 + (W_{rms}/W_1)^2]. \tag{6.10-31}$$

In making this selection, the output signal power in the receiver will be smaller by a factor K^2, since K is a voltage gain in the overall path. Thus, the improvement R_{FM} is not given by (6.10-25) when bandwidth constraints exist, but rather by (6.10-25) multiplied by the factor K^2:

$$R_{FM} = \frac{K^2(W_f/W_1)^3}{3[(W_f/W_1) - \tan^{-1}(W_f/W_1)]}. \tag{6.10-32}$$

Let us assume, for the sake of developing a numerical evaluation of R_{FM}, that a reasonable approximation for the power spectrum of some information signals is

$$\begin{aligned} \mathcal{S}_f(\omega) &= \mathcal{S}_{f0}[1 + (\omega/W_{f3})^2]^{-1}, &\quad |\omega| < W_f \\ &= 0, &\quad \text{elsewhere.} \end{aligned} \tag{6.10-33}$$

Here $\mathcal{S}_{f0}$ is a constant which determines the average power in $f(t)$, and W_{f3} is the 3-dB bandwidth of the power density spectrum. Using this signal description, K^2 evaluates to

$$K^2 = \frac{\tan^{-1}\left(\dfrac{W_f}{W_{f3}}\right)}{\left(\dfrac{W_{f3}}{W_1}\right)\left(\dfrac{W_f}{W_1}\right) + \left[1 - \left(\dfrac{W_{f3}}{W_1}\right)^2\right]\tan^{-1}\left(\dfrac{W_f}{W_{f3}}\right)}. \tag{6.10-34}$$

If the break frequency W_1 is made equal to the 3-dB bandwidth frequency W_{f3}, K^2 reduces to a particularly simple result:

ISBN 0-201-05758-1

$$K^2 = \left(\frac{W_1}{W_f}\right) \tan^{-1}\left(\frac{W_f}{W_1}\right). \tag{6.10-35}$$

Substitution of (6.10-35) into (6.10-32) leads to curve B of Fig. 6.10-6. The curve illustrates clearly that, under bandwidth constraints, the improvement in FM output signal-to-noise ratio may be much less than if no such constraint were present. It is to be noted, however, that improvement does occur, and as the next example illustrates, the improvement is very worthwhile.

Example 6.10-3

We again find the performance of a commercial FM broadcast system as described in example 6.10-2 except with bandwidth constraints. From (6.10-35) with $W_f/W_1 = 7.14$, we have $K^2 = 0.20$ corresponding to a loss due to bandwidth limiting of 6.98 dB. From (6.10-32), $R_{FM} = 4.26$ or 6.30 dB is the net improvement in $(S_o/N_o)_{FM}$.

A practical preemphasis network useful for broadcast FM is illustrated in Fig. 6.10-5(*b*). $|H_p(\omega)|^2$ is depicted in (*c*). The second break frequency W_2 is chosen such that $W_2 > W_f$ with $W_f/2\pi$ typically equal to 15 kHz. The network is used in conjunction with appropriate amplification so that the net low-frequency gain is K.

6.11. NOISE PERFORMANCE OF WIDEBAND PM

As noted in Section 6.7, demodulation of a wideband phase-modulated signal must be accomplished by using a frequency demodulator followed by an integrator. As a consequence, much of the foregoing FM performance analysis may be applied to the PM problem.

Signal and Noise Relationships

As in FM, the input signal and noise average powers are given by (6.10-1) through (6.10-3). The output signal from the discriminator is $K_D k_{PM}\, df(t)/dt$ where, as before, K_D is the discriminator gain. The integrator output becomes $K_D k_{PM} f(t)$, which has the output average signal power

$$S_o = K_D{}^2 k_{PM}{}^2\, \overline{f^2(t)}. \tag{6.11-1}$$

The only remaining task in finding signal-to-noise ratios is to determine output noise power. As before, we let $f(t) = 0$ to make this determination. The noise power density spectrum at the discriminator output is again given by (6.10-11) for PM. After considering that the square of the integrator transfer function magnitude is $1/\omega^2$, the final output noise power density spectrum is

$$\mathscr{S}_{n_d}(\omega) = \left(\frac{K_D}{A}\right)^2 \mathscr{S}_{n_s}(\omega), \qquad |\omega| < W_f \tag{6.11-2}$$

$$= 0, \qquad |\omega| > W_f.$$

where $\mathscr{S}_{n_s}(\omega)$ is given by (6.10-12).

Proper integration of (6.11-2) gives output noise power. Considering only the white noise case, we obtain

ISBN 0-201-05758-1

$$N_o = \frac{K_D{}^2 \mathcal{N}_0 W_f}{\pi A^2}. \quad (6.11\text{-}3)$$

Signal-to-noise ratios follow easily:

$$\left(\frac{S_i}{N_i}\right)_{PM} = \frac{A^2 \pi}{2\mathcal{N}_0 \, \Delta\omega}, \quad (6.11\text{-}4)$$

$$\left(\frac{S_o}{N_o}\right)_{PM} = \frac{A^2 \pi k_{PM}{}^2 \, \overline{f^2(t)}}{\mathcal{N}_0 W_f} = \frac{A^2 \pi}{\mathcal{N}_0 W_f} (\Delta\omega)^2 \frac{\overline{f^2(t)}}{|df(t)/dt|_{max}{}^2}. \quad (6.11\text{-}5)$$

As in FM for a given information signal, the output signal-to-noise ratio improves as the square of modulated signal bandwidth.

Comparisons with AM and FM

Comparing (6.11-5) with baseband, and therefore all AM systems indirectly, we use (5.8-7) to obtain

$$\left(\frac{S_o}{N_o}\right)_{PM} = (\Delta\omega)^2 \frac{\overline{f^2(t)}}{|df(t)/dt|_{max}{}^2} \left(\frac{S_o}{N_o}\right)_B. \quad (6.11\text{-}6)$$

Since the factor multiplying $(S_o/N_o)_B$ may be laıger than unity, $(S_o/N_o)_{PM} > (S_o/N_o)_B$ is possible. As a consequence PM may be superior to all AM methods.

We may compare the signal-to-noise ratio of (6.11-5) with that of FM by taking the ratio with (6.10-16). Using subscripts to distinguish $\Delta\omega$ in the two cases gives

$$\frac{(S_o/N_o)_{FM}}{(S_o/N_o)_{PM}} = 3\left(\frac{\Delta\omega_{FM}}{\Delta\omega_{PM}}\right)^2 \frac{|df(t)/dt|_{max}{}^2}{W_f^2 \, |f(t)|_{max}{}^2}. \quad (6.11\text{-}7)$$

A fair comparison assumes equal bandwidths under worst-case bandwidth conditions which, in turn, allows us to substitute (6.5-21). Hence,

$$\frac{(S_o/N_o)_{FM}}{(S_o/N_o)_{PM}} = 3, \quad (6.11\text{-}8)$$

showing FM to be three times as good as, or 4.8 dB better than, PM under conditions of equal transmission bandwidth. This fact accounts partly for why PM is less widely used than FM (except for certain digital modulation methods to be discussed in Chapter 8).

Emphasis in PM

Emphasis filters may also be added to a PM system. The improvement gained in output signal-to-noise ratio is not nearly so impressive as was achieved for FM, however. By retracing the techniques used in the FM analysis (without bandwidth constraint), and again assuming the form of the deemphasis filter is given by (6.10-24), the PM system improvement factor R_{PM}

ISBN 0-201-05758-1

for white input noise is found to be

$$R_{PM} = \frac{(W_f/W_1)}{\tan^{-1}(W_f/W_1)}. \tag{6.11-9}$$

We consider a case where $W_f/W_1 = 7.14$ so that a comparison with FM through example 6.10-2 can be made.

Example 6.11-1

If $W_f/W_1 = 7.14$, (6.11-9) gives $R_{PM} = 4.99$ or 6.98 dB. Compared to the FM improvement of example 6.10-2, the PM system improvement is smaller by a factor of 4.27, or 6.30 dB.

★ 6.12. THRESHOLD IN WIDEBAND FM

It was noted in Section 6.10 that $(S_o/N_o)_{FM}$ cannot be increased indefinitely by raising the deviation ratio D_{FM} because input signal-to-noise ratio decreases and a point is reached where assumptions leading to (6.10-16) are no longer valid. The most important assumption was that noise $n_i(t)$ at the demodulator input must nearly always be small in relation to the peak carrier level A. In effect, this requires $(S_i/N_i)_{FM}$ to be above a minimum value called the *threshold signal-to-noise ratio* $(S_i/N_i)_{FM,th}$. Thus, we require $(S_i/N_i)_{FM} \geqslant (S_i/N_i)_{FM,th}$ for (6.10-16) to be valid.

If $(S_i/N_i)_{FM}$ falls below the threshold value, it is found, both in theory and in practice, that $(S_o/N_o)_{FM}$ degrades to a value smaller than predicted by (6.10-16). Although a precise definition of threshold is somewhat arbitrary and not necessary at this point, it can be said that, if $(S_i/N_i)_{FM}$ is larger than about 10 dB to 12 dB, it is above threshold. A more careful definition of threshold will subsequently be given.

In general, there are two main points of interest in the FM threshold problem. The first is, How does it come about? This point will subsequently be considered first, and it will be seen that threshold corresponds to a sudden rise in effective output noise power. The second point relates to performance near or below threshold. Some specific, although approximate, results will be presented which lead to a definition of threshold itself.

In all work of this section we shall consider the FM system of Fig. 6.10-1. All filters are considered ideal, and input noise is assumed white with power density $\mathcal{N}_0/2$. Owing to the quite involved nature of the theory behind threshold effect, we give only a greatly simplified discussion. For additional detail the reader is referred to the literature [10-12].

Spike Noise as the Cause of Threshold

If one listens to the output of an FM receiver with large input signal-to-noise ratio $(S_i/N_i)_{FM}$ but no modulation present, the usual steady hiss of the noise will be obvious. This noise may typically appear as shown in Fig. 6.12-1(*a*) when displayed on an oscilloscope. If the input signal-to-noise ratio is reduced, a point is reached where snaps or clicks may be heard. The output noise may now appear as illustrated in (*b*). The clicks are due to "spikes" occurring at random times added to the usual background thermal-type noise. One click is observed for each spike. Further decrease in $(S_i/N_i)_{FM}$ increases the number of spikes occurring per second, and the clicks merge into a sort of crackling or sputtering sound.

Compared over an equal time interval the energy of a spike is much larger than that of the background noise. Thus, if the spike rate becomes significant, the overall output noise power

ISBN 0-201-05758-1

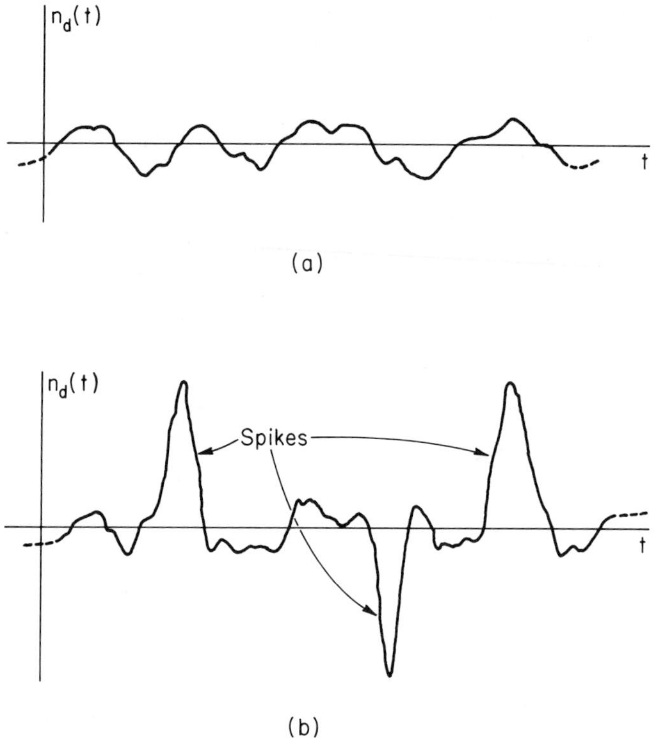

Fig. 6.12-1. Typical FM system output noise. (a) For input signal-to-noise ratio above threshold and (b) near threshold.

significantly increases. The added spike noise reduces output signal-to-noise ratio to a value lower than that computed earlier on the basis of a large input signal-to-noise ratio. Thus, spikes are the cause of FM threshold, and they begin to occur when the carrier level A ceases to be large relative to input noise $n_i(t)$ most of the time.

To explain how spikes come about it is helpful to resort to a phasor description of the input signal to the ideal demodulator of Fig. 6.10-1. For simplicity, assume that only the carrier is present so that the total input signal is the sum of carrier $A \cos(\omega_0 t + \theta_0)$ and noise $n_i(t)$. The band-pass input noise may be written as

$$n_i(t) = n_c(t) \cos(\omega_0 t + \theta_0) - n_s(t) \sin(\omega_0 t + \theta_0)$$

$$= r(t) \cos[\omega_0 t + \theta_0 + \theta(t)], \qquad (6.12\text{-}1)$$

where $r(t)$ is the envelope given by

$$r(t) = [n_c^2(t) + n_s^2(t)]^{1/2}, \qquad (6.12\text{-}2)$$

and $\theta(t)$ is a random phase angle:

$$\theta(t) = \tan^{-1}[n_s(t)/n_c(t)] \qquad (6.12\text{-}3)$$

(see Section 4.11). The total input waveform is now

ISBN 0-201-05758-1

$$A\cos(\omega_0 t + \theta_0) + r(t)\cos[\omega_0 t + \theta_0 + \theta(t)]$$

$$= \text{Re}\left\{e^{j\omega_0 t + j\theta_0}\,[A + r(t)e^{j\theta(t)}]\right\}$$

$$= \text{Re}\left\{e^{j\omega_0 t + j\theta_0}\,R(t)e^{j\psi(t)}\right\} = R(t)\cos[\omega_0 t + \theta_0 + \psi(t)]. \quad (6.12\text{-}4)$$

Here Re $\{\cdot\}$ implies the real part, and

$$R(t) = \left[\left\{A + r(t)\cos[\theta(t)]\right\}^2 + \left\{r(t)\sin[\theta(t)]\right\}^2\right]^{1/2}, \quad (6.12\text{-}5)$$

$$\psi(t) = \tan^{-1}\left\{\frac{r(t)\sin[\theta(t)]}{A + r(t)\cos[\theta(t)]}\right\}. \quad (6.12\text{-}6)$$

Figure 6.12-2 shows phasor diagrams for (6.12-4). We may suppress the factor $\exp(j\omega_0 t + j\theta_0)$ and "slow down" the rotation of all phasors. This means we may consider the carrier vector stationary. The noise phasor $r(t)\exp[j\theta(t)]$ is added to A as shown in (*a*). The resulting signal amplitude is $R(t)$, and its phase is $\psi(t)$. If $A \gg r(t)$ most of the time, then as the noise randomly varies with time the resultant may trace out a typical path as shown dashed in (*a*). For this case $R(t) \approx A$ and the phase angle $\psi(t)$, caused by noise, is always small. No spikes are generated in this high signal-to-noise ratio situation.

On the other hand, suppose noise is large as shown in (*b*). A typical path might lead to an encirclement of the origin as shown. When the resultant R is small and ψ is near π rad, then

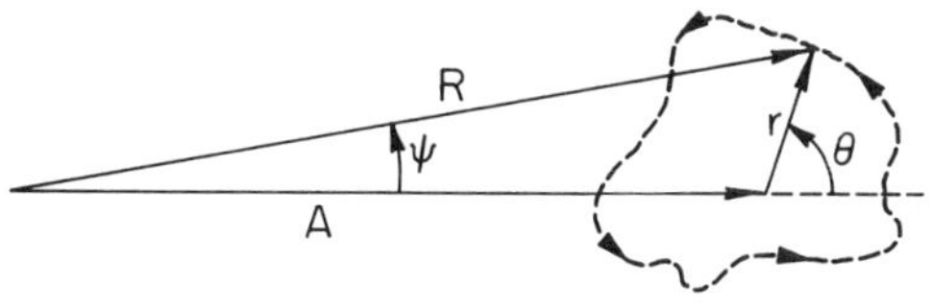

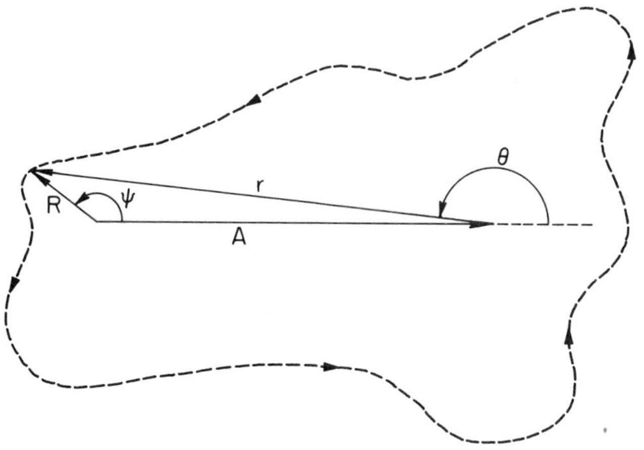

Fig. 6.12-2. Phasor diagrams applicable to the study of FM spikes. (a) Small noise case and (b) large noise case.

ISBN 0-201-05758-1

$\psi(t)$ may rapidly shift from small angles to angles near 2π rad. Because the FM system responds only to frequency, which is the rate of change of phase (the limiter prior to the discriminator removes any sensitivity to amplitude changes), it is only the behavior of $\psi(t)$ that is important. This behavior is illustrated in Fig. 6.12-3(a). Since the discriminator output is proportional to $d\psi(t)/dt$, the actual output with the resulting spike appears as in (b). The reader may verify that the area of any single spike is necessarily constant at either 2π or -2π.

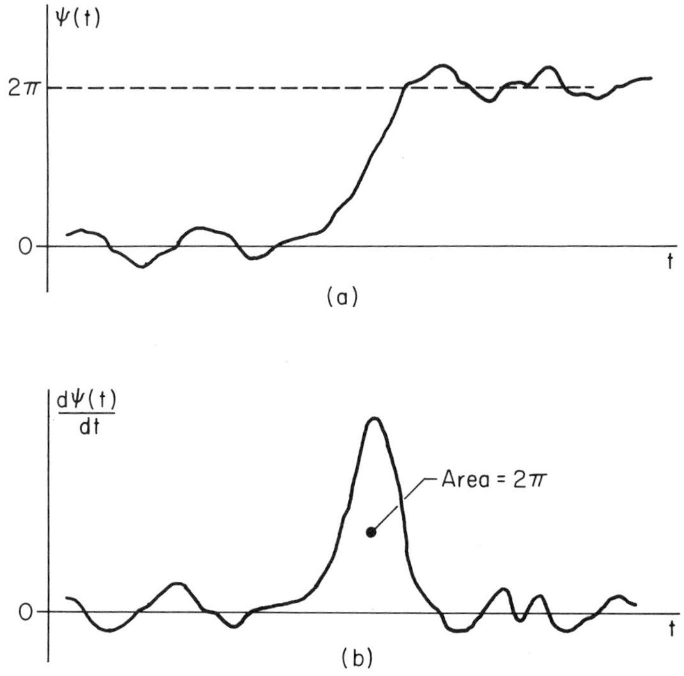

Fig. 6.12-3. (a) Typical noise phase, and (b) the time derivative of the noise phase, for times near a spike.

Thus, spikes occur when $R(t)$ rotates about the origin. This may happen by clockwise as well as counterclockwise rotation so that both negative and positive spikes are produced. When no modulation is present, the two spike polarities tend to occur with equal regularity. If modulation is present, positive spikes dominate when the instantaneous frequency is less than the carrier value ω_0, while negative spikes dominate for the opposite situation. Furthermore, presence of modulation causes the number of spikes per second to increase, leading to larger spike noise power when information is transmitted.

There are many other ways in which $\psi(t)$ may behave when near $\pm\,\pi$ rad. These in general are of much less importance than the behavior leading to spikes and are not considered further.

Average Spike Rate

We subsequently will find that output noise power due to spikes will increase linearly with the average spike rate $\overline{R}_s$. If there is no information transmitted, then it is found [12] that

$$\overline{R}_s = \frac{\Delta\omega}{\pi\sqrt{12}} \text{ erfc}\left(\sqrt{(S_i/N_i)_{\text{FM}}}\right). \tag{6.12-7}$$

ISBN 0-201-05758-1

Here erfc $(\cdot)$ is the *complementary error function* defined by

$$\operatorname{erfc}(x) = 1 - \frac{2}{\sqrt{\pi}} \int_0^x e^{-\xi^2}\, d\xi. \tag{6.12-8}$$

When modulation is present, the number of spikes per unit of time increases greatly. Although the general analysis of spike rate is too involved to include here, a simplified discussion will lead to approximate results. Suppose initially that a *constant* frequency offset $\delta\omega$ from the carrier is present. Rice [12, p. 413] has solved for the spike rate. It has two components, one of which is zero when $\delta\omega = 0$ and increases as $(|\delta\omega|/2\pi)\exp[-(S_i/N_i)_{\text{FM}}]$ with offset. The second component is maximum when $\delta\omega = 0$ and decreases to a negligible value rapidly as $|\delta\omega|$ increases. Its value for $\delta\omega = 0$ is given by (6.12-7). Thus, for all values of $|\delta\omega|$ of interest the spike rate is approximately given by

$$R_s = \frac{|\delta\omega|}{2\pi} \exp[-(S_i/N_i)_{\text{FM}}] \tag{6.12-9}$$

for a fixed offset $\delta\omega$.

The effect of modulation can now be approximated by a technique of Taub and Schilling [10] whereby modulation is considered to cause a continuously varying offset. The offset and modulating signal $f(t)$ are related by

$$\delta\omega = k_{\text{FM}} f(t). \tag{6.12-10}$$

On substitution of this expression into (6.12-9), and taking the average value, we find the average spike rate:

$$\overline{R_s} = \frac{k_{\text{FM}}}{2\pi} \overline{|f(t)|} \exp[-(S_i/N_i)_{\text{FM}}]. \tag{6.12-11}$$

For sinusoidal modulation it is readily shown that $k_{\text{FM}}\overline{|f(t)|}/2\pi = \Delta\omega/\pi^2$. For a Gaussian random message $k_{\text{FM}}\overline{|f(t)|}/2\pi = \Delta\omega_{\text{rms}}/(\pi\sqrt{2\pi})$, where $\Delta\omega_{\text{rms}}^2$ is the mean-squared frequency deviation $\Delta\omega_{\text{rms}}^2 = k_{\text{FM}}^2\overline{f^2(t)}$. Thus,

$$\overline{R_s} = \frac{\Delta\omega}{\pi^2} \exp[-(S_i/N_i)_{\text{FM}}], \qquad \text{sinusoidal modulation,} \tag{6.12-12}$$

$$\overline{R_s} = \frac{\Delta\omega_{\text{rms}}}{\pi\sqrt{2\pi}} \exp[-(S_i/N_i)_{\text{FM}}], \quad \text{Gaussian modulation.} \tag{6.12-13}$$

Average Power Due to Spikes

Our development of FM output signal-to-noise ratio near threshold can be completed once we relate average spike rate to average power due to spikes. We give an intuitive argument in the following which does not constitute a rigorous development for noise power. It does, however, have the advantage of getting the essential points across without undue labor.

Let K_{D} be the discriminator gain, as before, and let $Q_i(\omega)$ represent the Fourier transform of a single (the ith) spike. $Q_i(\omega)$ is the transform of $d\psi(t)/dt$ in a small time interval spanning

ISBN 0-201-05758-1

the spike. The energy density spectrum of the spike at the discriminator output is $K_D{}^2|Q_i(\omega)|^2$. Because spikes tend to occur independently, the energy density spectrum for N spikes is the sum of the individual energy density spectra, or

$$\text{energy density of } N \text{ spikes} = \sum_{i=1}^{N} K_D{}^2 |Q_i(\omega)|^2. \qquad (6.12\text{-}14)$$

Next, if N is large, (6.12-14) may be approximated as $NK_D{}^2$ times an *average* energy density spectrum $\overline{|Q(\omega)|^2}$ of a *typical* spike. Furthermore, if the N spikes occur over a time interval $2T$, the power density spectrum becomes

$$\text{power density of } N \text{ spikes} = \frac{N}{2T} K_D{}^2 \overline{|Q(\omega)|^2}. \qquad (6.12\text{-}15)$$

Taking the limit of this expression as $T \to \infty$ and observing that $N/2T$ becomes $\overline{R}_s$, the average spike rate, we have the power density spectrum $\mathscr{S}_s(\omega)$ for spike noise

$$\mathscr{S}_s(\omega) = \overline{R}_s K_D{}^2 \overline{|Q(\omega)|^2}. \qquad (6.12\text{-}16)$$

This expression indicates that the power density spectrum of spike noise at the discriminator output equals the product of average spike rate, the discriminator power gain, and the average energy density spectrum of a typical spike.

Now from the Fourier transform expression with $\omega = 0$, we know that $|Q(0)| = 2\pi$ because the area of any spike is either 2π or -2π. Thus, the power density spectrum for $\omega = 0$ equals $K_D{}^2 \overline{R}_s 4\pi^2$. The exact behavior of $\overline{|Q(\omega)|^2}$ for $|\omega| > 0$ is not too important to our problem. The reason is that we are only interested in the narrow frequency interval $|\omega| < W_f$ occupied by the demodulated information signal, while $\overline{|Q(\omega)|^2}$ is broadband. Because a broadband function of ω will not greatly change over a small frequency interval, we may consider $\overline{|Q(\omega)|^2}$ approximately constant, and equal to its value $4\pi^2$ at the origin. Hence,

$$\mathscr{S}_s(\omega) \approx 4\pi^2 K_D{}^2 \overline{R}_s, \qquad |\omega| < W_f. \qquad (6.12\text{-}17)$$

$\overline{|Q(\omega)|^2}$ is a broadband function, since spikes are caused by phase changes due to input noise, which possesses the full FM signal bandwidth, approximately $2\Delta\omega$.

Finally, the average output noise power N_s due to spikes is determined by integrating the power density spectrum:

$$N_s = \frac{1}{2\pi} \int_{-W_f}^{W_f} \mathscr{S}_s(\omega)\, d\omega = 4\pi K_D{}^2 W_f \overline{R}_s. \qquad (6.12\text{-}18)$$

The noise expression is completed by substitution of an appropriate average spike rate $\overline{R}_s$. An approximation for a zero-mean message having a symmetric probability density function results from substitution of (6.12-11). Specific examples for sinusoidal or Gaussian modulation would use (6.12-12) and (6.12-13), respectively. These examples are developed in the following discussion.

ISBN 0-201-05758-1

Signal-to-Noise Ratio near Threshold

Total output average noise power in the FM receiver is the sum of that due to receiver noise, given by (6.10-14), and that due to spikes, given by (6.12-18). Neglecting emphasis, we divide output average signal power, given by (6.10-4), by the sum of these two noise powers to obtain signal-to-noise ratio near threshold. For a sinusoidal modulating signal of frequency ω_f the ratio becomes

$$\left(\frac{S_o}{N_o}\right)_{FM} = \frac{3\beta_{FM}^{\ 3}\,(S_i/N_i)_{FM}}{1 + \dfrac{24\beta_{FM}^{\ 2}}{\pi}\,(S_i/N_i)_{FM}\,\exp[-(S_i/N_i)_{FM}]}. \qquad (6.12\text{-}19)$$

Here $\beta_{FM} = \Delta\omega/\omega_f$ is the modulation index. We observe that the numerator of (6.12-19) is equal to the signal-to-noise ratio above threshold which was found previously in (6.10-20). This should be the case, since the second denominator term is due to spike noise and it approaches zero for above-threshold operation. Figure 6.12-4 illustrates output versus input signal-to-noise ratios found from (6.12-19). The threshold effect is clearly demonstrated.

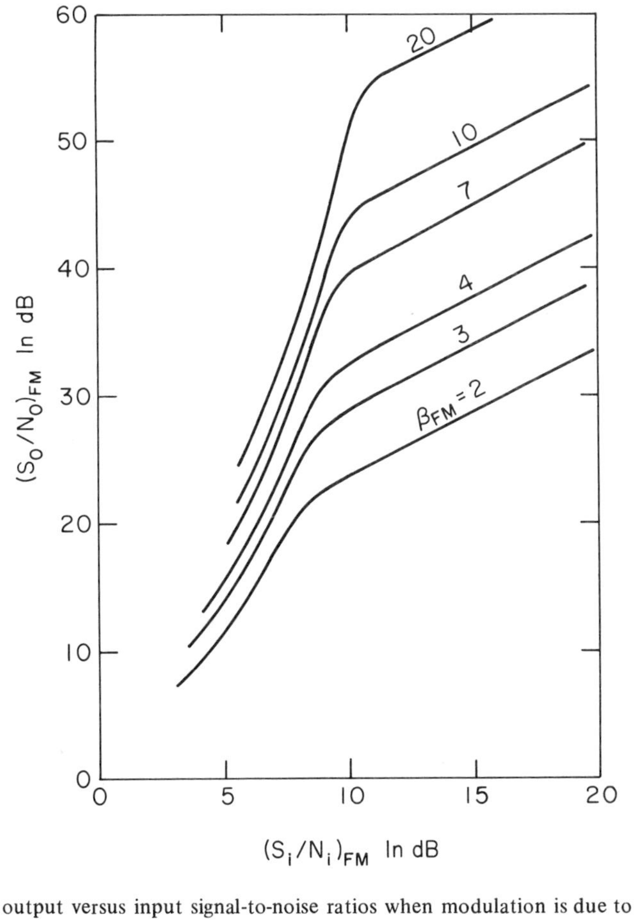

Fig. 6.12-4. FM output versus input signal-to-noise ratios when modulation is due to a sinusoidal information signal.

ISBN 0-201-05758-1

If the information signal is a Gaussian random message we substitute (6.12-13) into (6.12-18) to obtain

$$\left(\frac{S_o}{N_o}\right)_{FM} = \frac{6(\Delta\omega/\Delta\omega_{rms})(\Delta\omega_{rms}/W_f)^3(S_i/N_i)_{FM}}{1 + 12\sqrt{2/\pi}(\Delta\omega/\Delta\omega_{rms})(\Delta\omega_{rms}/W_f)^2(S_i/N_i)_{FM}\exp[-(S_i/N_i)_{FM}]}. \tag{6.12-20}$$

Figure 6.12-5 illustrates the behavior of this expression for several values of $\Delta\omega_{rms}/W_f$. In every case the value of $\Delta\omega$ was chosen to pass 98% of the FM signal power—that is, $\Delta\omega = 2.3\Delta\omega_{rms}$. Again, we find the threshold effect clearly demonstrated.

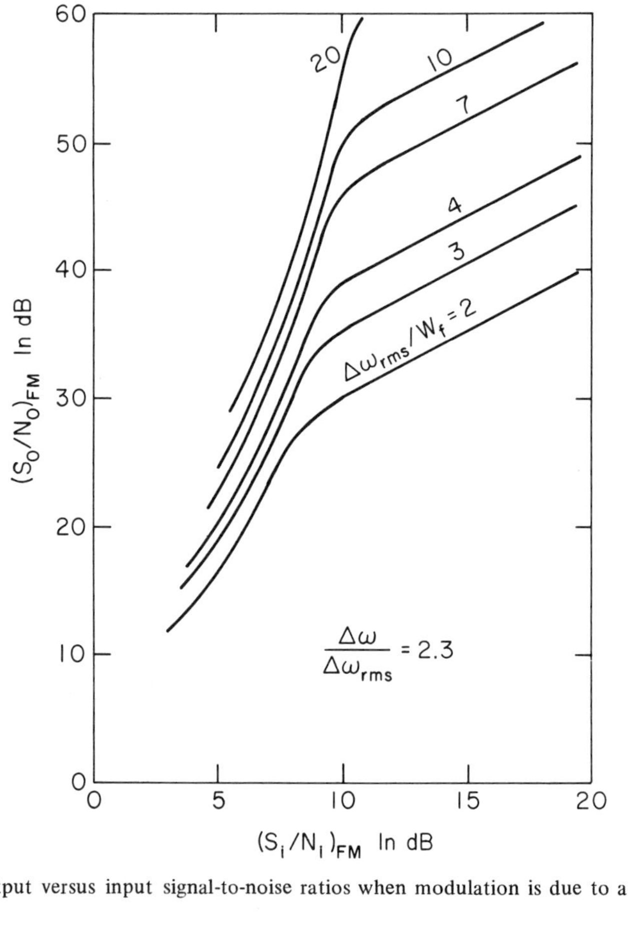

Fig. 6.12-5. FM output versus input signal-to-noise ratios when modulation is due to a Gaussian random information signal.

A comparison of figures 6.12-5 and 6.12-4 shows that the breakpoint or threshold occurs at an input signal-to-noise ratio of from 8 dB to 11 dB for either message type. Thus, threshold is nearly independent of the type of message and tends to occur near $(S_i/N_i)_{FM} = 10$.

In Section 6.9 the noise performance of a product-type demodulator for narrowband FM was analyzed. Of course, a discriminator can also be used for demodulation. If the receiver bandwidth ($2\Delta\omega$) is large in relation to the NBFM waveform bandwidth, the performance near threshold may be determined from substitution of (6.12-7) into (6.12-18), which relates

ISBN 0-201-05758-1

noise power due to spikes. By adding receiver noise power from (6.10-14) and dividing the sum into the signal power expression, given by (6.10-4), the output signal-to-noise ratio near threshold is determined. The reader is encouraged to carry out these steps as an exercise.

FM Threshold Evaluation

A precise definition of threshold signal-to-noise ratio $(S_i/N_i)_{FM,th}$ is arbitrary. We choose to define it as the value where $(S_o/N_o)_{FM}$ falls 1 dB below the value given by (6.10-16). Hence, threshold occurs where the second denominator term in either of the expressions (6.12-19) or (6.12-20) equals 0.26.

For example, assuming sinusoidal modulation and using (6.12-19), we have

$$(S_i/N_i)_{FM,th} \exp\left\{-(S_i/N_i)_{FM,th}\right\} = 0.034/\beta_{FM}^2. \tag{6.12-21}$$

The solution to this equation is plotted in Fig. 6.12-6. We see that threshold signal-to-noise ratio does not vary greatly with modulation index. A similar threshold expression may be obtained for a Gaussian modulating signal from (6.12-20). This calculation is left for the reader as an exercise.

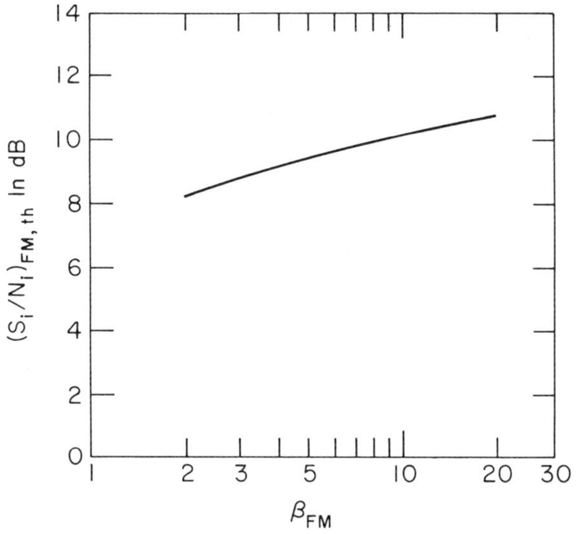

Fig. 6.12-6. FM threshold signal-to-noise ratio as a function of modulation index when modulation is due to a sinusoidal information signal.

Locked-loop demodulators of the types described in Section 6.8 are capable of reducing the value of the threshold by about 6 dB to 10 dB [13, p. 39] for $\Delta\omega_{rms}/W_f$ from about 2 to 10, respectively.* Such improvements are called *threshold extensions* and become very important in applications where transmitter power is at a premium. More sophisticated locked-loop demodulators also exist that are capable of even greater threshold extensions, as described by Klapper and Frankle [13]. The price paid for the improved performance is additional loop complexity.

Finally, we point out that the threshold effect is not unique to FM. It also obviously occurs in wideband PM, since message recovery must ultimately rely on use of an FM demodulator.

*These reductions apply to modulation by a speech signal modeled as a Gaussian random message having a peak-to-rms amplitude ratio of 10 dB.

ISBN 0-201-05758-1

PROBLEMS

6-1. An angle-modulated signal is $s(t) = A \cos [\omega_0 t + 100 \cos (\omega_f t)]$. (*a*) If the modulation is PM, what is $f(t)$ if $k_{PM} = 2$? (*b*) What is the peak frequency change $\Delta\omega$ caused by $f(t)$ in (*a*)?

6-2. Work problem 6-1 assuming modulation is FM and $k_{FM} = 2$.

6-3. Discuss what limits apply to the integral of (6.1-8).

6-4. If the approximations made in arriving at (6.2-2) are not valid, show that the first-order correction term to be added to the right side is

$$\frac{-k_{FM}^2}{6} \left[\int f(t)\, dt\right]^2 \left[s_{NBFM}(t) + 2A \cos (\omega_0 t + \theta_0)\right],$$

where $s_{NBFM}(t)$ is the uncorrected signal given by (6.2-2).

6-5. Equation (6.2-5) is valid if the transform $F(\omega)$ of $f(t)$ is zero for $\omega = 0$. Obtain a result valid without this restriction.

6-6. Work problem 6-4 as applied to (6.2-7) and show that the correction term for a PM signal is

$$\frac{-k_{PM}^2}{6} \left[f^2(t)\right] \left[s_{NBPM}(t) + 2A \cos (\omega_0 t + \theta_0)\right].$$

6-7. Find and sketch the magnitude of the spectrum of a narrowband FM signal if $f(t) = A_f \cos (\omega_f t)$ and $\theta_0 = 0$. Repeat for a PM wave. Make the sketches of the two spectra assuming $k_{PM} |f(t)|_{max} = \pi/6$ for PM and $k_{FM} |\int f(t)\, dt|_{max} = \pi/6$ for FM.

6-8. Assume carrier phase $\theta_0 \neq 0$. Obtain modified expressions for (6.3-2), (6.3-7), and (6.3-12).

6-9. A 100-MHz carrier is frequency-modulated by a sinusoidal signal of frequency 5 kHz. If the peak frequency deviation is 50 kHz, what is the approximate band of frequencies occupied by the FM waveform?

6-10. (*a*) What minimum value of β_{FM} is required in order that FM bandwidth as given by (6.3-14) will be within 10% of that given by (6.3-13)? (*b*) What is the minimum for 5%?

6-11. Verify (6.3-15).

6-12. (*a*) Find and sketch the power density spectrum corresponding to the spectrum of Fig. 6.3-3(*b*). (*b*) Find the percentage of total power in frequencies within the bandwidth W_{FM}. (Hint: Use tables for the Bessel coefficients.)

★ 6-13. Obtain and plot the magnitude of the spectrum of the FM wave having sawtooth frequency variations as shown in Fig. P6-13. Assume that the product $\Delta\omega T/2\pi$ is large and use the fact that

$$C(x) = \int_0^x \cos\left(\frac{\pi}{2}\xi^2\right) d\xi \approx \frac{1}{2},$$

$$S(x) = \int_0^x \sin\left(\frac{\pi}{2}\xi^2\right) d\xi \approx \frac{1}{2},$$

ISBN 0-201-05758-1

if $x >> 1$ where $C(-x) = -C(x)$, and $S(-x) = -S(x)$. $C(x)$ and $S(x)$ are called *Fresnel integrals.*

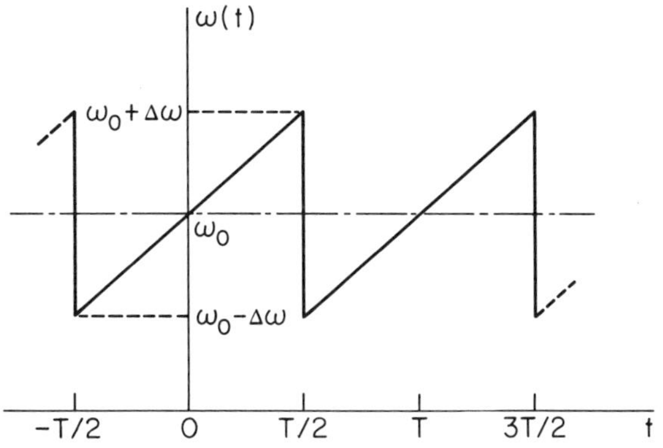

Fig. P6-13

6-14. Plot the magnitude of the spectrum of example 6.4-1 for β_{FM} large.

6-15. Use (6.4-7) to find the FM power density spectrum for square-wave modulation if the square-wave levels are $\Delta\omega/k_{FM}$ and $-\Delta\omega/k_{FM}$.

6-16. Use (6.4-7) to find the power density spectrum of an FM signal caused by Gaussian information defined by (6.5-13) through (6.5-15). Plot the result.

★6.17. Using the method of Section 6.4 determine the power density spectrum for FM by a sinusoidal signal.

6-18. A carrier is phase-modulated by a sinusoid having a peak voltage of 10 volts. What is k_{PM} if the modulation index is known to be 20? What is its unit? If peak frequency deviation is 300 kHz, what is the frequency of the modulating signal?

6-19. A capacitance is assumed to vary as

$$C = K_0[V_b + V]^{-1/2} = K_0[V_b + V_0 + f(t)]^{-1/2},$$

where $V \geqslant 0$ is the total voltage applied across the capacitor, K_0 and V_b are constants, and V_0 is the dc bias. (*a*) What will be the exact resonant frequency of an oscillator using C? (*b*) What is the maximum that $|f(t)|_{max}/(V_0 + V_b)$ can be if frequency is to vary linearly with $f(t)$ within 1%?

6-20. In the circuit of Fig. 6.7-2 explain the effect on an oscillator, using the circuit as a capacitance, if a nonnegligible capacitance exists between source and drain.

6-21. Analyze the circuit of Fig. 6.7-2 if Z_1 and Z_2 are as defined for a case 1 reactance modulator except that a capacitance C_g exists across R. Assume $|Z_1 + Z_2|$ is large so that (6.7-1) is valid and find Z_i. What is the effect of adding C_g?

6-22. Work problem 6-21 except assume a small inductance L is in series with R. What is its effect?

★6-23. In the circuit of Fig. 6.7-3 the VCO has a gain of 10^4 radians per second per volt. If the overall gain of the mixer-discriminator is 50 microvolts per radian per second, what low-pass filter gain is necessary to hold the VCO nominal frequency to within a 1-Hz error when it tries to drift by 1 kHz without the loop?

6-24. Find multiplier values N_1 and N_2 in an Armstrong FM generator using sinusoidal modulation at a frequency of 50 Hz if the narrowband modulator has a modulation index of

ISBN 0-201-05758-1

0.5. Assume $f_1 = 100$ kHz and $f_2 = 4$ MHz. The output frequency is to be 100 MHz with $\beta_{FM} = 1500$.

6-25. For the circuit of Fig. 6.8-1(*a*), if $\omega_0 = 2/RC$, what is the largest value of $\Delta\omega$ possible in FM demodulation such that the slope of the actual response departs less than 10% from the slope existing at ω_0?

6-26. Prove that the output of the circuit in Fig. 6.8-4 equals $(|E_A| - |E_B|)/2$ when the input FM signal amplitude is constant.

6-27. Show that, if the envelope detector of Fig. 6.8-5 is replaced by a square-law detector, defined by output = input-squared, followed by a suitable low-pass filter, demodulation of the FM signal is still possible except that a small inherent distortion is present.

6-28. A random message frequency modulates a carrier. The root-mean-squared frequency deviation is 30 kHz. (*a*) If a limiter-discriminator is used for demodulation, the mean-squared demodulated signal voltage is $\overline{s_d^2(t)} = 0.01$. What is the discriminator gain constant? (*b*) If the same limiter-discriminator is used with the network of Fig. 6.8-8, $\overline{s_d^2(t)} = 0.001$. What is the VCO gain constant?

6-29. A random message is transmitted by narrowband frequency modulation. The rms frequency deviation $\Delta\omega_{rms}$ is equal to a quarter of the signal's maximum frequency extent W_f. (*a*) If the receiver input power signal-to-noise ratio is 20 dB, what output signal-to-noise ratio can be achieved? (*b*) What output signal-to-noise ratio could be achieved by a simple baseband system?

6-30. An FM receiver uses a discriminator and has an input signal-to-white-noise power ratio of 20 dB and modulation index of 5 when using sinusoidal modulation. Find the output signal-to-noise power ratio.

6-31. If the modulation index of problem 6-30 is increased to 10 by increasing frequency deviation, find the output and input signal-to-noise power ratios. Assume the receiver is modified to accept the larger deviation.

6-32. (*a*) In problem 6-30 what performance improvement is possible if emphasis methods are added and if $W_f/W_1 = 7.5$? Assume no bandwidth restrictions. (*b*) What is the improvement in (*a*) if there are bandwidth restrictions? Assume (6.10-31) can be applied and $W_f = \omega_f$, the signal frequency.

6-33. An information signal has a power density spectrum

$$\mathcal{S}_f(\omega) = P_0 \cos\left(\frac{\pi\omega}{2W_f}\right), \qquad |\omega| \leqslant W_f$$
$$= 0, \qquad |\omega| > W_f.$$

Determine an expression for the degradation, due to bandwidth restrictions, in signal-to-noise ratio improvement when emphasis is added to an FM system using a discriminator. (Hint: Find W_{rms} for the message and solve for K^2.)

6-34. Two systems, one PM and one FM, transmit at the same carrier level A modulated with the same information signal. The corresponding receivers have identical white noise power densities $\mathcal{N}_0/2$. Find a condition that the signal must satisfy if the *input* signal-to-noise power ratios are to be equal in the two systems.

★ 6-35. By starting with (6.12-11) show that the average spike rates for (*a*) a sinusoidal information signal and (*b*) a Gaussian random signal are given by (6.12-12) and (6.12-13), respectively.

ISBN 0-201-05758-1

REFERENCES

[1] Abramowitz, M., and Stegun, I. A. (Editors), *Handbook of Mathematical Functions with Formulas, Graphs, and Mathematical Tables,* National Bureau of Standards Applied Mathematics Series 55, U.S. Government Printing Office, Washington, D.C., 1964.

[2] Carlson, A. B., *Communication Systems, An Introduction to Signals and Noise in Electrical Communication,* McGraw-Hill Book Co., New York, 1968. See also the Second Edition of the same title, 1975.

[3] Blachman, N. M., Limiting Frequency Modulation Spectra, *Information and Control,* Vol. 1, September, 1957, pp. 26–37.

[4] Blachman, N. M., and McAlpine, G. A., The Spectrum of a High-Index FM Waveform: Woodward's Theorem Revisited, *IEEE Transactions on Communication Technology,* Vol. COM-17, No. 2, April, 1969, pp. 201–208.

[5] Johns, P. B., and Rowbotham, T. R., *Communication Systems Analysis,* Van Nostrand-Reinhold Company, New York, 1972.

[6] Sunde, E. D., *Communication Systems Engineering Theory,* John Wiley & Sons, New York, 1969.

[7] Rowe, H. E., *Signals and Noise In Communication Systems,* D. Van Nostrand Co., Princeton, New Jersey, 1965.

[8] Kennedy, G., *Electronic Communication Systems,* McGraw-Hill Book Co., New York, 1970.

[9] Viterbi, A. J., *Principles of Coherent Communication,* McGraw-Hill Book Co., New York, 1966.

[10] Taub, H., and Schilling, D. L., *Principles of Communication Systems,* McGraw-Hill Book Co., New York, 1971.

[11] Schwartz, M., Bennett, W. R., and Stein, S., *Communication Systems and Techniques,* McGraw-Hill Book Co., New York, 1966.

[12] Rice, S. O., Noise in FM Receivers, Chapter 25, pp. 395–424, in *Proceedings of the Symposium on Time Series Analysis,* M. Rosenblatt (Editor), John Wiley & Sons, New York, 1963.

[13] Klapper, J., and Frankle, J. T., *Phase-Locked and Frequency-Feedback Systems,* Academic Press, New York, 1972.

ISBN 0-201-05758-1

Chapter 7

Pulse and Digital Modulation

7.0. INTRODUCTION

All the modulation methods discussed thus far have involved transmission of the entire information signal over the communication channel. Usually only one signal was continuously passed over the link, although it was also possible to send many simultaneous waveforms by resorting to frequency multiplexing. In this chapter we turn our attention to the transmission of only periodic *samples* of the information signal rather than the whole waveform.

One of the advantages to be gained by transmitting only samples is that it becomes possible to interlace samples from many different information signals in time. Thus, we have a process of *time-division multiplexing* analogous to frequency multiplexing. With the availability of modern high-speed switching circuits the practicality of time multiplexing is well established and, in many cases, may be preferred over frequency multiplexing.

Because only signal samples are to be transmitted, we consider the modulation of a pulse train to convey the information. We let each sample affect some characteristic of the pulse train.

In this chapter we discuss two methods of modulation. The first, called *pulse modulation,* is analog and involves changing a parameter of a pulse train such as amplitude, duration, or position of pulses. These cases are termed *pulse amplitude modulation* (PAM), *pulse duration modulation* (PDM), and *pulse position modulation* (PPM), respectively. The former is analogous to amplitude modulation. The latter two are analogous to angle modulation and are sometimes collectively called *pulse time modulation* (PTM). since they are both based on changing a time characteristic of the pulse train.

PAM, PDM, and PPM are not the only possible pulse modulations. They are, however, the most important ones and will be the only ones considered in this text. Figure 7.0-1 illustrates the principles of PAM, PDM, and PPM. Note that in pulse modulation the (usually linearly) modulated parameter may have *any* of a continuum of values within the range of allowed values. Thus, pulse modulation is discrete in time but continuous in the modulated parameter.

The second method of modulation we shall consider is *coded* or *digital modulation.* One of the most important of several types of digital modulation is *pulse code modulation* (PCM). Here a finite number or group of pulses in the train is affected by the information signal sample. Typically the sign (positive or negative) of pulses (*polar modulation*) or presence–absence of pulses (*unipolar modulation*) is related to the sample value. Since only a limited number of combinations of sign (or presence–absence) are possible in the pulse group, there is a limited number of values that may be assigned to the information signal sample. Hence, there must inherently be some round-off error associated with PCM systems. As a consequence, we see that PCM is discrete in both time and the modulated parameter.

In addition to PCM, several other types of digital modulation systems are discussed which may be considered variations of PCM. All work of this chapter is devoted to pulse and digital

ISBN 0-201-05758-1

References start on p. 409.

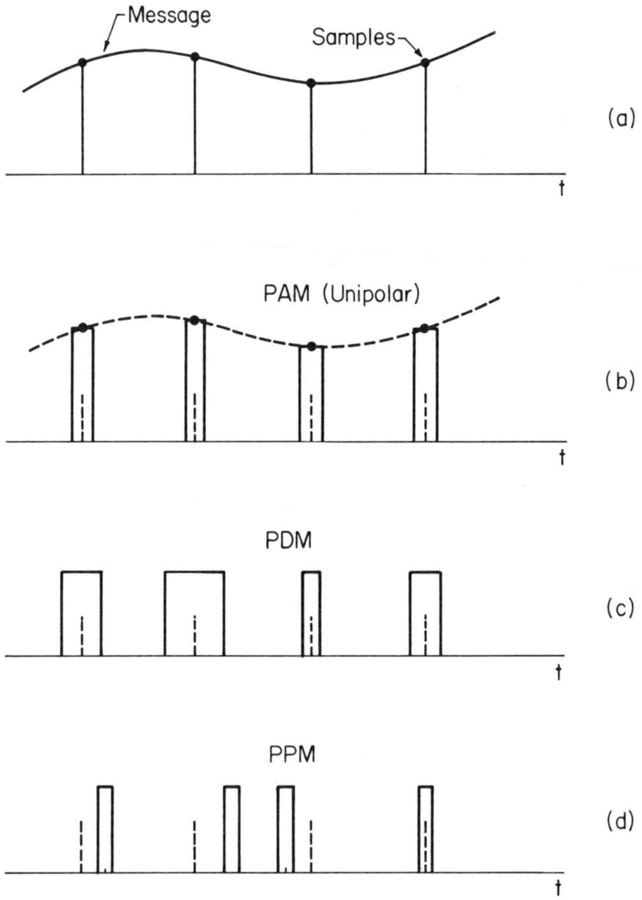

Fig. 7.0-1. Analog pulse-modulated signals. (a) A message and its samples, (b) unipolar PAM, (c) PDM, and (d) PPM.

systems that do not involve carrier modulation. We shall take up carrier modulation by digital signals in the following chapter.

In all systems with which we deal it is necessary that the receiver recover the information signal by using only the transmitted samples of the original waveform. The fundamental principle that makes this possible is the *sampling theorem.* Indeed, the sampling theorem is the very foundation of all pulse and digital modulation systems, and it is considered in some detail prior to discussion of individual modulation systems.

7.1. SAMPLING THEOREMS FOR LOW-PASS SIGNALS

There are various ways in which the sampling theorem for baseband or low-pass waveforms may be presented. We take these in turn.

Time-Domain Sampling of Nonrandom Signals

The low-pass sampling theorem may be stated as follows:

A low-pass signal $f(t)$, bandlimited such that it has no frequency components

ISBN 0-201-05758-1

above W_f rad/s, is uniquely determined by its values at equispaced points in time separated by $T_s \leqslant \pi/W_f$ seconds.*

This theorem is sometimes called the *uniform sampling theorem* owing to the equally spaced nature of the instantaneously taken samples.† It allows us to completely reconstruct a bandlimited signal from instantaneous samples taken at a rate $\omega_s = 2\pi/T_s$ of at least $2W_f$, which is twice the highest frequency present in the waveform. The *minimum* rate $2W_f$ is called the *Nyquist rate.*

Because the sampling theorem is such a remarkable result and so important, we shall include a proof.

A bandlimited signal $f(t)$ and its spectrum $F(\omega)$ are illustrated in Fig. 7.1-1(*a*). From our work in Chapter 2, the spectrum may be represented by a Fourier series developed by assum-

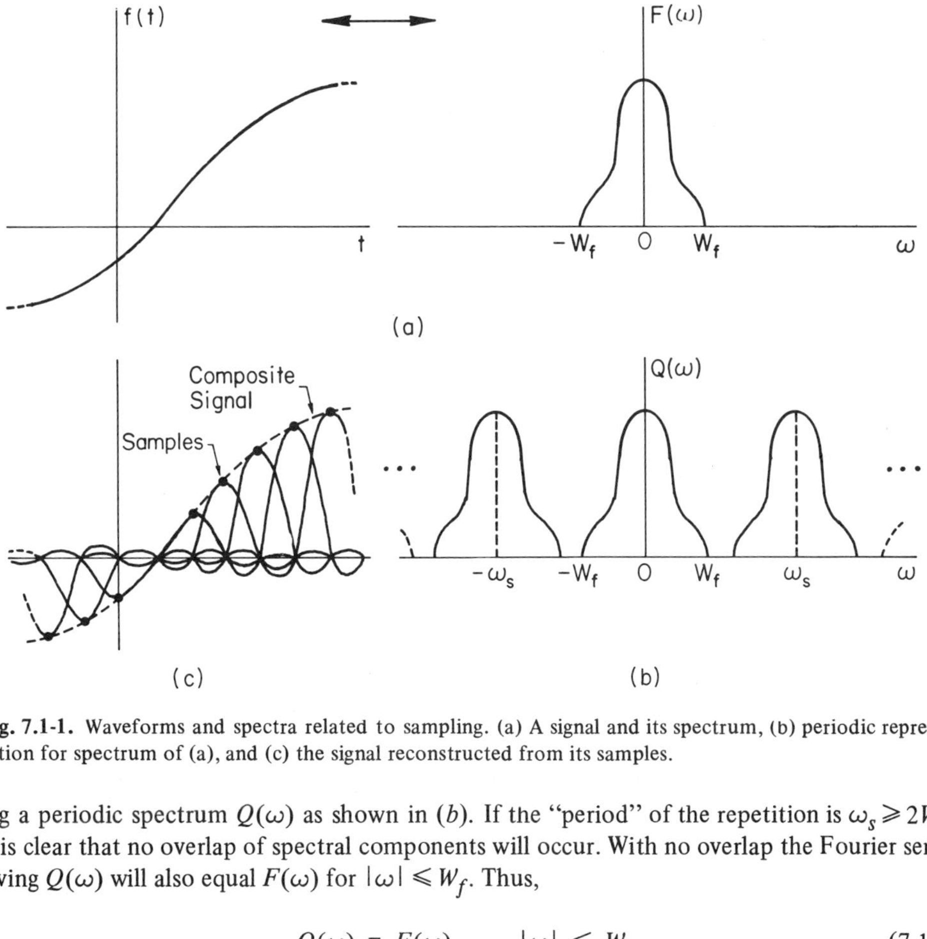

Fig. 7.1-1. Waveforms and spectra related to sampling. (a) A signal and its spectrum, (b) periodic representation for spectrum of (a), and (c) the signal reconstructed from its samples.

ing a periodic spectrum $Q(\omega)$ as shown in (*b*). If the "period" of the repetition is $\omega_s \geqslant 2W_f$, it is clear that no overlap of spectral components will occur. With no overlap the Fourier series giving $Q(\omega)$ will also equal $F(\omega)$ for $|\omega| \leqslant W_f$. Thus,

$$Q(\omega) = F(\omega), \qquad |\omega| \leqslant W_f. \tag{7.1-1}$$

Using (2.1-12) with proper variable change allows $Q(\omega)$ to be written as

*If a spectral impulse exists at $\omega = \pm W_f$, the equality is to be excluded [1].

†The theorem can be generalized to nonuniform samples taken one per interval anywhere within contiguous intervals $T < \pi/W_f$ [2] in duration.

ISBN 0-201-05758-1

$$Q(\omega) = \sum_{n=-\infty}^{\infty} C_n e^{jn2\pi\omega/\omega_s}, \tag{7.1-2}$$

where the series coefficients are given by (2.1-13):

$$C_n = \frac{1}{\omega_s} \int_{-\omega_s/2}^{\omega_s/2} Q(\omega) e^{-jn2\pi\omega/\omega_s} \, d\omega \tag{7.1-3}$$

for $n = 0, \pm 1, \pm 2, \ldots$. Since in the central period $Q(\omega) = F(\omega)$, the coefficients evaluate to

$$C_n = \frac{2\pi}{\omega_s} \frac{1}{2\pi} \int_{-W_f}^{W_f} F(\omega) e^{-jn2\pi\omega/\omega_s} \, d\omega = \frac{2\pi}{\omega_s} f\left(\frac{-n2\pi}{\omega_s}\right). \tag{7.1-4}$$

We see that the coefficients are proportional to instantaneous samples of $f(t)$ at times $-n2\pi/\omega_s = -nT_s$, where T_s is the time between samples:

$$T_s = 2\pi/\omega_s. \tag{7.1-5}$$

From (7.1-2) the periodic spectrum $Q(\omega)$ becomes

$$Q(\omega) = \sum_{n=-\infty}^{\infty} \frac{2\pi}{\omega_s} f\left(\frac{-n2\pi}{\omega_s}\right) e^{jn2\pi\omega/\omega_s}, \tag{7.1-6}$$

valid for all ω. Now we may use (7.1-6) with (7.1-1) to obtain $F(\omega)$. Inverse transforming the result gives the important expression

$$f(t) = \sum_{k=-\infty}^{\infty} \frac{2W_f}{\omega_s} f\left(\frac{k2\pi}{\omega_s}\right) \frac{\sin\left[W_f\left(t - \frac{k2\pi}{\omega_s}\right)\right]}{W_f\left(t - \frac{k2\pi}{\omega_s}\right)}, \tag{7.1-7}$$

where we have let $k = -n$. This equation constitutes a proof of the uniform sampling theorem, since it shows that $f(t)$ is known for all time from a knowledge of its samples. The samples determine the amplitudes of time signals in the sum having known form.

The time signals all have the form $\sin(x)/x$. This form was defined in Chapter 2 as a *sampling function* Sa (x)—that is,

$$\text{Sa}(x) = \frac{\sin(x)}{x}. \tag{7.1-8}$$

The sampling function is sometimes called an *interpolating function.* In terms of sampling functions, (7.1-7) can be restated as

$$f(t) = \frac{2W_f}{\omega_s} \sum_{k=-\infty}^{\infty} f\left(\frac{k2\pi}{\omega_s}\right) \text{Sa}\left[W_f\left(t - \frac{k2\pi}{\omega_s}\right)\right], \tag{7.1-9}$$

ISBN 0-201-05758-1

or equivalently,

$$f(t) = \frac{W_f T_s}{\pi} \sum_{k=-\infty}^{\infty} f(kT_s) \operatorname{Sa}\left[W_f T_s\left(\frac{t}{T_s} - k\right)\right]. \tag{7.1-10}$$

Recall that these results were derived under the restriction $\omega_s \geqslant 2W_f$, requiring a sampling interval

$$T_s \leqslant \pi/W_f, \tag{7.1-11}$$

as stated in the theorem. Important special cases of (7.1-9) and (7.1-10) occur for $T_s = \pi/W_f$ corresponding to sampling at the Nyquist (minimum) rate:

$$f(t) = \sum_{k=-\infty}^{\infty} f\left(\frac{k\pi}{W_f}\right) \operatorname{Sa}\left[W_f\left(t - \frac{k\pi}{W_f}\right)\right] \tag{7.1-12}$$

and

$$f(t) = \sum_{k=-\infty}^{\infty} f(kT_s) \operatorname{Sa}\left[\frac{\omega_s}{2}(t - kT_s)\right]. \tag{7.1-13}$$

Figure 7.1-1(c) illustrates a possible signal $f(t)$ as being the sum of time-shifted sampling functions as given by (7.1-13).

Example 7.1-1

As an example relating to sampling functions we show that

$$\int_{-\infty}^{\infty} \operatorname{Sa}\left[\frac{\omega_s}{2}(t - kT_s)\right] \operatorname{Sa}\left[\frac{\omega_s}{2}(t - mT_s)\right] dt = 0, \qquad m \neq k$$

$$= 2\pi/\omega_s, \qquad m = k.$$

The sampling functions are said to be *orthogonal* because they satisfy this relationship. We shall use Parseval's theorem, which may be written in the form

$$\int_{-\infty}^{\infty} x(t)y^*(t)\, dt = \frac{1}{2\pi} \int_{-\infty}^{\infty} X(\omega)Y^*(\omega)\, d\omega$$

for two signals $x(t)$ and $y(t)$ having Fourier transforms $X(\omega)$ and $Y(\omega)$, respectively. As always, the asterisk represents the complex conjugate operation.

Since

$$\operatorname{Sa}\left[\frac{\omega_s}{2}(t - mT_s)\right] \leftrightarrow \begin{cases} \dfrac{2\pi}{\omega_s} e^{-jm2\pi\omega/\omega_s}, & |\omega| < \omega_s/2 \\ 0, & |\omega| > \omega_s/2. \end{cases}$$

ISBN 0-201-05758-1

where the double-ended arrow represents a Fourier transform pair, we obtain the desired result by integration in the frequency domain.

$$\int_{-\infty}^{\infty} \text{Sa}\left[\frac{\omega_s}{2}(t - kT_s)\right] \text{Sa}\left[\frac{\omega_s}{2}(t - mT_s)\right] dt$$

$$= \frac{1}{2\pi}\int_{-\omega_s/2}^{\omega_s/2} \left(\frac{2\pi}{\omega_s}\right)^2 e^{-j(m-k)2\pi\omega/\omega_s}\, d\omega$$

$$= \frac{2\pi}{\omega_s}\frac{\sin[(m-k)\pi]}{(m-k)\pi} = 0, \qquad m \neq k$$

$$= 2\pi/\omega_s, \qquad m = k.$$

Our main results (7.1-9) and (7.1-10) are valid for complex as well as real waveforms. The waveforms are assumed to be nonrandom, however.

Time-Domain Sampling of Random Processes

There also exists a sampling theorem for random waveforms, such as noise, in the sense that the waveform is a sample function of a random process [3, p. 273]. Let $N(t)$ be a real, low-pass random process that is at least wide-sense stationary. If the process power density spectrum $\mathscr{S}_N(\omega)$, which is the Fourier transform of the autocorrelation function $R_N(\tau)$, is bandlimited to an interval $|\omega| \leqslant W_N$, then an approximation $\hat{N}(t)$ to $N(t)$ exists that has zero mean-squared error. That is,

$$E[\{\hat{N}(t) - N(t)\}^2] = 0. \tag{7.1-14}$$

The approximation is given by

$$\hat{N}(t) = \sum_{k=-\infty}^{\infty} N\left(\frac{k\pi}{W_N} + t_0\right) \text{Sa}\left[W_N\left(t - \frac{k\pi}{W_N} - t_0\right)\right], \tag{7.1-15}$$

where t_0 is an arbitrary constant and sampling is at the Nyquist rate.

A Network View of Sampling Theorems

There is a useful way in which low-pass sampling theorems may be viewed by using networks. Figure 7.1-2 illustrates the network. Either a nonrandom signal $f(t)$ or a random signal (noise) $N(t)$ may be instantaneously sampled by taking the product with a train of impulses as shown. By assuming the former, the output of the product is easily shown to be (see problem 7-4)

$$f_s(t) = f(t)\sum_{k=-\infty}^{\infty} \delta(t - kT_s) = \sum_{k=-\infty}^{\infty} f(kT_s)\delta(t - kT_s). \tag{7.1-16}$$

ISBN 0-201-05758-1

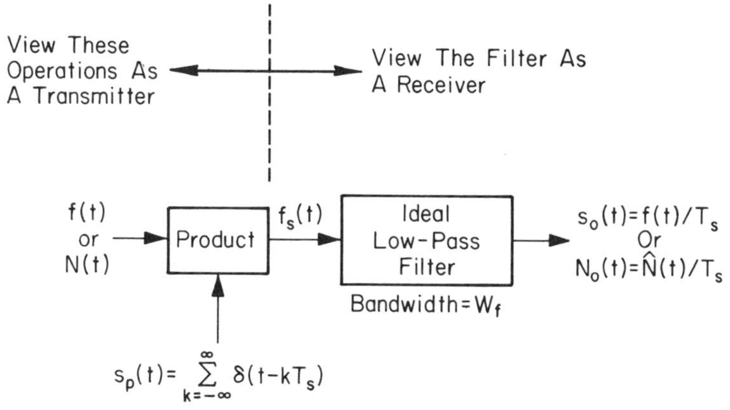

Fig. 7.1-2. Block diagram of a network useful in the interpretation of sampling theorems.

Next, assume the filter is ideal with a transfer function

$$\begin{aligned} H(\omega) &= 1, \qquad |\omega| < W_f \\ &= 0, \qquad |\omega| > W_f. \end{aligned} \tag{7.1-17}$$

Its impulse response, by inverse Fourier transformation of (7.1-17), is

$$h(t) = \frac{W_f}{\pi} \text{ Sa } (W_f t). \tag{7.1-18}$$

The filter output response to (7.1-16) now follows easily:

$$\begin{aligned} s_o(t) &= \frac{W_f}{\pi} \sum_{k=-\infty}^{\infty} f(kT_s) \text{ Sa } [W_f(t - kT_s)] \\ &= f(t)/T_s. \end{aligned} \tag{7.1-19}$$

Let us pause to summarize what has been developed. First, a train of instantaneous samples of $f(t)$, at the product device output, has been generated. The sample rate is $\omega_s = 2\pi/T_s \geqslant 2W_f$. These samples may be viewed as the output of a transmitter. Second, by use of an ideal low-pass filter having a bandwidth equal to the maximum frequency extent W_f of $f(t)$, the output is given by the middle term of (7.1-19). The filter may be viewed as the receiver which must recover $f(t)$. Finally, by application of the sampling theorem this output equals $f(t)/T_s$. Thus, we have shown that, within a constant factor, $f(t)$ is completely reconstructed from its samples by using an ideal low-pass filter. Reconstruction is valid for any sample rate $\omega_s \geqslant 2W_f$.

For a random (noise) process $N(t)$, a similar analysis gives the output process

$$\begin{aligned} N_o(t) &= \frac{W_n}{\pi} \sum_{k=-\infty}^{\infty} N(kT_s) \text{ Sa } [W_N(t - kT_s)] \\ &= \hat{N}(t)/T_s, \end{aligned} \tag{7.1-20}$$

ISBN 0-201-05758-1

where W_N is the maximum spectral extent of the noise and sampling is at a rate $\omega_s = 2\pi/T_s = 2W_N$. The filter bandwidth is also W_N.

Frequency-Domain Sampling Theorems

A frequency-domain sampling theorem may also be stated analogous to the time-domain theorem:

> A Fourier spectrum $F(\omega)$, corresponding to a signal $f(t)$, timelimited such that it is nonzero only in the interval $T_1 \leqslant t \leqslant T_2$, with $2T_f = T_2 - T_1$, is uniquely determined by its values at equispaced points in frequency separated by an amount $W_s \leqslant \pi/T_f$ rad/s.

By analogy, W_s and T_f here correspond to T_s and W_f, respectively, in the time-domain theorem. The result analogous to (7.1-10) may be found to be

$$F(\omega) = \frac{T_f W_s}{\pi} \sum_{k=-\infty}^{\infty} F(kW_s) \text{ Sa } [T_f(\omega - kW_s)] \tag{7.1-21}$$

for real or complex $f(t)$.

For noise having a timelimited autocorrelation function $R_N(\tau)$, and power density spectrum $\mathcal{S}_N(\omega)$, the corresponding result is

$$\mathcal{S}_N(\omega) = \frac{T_f W_s}{\pi} \sum_{k=-\infty}^{\infty} \mathcal{S}_N(kW_s) \text{ Sa } [T_f(\omega - kW_s)]. \tag{7.1-22}$$

Aliasing

All the time-domain sampling theorems are based on the signal's being strictly bandlimited. If it is not bandlimited there may be some overlap in the spectral components of the sampled signal. Such overlap, called spectral *aliasing,* is illustrated in Fig. 7.1-3(a), where the central portion of the spectrum about $\omega = 0$ is the unsampled message spectrum. Over the *maximum* allowable message spectral region of interest, $|\omega| \leqslant W_f \leqslant \omega_s/2$, there is clearly overlap from spectral "replicas," due to sampling, that are centered at ω_s and $-\omega_s$. Aliasing is undesirable because the message recovered from the sampled signal will contain uncorrectable distortion.

An obvious correction for the above aliasing is to filter the message prior to sampling to force it to be bandlimited. Of course, such filtering must not be so severe that the filtered message has an unacceptable distortion. Another correction consists in simply sampling at a higher rate such that $W_f \leqslant \omega_s/2$ is large enough to encompass the significant part of the message spectrum. A practical situation may involve both correction techniques.

Aliasing may also arise with bandlimited signals if the sample rate is too small. Figure 7.1-3(b) illustrates this situation. The obvious correction in this case is to sample at a higher rate.

7.2. PRACTICAL SAMPLING METHODS

The instantaneous sampling of the preceding section using impulses will be called *ideal sampling.* It must be approximated in practice by using large, vanishingly narrow pulses (in relation

ISBN 0-201-05758-1

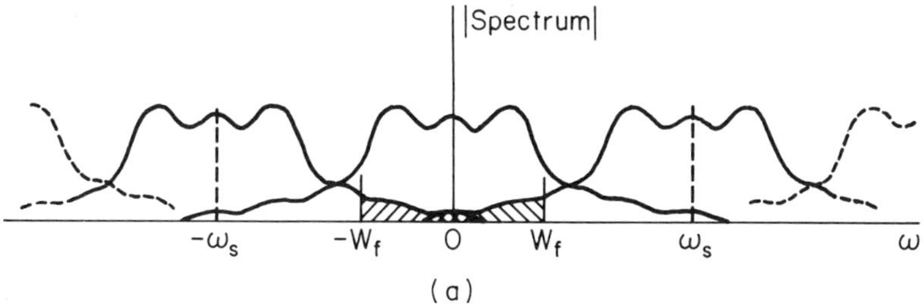

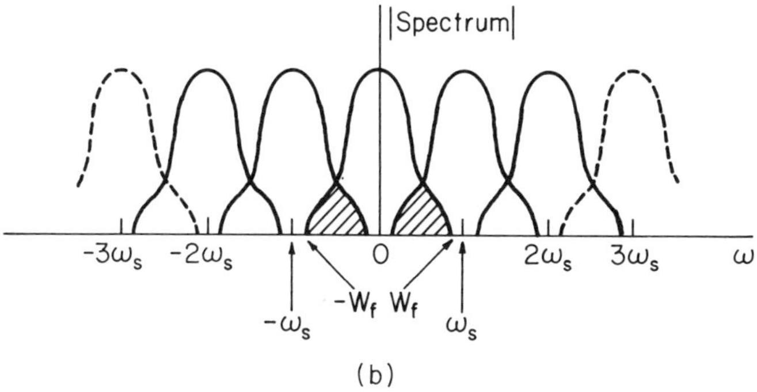

Fig. 7.1-3. (a) Sampled signal spectral overlap due to message not being bandlimited, and (b) spectral overlap caused by under sampling of a bandlimited message.

to π/W_f). On the other hand, it may be impractical or even undesirable to use very narrow pulses. Practical transmissions of interest will then use finite duration pulses. In the following paragraphs, we investigate several forms of such practical sampling.

Natural Sampling

Natural sampling involves a direct product of $f(t)$ and a train of rectangular pulses as shown in Fig. 7.2-1(a). The spectrum of $f(t)$ is defined as $F(\omega)$; it and $f(t)$ are sketched in (b). The pulse train $s_p(t)$ has amplitude K and has the Fourier series expansion

$$s_p(t) = \frac{K\tau}{T_s} \sum_{k=-\infty}^{\infty} \frac{\sin(k\omega_s\tau/2)}{k\omega_s\tau/2} e^{jk\omega_s t} \tag{7.2-1}$$

with τ being the pulse length and $\omega_s = 2\pi/T_s$ the pulse rate. The train and its spectrum

$$S_p(\omega) = \frac{2\pi K\tau}{T_s} \sum_{k=-\infty}^{\infty} \frac{\sin(k\omega_s\tau/2)}{k\omega_s\tau/2} \delta(\omega - k\omega_s) \tag{7.2-2}$$

are illustrated in (c). The product

$$f_s(t) = s_p(t)f(t) \tag{7.2-3}$$

is the sampled version of $f(t)$.

ISBN 0-201-05758-1

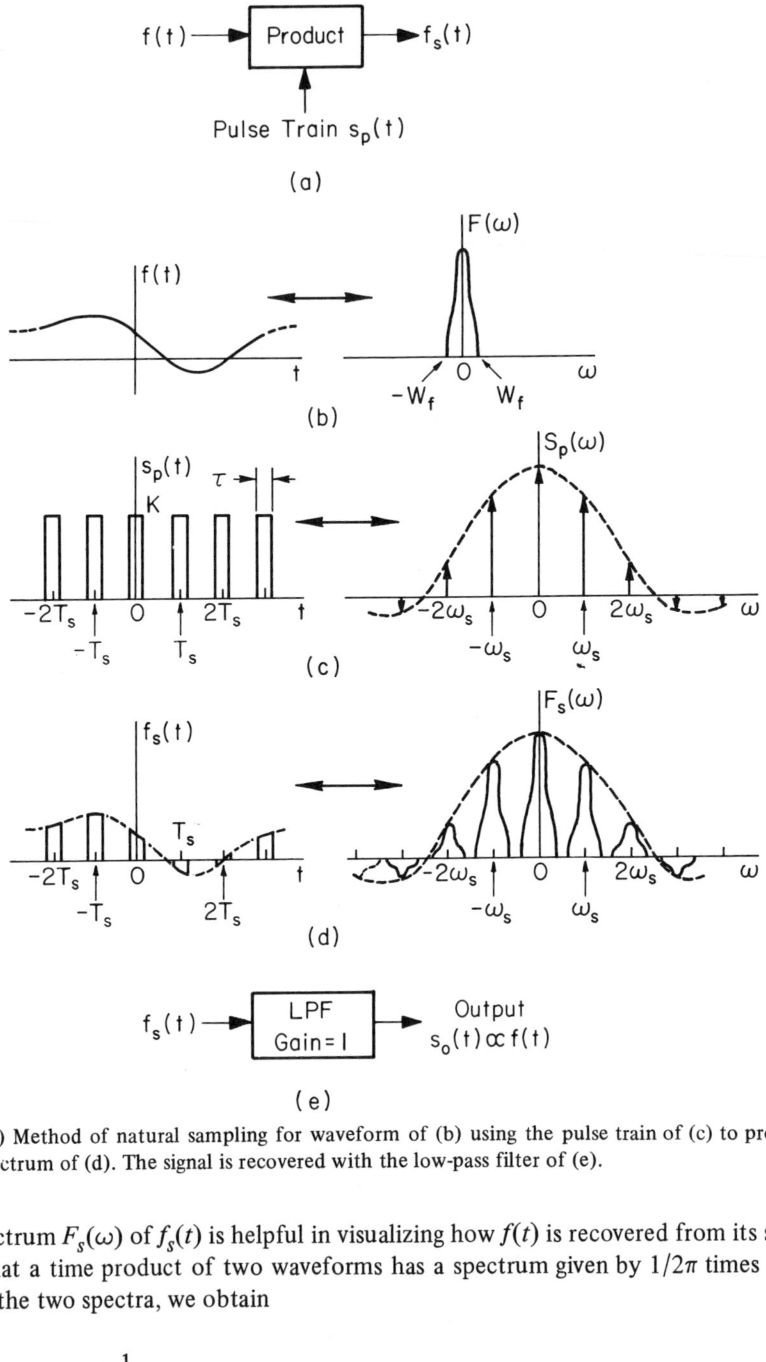

Fig. 7.2-1. (a) Method of natural sampling for waveform of (b) using the pulse train of (c) to produce the signal and spectrum of (d). The signal is recovered with the low-pass filter of (e).

The spectrum $F_s(\omega)$ of $f_s(t)$ is helpful in visualizing how $f(t)$ is recovered from its samples. Recalling that a time product of two waveforms has a spectrum given by $1/2\pi$ times the convolution of the two spectra, we obtain

$$F_s(\omega) = \frac{1}{2\pi} F(\omega) * S_p(\omega)$$

$$= \frac{K\tau}{T_s} \sum_{k=-\infty}^{\infty} \frac{\sin(k\omega_s\tau/2)}{k\omega_s\tau/2} \int_{-\infty}^{\infty} \delta(x - k\omega_s)F(\omega - x)\, dx$$

ISBN 0-201-05758-1

$$F_s(\omega) = \frac{K\tau}{T_s} \sum_{k=-\infty}^{\infty} \frac{\sin (k\omega_s \tau/2)}{k\omega_s \tau/2} F(\omega - k\omega_s). \tag{7.2-4}$$

The product and its spectrum are shown in Fig. 7.2-1(*d*).

From (7.2-4) it is seen that, so long as $\omega_s \geqslant 2W_f$, the spectrum of the sampled signal contains nonoverlapping, scaled, and frequency-shifted replicas of the information signal spectrum. By applying $f_s(t)$ to a low-pass filter* of bandwidth W_f, as shown in Fig. 7.2-1(*e*), $f(t)$ is easily recovered without distortion. The spectrum $S_o(\omega)$ of the output signal $s_o(t)$ from the filter will be

$$S_o(\omega) = \frac{K\tau}{T_s} F(\omega), \tag{7.2-5}$$

while the output signal is

$$s_o(t) = \frac{K\tau}{T_s} f(t). \tag{7.2-6}$$

Flat-Top Sampling

With this type of sampling the amplitude of each pulse in a pulse train is constant during the pulse but is determined by an instantaneous sample of $f(t)$ as illustrated in Fig. 7.2-2(*a*). The time of the instantaneous sample is chosen to occur at the pulse center for convenience. This

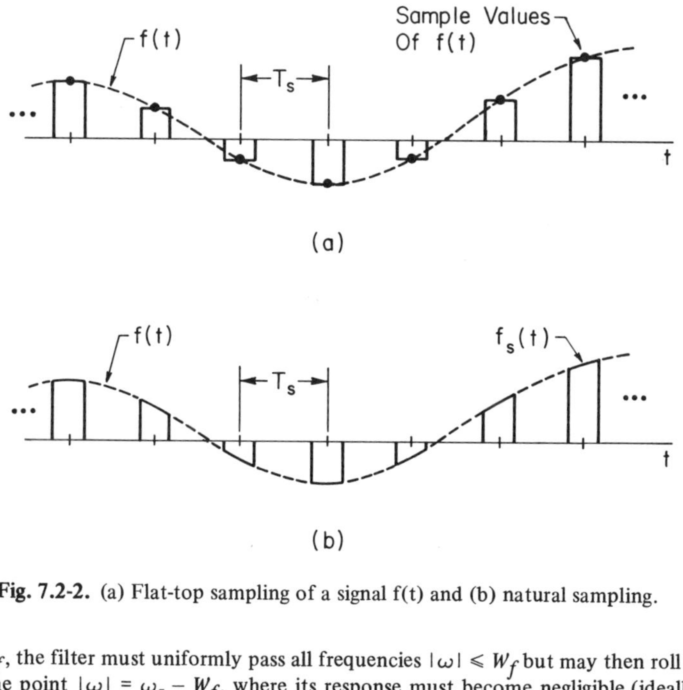

Fig. 7.2-2. (a) Flat-top sampling of a signal f(t) and (b) natural sampling.

*If $\omega_s > 2W_f$, the filter must uniformly pass all frequencies $|\omega| \leqslant W_f$ but may then roll off in a practical way out to the point $|\omega| = \omega_s - W_f$, where its response must become negligible (ideally zero). As ω_s approaches $2W_f$, the filter must approach the ideal low-pass filter.

ISBN 0-201-05758-1

choice is not necessary in general. Flat-top sampling differs from natural sampling which is illustrated for comparison in (*b*).

Assuming ideal rectangular pulses, the flat-top sampled signal is

$$f_s(t) = \sum_{k=-\infty}^{\infty} f(kT_s)\, K \operatorname{rect}\left(\frac{t - kT_s}{\tau}\right), \tag{7.2-7}$$

where K is a scale constant; it is the amplitude of a sampling pulse for a unit input (dc) signal. The function rect $(\cdot)$ is defined by (2.3-6).

To determine our ability to reconstruct $f(t)$ from the sampled signal (7.2-7), it is helpful to observe that the same expression derives from an ideal sampler followed by an amplifier with gain $K\tau$ and a filter. The filter must have an impulse response

$$q(t) = \frac{1}{\tau} \operatorname{rect}\left(\frac{t}{\tau}\right) \tag{7.2-8}$$

and transfer function

$$Q(\omega) = \frac{\sin(\omega\tau/2)}{\omega\tau/2}. \tag{7.2-9}$$

The network is illustrated* in Fig. 7.2-3(*a*). It is straightforward to show that the spectrum of $f_s(t)$ is

$$F_s(\omega) = \frac{K\tau}{T_s} \sum_{k=-\infty}^{\infty} \frac{\sin(\omega\tau/2)}{\omega\tau/2} F(\omega - k\omega_s). \tag{7.2-10}$$

$F_s(\omega)$ of (7.2-10) appears on the surface to be similar to (7.2-4) for natural sampling. There is an important difference, however. A low-pass filter operating on (7.2-10) will not give a distortion-free output proportional to $f(t)$. To see this fact, suppose a low-pass filter alone were used. The output spectrum would be $(K\tau/T_s)F(\omega)\sin(\omega\tau/2)/(\omega\tau/2)$, which is clearly not proportional to $F(\omega)$ as needed. The factor $Q(\omega) = \sin(\omega\tau/2)/(\omega\tau/2)$ represents distortion which may be corrected by adding a second filter, called an *equalizing filter.* It must have a transfer function $H_{eq}(\omega) = 1/Q(\omega)$ for $|\omega| \leqslant W_f$, that is,

$$\begin{aligned} H_{eq}(\omega) &= \frac{\omega\tau/2}{\sin(\omega\tau/2)}, \qquad |\omega| \leqslant W_f \\ &= \text{arbitrary elsewhere.} \end{aligned} \tag{7.2-11}$$

The equalized output spectrum is

$$S_o(\omega) = \frac{K\tau}{T_s} F(\omega), \tag{7.2-12}$$

and the output signal is

*This model is only for mathematical and conceptual convenience. Actual implementation in practice may be quite different. The amplifier is implied simply so that $Q(\omega)$ as defined may have a low-frequency gain of unity. The reasoning will become clearer as the reader proceeds to subsequent developments.

ISBN 0-201-05758-1

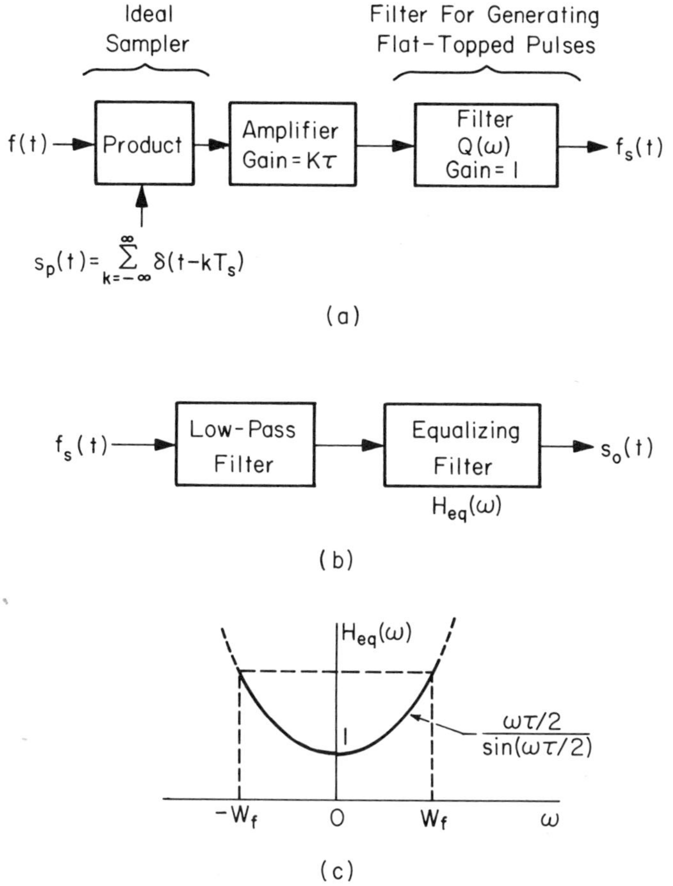

Fig. 7.2-3. (a) Generation method for flat-top samples. (b) Signal recovery method requiring the equalization filter with the transfer function of (c).

$$s_o(t) = \frac{K\tau}{T_s} f(t). \tag{7.2-13}$$

Our analysis has shown, in summary, that flat-top sampling still allows distortion-free reconstruction of the information signal from its samples as long as a proper equalization filter is added to the reconstruction (low-pass) filter path. These operations are given in block diagram form in Fig. 7.2-3(*b*). The equalizing filter transfer function is shown in (*c*).

Generalization to Arbitrary Pulse Shape

In the real world pulses are never perfectly rectangular as was assumed above for natural and flat-top sampling. To have such pulses would require infinite bandwidth in all circuits, an obviously unrealistic situation. It becomes appropriate to then ask: What can be done in a more realistic system? To answer this question, let us assume an arbitrarily shaped sampling pulse $p(t)$. The sampling pulse train will be a sum of such pulses occurring at the sampling rate ω_s. The spectrum of $p(t)$ is defined as $P(\omega)$.

Consider first a generalization of natural sampling. The sampling pulse train now becomes

ISBN 0-201-05758-1

$$s_p(t) = \sum_{k=-\infty}^{\infty} p(t - kT_s), \tag{7.2-14}$$

where $p(t)$ is some arbitrary pulse shape as illustrated in Fig. 7.2-4(a). Expanding $s_p(t)$ into its complex Fourier series having coefficients C_k, we develop the sampled signal as

$$f_s(t) = f(t) \sum_{k=-\infty}^{\infty} p(t - kT_s) = \sum_{k=-\infty}^{\infty} C_k f(t) e^{jk\omega_s t}. \tag{7.2-15}$$

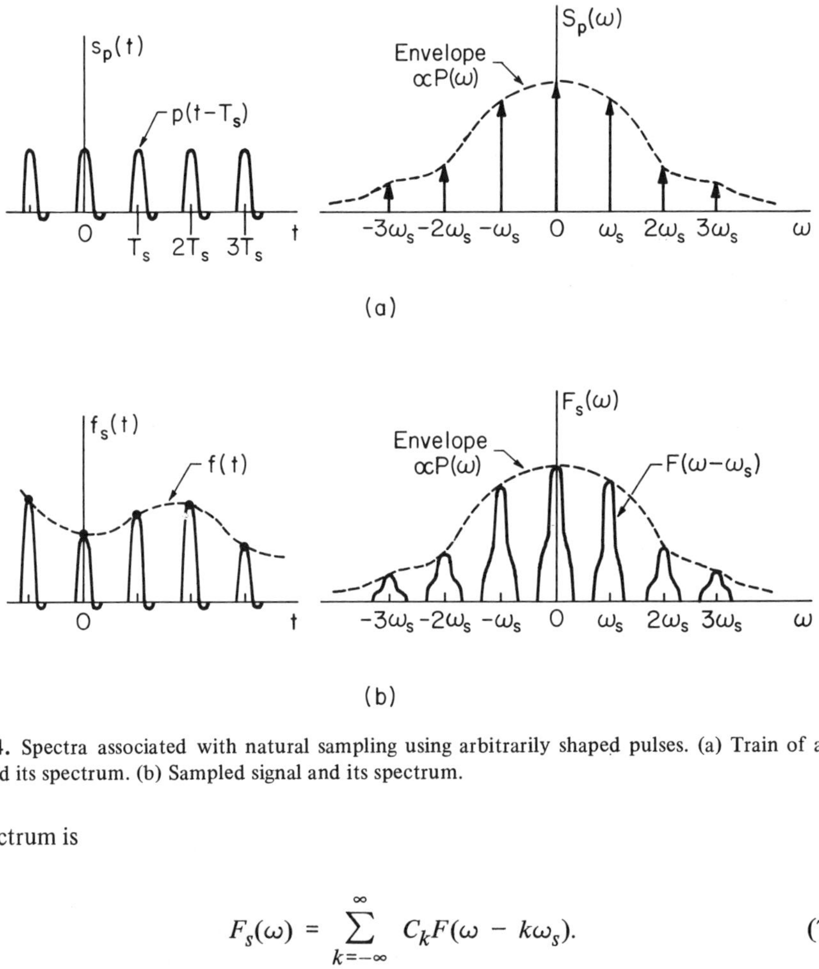

Fig. 7.2-4. Spectra associated with natural sampling using arbitrarily shaped pulses. (a) Train of arbitrary pulses and its spectrum. (b) Sampled signal and its spectrum.

The spectrum is

$$F_s(\omega) = \sum_{k=-\infty}^{\infty} C_k F(\omega - k\omega_s). \tag{7.2-16}$$

Both $f_s(t)$ and $F_s(\omega)$ are sketched in Fig. 7.2-4(b).

Again we see that by using a low-pass filter the signal $f(t)$ may be recovered without distortion. The filter will pass the term of (7.2-16) for $k = 0$. The output is easily found to be

$$s_o(t) = C_0 f(t). \tag{7.2-17}$$

Here C_0 is the dc component of the sampling pulse train. We may define parameters K and

ISBN 0-201-05758-1

τ such that $C_0 = K\tau/T_s$, where K is defined as the actual amplitude of $p(t)$ at $t = 0$. We may think of τ as the duration of an *equivalent rectangular pulse* [see (3.4-27)] that gives the same pulse train dc level and has the same amplitude at $t = 0$ as the actual pulse. It is easily determined that τ is given by

$$\tau = \frac{1}{p(0)} \int_{-T_s/2}^{T_s/2} p(t)\, dt \tag{7.2-18}$$

(see problem 7-6). With these definitions

$$s_o(t) = \frac{K\tau}{T_s} f(t). \tag{7.2-19}$$

The analysis above has shown that the only effect of using an arbitrary pulse shape in the sampling pulse train, as far as signal recovery in natural sampling goes, is to produce a reconstructed signal scaled by a factor equal to the dc component of the pulse train.

Flat-top (instantaneous) sampling may also be generalized. Following the points discussed above we define an arbitrary pulse shape $q(t)$, having a spectrum $Q(\omega)$, which is related to the arbitrary pulse $p(t)$ by

$$p(t) = K\tau q(t). \tag{7.2-20}$$

It is a normalized version of $p(t)$ having amplitude $1/\tau$ at $t = 0$ and unit spectral magnitude* at $\omega = 0$. As before, K is the amplitude of $p(t)$ at $t = 0$ and τ is the equivalent rectangular pulse of $p(t)$ found from (7.2-18). The sampled signal becomes

$$f_s(t) = \sum_{k=-\infty}^{\infty} f(kT_s) K\tau q(t - kT_s). \tag{7.2-21}$$

This is a generalization of (7.2-7). We now recognize that the block diagram of Fig. 7.2-3(*a*) applies to (7.2-21) if the filter impulse response is $q(t)$. The spectrum at the filter output is now found to be

$$F_s(\omega) = \frac{K\tau}{T_s} \sum_{k=-\infty}^{\infty} Q(\omega) F(\omega - k\omega_s). \tag{7.2-22}$$

Again, in reconstruction of $f(t)$, only the term for $k = 0$ is of interest, since it is the only one passed by the low-pass filter. The output spectrum is $K\tau Q(\omega)F(\omega)/T_s$, and the factor $Q(\omega)$ represents distortion. As in the flat-top case we may use an equalizing filter to remove the distortion. The required filter transfer function is

$$H_{eq}(\omega) = 1/Q(\omega). \tag{7.2-23}$$

The overall output with equalization becomes

*This is seen by calculating the Fourier series coefficient for both sides of (7.2-20), substituting the Fourier transform of $q(t)$ for $\omega = 0$, and using (7.2-18).

ISBN 0-201-05758-1

$$S_o(\omega) = \frac{K\tau}{T_s} F(\omega), \tag{7.2-24}$$

$$s_o(t) = \frac{K\tau}{T_s} f(t). \tag{7.2-25}$$

In all our practical sampling methods the output of the reconstruction filter in the receiver is $f(t)$ scaled by the dc component of the unmodulated pulse train.

★ 7.3. OTHER SAMPLING THEOREMS

Although there will be no need to invoke them in the following work, we shall briefly study some other sampling theorems.

It will be helpful to recall and use the network view of the sampling process. For ideal sampling of a low-pass signal we found that the sampling theorem representation for a signal $f(t)$ could be modeled as shown in Fig. 7.1-2. We interpreted the filter as a reconstruction filter in the receiver which used instantaneous (impulsive) samples as its input.

Even in cases where impulses were not transmitted, such as with natural or flat-topped pulses or the generalization to an arbitrary sample pulse $q(t)$, we still found that an ideal low-pass filter was required.* The effect of the shape of $q(t)$ only caused the need for an equalizing filter with transfer function $1/Q(\omega)$ with $q(t) \leftrightarrow Q(\omega)$. These comments are incorporated in the network interpretation of the low-pass sampling theorem shown in Fig. 7.3-1(a). Since $Q(\omega)H_{eq}(\omega) = 1$, there is no overall effect on the recovered output due to arbitrary pulse shape, and the equivalent representation of (b) applies to any of these sampling methods based on instantaneous samples.†

Higher-Order Sampling Defined

The low-pass sampling theorem may be summarized by writing $f(t)$ in the form

$$f(t) = \sum_{n=-\infty}^{\infty} f(nT_s)h(t - nT_s), \tag{7.3-1}$$

where we use the model of Fig. 7.3-1(b). The filter impulse response must be

$$h(t) = \frac{W_f T_s}{\pi} \text{ Sa } [W_f t] \tag{7.3-2}$$

in order for (7.3-1) to equal (7.1-10).

More generally, we may allow $f(t)$ to be either low-pass or band-pass, but still a band-limited, function. We now define $g(t)$, equal to the right side of (7.3-1), as a *first-order sampling* of $f(t)$ [4]. Thus, first-order sampling involves a single train of uniformly separated samples of $f(t)$. Extending the idea we define a p*th-order sampling* of $f(t)$ as

$$f(t) = \sum_{i=1}^{p} g_i(t) = \sum_{i=1}^{p} \sum_{n=-\infty}^{\infty} f(nT_{s_i} + \tau_i)h_i(t - nT_{s_i} - \tau_i). \tag{7.3-3}$$

*This fact is strictly true only if $\omega_s = 2W_f$. See footnote, p. 297.

†Note that we have assigned a gain T_s to the filter for convenience.

N 0-201-05758-1

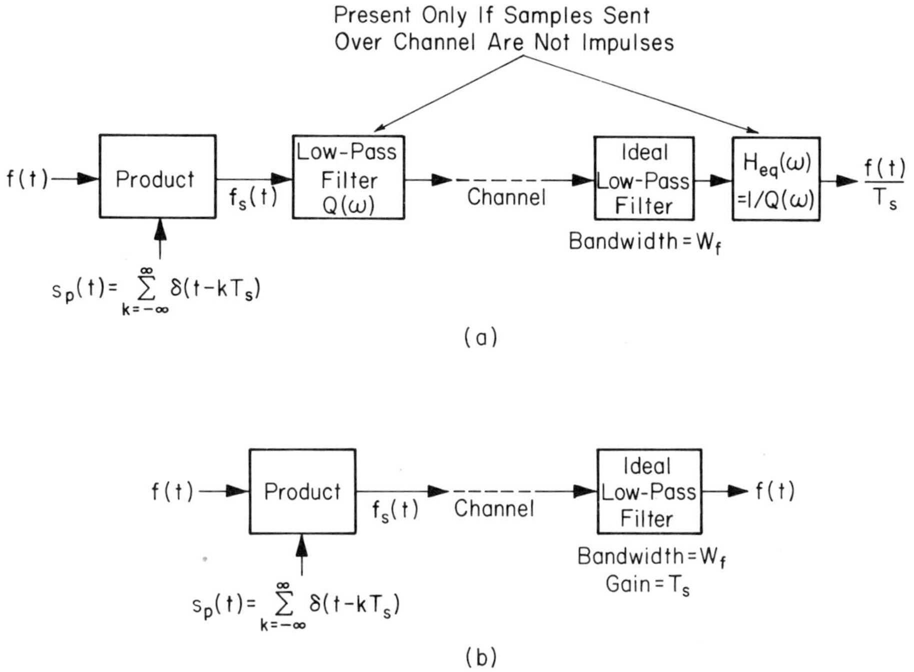

Fig. 7.3-1. Transmission and reception of sampled signals. (a) System using samples having arbitrary sampling waveform and (b) equivalent system.

Here we have p functions like (7.3-1). Each has uniformly separated samples T_{s_i} seconds apart. Each train has a time displacement τ_i relative to the chosen origin.

The general problem in higher-order sampling is to find the $h_i(t)$ which make (7.3-3) valid [4]. Only certain special cases are of usual interest. These are: (1) first-order low-pass signal sampling, (2) second-order low-pass signal sampling, (3) first-order band-pass signal sampling, and (4) second-order band-pass signal sampling. We have already discussed (1) extensively. In the following paragraphs we discuss (2), (3), and (4). In addition we shall also consider *quadrature sampling* of a band-pass waveform (it is a special form of sampling which involves preprocessing of the signal) and low-pass sampling in two dimensions.

Second-Order Sampling of Low-Pass Signals

Here $p = 2$. Let $\tau_1 = 0, \tau_2 = \tau$, $T_{s_2} = T_{s_1} = 2\pi/W_f$, the maximum sample interval (smallest sample rate allowable for each pulse train). The sample times of the first train are then $n2\pi/W_f$, while those of the second train are $(n2\pi/W_f) + \tau$. Figure 7.3-2(*a*) and (*b*) show the two trains of sampling impulses where $T_s = T_{s_1}/2$. The composite sampled signal is

$$f_s(t) = \sum_{n=-\infty}^{\infty} f(nT_{s_1})\delta(t - nT_{s_1}) + \sum_{n=-\infty}^{\infty} f(nT_{s_1} + \tau)\delta(t - nT_{s_1} - \tau), \tag{7.3-4}$$

which is illustrated in (*c*).

ISBN 0-201-05758-1

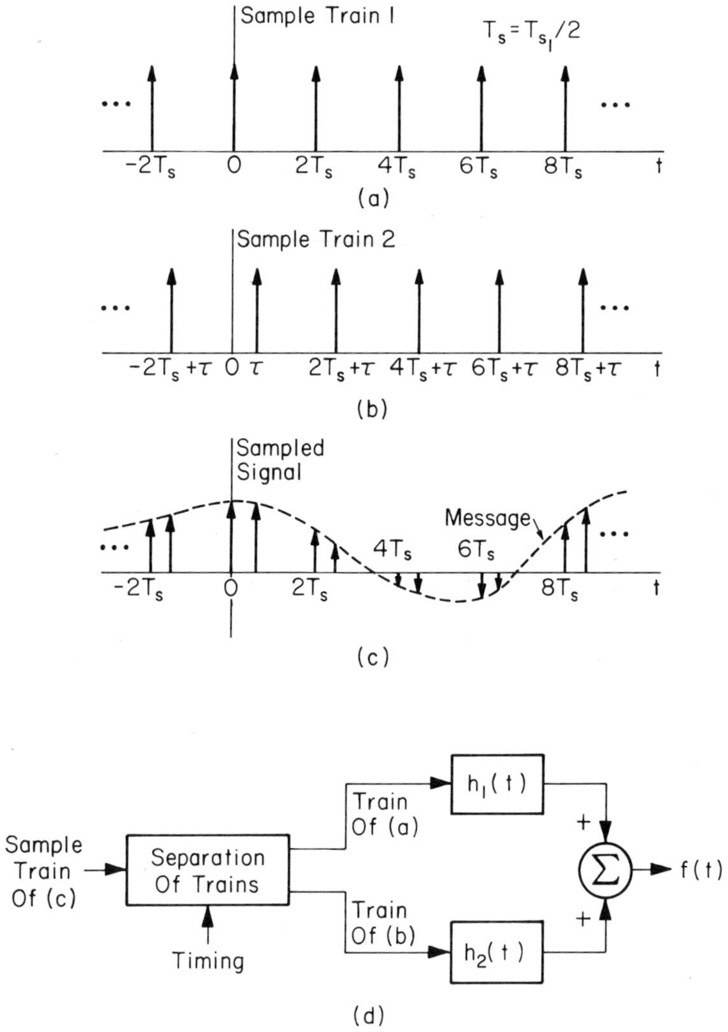

Fig. 7.3-2. (a) and (b) are sampling pulse trains used to produce the second-order sampled signal of (c). The message reconstruction method is shown in (d).

For a real time function $f(t)$ the necessary reconstruction filter impulse responses are [1]

$$h_1(t) = \frac{\cos\left[W_f\left(t - \frac{\tau}{2}\right)\right] - \cos(W_f\tau/2)}{W_f t \sin(W_f\tau/2)}, \tag{7.3-5}$$

$$h_2(t) = h_1(-t). \tag{7.3-6}$$

The receiver reconstruction of $f(t)$ is illustrated in Fig. 7.3-2(d). Note that, if $\tau = T_{s_1}/2 = T_s = \pi/W_f$, (7.3-5) and (7.3-6) reduce to

$$h_1(t) = h_2(t) = \frac{\sin(W_f t)}{W_f t} = \mathrm{Sa}(W_f t), \tag{7.3-7}$$

ISBN 0-201-05758-1

and (7.3-3) reduces to the first-order sampling result (7.1-12), as it should.

Probably the main advantage of second-order sampling of real low-pass time functions is that a nonuniform sampling may be used.

First-Order Sampling of Band-Pass Signals

Let $f(t)$ be a band-pass signal having spectral components only in the range $W_0 \leqslant |\omega| \leqslant W_0 + W_f$. It results that the Nyquist (minimum) sampling rate $2W_f$ can now be realized only for certain discrete values of $(W_0 + W_f)/W_f$. For other ratios the minimum rate will be larger than $2W_f$ but never larger than $4W_f$.

We let $\tau_1 = \tau = 0$ for convenience and also let $T_{s_1} = T_s$. The reconstruction filter impulse response to be used with (7.3-1) is [4]

$$h(t) = \frac{T_s}{\pi t} \left\{ \sin\,[(W_0 + W_f)t] - \sin\,(W_0 t) \right\}. \tag{7.3-8}$$

Only certain values of $\omega_s = 2\pi/T_s$ are allowed. Assuming $W_0 \neq 0$ and using results of Kohlenberg [4], the minimum allowable sample rate is

$$\omega_{s(\min)} = \frac{2}{M + 1} \left(1 + \frac{W_0}{W_f} \right) W_f, \tag{7.3-9}$$

where M is the largest nonnegative integer satisfying

$$M \leqslant W_0/W_f. \tag{7.3-10}$$

A plot of (7.3-9) is given in Fig. 7.3-3. The values of the peaks are

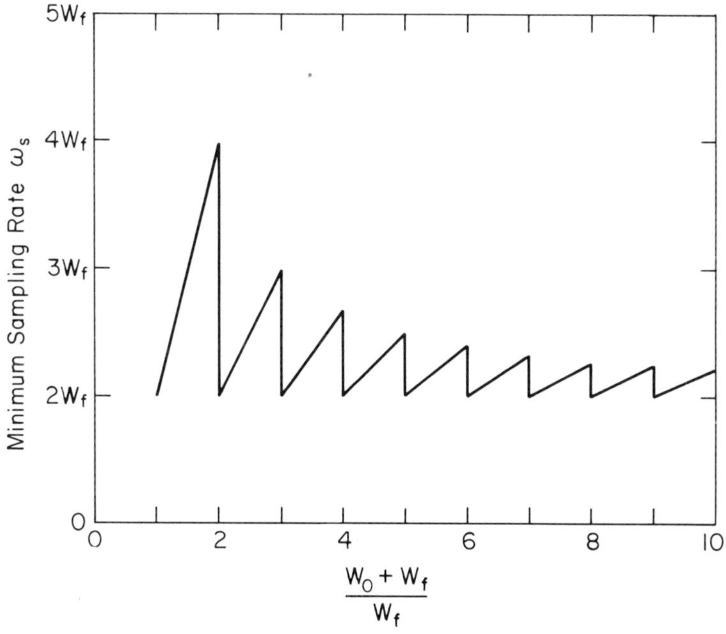

Fig. 7.3-3. Minimum sampling rate for first-order sampling of band-pass signals.

ISBN 0-201-05758-1

$$\omega_{s(\text{peaks})} = \frac{2(M+2)}{M+1} W_f, \qquad M = 0, 1, 2, \ldots . \tag{7.3-11}$$

We see that only when W_0/W_f equals an integer do we realize the Nyquist rate.

In general the samples of $f(t)$ are not independent, even when sampling is at the minimum rate given by (7.3-9), except when $W_0/W_f = 0, 1, \ldots$. As $W_0/W_f \to \infty$ they approach independence, however. Thus, in narrowband systems where $W_f \ll \omega_0 = W_0 + (W_f/2)$, samples are approximately independent.

Second-Order Sampling of Band-Pass Signals

The advantage of second-order versus first-order sampling of band-pass waveforms is that the minimum (Nyquist) rate of sampling is allowed for any choices of W_0 and W_f. Furthermore, the samples of $f(t)$ are independent.

Again letting $\tau_1 = 0$, $\tau_2 = \tau$, $T_{s_1} = T_{s_2} = 2T_s = 2\pi/W_f$, the reconstructed signal is given by (7.3-3) with [1]

$$h_1(t) = \frac{\cos[mW_f\tau - (W_0 + W_f)t] - \cos\{mW_f\tau - [(2m-1)W_f - W_0]t\}}{W_f t \sin(mW_f\tau)}$$
$$+ \frac{\cos\{[(2m-1)W_f\tau/2] - [(2m-1)W_f - W_0]t\} - \cos\{[(2m-1)W_f\tau/2] - W_0 t\}}{W_f t \sin[(2m-1)W_f\tau/2]}, \tag{7.3-12}$$

$$h_2(t) = h_1(-t). \tag{7.3-13}$$

Here m is the largest integer for which $(m-1)W_f < W_0$ and τ is arbitrary, except that it may not be an integral multiple of π/W_f unless $(m-1)W_f = W_0$ [1]. In the latter case W_0/W_f is an integer, and a development based on first-order sampling can be used to give

$$f(t) = \sum_{n=-\infty}^{\infty} f(nT_s)h(t - nT_s), \tag{7.3-14}$$

$$h(t) = \frac{T_s}{\pi t}\{\sin(mW_f t) - \sin[(m-1)W_f t]\}. \tag{7.3-15}$$

where samples are independent.

We observe that (7.3-15) agrees with (7.3-8) if we allow for the fact that $(m-1)W_f = W_0$.

Quadrature Sampling of Band-Pass Signals

We conclude the discussions on band-pass signal sampling by observing that it is not necessary that a band-pass signal be directly sampled. It is possible to precede sampling by preparatory processing of the waveform. Such operations will always allow sampling at the Nyquist rate if the signal is bandlimited.

Let

$$f(t) = A(t)\cos[\omega_0 t + \psi(t)] \tag{7.3-16}$$

ISBN 0-201-05758-1

be a band-pass signal having all its spectral components in the band $\omega_0 - (W_f/2) \leqslant |\omega| \leqslant \omega_0 + (W_f/2)$. Suppose we process $f(t)$ as illustrated in Fig. 7.3-4(*a*) prior to sampling. The product can be implemented with a balanced modulator. The low-pass filters are to remove spectral components centered around $2\omega_0$ while passing components in the band $-W_f/2 \leqslant \omega \leqslant W_f/2$. The "in-phase" and "quadrature" signals

$$f_I(t) = A(t) \cos [\psi(t)], \qquad (7.3\text{-}17)$$

$$f_Q(t) = A(t) \sin [\psi(t)], \qquad (7.3\text{-}18)$$

are bandlimited low-pass signals. *Each* may then be sampled at a minimum rate of $W_f/2\pi$ samples per second and reconstructed as shown in (*b*). By forming the products indicated in (*b*) we may recover $f(t)$. Thus, we may sample an arbitrary bandlimited band-pass signal at a *total* rate of W_f/π samples per second, using preprocessing, regardless of the ratio of $\omega_0 + (W_f/2)$ and $\omega_0 - (W_f/2)$, and recover $f(t)$. Notice that this is different from the sampling discussed before because we now have *two* samples being transmitted, each at a rate $W_f/2\pi$ samples per second, instead of one sample at a rate $2W_f/2\pi$.

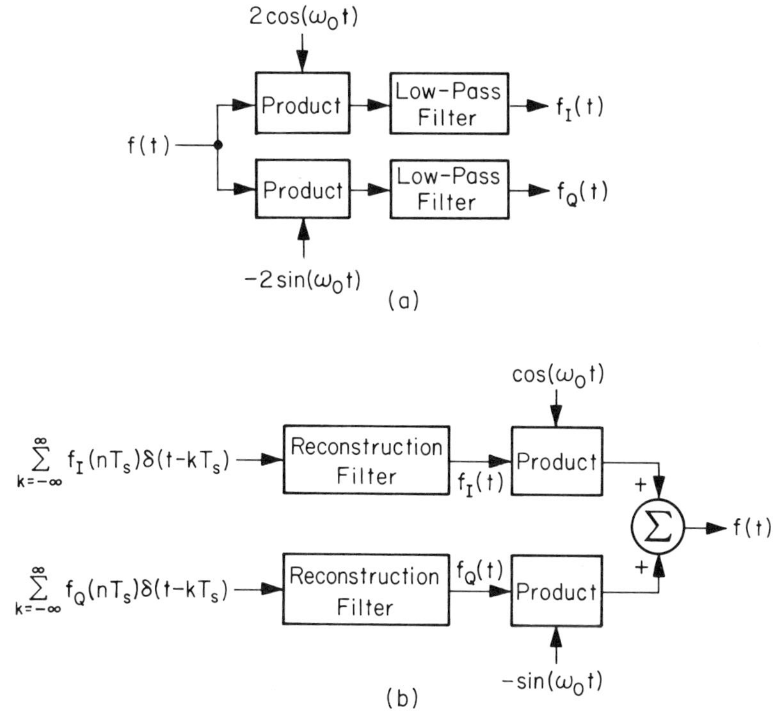

Fig. 7.3-4. Quadrature sampling of a band-pass signal f(t). (a) Quadrature signal generation prior to sampling and (b) waveform recovery from samples.

In quadrature sampling a means must be provided in the receiver to separate the two sample trains. Finally, we observe that the samples of $f_I(t)$ and $f_Q(t)$ do not have to occur at the same times. They may be interlaced forming a composite sample train at a rate W_f/π samples per second which alternately carries samples of $f_I(t)$ and $f_Q(t)$. Viewed in this manner we might think of quadrature sampling as being very similar to first-order band-pass sampling but having the advantage of always being able to sample at the Nyquist (minimum) rate.

ISBN 0-201-05758-1

Low-Pass Sampling Theorem in Two Dimensions

As a final topic in sampling theory we shall consider the uniform sampling theorem in two dimensions for low-pass functions [5]. Although the theorem has little application to ordinary radio communication systems, it is important in optical communication systems, antenna theory, picture processing, image enhancement, pattern recognition, and other areas.

It is difficult to state a completely general theorem owing to factors which we subsequently discuss.* However, we state the most useful and widely applied theorem:

> A low-pass function $f(t, u)$, bandlimited such that its Fourier transform $F(\omega, \sigma)$ is nonzero only within, at most, the region bounded by $|\omega| \leqslant W_{f\omega}$ and $|\sigma| \leqslant W_{f\sigma}$, may be completely reconstructed from uniform samples separated by an amount $T_{st} \leqslant \pi/W_{f\omega}$ in t and an amount $T_{su} \leqslant \pi/W_{f\sigma}$ in u.

In following paragraphs we shall prove this theorem and discuss some additional fine points. In particular we shall find that

$$f(t, u) = \sum_{k=-\infty}^{\infty} \sum_{n=-\infty}^{\infty} f(kT_{st}, nT_{su}) \operatorname{Sa}\left[\pi\left(\frac{t}{T_{st}} - k\right)\right] \operatorname{Sa}\left[\pi\left(\frac{u}{T_{su}} - n\right)\right], \tag{7.3-19}$$

which is the two-dimensional extension of (7.1-12). Although we only consider two dimensions and sampling on a rectangular lattice, the theorem can be generalized to N-dimensional Euclidean space with sampling on an N-dimensional parallelepiped [5].

The proof of (7.3-19) amounts to postulating the function

$$f(t, u) = \sum_{k=-\infty}^{\infty} \sum_{n=-\infty}^{\infty} f(t_k, u_n) h(t - t_k, u - u_n), \tag{7.3-20}$$

where

$$t_k = kT_{st}, \tag{7.3-21}$$

$$u_n = nT_{su}, \tag{7.3-22}$$

and finding an appropriate function $h(t, u)$ which will make (7.3-20) true for bandlimited $f(t, u)$. The function $h(t, u)$ is called an *interpolating function* and, as will be found, its solution is not unique.

We first extend the definition of the delta function to two dimensions as follows

$$\int_{-\infty}^{\infty} \int_{-\infty}^{\infty} \phi(\xi, \eta) \delta(\xi - x, \eta - y)\, d\xi\, d\eta = \phi(x, y), \tag{7.3-23}$$

where $\phi(\xi, \eta)$ is an arbitrary function, continuous at (x, y). Now using the fact that

*Recall that, even in the one-dimensional case, the low-pass theorem was not the most general theorem which could have been stated.

ISBN 0-201-05758-1

$$\delta(\xi - x, \eta - y) = \delta(\xi - x)\delta(\eta - y), \tag{7.3-24}$$

we may apply (7.3-23) to write (7.3-20) in the form

$$f(t, u) = \sum_{k=-\infty}^{\infty} \sum_{n=-\infty}^{\infty} \int_{-\infty}^{\infty} \int_{-\infty}^{\infty} f(\xi, \eta) h(t - \xi, u - \eta) \delta(\xi - t_k, \eta - u_n)\, d\xi\, d\eta$$

$$= \int_{-\infty}^{\infty} \int_{-\infty}^{\infty} f(\xi, \eta) h(t - \xi, u - \eta) \sum_{k=-\infty}^{\infty} \delta(\xi - t_k) \sum_{n=-\infty}^{\infty} \delta(\eta - u_n)\, d\xi\, d\eta, \tag{7.3-25}$$

if it is assumed that the order of integrations and summations may be reversed. From problem 2-16 it is known that

$$\sum_{k=-\infty}^{\infty} \delta(\xi - t_k) = \sum_{k=-\infty}^{\infty} \delta(\xi - kT_{st}) = \frac{1}{T_{st}} \sum_{k=-\infty}^{\infty} e^{jk\omega_{st}\xi}, \tag{7.3-26}$$

where the sample rate in t is

$$\omega_{st} = 2\pi/T_{st}. \tag{7.3-27}$$

Defining the sample rate in u as

$$\omega_{su} = 2\pi/T_{su}, \tag{7.3-28}$$

an expression similar to (7.3-26) can be written. Substitution of the two expressions into (7.3-25) gives

$$f(t, u) = \int_{-\infty}^{\infty} \int_{-\infty}^{\infty} h(t - \xi, u - \eta) \frac{1}{T_{st}T_{su}} \sum_{k=-\infty}^{\infty} \sum_{n=-\infty}^{\infty} f(\xi, \eta) e^{jk\omega_{st}\xi + jn\omega_{su}\eta}\, d\xi\, d\eta. \tag{7.3-29}$$

This expression is recognized as the two-dimensional convolution of the two functions $h(t, u)/T_{st}T_{su}$ and

$$f_s(t, u) = \sum_{k=-\infty}^{\infty} \sum_{n=-\infty}^{\infty} f(t, u) e^{jk\omega_{st}t + jn\omega_{su}u}. \tag{7.3-30}$$

If the Fourier transform of $f_s(t, u)$ is defined as $F_s(\omega, \sigma)$, the frequency shifting property of Fourier transforms gives

$$F_s(\omega, \sigma) = \sum_{k=-\infty}^{\infty} \sum_{n=-\infty}^{\infty} F(\omega - k\omega_{st}, \sigma - n\omega_{su}). \tag{7.3-31}$$

ISBN 0-201-05758-1

Next, we recognize that the Fourier transform of a convolution of two functions in the t, u domain is the product of individual transforms in the ω, σ domain. The transform of (7.3-29) is then

$$F(\omega, \sigma) = \frac{H(\omega, \sigma)}{T_{st}T_{su}} \sum_{k=-\infty}^{\infty} \sum_{n=-\infty}^{\infty} F(\omega - k\omega_{st}, \sigma - n\omega_{su}), \tag{7.3-32}$$

where $H(\omega, \sigma)$ is defined as the transform of $h(t, u)$.

This expression allows us to determine the required properties of the interpolating function $h(t, u)$. Figure 7.3-5 will aid in its interpretation. The bandlimited spectrum $F(\omega, \sigma)$ is illustrated in (a). The double sum of terms in (7.3-32) is illustrated in (b). Clearly, if the right side of (7.3-32) must equal $F(\omega, \sigma)$, two requirements must be satisfied. First, $\omega_{st} \geqslant 2W_{f\omega}$ and $\omega_{su} \geqslant 2W_{f\sigma}$ are necessary so that the replicas in (b) do not overlap, and second, the function $H(\omega, \sigma)/T_{st}T_{su}$ must equal unity over the *aperture* region in the ω, σ plane occupied by $F(\omega, \sigma)$ and must be zero in all regions occupied by the replicas. In the space between these regions $H(\omega, \sigma)/T_{st}T_{su}$ may be arbitrary. Hence, there is no unique interpolating function in general. The first requirement establishes the sampling intervals stated in the original theorem.

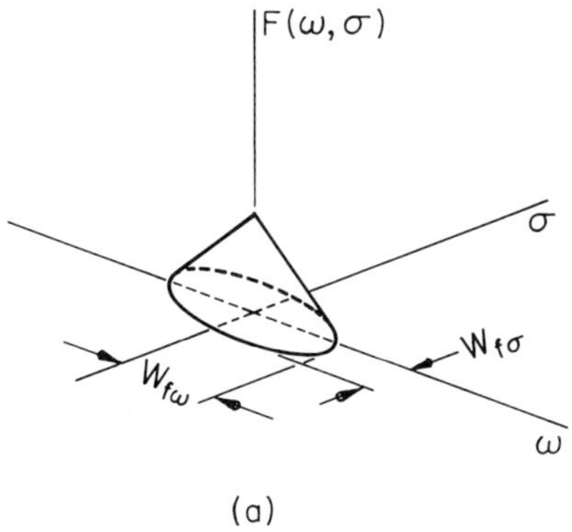

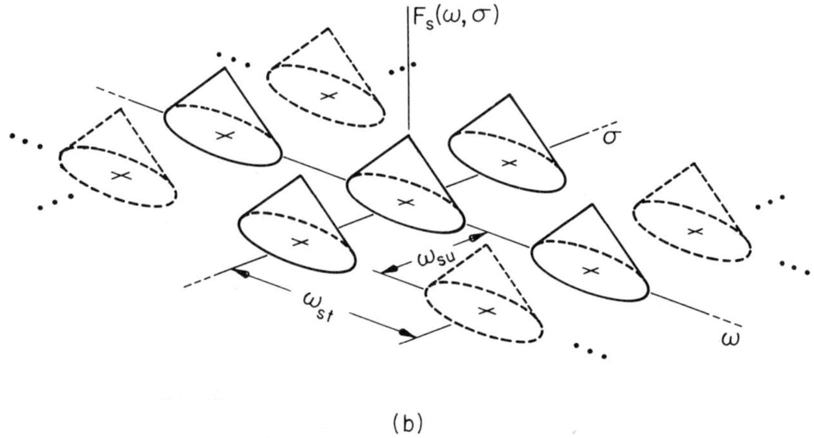

Fig. 7.3-5. A two-dimensional, bandlimited Fourier transform (a) and its periodic version (b) representing the spectrum of the sampled signal.

ISBN 0-201-05758-1

Regarding the second condition, we may ask, What interpolating function should be used? There is no one correct answer. However, suppose we select

$$H(\omega, \sigma) = T_{st}T_{su} \operatorname{rect}\left(\frac{\omega}{\omega_{st}}\right) \operatorname{rect}\left(\frac{\sigma}{\omega_{su}}\right). \tag{7.3-33}$$

This choice has the advantage of admitting *all* aperture shapes within the prescribed rectangular boundary. By inverse transformation

$$h(t, u) = \frac{1}{(2\pi)^2} \int_{-\infty}^{\infty} \int_{-\infty}^{\infty} H(\omega, \sigma) e^{j\omega t + j\sigma u} \, d\omega \, d\sigma$$

$$= \frac{T_{st}}{2\pi} \int_{-\omega_{st}/2}^{\omega_{st}/2} e^{j\omega t} \, d\omega \, \frac{T_{su}}{2\pi} \int_{-\omega_{su}/2}^{\omega_{su}/2} e^{j\sigma u} \, d\sigma$$

$$= \operatorname{Sa}(\pi t/T_{st}) \operatorname{Sa}(\pi u/T_{su}). \tag{7.3-34}$$

By substituting this expression back into (7.3-20), we have finally proved the original theorem embodied in (7.3-19).

The interpolating function defined by either (7.3-33) or (7.3-34) is called the *canonical interpolating function* [5]. It is *orthogonal* in the t, u space, which means that samples of $f(t, u)$ are linearly independent.

For the rectangular sampling plan the canonical interpolating function may give 100% *sampling efficiency*. Let sampling efficiency η be defined as the ratio of the area A_a in the ω, σ plane over which $F(\omega, \sigma)$ is nonzero to the area A_s of the rectangle defining the repetitive "cell" due to sampling [5]. Since $A_s = \omega_{st}\omega_{su}$, we have

$$\eta = A_a/A_s = A_a/\omega_{st}\omega_{su}. \tag{7.3-35}$$

Efficiency is maximized by sampling at the minimum rates. Assuming this to be the case, A_a is maximum for $F(\omega, \sigma)$ existing over a rectangular aperture. The corresponding efficiency is $\eta = 1.0$ or 100%. For comparison, it can be shown that maximum efficiency is 78.5% for either a circular aperture, or an elliptical aperture with major axis in either ω or σ directions.

7.4. PULSE AMPLITUDE MODULATION

We are now in a position to apply the foregoing sampling principles to the practical transmission of information. We begin with the transmission of a single information message using *pulse amplitude modulation* (PAM). In the following section time multiplexing is introduced, which will allow us to study multimessage transmission using PAM.

PAM Signal and Spectrum

ISBN 0-201-05758-1

The usual PAM signal is of the generalized form described in Section 7.2. The sampled signal, assuming bipolar modulation, is given by (7.2-21). It will be recalled that K is the amplitude of the sampling pulses $p(t - kT_s) = K\tau q(t - kT_s)$; $q(t)$ determined the shape of the sampling pulses and had an amplitude $1/\tau$, where τ was given by (7.2-18).* However, unipolar modulation is more typical and can be considered by adding a dc level A to $f(t)$ that is sufficiently large to assure $A + f(t) \geqslant 0$. The sampled signal in either case can then be expressed by

$$s_{\text{PAM}}(t) = K\tau \sum_{n=-\infty}^{\infty} [A + f(nT_s)]\, q(t - nT_s). \tag{7.4-1}$$

The functional operations leading to the generation of $s_{\text{PAM}}(t)$ are shown in Fig. 7.4-1(*a*). The PAM signal is shown in (*b*). To find the spectrum $S_{\text{PAM}}(\omega)$ of $s_{\text{PAM}}(t)$, it is easiest to take the product of the spectrum of $f_s(t)$ shown in (*a*) with $K\tau$ times the filter transfer function $Q(\omega)$ where

$$q(t) \leftrightarrow Q(\omega). \tag{7.4-2}$$

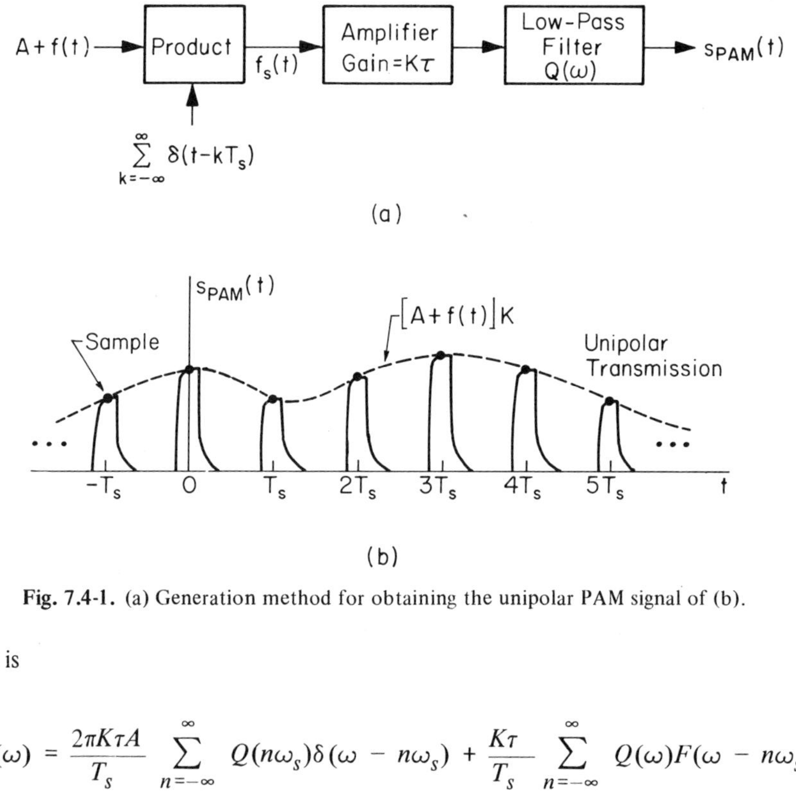

Fig. 7.4-1. (a) Generation method for obtaining the unipolar PAM signal of (b).

The result is

$$S_{\text{PAM}}(\omega) = \frac{2\pi K\tau A}{T_s} \sum_{n=-\infty}^{\infty} Q(n\omega_s)\delta(\omega - n\omega_s) + \frac{K\tau}{T_s} \sum_{n=-\infty}^{\infty} Q(\omega)F(\omega - n\omega_s), \tag{7.4-3}$$

where $\omega_s = 2\pi/T_s$ is the sampling rate and $F(\omega)$ is the spectrum of $f(t)$. A sketch of (7.4-3) is given in Fig. 7.4-2(*a*). The discrete spectral lines (impulses) are due to the assumption of uni-

*These definitions necessarily result in $\int_{-T_s/2}^{T_s/2} q(t)\, dt = 1$; that is, the area of $q(t)$ over a sample period is unity. The area of $p(t)$ will be $K\tau$.

ISBN 0-201-05758-1

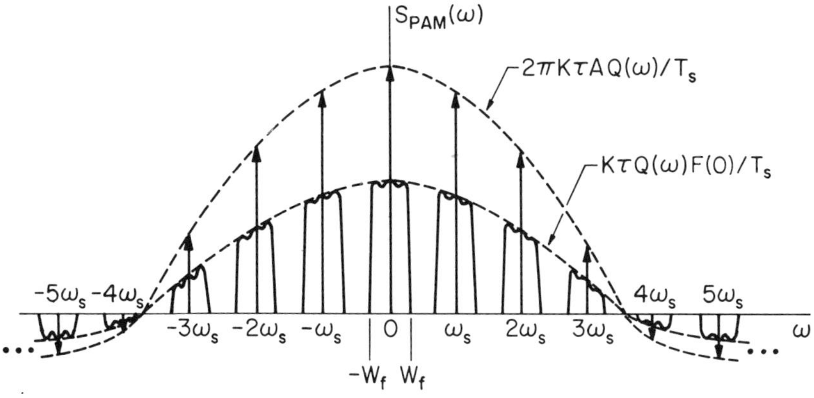

(a)

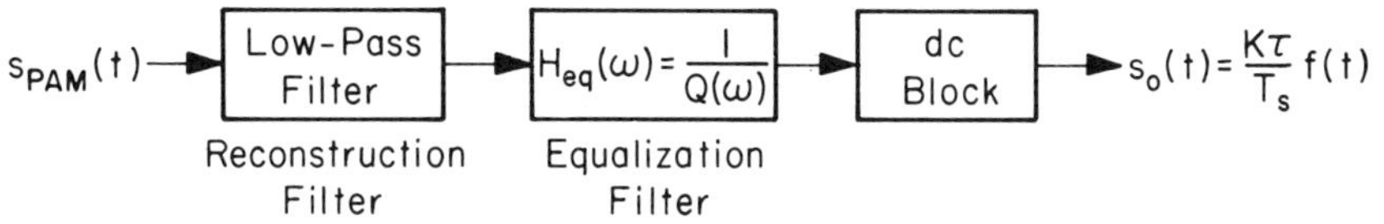

(b)

Fig. 7.4-2. (a) Spectrum of PAM signal using arbitrarily shaped sampling pulses, and (b) required system to recover the message.

polar transmission ($A \neq 0$). For bipolar transmission $A = 0$, and there are no discrete lines. As usual, we assume that $f(t)$ is bandlimited to $|\omega| < W_f$, so that so long as $\omega_s \geqslant 2W_f$ there is no spectral overlap in shifted replicas of $F(\omega)$.

Example 7.4-1

Let us take an example of PAM using a triangular sampling pulse defined by

$$p(t) = K\left(1 - \frac{|t|}{\tau_0}\right), \qquad 0 \leqslant |t| \leqslant \tau_0$$

$$= 0, \qquad \tau_0 < |t|.$$

The example will illustrate the calculation of τ, $q(t)$, and $Q(\omega)$ required to define (7.4-1) and (7.4-3). To find τ we use (7.2-18):

$$\tau = \frac{\displaystyle\int_{-\tau_0}^{0} K\left(1 + \frac{t}{\tau_0}\right) dt + \int_{0}^{\tau_0} K\left(1 - \frac{t}{\tau_0}\right) dt}{K}$$

which evaluates easily to

$$\tau = \tau_0.$$

ISBN 0-201-05758-1

Next, we write $p(t)$ in the form

$$p(t) = K\tau_0 \left(\frac{1 - \dfrac{|t|}{\tau_0}}{\tau_0} \right) = k\tau_0 q(t), \qquad 0 \leqslant |t| \leqslant \tau_0,$$

which defines $q(t)$ as

$$q(t) = \frac{1}{\tau_0}\left(1 - \frac{|t|}{\tau_0}\right), \qquad 0 \leqslant |t| \leqslant \tau_0$$

$$= 0, \qquad \tau_0 < |t|.$$

Finally, we use the results of example 2.3-2 of Chapter 2 to obtain

$$Q(\omega) = \mathrm{Sa}^2\ (\omega\tau_0/2).$$

Observe that the sample pulses have amplitude K, while the amplitude of $q(t)$ is $1/\tau_0$ and $Q(0) = 1$, all as required by the definitions of $p(t)$ and $q(t)$.

Signal Recovery by Low-Pass Filtering

A low-pass filter in the receiver will remove all spectral components in (7.4-3) except those for $n = 0$ if its response is zero for $|\omega| > \omega_s - W_f$. The spectrum at its output will then contain a dc component $2\pi KA\tau Q(0)\delta(\omega)/T_s$ and the information component $K\tau Q(\omega)F(\omega)/T_s$. By incorporating a dc block and equalization filter to remove the distortion due to the factor $Q(\omega)$, we recover

$$s_o(t) = \frac{K\tau}{T_s} f(t), \tag{7.4-4}$$

as illustrated in Fig. 7.4-2(b).

If $\omega_s > 2W_f$, the low-pass filter does not have to be ideal. There will be a frequency interval of width $\omega_s - 2W_f$, where a practical filter may roll off to a negligible transmission level. As $\omega_s \to 2W_f$ the filter must approach ideal.

Signal Recovery by Zero-Order Sample-Hold

Because of the factor τ/T_s the output of the receiver using the low-pass filter reconstruction method is not as large as we might like. A simple sample-hold circuit may be used to increase the output by a factor T_s/τ. The circuit is shown in Fig. 7.4-3(a). The amplifier gain is arbitrary and assumed to equal unity; it only needs to provide a very low output impedance when driving the capacitor.

For purposes of description, assume the input to the sample-hold circuit is a flat-top PAM signal as illustrated in (b). At the time sample points (shown as heavy dots) the switch instantaneously* closes and the output level equals the input sample amplitude. While the switch is open the capacitor holds the voltage, as shown dashed, until the next closure. The

*In practice it would close just after the start of a sample pulse and open just before it ends.

ISBN 0-201-05758-1

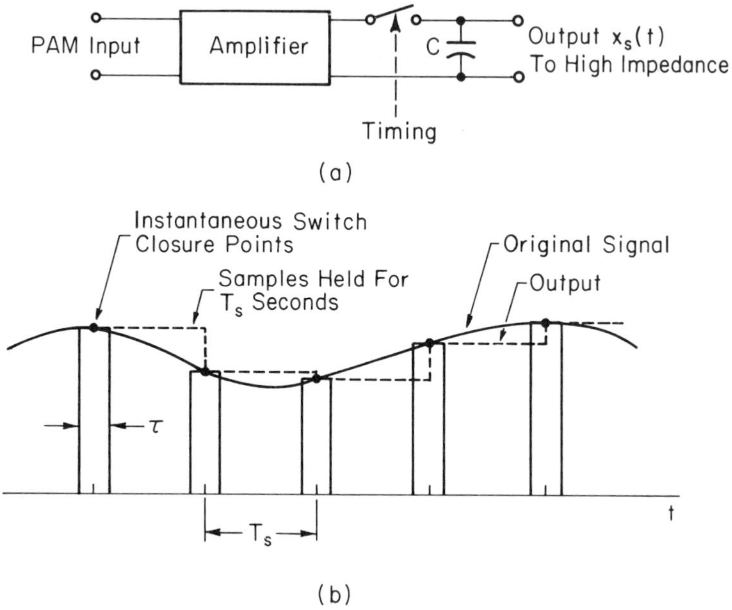

Fig. 7.4-3. (a) Sample-hold circuit, and (b) applicable waveforms.

output still looks like a flat-top sampled PAM signa!, but its pulse width is now T_s instead of τ seconds.

Using (7.4-1) as the PAM input, the sample amplitudes are $K[A + f(nT_s)]$ [recall that $q(nT_s) = 1/\tau$]. The new sampled and held PAM signal becomes

$$x_s(t) = K \sum_{n=-\infty}^{\infty} [A + f(nT_s)]\, h(t - nT_s). \tag{7.4-5}$$

Here

$$\begin{aligned} h(t) &= 1, \qquad 0 < t < T_s \\ &= 0, \qquad \text{elsewhere} \end{aligned} \tag{7.4-6}$$

is due to the action of the sample-hold circuit; it is its impulse response.

If $H(\omega)$ is the Fourier transform of $h(t)$, we find the spectrum $X_s(\omega)$ of the output of the sample-hold circuit to be

$$\begin{aligned} X_s(\omega) = {} & \frac{2\pi KA}{T_s} \sum_{n=-\infty}^{\infty} H(n\omega_s)\delta(\omega - n\omega_s) \\ & + \frac{K}{T_s} \sum_{n=-\infty}^{\infty} H(\omega)F(\omega - n\omega_s), \end{aligned} \tag{7.4-7}$$

where

ISBN 0-201-05758-1

$$H(\omega) = T_s \frac{\sin(\omega T_s/2)}{\omega T_s/2} \exp(-j\omega T_s/2). \quad (7.4\text{-}8)$$

From (7.4-7) we see that a low-pass filter, equalizer, and dc block are required to recover $f(t)$. The overall receiver block diagram is shown in Fig. 7.4-4(*a*). The equalizer transfer function required is

$$\begin{aligned} H_{eq}(\omega) &= T_s/H(\omega), \qquad |\omega| \leqslant W_f \\ &= \text{arbitrary}, \qquad \text{elsewhere}. \end{aligned} \quad (7.4\text{-}9)$$

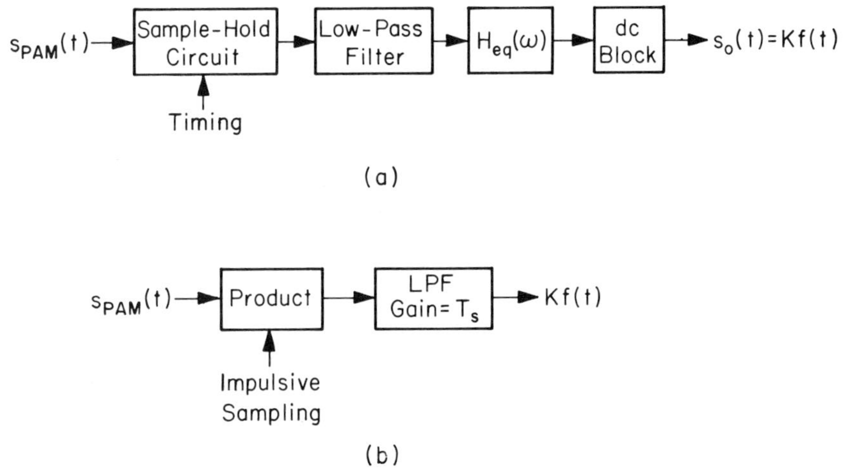

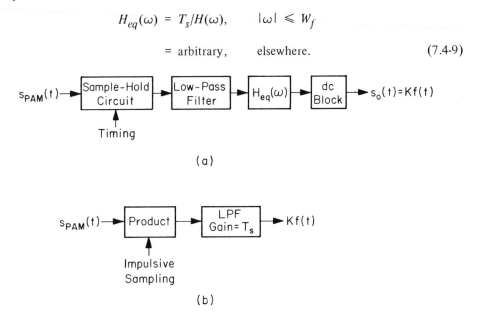

Fig. 7.4-4. (a) PAM receiver using a sample-hold circuit for signal recovery. (b) Equivalent receiver.

The output recovered message will be $Kf(t)$. From the standpoint of message reconstruction, the overall sample-hold receiving system can be replaced by the equivalent system illustrated in (*b*). It is made up of an ideal sampler followed by an ideal filter† having a gain T_s and bandwidth W_f.

The sample-hold circuit described above is called zero-order because the "held" voltage may be described by a polynomial of order zero.

First-Order Sample-Hold

Figure 7.4-5 illustrates the process of first-order sampling and holding. At a particular instant (say nT_s) a sample of the PAM signal is held until the next sample. Now, however, rather than being a constant level held, the level between samples varies linearly. The slope is determined by the present sample (at time nT_s) and the immediately past sample [at $(n-1)T_s$].

The output of the sample-hold circuit again has the spectrum of (7.4-7) where the transfer function $H(\omega)$ of the sample-hold circuit is now [6]

$$H(\omega) = T_s(1 + j\omega T_s)\left[\frac{\sin(\omega T_s/2)}{\omega T_s/2}\right]^2 \exp(-j\omega T_s). \quad (7.4\text{-}10)$$

The block diagram of Fig. 7.4-4 also applies to first-order sampling and holding. The equalization filter transfer function to be used is

†The filter is assumed to include also a dc block.

ISBN 0-201-05758-1

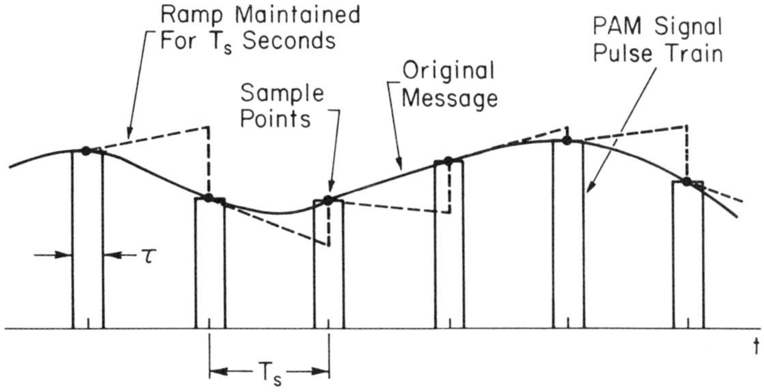

Fig. 7.4-5. Waveforms applicable to first-order sample-hold.

$$H_{eq}(\omega) = T_s/H(\omega), \qquad |\omega| \leqslant W_f$$
$$= \text{arbitrary}, \qquad \text{elsewhere.} \tag{7.4-11}$$

Higher-Order Sample-Holds

Sample-hold operations in general may be fractional or higher integer-order. These are discussed in reference [6]. A zero-order operation is capable of reproducing a constant (zero-order polynomial) signal $f(t)$ perfectly. A first-order sample-hold operation can reproduce exactly a constant or ramp (first-order polynomial) signal $f(t)$. Thus, an nth-order sample-hold can reproduce exactly a polynomial signal of order n. Sample-hold circuits above first-order are rarely used in practice for various reasons including economy. Fractional-order data holds are sometimes preferred in control systems [6].

7.5. TIME-DIVISION MULTIPLEXING

The concept of time-division multiplexing (TDM) as a means of combining several information signals in time by sampling was briefly introduced at the start of this chapter. The concept will now be discussed in more detail. We shall assume that PAM sampling is involved in subsequent work. This choice is not to imply that TDM cannot be implemented for other types of modulation. On the contrary, all other types of modulation (PDM, PPM, PCM, etc.) may be time-multiplexed. However, since only PAM has been introduced at this point, we elect to include TDM as part of discussions of other modulation methods as they are introduced.

The Basic Concept

The conceptual implementation of a TDM system is shown in Fig. 7.5-1(a) for multiplexing N nearly similar information sources (such as telephone signals). Assuming unipolar PAM, a dc level A plus a message is passed through a low-pass filter to assure that the message to be sampled is bandlimited. The filtered signal is applied to a sampling switch or *commutator* which makes one revolution every T_s seconds, thereby sampling all N signals once. The sampled version of message 1 is shown in (b) while that for message 2 is given in (c). The composite multiplexed signal is the sum of N sampled messages as illustrated in (d). A group of pulses comprising one sample from each message is called a *frame*. The time within one frame for each message sample is a *time slot*. The portion of the time slot not used by the sample pulse is called *guard* time.

ISBN 0-201-05758-1

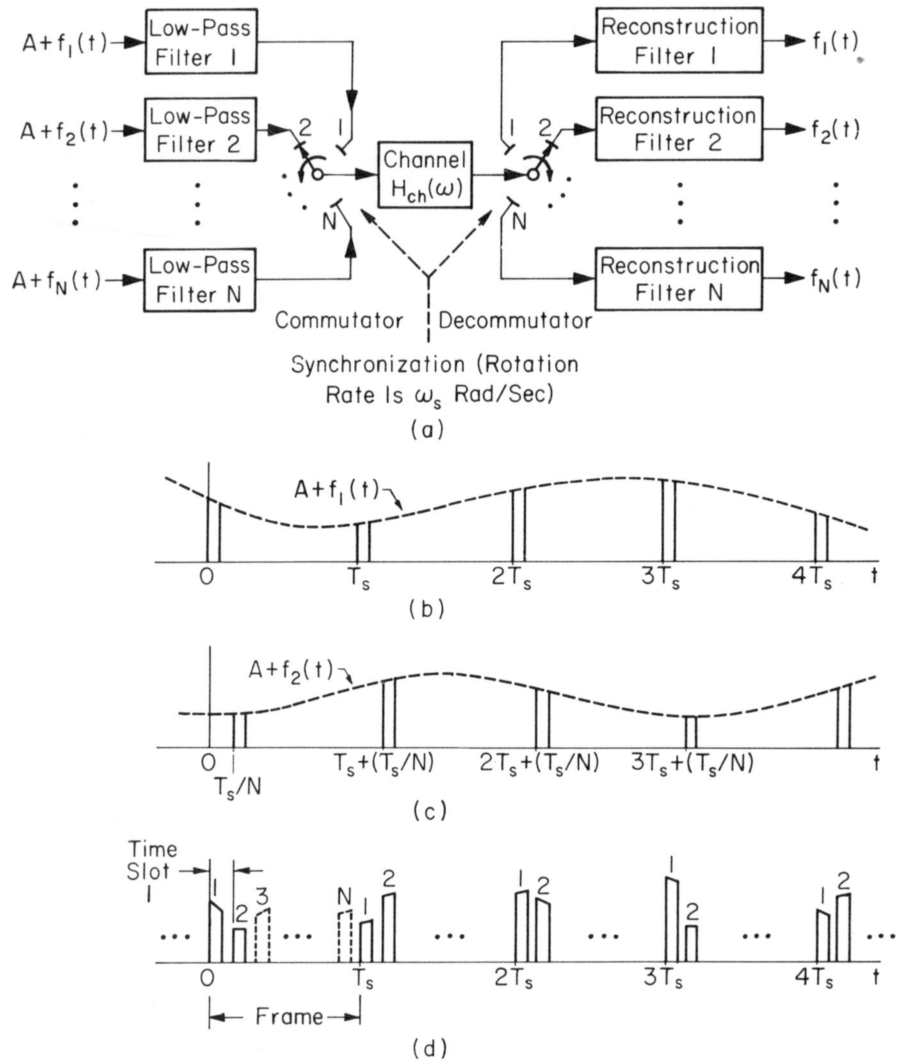

Fig. 7.5-1. (a) Time-division multiplexing system. Sampled messages, (b) number 1, and (c) number 2. (d) Composite waveform comprising N multiplexed PAM signals.

The multiplexed signal is sent over the channel and arrives at a *decommutator switch* or *distributor* which separates the samples and applies each sample train to its appropriate reconstruction filter. This filter may be any of those discussed earlier and varies from a simple low-pass filter when natural sampling is used to a sample-hold low-pass filter and equalizing network combination for more general instantaneous sampling methods.

Example 7.5-1

Suppose we want to time-multiplex $N = 50$ similar telephone messages using PAM. Assume each message is bandlimited to 3.3 kHz. Thus, $W_f = 2\pi(3.3)10^3$, and we must sample each message at a rate of at least $6.6(10^3)$ samples per second. From practical considerations let us be limited to a sampling rate of 8 kHz and use a guard time equal to the sample pulse duration τ. We find the required value of τ.

ISBN 0-201-05758-1

Equating 2τ, the slot time per sample, to T_s/N, the allowed slot time, gives $\tau = T_s/(2N)$. But $T_s = 1/8$ ms, so $\tau = 10^{-3}/(8 \cdot 100) = 1.25\ \mu s$.

Synchronization

It is necessary that the commutator and decommutator switches be synchronized. Synchronization is probably the most critical part of TDM. Since the switches are usually electronic, there is typically a *clock* pulse train at both transmitter and receiver which regulates switching. Synchronization amounts to keeping the two clock trains together in time, a problem analogous to keeping local sub-carrier signals in phase synchronism for FDM systems.

In TDM-PAM a simple technique is to allocate one (or more) time slots per frame for synchronization. A "sync" pulse is placed in the slot and is chosen to have a larger amplitude than the largest expected amplitude of any message pulse. In the receiver a threshold circuit may be used to determine when the sync pulse occurs. Thus, amplitude is the characteristic typically used to synchronize TDM-PAM. Similarly, time is the characteristic which relates to synchronization of TDM-PTM systems.

Bandwidth of Ideal Channel

A typical channel over which the TDM waveform passes does not have infinite bandwidth. Indeed, for purposes of noise and interference reduction, one may not even want infinite bandwidth. What then is the minimum bandwidth W_{ch} (radians per second) necessary to transmit the TDM-PAM signal without distortion in the recovered messages? To consider this question, let the channel be idealized by representing it as an ideal low-pass filter having the transfer function

$$H_{ch}(\omega) = 1, \qquad |\omega| < W_{ch}$$
$$= 0, \qquad \text{elsewhere.} \tag{7.5-1}$$

Its impulse response by inverse Fourier transformation is

$$h_{ch}(t) = \frac{W_{ch}}{\pi} \frac{\sin(W_{ch}t)}{W_{ch}t}. \tag{7.5-2}$$

$H_{ch}(\omega)$ and $h_{ch}(t)$ are sketched in Fig. 7.5-2(a).

For simplicity of discussion we represent the sampling process by ideal sampling with impulses. The sampled TDM waveform for N messages is

$$s_{\text{PAM}}(t) = \sum_{n=1}^{N} \left\{ \sum_{k=-\infty}^{\infty} \left\langle A + f_n\left[kT_s + \frac{(n-1)T_s}{N}\right]\right\rangle \delta\left[t - kT_s - \frac{(n-1)T_s}{N}\right]\right\}. \tag{7.5-3}$$

The sum within the braces is just the sampled version of message $A + f_n(t)$. The response $s_{ch,\text{PAM}}(t)$ of the channel to which (7.5-3) is applied follows easily:

ISBN 0-201-05758-1

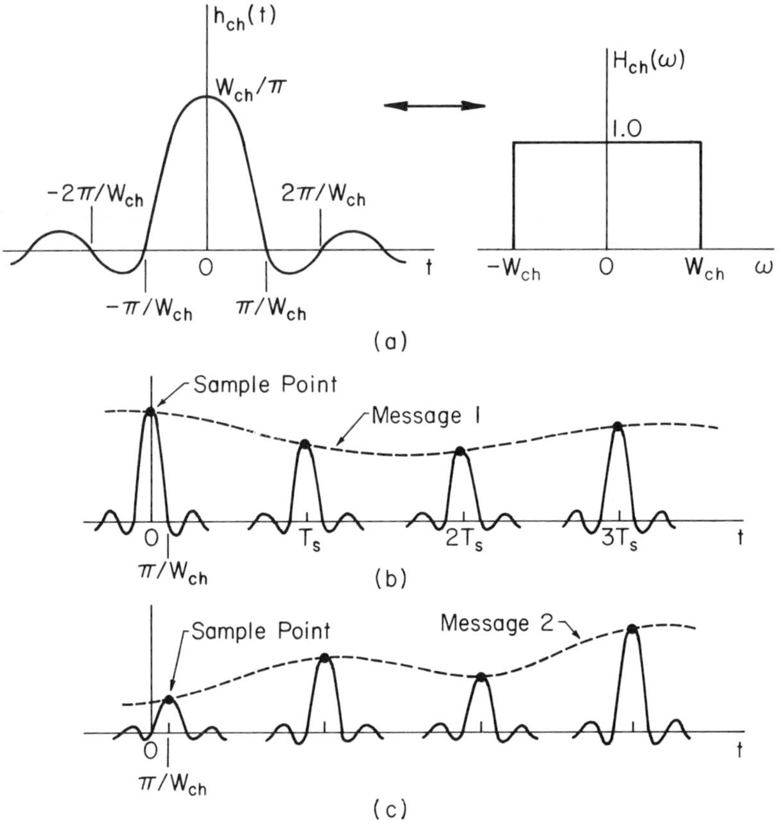

Fig. 7.5-2. (a) Ideal channel impulse response and transfer function. (b) and (c) Channel responses to two impulse sample pulse trains separated by π/W_{ch} seconds in time.

$$s_{ch,\text{PAM}}(t) = \sum_{n=1}^{N} \left\{ \sum_{k=-\infty}^{\infty} \left\langle A + f_n\left[kT_s + \frac{(n-1)T_s}{N}\right]\right\rangle \frac{W_{ch}}{\pi} \cdot \frac{\sin\left\langle W_{ch}\left[t - kT_s - \frac{(n-1)T_s}{N}\right]\right\rangle}{W_{ch}\left[t - kT_s - \frac{(n-1)T_s}{n}\right]}\right\}. \tag{7.5-4}$$

Figure 7.5-2(b) gives a sketch of (7.5-4) for the first message only. Note that the closest that a similar train [say for message 2 as shown in (c)] can be added and still have sample points independent is π/W_{ch} seconds. This happens because, at a given message sample time, the channel response to *all* other messages is passing through a null.

Since N messages are involved, we equate the minimum train separation time to the width T_s/N of a time slot:

$$\pi/W_{ch} = T_s/N. \tag{7.5-5}$$

Solving for W_{ch} and using the fact that $\omega_s = 2\pi/T_s \geqslant 2W_f$ is required, we obtain

ISBN 0-201-05758-1

$$W_{ch} = N\omega_s/2 \geqslant NW_f. \tag{7.5-6}$$

The minimum required channel bandwidth is then NW_f.

Because the message samples all remain independent, the reciever may have *perfect* message reconstruction by resampling and low-pass filtering the sample train at the channel output.

Practical Channel Bandwidth, Crosstalk

If the channel is not ideal there may be *crosstalk*—the interference of one or more message channels with another. In telephone systems the reader has no doubt encountered crosstalk occasionally when another line is heard in the background while he is using his own line.

The effect of a practical channel on crosstalk may be studied in one of its simplest forms by modeling it as a simple low-pass filter with 3-dB bandwidth W_{ch}. Its transfer function is

$$H_{ch}(\omega) = \frac{1}{1 + j(\omega/W_{ch})}. \tag{7.5-7}$$

The impulse response is the inverse Fourier transform of $H_{ch}(\omega)$:

$$\begin{aligned} h_{ch}(t) &= W_{ch} \exp(-W_{ch}t), \qquad 0 < t \\ &= 0, \qquad t < 0. \end{aligned} \tag{7.5-8}$$

Let us assume impulsive or ideal sampling at the transmitter as done earlier. The channel, defined by (7.5-8), is excited by the PAM signal of (7.5-3) to produce the response

$$s_{ch,\mathrm{PAM}}(t) = \sum_{n=1}^{N} \left\{ \sum_{k=-\infty}^{\infty} \left\langle A + f_n \left[kT_s + \frac{(n-1)T_s}{N}\right]\right\rangle \cdot u\left[t - kT_s - \frac{(n-1)T_s}{N}\right] W_{ch} e^{-W_{ch}\left[t - kT_s - \frac{(n-1)T_s}{N}\right]} \right\} \tag{7.5-9}$$

at the receiver. Here $u(\cdot)$ is the unit step function. A resampling of this waveform at time $kT_s + [(n-1)T_s/N]$ will no longer be proportional to the original message level $A + f_n\{kT_s + [(n-1)T_s/N]\}$. There is now a component of voltage in the sample due to crosstalk from other messages in the TDM signal. *A crosstalk factor* K_{ct} may be defined to describe the level of interference of message n with message m $(m > n)$. It is the ratio of the rms level V_1 of message n, that occurs at the sample time of message m, to the rms level V_2 of message m. From (7.5-9) these levels are

$$V_2 = \left\langle \overline{A^2 + f_m^2\{kT_s + [(m-1)T_s/N]\}} \right\rangle^{1/2} W_{ch}, \tag{7.5-10}$$

$$V_1 = \left\langle \overline{A^2 + f_n^2\{kT_s + [(n-1)T_s/N]\}} \right\rangle^{1/2} W_{ch} e^{-W_{ch}(m-n)T_s/N}, \qquad m > n. \tag{7.5-11}$$

Thus, for similar messages having nearly the same power,

ISBN 0-201-05758-1

$$K_{ct} = e^{-W_{ch}(m-n)T_s/N}, \qquad m > n. \tag{7.5-12}$$

The worst case corresponds to adjacent time slots where $m = n + 1$. Using (7.5-6), the largest factor is

$$K_{ct} = \exp(-W_{ch}T_s/N) \geqslant \exp[-\pi W_{ch}/(NW_f)]. \tag{7.5-13}$$

Figure 7.5-3(*a*) illustrates crosstalk between adjacent time slots. Minimum K_{ct} corresponds to sampling at the Nyquist rate; it is plotted in (*b*). If $W_{ch} = NW_f$, the ideal channel value, minimum crosstalk factor is 4.3%. It decreases rapidly for increasing W_{ch}.

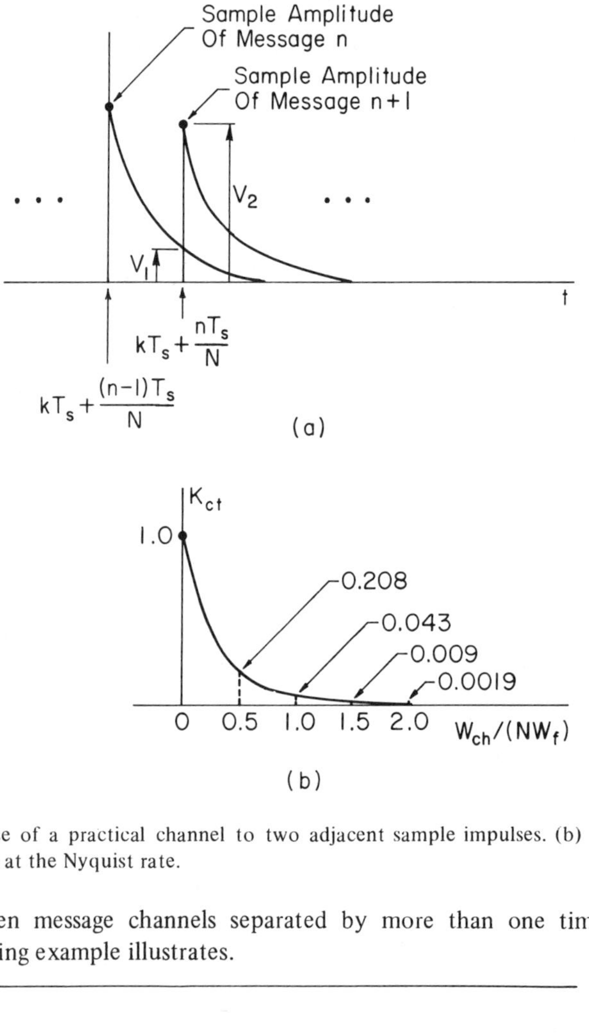

Fig. 7.5-3. (a) Response of a practical channel to two adjacent sample impulses. (b) Minimum crosstalk factor when sampling is at the Nyquist rate.

Crosstalk between message channels separated by more than one time slot decreases rapidly as the following example illustrates.

Example 7.5-2

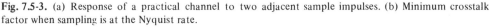

Assume $W_{ch} = NW_f$ and find crosstalk factor for message channels separated by two time slots. From (7.5-12) with $m = n + 2$, K_{ct} is the *square* of that given by (7.5-13). Thus, $K_{ct} = (0.0432)^2 = 0.0019$.

ISBN 0-201-05758-1

If samples have finite pulse width, such as when flat-top samples are used, a crosstalk factor may be defined based on interference "area" [7]. Figure 7.5-4(*b*) illustrates channel response to a sample in one time slot assuming the channel is described by (7.5-7) or (7.5-8). The channel input (average) is shown in (*a*). The time between the end of one sample pulse and the start of another is the guard time τ_g. Large guard time corresponds to small crosstalk.* Crosstalk factor K_{ct} is now defined by the ratio of areas A_1/A_2 shown in (*b*). For $W_{ch} >> 1/\tau$, as must be the case if K_{ct} is to be small, an approximation for K_{ct} is

$$K_{ct} = \frac{1}{W_{ch}\tau} \exp(-W_{ch}\tau_g). \tag{7.5-14}$$

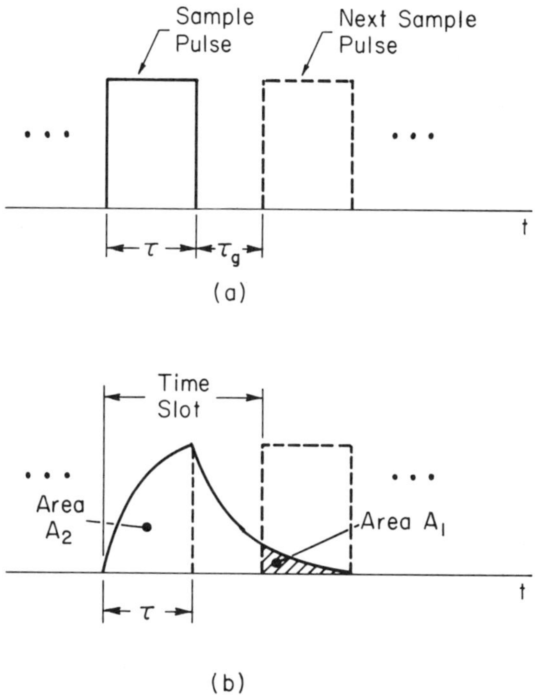

Fig. 7.5-4. (a) Finite duration pulse in a typical time slot. The pulse is applied to a practical low-pass channel to produce the response of (b).

Sometimes this expression is written in terms of *channel time constant* τ_{ch} defined by [7]

$$\tau_{ch} = 1/W_{ch}. \tag{7.5-15}$$

Example 7.5-3

Suppose the TDM-PAM signal of example 7.5-1 is transmitted over a simple low-pass channel defined by (7.5-7) that has a bandwidth selected according to the expression (7.5-5). Find the approximate crosstalk between adjacent message time slots. Here $\tau = \tau_g = T_s/(2N)$ and $W_{ch} = \pi N/T_s$. Thus, $W_{ch}\tau = W_{ch}\tau_g = \pi/2$ and $K_{ct} = (2/\pi)\exp(-\pi/2) = 0.132$ or 13.2%.

---

*The principle reason for including guard time is the reduction of crosstalk.

ISBN 0-201-05758-1

Crosstalk Due to Low-Frequency Cutoff of Channel

The approximate effect of low-frequency cutoff of the channel may be found by modeling the channel as a simple high-pass RC filter having a cutoff frequency equal to the channel's low-end cutoff frequency W_L. Hence, $W_L = 1/RC$. If we assume that rectangular sampling pulses are transmitted, similar messages are multiplexed, and reconstruction utilizes instantaneous resampling, it can readily be shown that the crosstalk factor for adjacent message slots is approximately

$$K_{ct} = W_L \tau \exp(-W_L \tau_g). \tag{7.5-16}$$

In the derivation of (7.5-16) the assumption $W_L \tau \ll 1$ was made where τ is the pulse duration. Such an assumption is nearly always valid.

If the above high-pass filter channel model is again employed, but an analysis based on interference area is used, it is found [7] that the crosstalk factor for two adjacent message slots is approximately

$$K_{ct} = W_L \tau. \tag{7.5-17}$$

In a practical situation guard time τ_g will not be markedly different from pulse duration τ, so $W_L \tau_g \ll 1$ is likely to be true. Thus, (7.5-17) and (7.5-16) show that crosstalk factors in the two cases discussed are nearly the same.

From either (7.5-16) or (7.5-17) crosstalk is made small if $W_L \tau \ll 1$. Unfortunately, a more detailed analysis shows that, while this condition tends to make crosstalk small between adjacent channels, crosstalk due to low-frequency cutoff of the channel is a slowly decreasing function and can extend over many message time slots.

7.6. NOISE PERFORMANCE OF PAM

The noise behavior of a PAM time-multiplexed system may be determined by analysis of the system illustrated in Fig. 7.6-1 for a typical message path. N similar messages are multiplexed. These signals are assumed bandlimited with maximum frequency extent W_f and are transmitted by unipolar sampling. Bipolar transmission may be handled as a special case where $A = 0$.

For simplicity, ideal sampling is assumed in the transmitter. All networks between the transmitter and message reconstruction samplers in the receiver are lumped into the one filter

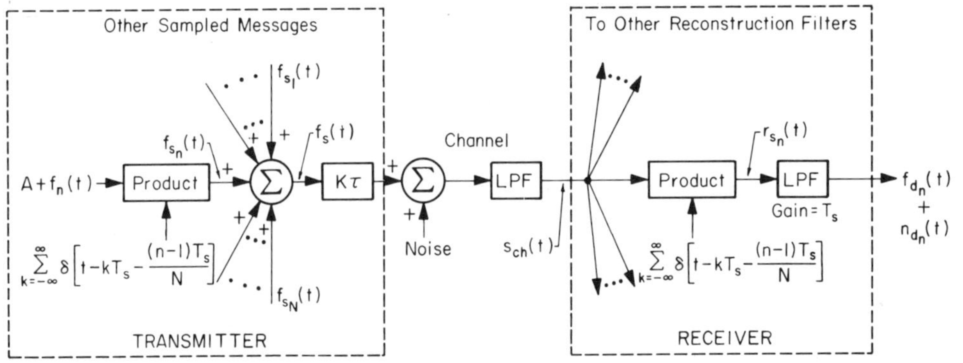

Fig. 7.6-1. Time multiplexing and demultiplexing of PAM signals.

ISBN 0-201-05758-1

representing a bandlimited channel. This filter is assumed ideal with a bandwidth given by (7.5-6), which was derived for the model assumed here. A gain $K\tau$ is shown in the transmitter to be consistent with earlier work. That is, τ is made equal to the equivalent duration of the impulse response of the channel filter; this response will have a peak magnitude of $1/\tau$. Therefore, the constant K is simply the amplitude of sampling pulses at the input to the receiver if the information signals were just dc levels of unity $[A + f_n(t) = 1]$. For the channel impulse response, that is given by (7.5-2), it is subsequently shown that $\tau = \pi/W_{ch}$. These considerations lead us to write the input waveform at the receiver as:

$$s_{ch}(t) = \sum_{n=1}^{N} K \sum_{k=-\infty}^{\infty} \left\{A + f_n\left[kT_s + \frac{(n-1)T_s}{N}\right]\right\} \text{Sa}\left\{W_{ch}\left[t - kT_s - \frac{(n-1)T_s}{N}\right]\right\} + n_i(t). \tag{7.6-1}$$

For a more generalized system, only the shape of the sampling pulses would be different.

For message recovery the receiver is assumed to use instantaneous resampling followed by an ideal low-pass filter with bandwidth W_f and gain T_s.* Noise $n_i(t)$ at the receiver input is assumed to be stationary and to have a power density $\mathcal{N}_0/2$. It is the result of the channel filter's acting on white channel noise.

Signal and Noise Powers

Since all signals are similar, we examine only a single typical reconstruction path. Consider first the receiver input. Noise power is easily determined to be

$$N_i = \mathcal{N}_0 W_{ch}/2\pi. \tag{7.6-2}$$

For the input signal power, due to, say, message n, we need only find the power in the signal:

$$s_{i_n}(t) = K \sum_{k=-\infty}^{\infty} \left\{A + f_n\left[kT_s + \frac{(n-1)T_s}{N}\right]\right\} \text{Sa}\left\{W_{ch}\left[t - kT_s - \frac{(n-1)T_s}{N}\right]\right\}. \tag{7.6-3}$$

By treating the messages $f_n(t)$ as sample functions of jointly statistically independent stationary random processes, we may show that the average power in $s_{i_n}(t)$ is

$$S_i = \frac{\pi K^2}{W_{ch}T_s} [A^2 + \overline{f_n^2(t)}]. \tag{7.6-4}$$

The procedure is to take the expected value of $s_{i_n}^2(t)$, observe that the resulting expression is time-dependent, and form the time average. The reader may wish to carry out these steps to verify (7.6-4).

Let us next show, by means of an example, that the equivalent duration of the channel impulse response is $\tau = \pi/W_{ch}$. With this fact, (7.6-4) may be written as

$$S_i = K^2 [A^2 + \overline{f_n^2(t)}] (\tau/T_s). \tag{7.6-5}$$

This is an intuitively satisfying way of interpreting (7.6-4), since $K^2 [A^2 + \overline{f_n^2(t)}]$ would be

*From the standpoint of the message this is equivalent to a sample-hold receiver (Fig. 7.4-4).

ISBN 0-201-05758-1

the message power if no sampling took place, and, with sampling, the average power due to message n would be reduced by the *duty cycle* τ/T_s.

Example 7.6-1

The equivalent duration of the channel impulse response of (7.5-2) is found from application of (3.4-27). Thus,

$$\tau = \frac{\pi}{W_{ch}} \int_{-\infty}^{\infty} \frac{W_{ch}}{\pi} \text{ Sa } (W_{ch}t)\, dt = \int_{-\infty}^{\infty} \text{Sa } (W_{ch}t)\, dt$$

$$= \frac{1}{W_{ch}} \int_{-\infty}^{\infty} \text{Sa } (x)\, dx = \frac{\pi}{W_{ch}}$$

after using the fact that $\int_{-\infty}^{\infty} \text{Sa } (x)\, dx = \pi$ from a table of integrals (Appendix B).

Turning attention to the output, the output of the receiver sampler for message path n is

$$r_{s_n}(t) = \sum_{k=-\infty}^{\infty} K\left\{A + f_n\left[kT_s + \frac{(n-1)T_s}{N}\right]\right\} \delta\left[t - kT_s - \frac{(n-1)T_s}{N}\right]$$

$$+ \sum_{k=-\infty}^{\infty} n_i\left[kT_s + \frac{(n-1)T_s}{N}\right] \delta\left[t - kT_s - \frac{(n-1)T_s}{N}\right]. \tag{7.6-6}$$

The first sum is the sampler output message component. The second sum is the noise component. We evaluate the signal component by substituting the impulse response $(W_f T_s/\pi)$ Sa $(W_f t)$ of the low-pass filter.

$$f_{d_n}(t) = \frac{W_f T_s}{\pi} \sum_{k=-\infty}^{\infty} K\left\{A + f_n\left[kT_s + \frac{(n-1)T_s}{N}\right]\right\} \text{Sa}\left\{W_f\left[t - kT_s - \frac{(n-1)T_s}{N}\right]\right\}$$

$$= K\left\{A + f_n(t)\right\} \tag{7.6-7}$$

from the sampling theorem embodied in (7.1-10). The average power in the message portion of the reconstructed output becomes

$$S_o = K^2\overline{f_n^{\,2}(t)} = \overline{f_{d_n}^{\,2}(t)}. \tag{7.6-8}$$

In a similar fashion, the output noise is found from the filter noise response:

$$n_{d_n}(t) = \frac{W_f T_s}{\pi} \sum_{k=-\infty}^{\infty} n_i\left[kT_s + \frac{(n-1)T_s}{N}\right] \text{Sa}\left\{W_f\left[t - kT_s - \frac{(n-1)T_s}{N}\right]\right\}. \tag{7.6-9}$$

ISBN 0-201-05758-1

Average noise power is less simple to evaluate. The procedure is to take the time average of the expected value of $n_{d_n}^2(t)$. Such steps lead to

$$N_o = \frac{W_f T_s}{\pi} N_i = \frac{W_f T_s \mathcal{N}_0 W_{ch}}{2\pi^2} . \tag{7.6-10}$$

Sampling near the Nyquist rate is to be expected, so $N_o \approx N_i$. It might appear at first glance that output average noise power should be smaller than the input noise power by a bandwidth reduction factor W_f/W_{ch}, since the input noise is spread over a band $W_{ch} \geqslant NW_f$ while output noise is spread over a band W_f. Such a reduction *would* be possible if the sampling rate were high enough to reconstruct the noise along with the message. A sampling rate $\omega_s \geqslant 2W_{ch}$ would be required. The reconstructed output could then be passed through a low-pass filter of bandwidth W_f and $N_o << N_i$ would result. Indeed, assuming minimum channel bandwidth and sampling at the rate $\omega_s = 2W_{ch}$, (7.6-10) becomes $N_o = N_i(W_f/W_{ch})$, and a bandwidth reduction factor is achieved. However, the sampling rate is usually *not* designed to reconstruct the noise because such a choice allows only a single message to be transmitted. Rather, ω_s is chosen as small as possible for message reconstruction so that time multiplexing becomes possible.

How then can we intuitively justify $N_o \approx N_i$ when different noise bandwidths are involved? The answer lies in the fact that, with a sampling rate designed for message reconstruction, the (broadband) input noise is being *undersampled* by a factor $W_{ch}/W_f = N$ below its Nyquist rate. As found in Section 7.1, undersampling a bandlimited signal leads to aliasing—in this case severe aliasing, such that N replica spectra in the sampled signal overlap in the region $|\omega| < W_f$ (see Fig. 7.1-3). This leads to a noise power increase of N times due to aliasing and justifies $N_o \approx N_i$.

Performance

Output signal-to-noise power ratio for message n of N similar messages follows from the ratio of (7.6-8) and (7.6-10):

$$\left(\frac{S_o}{N_o}\right)_{\text{PAM}} = \frac{2\pi^2 K^2 \overline{f_n^2(t)}}{W_f T_s W_{ch} \mathcal{N}_0} . \tag{7.6-11}$$

The ratio of input average signal power, due to message n, and total input average noise power is

$$\left(\frac{S_i}{N_i}\right)_{\text{PAM}} = \frac{2\pi^2 K^2 [A^2 + \overline{f_n^2(t)}]}{W_{ch}^2 T_s \mathcal{N}_0} \tag{7.6-12}$$

from (7.6-4) and (7.6-2). More useful forms of (7.6-11) are

$$\left(\frac{S_o}{N_o}\right)_{\text{PAM}} = \frac{\overline{f_n^2(t)}}{A^2 + \overline{f_n^2(t)}} \left(\frac{W_{ch}}{W_f}\right) \left(\frac{S_i}{N_i}\right)_{\text{PAM}} , \tag{7.6-13}$$

$$\left(\frac{S_o}{N_o}\right)_{\text{PAM}} = \frac{\overline{f_n^2(t)}}{A^2 + \overline{f_n^2(t)}} \left(\frac{S_o}{N_o}\right)_{\text{B}} . \tag{7.6-14}$$

ISBN 0-201-05758-1

The last expression derives from comparing the PAM system to a baseband system [see (5.8-7)] having the same input average signal power and noise power density $\mathcal{N}_0/2$.

From a consideration of (7.6-14), PAM gives a performance never exceeding simple baseband transmission. The two become equal for bipolar PAM where $A = 0$. Furthermore, by comparing (7.6-14) with the corresponding results for AM systems, as given by (5.8-35) and (5.8-36), PAM is found to give similar results. An off hand comparison of (7.6-13) with the similar result (5.8-34) for AM seems to imply that demodulation gain in PAM far exceeds that for AM. Indeed, for the way in which $(S_i/N_i)_{\text{PAM}}$ has been defined, it does. However, a careful comparison of gain based on equal input signal-to-noise ratios shows PAM and AM to be the same.

Example 7.6-2

Let the message be a sinusoidal signal. We determine how much poorer the PAM system is than a baseband system. If we give the PAM system the best possible advantage by letting A be as small as possible, but still maintaining unipolar transmission, then $A = |f_n(t)|_{\text{max}}$. For a sinusoid $\overline{f_n^2(t)}/|f_n(t)|_{\text{max}}^2 = 1/2$, so

$$(S_o/N_o)_{\text{PAM}} = (S_o/N_o)_{\text{B}}/3 \qquad \text{(or 4.8 dB poorer)}$$

from (7.6-14).

7.7. PULSE DURATION MODULATION

In *pulse duration modulation* (PDM) the width of a pulse in the train is made to be a linear function of the message signal sample. As illustrated in Fig. 7.7-1, either one or both pulse edges may be modulated. Trailing-edge modulation is the more common in practice, but most analyses become simpler if one assumes both edges are modulated.

PAM was found to be analogous to AM modulation. It will subsequently be found that PDM is analogous to angle modulation. We would then expect noise performance to be better than PAM in a similar way that angle modulation is better than AM. This is indeed the case, and the resulting noise immunity is one of the advantages of using PDM.

Signal Generation and Spectrum

There are two basic ways of generating PDM signals. These are called *uniform* and *nonuniform* sampling. Figure 7.7-2 illustrates generation of trailing-edge modulation using uniform sampling. N messages are first multiplexed to form a PAM waveform with flat-top samples as shown in (b) for four messages within one frame. Next the ramp illustrated in (c) is added to the PAM signal to obtain the waveform of (d). This signal is compared to a threshold voltage. The portion of the waveform exceeding the threshold triggers a bistable circuit which generates the pulse train of (e). The overall block diagram of (a) applies. By changing the slope of the ramp, leading-edge modulation results. Similarly, by replacing the ramp with a triangular pulse, two-edge variation may be achieved.

Uniform sampling derives its name from the fact that the widths of all pulses are determined by instantaneous samples taken uniformly in time. On the other hand, nonuniform sampling implies pulse widths are related to samples of the messages *not* taken uniformly in time.

ISBN 0-201-05758-1

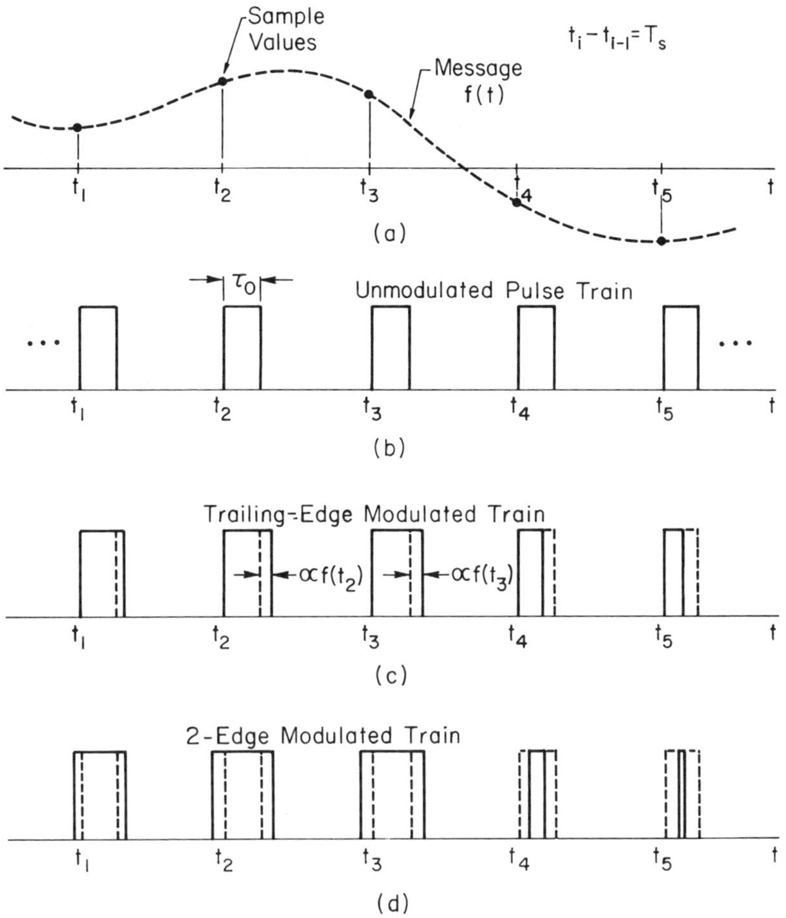

Fig. 7.7-1. (a) A message showing its sample values. (b) An unmodulated pulse train. (c) A trailing-edge modulated PDM signal, and (d) two-edge modulated waveform.

Figure 7.7-3 illustrates a simple method for generating nonuniform samples. The block diagram is given in (*a*). A typical message is shown in (*b*). A ramp is added to the message as in (*c*) and compared to a threshold level. Again a bistable circuit generates a pulse with duration equal to the time the ramp-plus-message exceeds the threshold. Pulse widths shown in (*c*) and (*d*) have been exaggerated for clarity; they would actually be much narrower if N is large. In this method of generation, the duration of pulses is determined by message values (samples) at the *modulated edge.* These "sample" times then vary and are not fixed as in uniform sampling.

The uniform sampling theorem does not strictly apply to the nonuniform method. However, if pulse widths are small relative to T_s, as they would be for N large, the error made by assuming uniform sampling is insignificant [8]. This fact is useful in approximating the spectrum of PDM waveforms.

To gain some insight into the behavior of the PDM signal spectrum, let us assume two-edge modulation and nonuniform sampling. Our general conclusions from such assumptions will not be grossly in error for other forms of modulation and sampling, if average pulse duration τ_0 is small relative to T_s, as would be true for N large. These assumptions mean that pulse

ISBN 0-201-05758-1

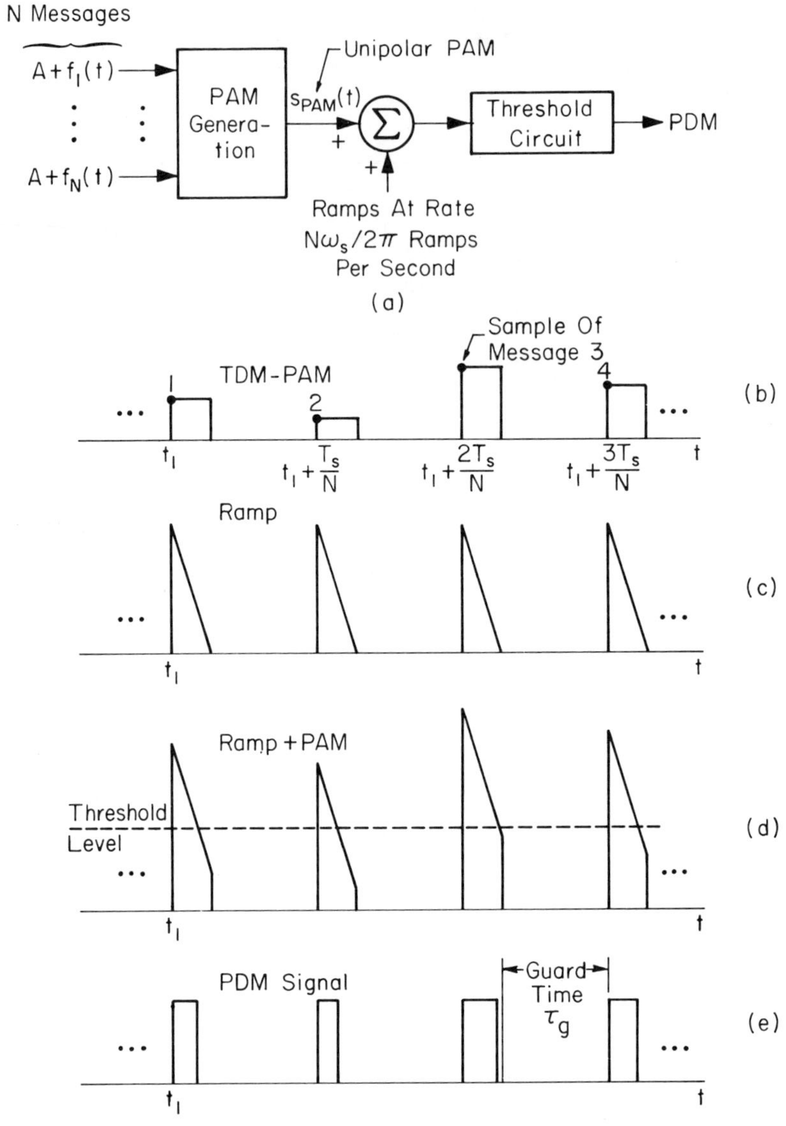

Fig. 7.7-2. (a) Generation method for TDM–PDM using uniform sampling. (b) TDM–PAM signal. (c) A train of ramps. (d) Sum of train of ramps and TDM–PAM signal, and (e) the PDM signal.

spacing is approximately constant and a sort of quasi-static method can be used to approximate the spectrum. First, sampling can be taken into account by expanding the pulse train into its Fourier series representation. Next, the effect of modulation is introduced as a slow, continuously changing pulse duration. Finally, the spectrum is found, in principle, through Fourier transformation.

Consider a single typical message. From (7.2-1) the Fourier series for a train of pulses with amplitude A, duration τ, and rate $\omega_s = 2\pi/T_s$ can be put in the form

$$s_p(t) = \frac{A\tau}{T_s} + \frac{A}{\pi}\sum_{k=1}^{\infty}\frac{1}{k}\left\{\sin\left[k\omega_s\left(t+\frac{\tau}{2}\right)\right] - \sin\left[k\omega_s\left(t-\frac{\tau}{2}\right)\right]\right\}. \quad (7.7\text{-}1)$$

ISBN 0-201-05758-1

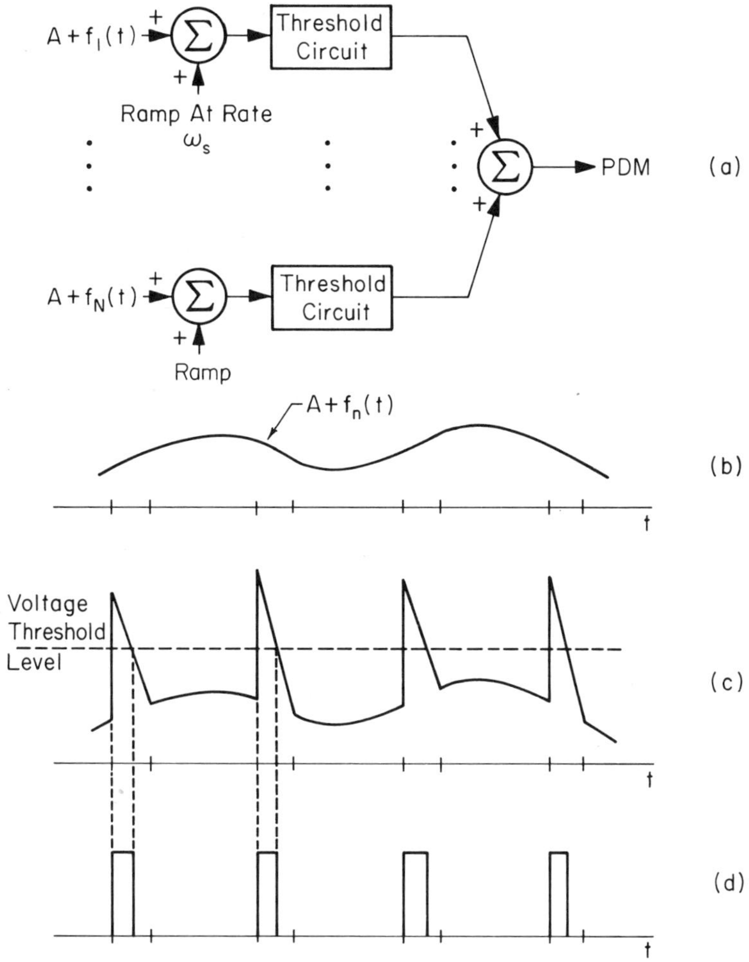

Fig. 7.7-3. (a) Block diagram for PDM generation using nonuniform sampling. (b) Message, and (c) message plus ramp. (d) PDM signal generated by threshold crossing of the waveform of (c).

Pulse length can now be approximated as a linear continuous function of the message

$$\tau = \tau_0 + k_{PDM} f_n(t), \tag{7.7-2}$$

where τ_0 is the average width and k_{PDM} is a positive constant. Substitution results in the PDM signal (only one message component):

$$\begin{aligned} s_{PDM}(t) &= \frac{A\tau_0}{T_s} + \frac{Ak_{PDM} f_n(t)}{T_s} \\ &+ \frac{A}{\pi}\sum_{k=1}^{\infty}\frac{1}{k}\sin\left[k\omega_s t + \frac{k\omega_s\tau_0}{2} + \frac{k\omega_s k_{PDM} f_n(t)}{2}\right] \\ &- \frac{A}{\pi}\sum_{k=1}^{\infty}\frac{1}{k}\sin\left[k\omega_s t - \frac{k\omega_s\tau_0}{2} - \frac{k\omega_s k_{PDM} f_n(t)}{2}\right]. \end{aligned} \tag{7.7-3}$$

ISBN 0-201-05758-1

This expression shows that the PDM signal comprises a dc component, the message itself, and *phase-modulated* waves at "carrier" frequencies $k\omega_s$.

As with any general phase-modulated wave, the spectrum of (7.7-3) is difficult to find. The methods of Section 6.4 may be applied to components where the peak phase deviation

$$\Delta\theta = k\omega_s k_{PDM} |f_n(t)|_{max}/2 \tag{7.7-4}$$

is large. The rough form of the spectrum may be sketched in Fig. 7.7-4. Note that spectral components centered at frequencies $k\omega_s$ decrease at least as fast as $1/k$ but tend to have broader bandwidths owing to the factor k in (7.7-4). In (7.7-4) $\Delta\theta$ does not tend to become large until k is about N or larger.

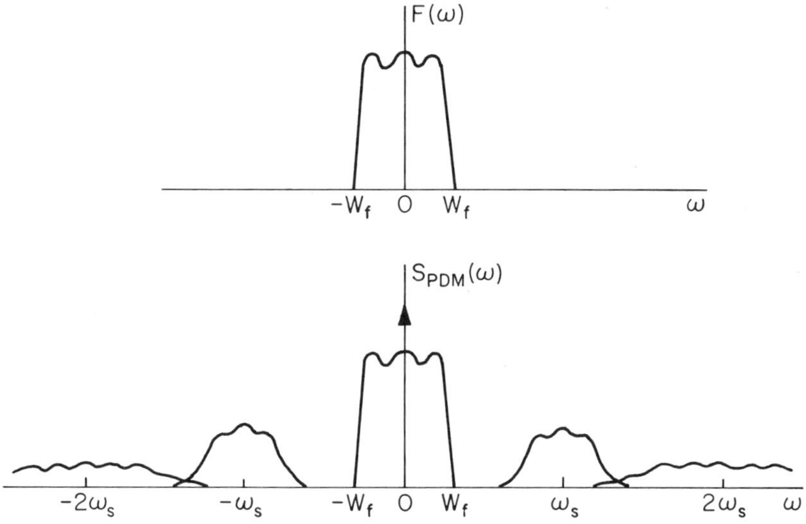

Fig. 7.7-4. A rough interpretation of the spectrum of a PDM waveform.

Channel Bandwidth

Although the precise spectrum of the multiplexed PDM signal is difficult to obtain, it is possible to derive an approximate value for the minimum channel bandwidth required to support the signal. Consider trailing-edge modulation with N similar messages being multiplexed. The messages all have the same maximum spectral extent W_f, and are sampled at a rate $\omega_s \geqslant 2W_f$. The composite multiplexed signal pulse rate is $N\omega_s/2\pi$ pulses per second. Separation time between pulses is $2\pi/N\omega_s = T_s/N$. Now suppose that, when the sample pulse in a given message time slot is at its maximum duration τ_{max}, its trailing edge remains separated by τ_g seconds from the leading edge of the adjacent sample pulse. We may call this separation *guard time.* It is useful for reducing crosstalk between message slots. Next, note that pulse separation time must equal $\tau_{max} + \tau_g$; that is, $2\pi/N\omega_s = \tau_{max} + \tau_g$. However, $\tau_{max} = \tau_{min} + 2\Delta\tau$, where τ_{min} is the minimum pulse duration and $\Delta\tau$ is the peak duration deviation of the sample pulses from an average value τ_0. It is given by

$$\Delta\tau = k_{PDM} |f_n(t)|_{max} \tag{7.7-5}$$

ISBN 0-201-05758-1

for a typical message $f_n(t)$. Thus, we have

$$\frac{2\pi}{N\omega_s} = \tau_{\min} + 2\Delta\tau + \tau_g. \tag{7.7-6}$$

We may now observe that channel bandwidth determines how well the *narrowest* pulse is received. To express $\tau_{\min}$ in terms of W_{ch}, suppose that a rectangular pulse of duration $\tau_{\min}$ is passed over a realizable low-pass channel having the transfer function of (7.5-7). The channel response pulses will maintain reasonable quality if $\tau_{\min} \geqslant \pi/W_{ch}$. On the other hand, some applications may require slightly better preservation of pulse shape, as achieved when $\tau_{\min} \geqslant 2\pi/W_{ch}$ or twice the channel rise time.* Recognizing these facts, let us use the minimum value π/W_{ch} in (7.7-6) to solve for minimum channel bandwidth:

$$W_{ch} \geqslant \frac{N\omega_s}{2}\left(1 + \frac{2\Delta\tau + \tau_g}{\tau_{\min}}\right). \tag{7.7-7}$$

Since $\omega_s \geqslant 2W_f$, an alternate form is

$$W_{ch} \geqslant NW_f\left(1 + \frac{2\Delta\tau + \tau_g}{\tau_{\min}}\right). \tag{7.7-8}$$

On comparing this expression with W_{ch} for multiplexed PAM, as given by (7.5-6), it is clear that PDM requires a larger channel bandwidth. Indeed, since $\Delta\tau$ may be several times as large as $\tau_{\min}$, the ratio of the two bandwidths may be quite large.

Although the discussion leading to (7.7-8) assumed single-edge modulation, the reader may retrace the steps and verify its applicability to two-edge modulation as well, if messages are similar.

Signal Recovery by Low-Pass Filtering

Equation (7.7-3) and the form of its transform suggest that $f_n(t)$ may be recovered by low-pass filtering. The filter response will be approximately

$$s_o(t) = \frac{A\tau_0}{T_s} + \frac{Ak_{\text{PDM}}}{T_s} f_n(t). \tag{7.7-9}$$

Signal recovery is not exact and some distortion is caused by spectral overlap from spectra centered about frequencies $k\omega_s$ (a kind of aliasing). For voice communication this distortion can be maintained within acceptable limits [7]. More generally, the distortion can be reduced by raising the sampling rate. Recall that the output (7.7-9) was developed by assuming non-uniform sampling. If uniform sampling were used instead, additional spectral terms may be generated at baseband [2], causing the quality of the recovered signal to suffer somewhat [7]. Of course, the chief advantage of message recovery by low-pass filtering is its simplicity. However, one of its main disadvantages is low output signal-to-noise ratio as compared to other recovery methods.

*These expressions are sufficient to determine W_{ch}. However, we go one step further to develop expressions containing other parameters in the PDM system.

ISBN 0-201-05758-1

Signal Recovery by Reconstruction

Although message recovery using a low-pass filter is simple, it involves inherent distortion. By implementing a more complicated receiver, exact message recovery is possible in principle. The process will be called *reconstruction.* Assume uniform sampling is used to generate the PDM wave. The reconstruction process amounts to first converting the PDM wave to a PAM signal and then performing demodulation by any of the PAM methods already discussed.

A straightforward method of converting PDM to PAM is illustrated in Fig. 7.7-5. At the start of a sample pulse in the train (*a*) of a typical message, a ramp is generated (*b*) which stops at the sample pulse trailing edge. By adding a narrow pulse to the ramp, and preserving only the portion exceeding a prescribed clipping level as in (*c*), a PAM wave (*d*) is generated.

As will be shown in Section 7.8, message reconstruction leads to a larger output signal-to-noise ratio than for simple low-pass filtering, a fact which may justify the added receiver complexity in some cases.

If nonuniform sampling is employed to generate the PDM signal in place of uniform sampling, the message obtained by reconstruction will contain distortion. The distortion be-

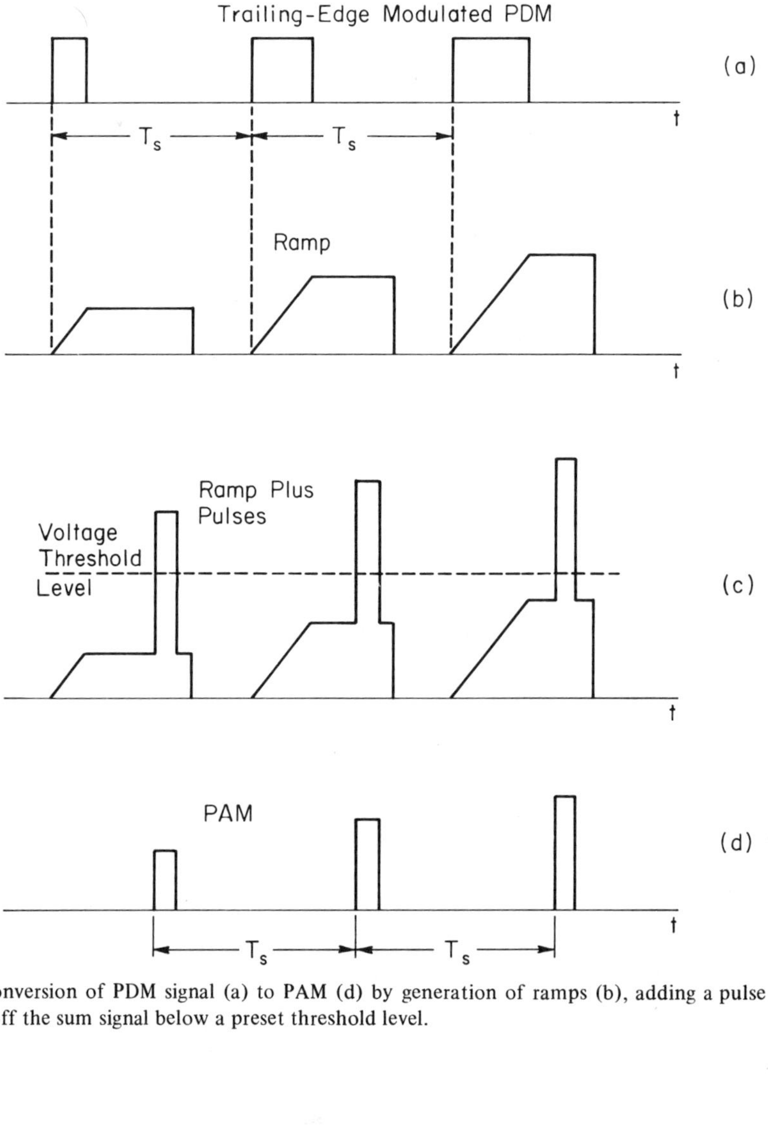

Fig. 7.7-5. Conversion of PDM signal (a) to PAM (d) by generation of ramps (b), adding a pulse train (c), and clipping off the sum signal below a preset threshold level.

ISBN 0-201-05758-1

comes smaller if a large number of messages is multiplexed. In any case, the distortion can be reduced by decreasing the magnitude of the modulating message, a measure which naturally reduces the peak deviation $\Delta\tau = k_{PDM}\ |f_n(t)|_{max}$ of pulse duration.

7.8. NOISE PERFORMANCE OF PDM

The two receiving systems discussed in the last section will next be analyzed to determine output signal-to-noise ratios.

Performance Using Low-Pass Filter

The appropriate receiver is illustrated in Fig. 7.8-1. The channel is represented by an ideal filter with bandwidth W_{ch}, and channel output noise is assumed to be bandlimited white noise with power density $\mathscr{N}_0/2$. The receiver *gates* are simply devices which connect the receiver input to a given message low-pass filter for only the times that a particular message sample pulse train exists. The gates provide the decommutation or signal separation operation. Each may be modeled as a product device. The output is the product of the input times a unit-amplitude positive train of pulses each having a duration just long enough to admit the widest sample pulse for a given message. Again, for simplicity, assume that all messages are similar so that all message channels are the same with gate pulse durations equal to τ_{max}.

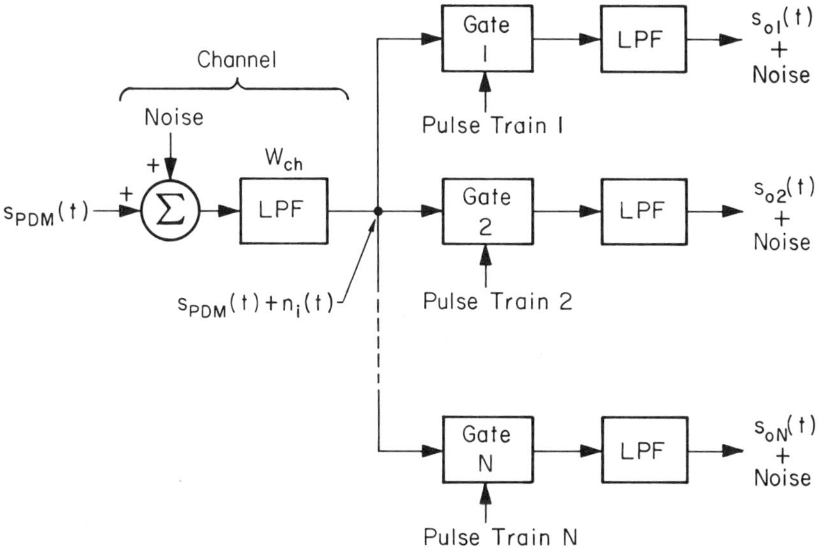

Fig. 7.8-1. Block diagram of a PDM receiver which uses low-pass filters for message recovery.

Average signal power per message and average noise power at the receiver input are readily determined to be

$$S_i = A^2\tau_0/T_s, \tag{7.8-1}$$

$$N_i = \overline{n_i^2(t)} = \mathscr{N}_0 W_{ch}/2\pi, \tag{7.8-2}$$

where A is the amplitude of the input PDM pulses, that have an average duration τ_0, and T_s

ISBN 0-201-05758-1

is the sample period per message. Output signal power for a typical message channel is also readily calculated from (7.7-9):

$$S_o = \overline{s_{on}^2(t)} = A^2 k_{\text{PDM}}^2 \overline{f_n^2(t)}/T_s^2. \tag{7.8-3}$$

Output average noise power is not as easily found, and we shall only give the result, omitting the derivation:

$$N_o = \frac{\tau_{\max}}{T_s} \frac{\mathcal{N}_0 W_f}{2\pi}. \tag{7.8-4}$$

The procedure for arriving at this expression is to expand the gate pulse train into its Fourier series representation, use (4.11-6) for the noise power density spectrum at the output of a product device, and consider the output spectral density in the region $|\omega| < W_f$. It will be found that severe aliasing occurs in the power spectrum replicas associated with the "sampled" gate output noise. Approximate evaluation shows the power density to be constant over the region $|\omega| < W_f$ (owing to the channel output noise density's being constant) at a level $\tau_{\max}\mathcal{N}_0/2T_s$.

The preceding four expressions allow input and output signal-to-noise ratios to be stated:

$$\left(\frac{S_i}{N_i}\right)_{\text{PDM}} = \frac{2\pi A^2 \tau_0}{\mathcal{N}_0 W_{ch} T_s}, \tag{7.8-5}$$

$$\left(\frac{S_o}{N_o}\right)_{\text{PDM}} = \frac{2\pi A^2 k_{\text{PDM}}^2 \overline{f_n^2(t)}}{\mathcal{N}_0 W_f \tau_{\max} T_s}. \tag{7.8-6}$$

Other forms of $(S_o/N_o)_{\text{PDM}}$ are instructive. For example, in terms of the output performance of a baseband system having the same input power and noise density,

$$\left(\frac{S_o}{N_o}\right)_{\text{PDM}} = \frac{k_{\text{PDM}}^2 \overline{f_n^2(t)}}{\tau_{\max}\tau_0} \left(\frac{S_o}{N_o}\right)_{\text{B}}. \tag{7.8-7}$$

By recognizing $k_{\text{PDM}}^2 \overline{f_n^2(t)}$ as the mean-squared deviation $\Delta\tau_{\text{rms}}^2$ of pulse duration, and noting that $\tau_{\max} \approx 2\Delta\tau$ and $\tau_0 \approx \Delta\tau$ for a heavily deviated PDM system, we have

$$\left(\frac{S_o}{N_o}\right)_{\text{PDM}} = \left(\frac{\Delta\tau_{\text{rms}}}{\Delta\tau}\right)^2 \frac{1}{2}\left(\frac{S_o}{N_o}\right)_{\text{B}}. \tag{7.8-8}$$

This expression proves that PDM is generally inferior to a baseband system. For sinusoidal modulation, $\Delta\tau_{\text{rms}}/\Delta\tau = 1/\sqrt{2}$ and performance is 6.02 dB poorer than baseband.

Equation (7.8-6) may be couched in still another form:

$$\left(\frac{S_o}{N_o}\right)_{\text{PDM}} = \frac{\Delta\tau_{\text{rms}}^2}{\tau_{\max}\tau_0} \frac{W_{ch}}{W_f} \left(\frac{S_i}{N_i}\right)_{\text{PDM}}$$

ISBN 0-201-05758-1

$$\approx \left(\frac{\Delta\tau_{\text{rms}}}{\Delta\tau}\right)^2 \frac{W_{ch}}{2W_f} \left(\frac{S_i}{N_i}\right)_{\text{PDM}}, \tag{7.8-9}$$

which shows that demodulation gain can be large, because $W_{ch} >> 2W_f$ when N is large.

Performance Using Reconstruction

Here average input signal and noise powers are still given by (7.8-1) and (7.8-2). If there were no noise the receiver would reconstruct a typical message $f_n(t)$ exactly. With noise the reconstruction will involve an error $\epsilon_n(t)$:

$$s_{on}(t) = f_n(t) + \epsilon_n(t). \tag{7.8-10}$$

The output signal power is then

$$S_o = \overline{f_n^2(t)}, \tag{7.8-11}$$

while the "noise" power is

$$N_o = \overline{\epsilon_n^2(t)}. \tag{7.8-12}$$

To determine $\epsilon_n(t)$, we observe that, neglecting equipment imperfections, the error would be zero if the receiver made perfect determinations of the times of the two edges of the sample pulses in the PDM train (see Fig. 7.7-5). Hence $\overline{\epsilon_n^2(t)}$ is related to the average error in determining the PDM pulse widths. Since we have pulse width given by (7.7-2), we take differentials and relate small message changes (errors) to small time errors:

$$\delta\tau = k_{\text{PDM}}\,\delta\,[f_n(t)] = k_{\text{PDM}}\epsilon_n(t). \tag{7.8-13}$$

Averaging the square of (7.8-13) and substituting into (7.8-12) produces

$$N_o = \overline{(\delta\tau)^2}/k_{\text{PDM}}^2. \tag{7.8-14}$$

It only remains to relate errors $\delta\tau$ in pulse width to input noise $n_i(t)$. This amounts to finding the error in measuring the time-position of the trailing edge in one-edge modulation (the leading edge is known from synchronization). Figure 7.8-2 is helpful in the derivation. If the trailing edge is said to occur when the pulse-plus-noise waveform falls below some prescribed threshold voltage, the error $\delta\tau$ in pulse length is related to noise approximately as $\delta\tau \approx t_r n_i(t)/A$, where t_r is the pulse fall-time* and $t_r \approx \pi/W_{ch}$. Squaring and averaging $\delta\tau$, and using in (7.8-14), gives the output noise power, valid for the case of small noise:

$$N_o = \frac{\pi^2\,\overline{n_i^2(t)}}{A^2 k_{\text{PDM}}^2 W_{ch}^2} = \frac{\pi\mathcal{N}_0}{2A^2 k_{\text{PDM}}^2 W_{ch}}. \tag{7.8-15}$$

*More precisely, A/t_r represents approximately the magnitude of the slope of the pulse in the vicinity of the threshold voltage, which is often placed at about $A/2$.

ISBN 0-201-05758-1

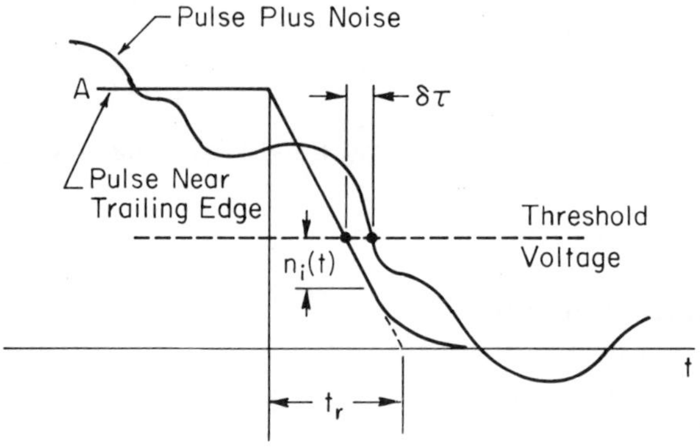

Fig. 7.8-2. Detailed view of a pulse with noise as seen near the trailing edge.

Input signal-to-noise ratio is given by (7.8-5). For the output, the signal-to-noise ratio is found from (7.8-11) and (7.8-15):

$$\left(\frac{S_o}{N_o}\right)_{\text{PDM}} = \frac{2A^2 k_{\text{PDM}}{}^2 \overline{f_n{}^2(t)} W_{ch}}{\pi \mathcal{N}_0}. \tag{7.8-16}$$

Other useful forms follow by substituting $\Delta\tau_{\text{rms}}{}^2 = k_{\text{PDM}}{}^2 \overline{f_n{}^2(t)}$:

$$\left(\frac{S_o}{N_o}\right)_{\text{PDM}} = \left(\frac{\Delta\tau_{\text{rms}}}{\tau_0}\right)^2 \frac{W_{ch}{}^2 T_s \tau_0}{\pi^2} \left(\frac{S_i}{N_i}\right)_{\text{PDM}}, \tag{7.8-17}$$

$$\left(\frac{S_o}{N_o}\right)_{\text{PDM}} = \left(\frac{\Delta\tau_{\text{rms}}}{\tau_0}\right)^2 \left(\frac{W_{ch}\tau_0}{\pi}\right) \left(\frac{W_f T_s}{\pi}\right) \left(\frac{S_o}{N_o}\right)_{\text{B}}. \tag{7.8-18}$$

For a system with two-edge modulation, $(S_o/N_o)_{\text{PDM}}$, as given by the above expressions, must be reduced by a factor of 2.

Performance of a PDM system with signal recovery by reconstruction may be worse than, equal to, or better than that of a baseband system, depending on specific system parameters. Two examples serve to demonstrate these facts.

Example 7.8-1

A PDM system multiplexes N similar messages having zero mean values using the minimum (Nyquist) sample rate. If the time slot is $3\tau_0$ wide, the maximum sample pulse width is $1.9\tau_0$, and $\Delta\tau_{\text{rms}} = 0.225\tau_0$, solve (7.8-18) and (7.8-17) assuming $\tau_{\text{min}} = 2\pi/W_{ch}$ for an idealized (rectangular passband) channel.

Because sampling is at the Nyquist rate, $W_f T_s/\pi = 1$. The minimum pulse width is $\tau_0 - (\tau_{\text{max}} - \tau_0) = 0.1\tau_0$, which must equal twice the idealized channel rise time. Hence $0.1\tau_0 \approx 2\pi/W_{ch}$, so $W_{ch}\tau_0/\pi \approx 20$, giving $(S_o/N_o)_{\text{PDM}} = (0.225)^2 20 (S_o/N_o)_{\text{B}} = 1.01 (S_o/N_o)_{\text{B}}$. In this case PDM and baseband system performances are about equal.

To solve (7.8-17) we relate T_s to τ_0 through $T_s/N = 3\tau_0$. Thus, $(S_o/N_o)_{\text{PDM}} = (0.225)^2 (20)^2 3N(S_i/N_i)_{\text{PDM}} = 60.8N(S_i/N_i)_{\text{PDM}}$. We see that a large demodulation gain occurs in PDM.

ISBN 0-201-05758-1

Example 7.8-2

If we increase the rms and peak duration deviations in the problem of example 7.8-1 so that the minimum pulse is $0.05\tau_0$, with all other parameters fixed, then $(S_o/N_o)_{PDM} = 2.26(S_o/N_o)_B$ or 3.53 dB *better* than baseband.

Naturally, reducing the peak duration deviation will cause the system to be poorer than baseband.

A comparison of (7.8-18) with the output signal-to-noise ratio for low-pass filter recovery, as given by (7.8-7), shows that reconstruction produces a larger output ratio:

$$\frac{(S_o/N_o)_{PDM}\ \text{(reconstruction)}}{(S_o/N_o)_{PDM}\ \text{(low-pass filter)}} = \left(1 + \frac{2\Delta\tau}{\tau_{min}}\right)\left(\frac{W_{ch}\tau_{min}}{\pi}\right)\left(\frac{W_f T_s}{\pi}\right). \tag{7.8-19}$$

Since the smallest value of either $W_{ch}\tau_{min}/\pi$ or $W_f T_s/\pi$ is on the order of unity, the signal-to-noise ratios are different by at least a factor of $[1 + (2\Delta\tau/\tau_{min})]$.

Maximum Possible Output Signal-to-Noise Ratio

It is instructive to determine the largest possible value of (7.8-18). This situation will occur when the minimum sample rate is used (largest T_s) and only one message ($N = 1$) is conveyed (largest τ_0). These facts allow $W_f T_s/\pi = 1$ and $\tau_0 = T_s/2 = \pi/(2W_f)$, giving a comparison of (7.8-18) with baseband performance:

$$\left(\frac{S_o}{N_o}\right)_{PDM\,(max)} = \frac{1}{2}\left(\frac{\Delta\tau_{rms}}{T_s/2}\right)^2 \frac{W_{ch}}{W_f}\left(\frac{S_o}{N_o}\right)_B. \tag{7.8-20}$$

Another expression, useful for comparing with FM, follows by noting that $k_{PDM}|f_n(t)|_{max} = \Delta\tau \approx T_s/2$:

$$\left(\frac{S_o}{N_o}\right)_{PDM\,(max)} = \frac{1}{2}\left(\frac{W_{ch}}{W_f}\right)\frac{\overline{f_n^2(t)}}{|f_n(t)|_{max}^2}\left(\frac{S_o}{N_o}\right)_B. \tag{7.8-21}$$

This equation is to be compared with (6.10-17) with $W_{ch} = 2\Delta\omega$ for wideband FM:

$$\left(\frac{S_o}{N_o}\right)_{FM} = \frac{3}{4}\left(\frac{W_{ch}}{W_f}\right)^2\frac{\overline{f_n^2(t)}}{|f_n(t)|_{max}^2}\left(\frac{S_o}{N_o}\right)_B. \tag{7.8-22}$$

Comparison shows that, for equal average transmitted power and channel bandwidths, PDM is much inferior to WBFM even under the most favorable performance conditions, a point which the following example emphasizes.

Example 7.8-3

A single message is conveyed by using PDM, and sampling is at the Nyquist rate. The minimum pulse duration during modulation is assumed to be $\tau_{min} = T_s/10$. We shall find the best

ISBN 0-201-05758-1

possible PDM performance as compared with wideband FM having the same channel bandwidth and same transmitted average power.

From (7.8-22) and (7.8-21),

$$\frac{(S_o/N_o)_{\text{FM}}}{(S_o/N_o)_{\text{PDM (max)}}} = \frac{3}{2}\left(\frac{W_{ch}}{W_f}\right).$$

Now because of sampling at the Nyquist rate, $W_f = \pi/T_s$. If we assume that the minimum pulse duration equals twice the channel rise time π/W_{ch}, then $W_{ch} = 2\pi/\tau_{\min}$, so $W_{ch}/W_f = 2T_s/\tau_{\min} = 20$, and

$$\frac{(S_o/N_o)_{\text{FM}}}{(S_o/N_o)_{\text{PDM (max)}}} = 3\left(\frac{T_s}{\tau_{\min}}\right) = 30 \qquad (14.8 \text{ dB}).$$

Performance near Threshold

All the above relationships for output signal-to-noise ratio are valid only for large input signal-to-noise ratio. When this constraint is not satisfied, a threshold effect takes place similar to that observed in Chapter 6 for FM. To obtain a picture of how the effect develops, let us assume message demodulation is achieved by reconstruction with trailing-edge PDM. Since the leading-edge position of a received pulse is known (from synchronization), the receiver determines the trailing-edge position by observing the time when the received voltage (which is the sum of the pulse and noise) decreases through a preset threshold voltage level. When the input signal-to-noise ratio becomes small enough, the threshold voltage crossing can occur from a negative-going noise peak. Such a noise peak may occur at any instant following the pulse leading-edge time, giving an indication of pulse trailing edge that is considerably in error.

Analyses of the threshold effect for both PDM and pulse position modulation (PPM) systems are nearly identical. We shall defer the details until our later discussion of PPM systems (Section 7.10) and only summarize the PDM results. For received pulses of peak amplitude A, arriving from a channel of bandwidth W_{ch}, the added output noise due to the threshold effect is

$$N_{rec} = \frac{W_{\text{rms}}\Delta\tau^3}{2\pi k_{\text{PDM}}^2}\left[\frac{\overline{f_n^2(t)}}{|f_n(t)|_{\max}^2} + \frac{1}{3}\right]\exp(-A^2/8N_i). \qquad (7.8\text{-}23)$$

Here N_i is the system input average noise power at the point where the pulse edge position decisions are made, $\Delta\tau$ is the peak duration deviation, and W_{rms} is the rms bandwidth of the power density spectrum of the system input noise. W_{rms} is determined from (4.9-35).

The expression (7.8-23) is valid for any pulse shape,* any Gaussian input noise power density spectrum, and any zero-mean random message having a symmetric probability density function. A more useful form derives from the assumptions that guard time is small relative to $\Delta\tau$ and sampling is at the Nyquist rate. For these assumptions the output signal-to-noise ratio becomes

$$\left(\frac{S_o}{N_o}\right)_{\text{PDM}} = \frac{\dfrac{1}{4N^2}\left(\dfrac{W_{ch}}{W_f}\right)^2 \dfrac{\overline{f_n^2(t)}}{|f_n(t)|_{\max}^2}\left(\dfrac{\hat{S}_i}{N_i}\right)_{\text{PDM}}}{1 + \dfrac{1}{16N^3}\left(\dfrac{W_{\text{rms}}}{W_f}\right)\left(\dfrac{W_{ch}}{W_f}\right)^2\left[\dfrac{\overline{f_n^2(t)}}{|f_n(t)|_{\max}^2} + \dfrac{1}{3}\right]\left(\dfrac{\hat{S}_i}{N_i}\right)_{\text{PDM}} \exp\left[-\dfrac{1}{8}\left(\dfrac{\hat{S}_i}{N_i}\right)_{\text{PDM}}\right]}, \qquad (7.8\text{-}24)$$

*The minimum pulse duration is assumed to be small relative to the maximum pulse duration.

ISBN 0-201-05758-1

where we have denoted the input *peak* signal-to-average noise ratio by

$$\left(\frac{\hat{S}_i}{N_i}\right)_{\text{PDM}} = \frac{A^2}{N_i}, \tag{7.8-25}$$

and, as always, W_f is the maximum spectral extent of the typical message $f_n(t)$. Peak and average input signal-to-noise ratios are related approximately by

$$\left(\frac{S_i}{N_i}\right)_{\text{PDM}} = \frac{1}{2N}\left(\frac{\hat{S}_i}{N_i}\right)_{\text{PDM}}. \tag{7.8-26}$$

In terms of a baseband system having the same average transmitted power and input noise power density, we may also show that

$$\left(\frac{S_o}{N_o}\right)_{\text{B}} = \frac{1}{2N}\left(\frac{W_{ch}}{W_f}\right)\left(\frac{\hat{S}_i}{N_i}\right)_{\text{PDM}}. \tag{7.8-27}$$

Our main result (7.8-24) is identical in form to (7.10-16) which is derived below for pulse position modulation. By assuming a single message for which $|f_n(t)|_{\max}^2 = 2\overline{f_n^2(t)}$, and white noise uniformly distributed over the channel passband, we find that Fig. 7.10-2 applies with appropriate subscript changes.

7.9. PULSE POSITION MODULATION

In PDM part of the signal power carries no information and is wasted. Recall that information in PDM is contained in the pulse *edge locations* in time and not in the pulses themselves. Thus, the narrow pulses contain as much information as the wider pulses do, leading to the conclusion that power is wasted in the longer pulses. A system using *pulse position modulation* (PPM) overcomes this difficulty.

PPM has the additional advantage over PDM of superior noise performance if the comparison is based on equal transmitted average powers. A disadvantage is the fact that it is still not as good as wideband FM, and in some cases not as good as pulse code modulation to be studied later. PPM also requires a large channel bandwidth. In some situations this fact may be a disadvantage in relation to pulse code modulation.

Signal Generation and Spectrum

A PPM signal consists of fixed-duration pulses in a train. Each pulse is displaced from a nominal position by an amount linearly related to a message sample value. Figure 7.0-1 illustrates a typical PPM signal.

As with PDM, PPM can be generated by either the uniform or nonuniform sampling methods. One direct way is to first generate and then differentiate a trailing-edge modulated PDM waveform, invert the derivative waveform, and shape the positive spikes to obtain the PPM signal. Equivalently, the negative spikes in the derivative waveform may be used to trigger a monostable circuit (sometimes called a *one-shot* circuit) to generate a pulse for each negative spike.

Another generation method using uniform sampling is shown in Fig. 7.9-1 for a typical message $f_n(t)$ of N messages. The message (a) is sampled to generate the PAM signal of (b).

ISBN 0-201-05758-1

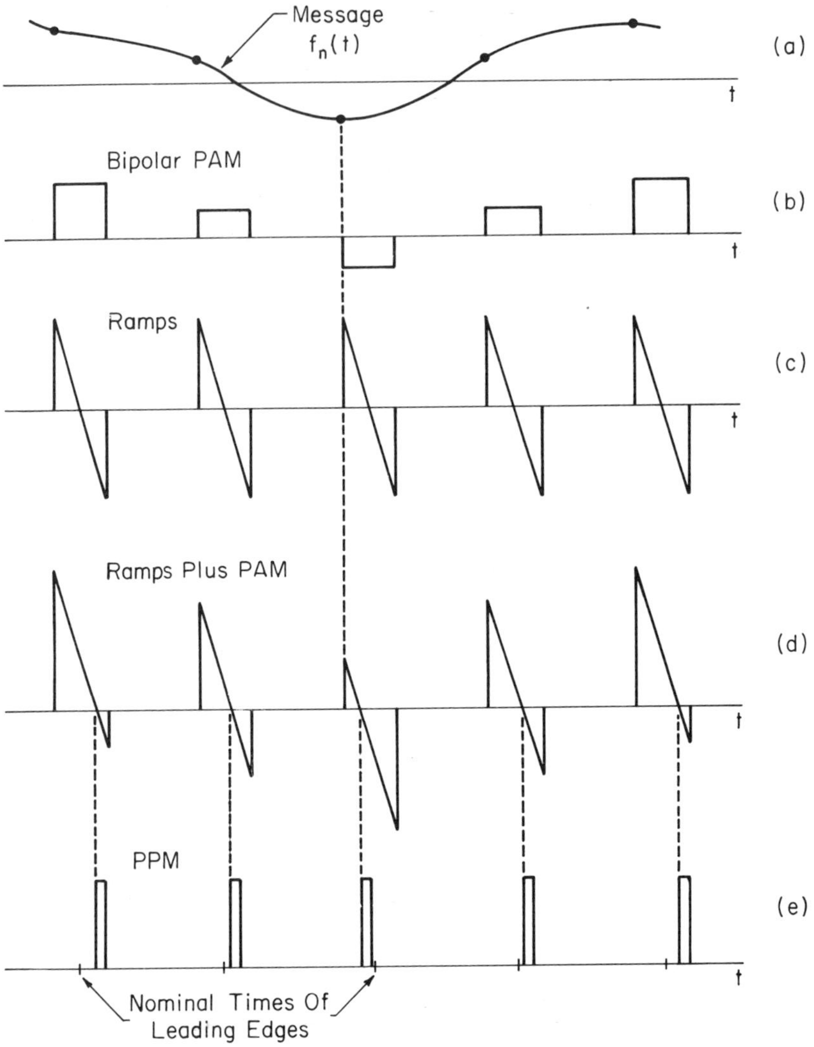

Fig. 7.9-1. Generation of a PPM signal by uniform sampling. (a) A message and (b) its conversion to bipolar PAM. (c) A train of ramps. (d) The addition of the signals of (b) and (c). (e) The PPM signal with pulses generated as the waveform of (d) goes negative.

To this is added the ramp train of (*c*) to obtain the waveform of (*d*). A monostable circuit is then used to generate a pulse each time this waveform crosses zero in a negative-going direction. If the message were absent the pulses would all start at the nominal positions shown in (*e*), which are the same as the zero-crossing points in the waveform of (*c*). The sign of the displacement can be changed by reversing the slope of the ramp.

A simple means of generating PPM by nonuniform sampling is to eliminate the PAM step above. The message is added directly to the ramp waveform during its time of occurrence. The remaining threshold and monostable circuit operations are the same as in uniform sampling. Figure 7.9-2 summarizes the various waveforms. As in PDM, nonuniform and uniform generation methods produce different results in general and the uniform sampling theorem does not strictly apply to the former. However, if the maximum position modulation of pulses is small compared with the period T_s, the differences will be small [9, p. 260].

ISBN 0-201-05758-1

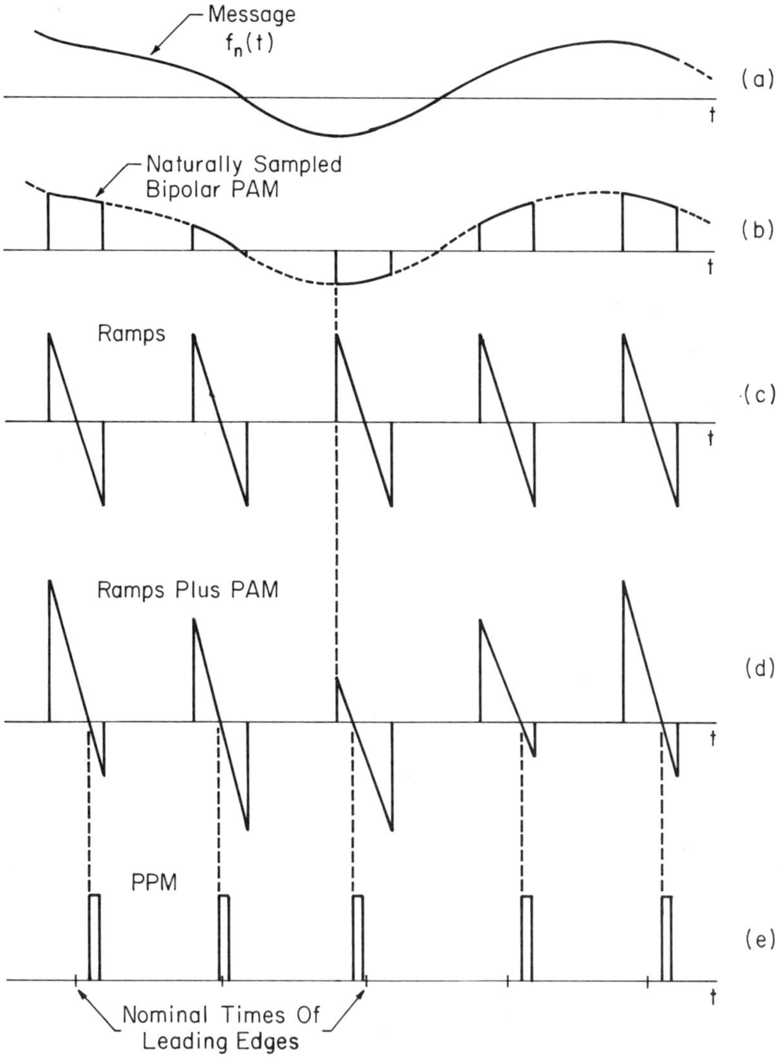

Fig. 7.9-2. Generation of PPM by nonuniform sampling. (a) A message and (b) its conversion to naturally sampled PAM. (c) A train of ramps. (d) The addition of the signals of (b) and (c). (e) The PPM signal with pulses generated when the waveform of (d) goes negative.

The small-difference conclusion may be made more convincing by comparing the waveform of Fig. 7.9-1(*b*) for uniform sampling with that of Fig. 7.9-2(*b*) for nonuniform sampling. Since the subsequent processing steps of ramp addition, threshold, and final pulse generation in the two cases are the same, the final results will be nearly the same if these two waveforms are nearly equal. Such near-equality will occur if the message waveform does not change appreciably over the duration of a ramp. In a typical application where a large number of channels is multiplexed, the ramp duration is small in relation to the sampling period T_s, and $f_n(t)$ will not vary greatly during such an interval.

Example 7.9-1

Let ten similar messages be time-multiplexed using PPM. We determine the maximum varia-

ISBN 0-201-05758-1

tion possible for a pulse duration of τ. To maintain distinct (single) pulses in any given time slot we restrict the maximum extent of the pulse variation to be less than 80% of the slot duration. A pulse duration τ of one-tenth of the slot duration will also be assumed.

The peak deviation is $\Delta\tau = (0.4T_s/N) - (\tau/2)$, where the time slot duration is T_s/N and $N = 10$. Thus

$$\frac{\Delta\tau}{T_s} = 0.04 - \frac{\tau}{2T_s}.$$

But τ/T_s is 1/100, so $\Delta\tau/T_s = 0.035$. The assumption in the text of $\Delta\tau \ll T_s$ is thus justified.

To examine the spectrum of PPM we take the case of nonuniform sampling. Results for uniform sampling will be similar for a typical multiplexed system. Consider only a typical message $f_n(t)$, and let t_k be the times of occurrence of the pulses in the PPM signal. The signal is

$$s_{\text{PPM}}(t) = \sum_{k=-\infty}^{\infty} A\tau q\left[t - kT_s - k_{\text{PPM}} f_n(t_k)\right] \tag{7.9-1}$$

with

$$t_k = kT_s + k_{\text{PPM}} f_n(t_k). \tag{7.9-2}$$

Here A is the amplitude of pulses having the form $q(t)$; $q(0) = 1/\tau$, τ is the equivalent rectangular pulse width* of $q(t)$, and k_{PPM} is a positive proportionality constant for the PPM modulation process. Equation (7.9-1) may also be written as

$$s_{\text{PPM}}(t) = q(t) * A\tau \sum_{k=-\infty}^{\infty} \delta(t - t_k), \tag{7.9-3}$$

where the asterisk denotes convolution. Although somewhat involved, it may be shown, using methods of Friedman [10],† that

$$A\tau \sum_{k=-\infty}^{\infty} \delta(t - t_k) = \frac{A\tau}{T_s}\left|1 - k_{\text{PPM}}\frac{df_n(t)}{dt}\right|\left\{1 + 2\sum_{k=1}^{\infty} \cos\left[k\omega_s t - k\omega_s k_{\text{PPM}} f_n(t)\right]\right\}. \tag{7.9-4}$$

In trying to interpret (7.9-3) using (7.9-4) it is helpful to resort to the frequency domain. The spectrum $S_{\text{PPM}}(\omega)$ of the PPM signal will be the product of $Q(\omega)$, the spectrum of $q(t)$, and the Fourier transform of (7.9-4). This is seen by Fourier transforming the right side of (7.9-3). Hence, the main behavior of $S_{\text{PPM}}(\omega)$ is determined by (7.9-4), which indicates the presence of spectra centered about frequencies $k\omega_s$ due to both amplitude and phase modulation. More detailed interpretation of the spectrum is quite difficult and not necessary here.

*See Section 7.2.

†For a less detailed development see [8].

ISBN 0-201-05758-1

Channel Bandwidth

As in PDM, even though the detailed form of the PPM signal spectrum is difficult to determine, we may develop an expression for the minimum channel bandwidth required to support the PPM signal with reasonable fidelity. All the steps involved are direct parallels of those used in Section 7.7 to find the corresponding bandwidth in PDM. Thus, our discussion is brief, and the reader should refer to the earlier work for detail.

Let τ be the duration of the pulses. Peak position deviation $\Delta\tau$ is given by

$$\Delta\tau = k_{\mathrm{PPM}} \, |f_n(t)|_{\max} \tag{7.9-5}$$

for a typical message $f_n(t)$ of N similar multiplexed messages. The messages are sampled at a rate $\omega_s \geqslant 2W_f$, where, as always, W_f is the maximum signal spectral extent. When sample pulses in two adjacent time slots are at their closest possible separation, assume the trailing edge of one remains τ_g seconds from the leading edge of the other. The separation is the guard time needed to reduce crosstalk. With these definitions

$$W_{ch} \geqslant \frac{N\omega_s}{2}\left(1 + \frac{2\Delta\tau + \tau_g}{\tau}\right). \tag{7.9-6}$$

Alternatively,

$$W_{ch} \geqslant NW_f\left(1 + \frac{2\Delta\tau + \tau_g}{\tau}\right). \tag{7.9-7}$$

These expressions are to be compared with (7.7-7) and (7.7-8).

Signal Recovery Using a Filter-Integrator

We now show that an ideal low-pass filter with a dc blocking circuit followed by an integrator can recovery the message $f_n(t)$. From (7.9-3), by using (7.9-4), we find that the baseband component of $s_{\mathrm{PPM}}(t)$ which is passed by the low-pass filter is

$$q(t) * \frac{A\tau}{T_s}\left|1 - k_{\mathrm{PPM}} \frac{df_n(t)}{dt}\right|. \tag{7.9-8}$$

Restricting analysis to the case where*

$$k_{\mathrm{PPM}} \left|\frac{df_n(t)}{dt}\right|_{\max} < 1, \tag{7.9-9}$$

and dropping the first term (dc component) of (7.9-8), we obtain the output $s_1(t)$ of the low-pass filter and dc blocking circuit:

$$s_1(t) \approx \frac{-A\tau k_{\mathrm{PPM}}}{T_s}\, q(t) * \frac{df_n(t)}{dt}$$

*This condition is satisfied for $N > 1$.

ISBN 0-201-05758-1

$$= \frac{-A\tau k_{\text{PPM}}}{T_s} \int_{-\infty}^{\infty} q(x)\dot{f}_n(t - x)\, dx, \tag{7.9-10}$$

where $\dot{f}_n(x) = df_n(x)/dx$. The integrator response $s_d(t)$ becomes

$$s_d(t) \approx \frac{-A\tau k_{\text{PPM}}}{T_s} \int_{-\infty}^{t} \int_{-\infty}^{\infty} q(x)\dot{f}_n(u - x)\, dx\, du$$

$$= \frac{-A\tau k_{\text{PPM}}}{T_s} \int_{-\infty}^{\infty} q(x) f_n(t - x)\, dx. \tag{7.9-11}$$

Since $q(t)$ is a pulse of small duration τ and amplitude $1/\tau$, we approximate $q(t)$ by an impulse $\delta(t)$ giving

$$s_d(t) \approx -A\tau k_{\text{PPM}} f_n(t)/T_s. \tag{7.9-12}$$

The filter-integrator recovery method is simple and straightforward. Its largest disadvantage is that it fails to take full advantage of the noise reduction properties of PPM. A more efficient recovery method is modulation conversion to PAM followed by low-pass filtering.

Signal Recovery by Reconstruction

It was found in Section 7.4 that a bandlimited information signal could be recovered without error using PAM. Because PPM can be generated from PAM it becomes logical to employ the reverse process, PPM-to-PAM conversion, and recover the message error-free (except for noise effects) from the PAM waveform. This process of demodulation is called *reconstruction.* One method first uses an intermediate conversion to PDM; the PDM-to-PAM method implied in Fig. 7.7-5 may then be used to complete the process.

The PPM-to-PDM conversion may be accomplished as illustrated in Fig. 7.9-3(*a*). A low-pass filter at the input serves to pass the received pulses with acceptable fidelity while restricting the amount of channel noise admitted to the rest of the system. It is helpful to think of the filter as ideal and broadband enough so that rectangular input pulses, with amplitudes A, emerge approximately undistorted as shown in (*c*). The converter uses a train of clock pulses (*b*) to set a bistable circuit to one of its stable levels, such as level L in (*d*). Clock train pulses are offset (advanced) relative to the unmodulated position of the PPM pulses sufficiently to prevent overlap when maximum pulse position deviation takes place. Internal to the converter is a preset voltage threshold with which the input PPM signal is compared. When a pulse exceeds the voltage threshold the bistable circuit is triggered to its second stable level (zero volts). The overall effect is to construct the PDM waveform of (*d*) which is free of noise. Of course, the *effects* of noise remain in the form of errors in the positions of pulse trailing edges.

7.10. NOISE PERFORMANCE OF PPM

Because demodulation by reconstruction takes advantage of the noise reduction properties of PPM, we consider only the performance of this method.

ISBN 0-201-05758-1

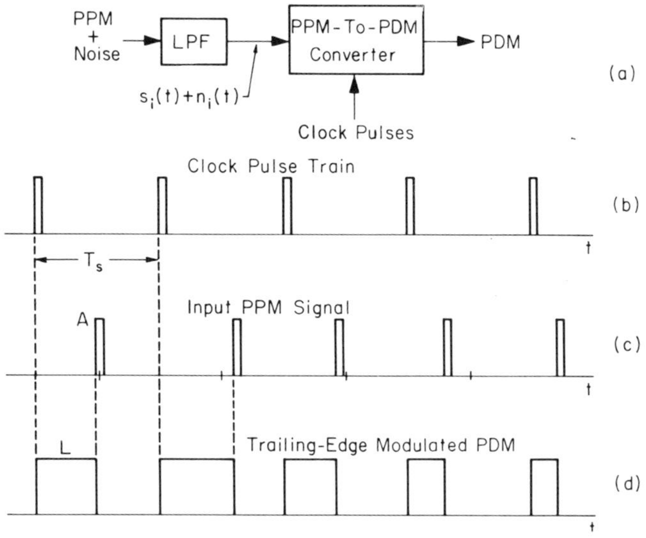

Fig. 7.9-3. Intermediate operation in the demodulation of PPM. (a) Applicable block diagram. (b) – (d) Applicable waveforms. The output PDM signal (d) is then demodulated via a PDM-to-PAM conversion.

Performance Using Reconstruction

Analysis exactly parallels that developed for PDM performance using reconstruction. The reason is that both cases ultimately depend on determining a time property of the pulses. In PDM it is the time of the trailing edge (for single-edge modulation) that is important; in PPM it is pulse position, which is typically found from an edge position measurement also.

Omitting the detailed steps, which the reader may fill in as an exercise, we summarize the principal findings. The input multiplexed PPM signal consists of pulses of amplitude A and width τ. Input average signal power for a typical message component is given by (7.8-1) with τ_0 replaced by τ. Output signal average power is given by (7.8-11). For white noise with power density $\mathcal{N}_0/2$ and a channel bandwidth W_{ch}, the input average noise power is again given by (7.8-2). Output average noise power becomes (7.8-15) with k_{PDM} replaced by k_{PPM}. The important input and output signal-to-noise ratios are stated for completeness:

$$\left(\frac{S_i}{N_i}\right)_{PPM} = \frac{2\pi A^2 \tau}{\mathcal{N}_0 W_{ch} T_s}, \tag{7.10-1}$$

$$\left(\frac{S_o}{N_o}\right)_{PPM} = \frac{2A^2 k_{PPM}^2 \overline{f_n^2(t)} W_{ch}}{\pi \mathcal{N}_0}. \tag{7.10-2}$$

Alternatively, noting that $\Delta\tau_{rms}^2 = k_{PPM}^2 \overline{f_n^2(t)}$, we have

$$\left(\frac{S_o}{N_o}\right)_{PPM} = \left(\frac{\Delta\tau_{rms}}{\tau}\right)^2 \frac{W_{ch}^2 T_s \tau}{\pi^2} \left(\frac{S_i}{N_i}\right)_{PPM} \tag{7.10-3}$$

and

ISBN 0-201-05758-1

$$\left(\frac{S_o}{N_o}\right)_{\text{PPM}} = \left(\frac{\Delta\tau_{\text{rms}}}{\tau}\right)^2 \left(\frac{W_{ch}\tau}{\pi}\right)\left(\frac{W_f T_s}{\pi}\right)\left(\frac{S_o}{N_o}\right)_{\text{B}}, \tag{7.10-4}$$

which are analogous to (7.8-17) and (7.8-18).

Example 7.10-1

Let us rework the PDM example 7.8-1 for a PPM case, except we now let τ equal the minimum pulse length in the PDM case, since this establishes equal channel bandwidths. Hence τ here equals $\tau_0/10$ in the PDM example. Furthermore, now $\Delta\tau_{\text{rms}} = 0.225(10\tau) = 2.25\tau$, and $\tau = 2\pi/W_{ch}$, so from (7.10-4), $(S_o/N_o)_{\text{PPM}} = (2.25)^2 2(1)(S_o/N_o)_{\text{B}} = 10.13(S_o/N_o)_{\text{B}}$. This is about 10 dB better than PDM. Based on (7.10-3) the result is also 10 dB better than for PDM. Both comparisons assume equal average signal powers at the receiver inputs (pulse amplitudes are different).

Comparison with PDM

The above example shows the supriority of PPM over PDM that results from being able to use narrow pulses all the time. Assuming equal noise densities, channel bandwidths, average receiver input signal power, signal characteristics, and values of $\Delta\tau_{\text{rms}}$, we find that

$$\left(\frac{S_o}{N_o}\right)_{\text{PPM}} = \frac{\tau_0}{\tau}\left(\frac{S_o}{N_o}\right)_{\text{PDM}}. \tag{7.10-5}$$

Thus, PPM is generally superior to PDM in performance because $\tau_0 > \tau$. Note that the assumption of equal *input* powers requires pulse amplitude A in the PPM system to be larger than that in the PDM system by a factor $\sqrt{\tau_0/\tau}$. If the two amplitudes are equal, both PPM and PDM give the same output signal-to-noise ratio, all other parameters being equal.

Maximum Possible Output Signal-to-Noise Ratio

The maximum value that $(S_o/N_o)_{\text{PPM}}$ may take on occurs, as in PDM, when sampling is at the Nyquist rate $(W_f T_s/\pi = 1)$ and only one message is transmitted. Assuming a peak position deviation that is large relative to a pulse length allows the approximation $\Delta\tau \approx T_s/2$. The additional assumption, that pulse length is approximately related to the channel bandwidth by $\tau \approx \pi/W_{ch}$, will allow (7.10-4) to be put in the form

$$\left(\frac{S_o}{N_o}\right)_{\text{PPM (max)}} = \frac{1}{4}\left(\frac{W_{ch}}{W_f}\right)^2 \frac{\overline{f_n^{\,2}(t)}}{|f_n(t)|_{\max}^2}\left(\frac{S_o}{N_o}\right)_{\text{B}}. \tag{7.10-6}$$

This expression implies that maximum signal-to-noise ratio increases as the square of the channel bandwidth. Recall that wideband FM behaves in a similar manner. Thus, a comparison of the two modulation methods is in order.

Comparison with FM Performance

Suppose (7.10-6) is compared with (7.8-22) on the basis of equal channel bandwidth and equal transmitted power, which means that (S_o/N_o) is the same in the two expressions. We have

ISBN 0-201-05758-1

$$\left(\frac{S_o}{N_o}\right)_{\text{FM}} = 3\left(\frac{S_o}{N_o}\right)_{\text{PPM (max)}}, \tag{7.10-7}$$

which shows FM to be better by a factor of 3 (4.8 dB).

All the above performance evaluations assume that input signal-to-noise ratio is large. When this condition is not true, a threshold effect occurs similar to that studied in FM and PDM.

★ Threshold Noise Power in PPM

When input signal-to-noise ratio is not sufficiently large, noise pulses will occasionally cross the voltage threshold in the PPM reconstruction process and produce pulse position estimates which are greatly in error. We may think of the overall effect as adding a new error term to the reconstructed message that is due to receiver noise. If N_{rec} is the added receiver noise power, the ratio of the output signal power S_o to the noise power is

$$\left(\frac{S_o}{N_o}\right)_{\text{PPM}} = \frac{S_o}{N_o + N_{rec}} = \frac{(S_o/N_o)}{1 + (N_{rec}/N_o)}, \tag{7.10-8}$$

where S_o and N_o are the output signal and noise powers when operating above the threshold.

The problem of finding N_{rec} is generally difficult [3, 11]. Blachman has presented a simple approach [12] that applies to a carrier modulated by PPM. We here adopt Blachman's method but extend it to the case of a baseband PPM system. Our goal is only to develop approximate results to illustrate the threshold effect. Thus, the simple method is especially attractive because it leads rapidly to an instructive expression for N_{rec}.

Assume the receiver to be specified by Fig. 7.9-3(a). The time interval applicable to a typical pulse is sketched in Fig. 7.10-1, where the time origin is taken as the position of an unmodulated pulse for convenience. The received pulse occurs at time* u with $-\Delta\tau \leqslant u = k_{\text{PPM}} f_n(t_k) \leqslant \Delta\tau$. Now if a noise pulse crosses the voltage threshold at some time $v < u$, an error $(v - u)$ is made in the true time-position of the pulse. Noise crossings for $v > u$ are of little or no interest because of the high probability of the pulse's crossing the threshold at time u (v occurring above u requires the *joint* occurrence of a missed pulse and a noise crossing, an event with small probability). Since noise crossings are equally likely to occur at all times, the crossing time v is a uniformly distributed random quantity. Similarly, u is a random quantity, if a random message is presumed; its distribution is described by the message probability density function.

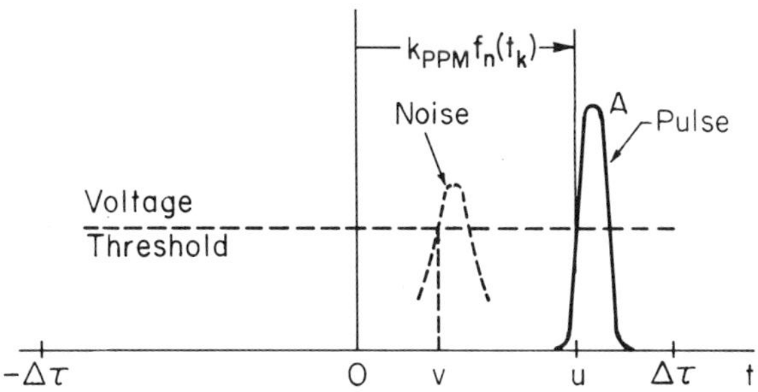

Fig. 7.10-1. Waveforms applicable to the threshold effect in PPM.

*The sample time is t_k for the kth sample.

ISBN 0-201-05758-1

The first step in the determination of N_{rec} is to find the mean-squared value of the error $(v - u)$. Denote this mean-squared error by $\overline{\epsilon_i^2}$. Initially we consider a particular value of u and observe that there will be some average time $\overline{T}_n$ between noise crossings of the threshold voltage level. Now if we assume that $\overline{T}_n >> 2\Delta\tau$ so that it is unlikely that more than one noise pulse is involved, and if we assume that the pulse duration τ is small in relation to $\Delta\tau$ so that end effects may be neglected, the probability of a false pulse is $(u + \Delta\tau)/\overline{T}_n$. The corresponding mean-squared value of the error, which is distributed uniformly over the interval $-\Delta\tau < v < u$, is $(u + \Delta\tau)^2/3$. Multiplying these two together, we get the expected mean-squared error $(u + \Delta\tau)^3/3\overline{T}_n$, which is conditional on a particular value of u. We account for all values of u by averaging over the distribution of u, denoted by $p_u(u)$. Thus we have

$$\overline{\epsilon_t^2} = \int_{-\Delta\tau}^{\Delta\tau} \frac{(u + \Delta\tau)^3}{3\overline{T}_n} p_u(u)\, du. \tag{7.10-9}$$

This expression easily reduces to

$$\overline{\epsilon_t^2} = \frac{\Delta\tau}{\overline{T}_n}\left(\Delta\tau_{\text{rms}}^2 + \frac{\Delta\tau^2}{3}\right), \tag{7.10-10}$$

where $\overline{u^2} = \Delta\tau_{\text{rms}}^2$ is the mean-squared position deviation. In obtaining (7.10-10) the message was assumed symmetrically distributed about a zero mean.

Final determination of N_{rec} requires evaluation of $\overline{T}_n$. To this end we follow Blachman's procedure [12]. Let $n_i(t)$ be the input noise that occasionally may cross the voltage threshold* $A/2$. The slope of this noise is $\dot{n}_i(t) = dn_i(t)/dt$. A noise crossing will occur if at some time n_i is just below $A/2$ and $\dot{n}_i$ is positive—that is, n_i is increasing with time. Because n_i and $\dot{n}_i$ are statistically independent random variables at any given instant if the noise n_i is Gaussian,† the joint probability density function of n_i and $\dot{n}_i$ becomes the product of individual density functions, defined as $p_{n_i}(n_i)$ and $p_{\dot{n}_i}(\dot{n}_i)$, respectively. Thus, the joint probability that n_i will lie in the interval $(A/2) - dn_i$ to $A/2$ and $\dot{n}_i$ will fall in any interval $d\dot{n}_i$ is $p_{n_i}(A/2)\, dn_i p_{\dot{n}_i}(\dot{n}_i)\, d\dot{n}_i$. Now because $dn_i = \dot{n}_i\, dt$, this probability also equals $p_{n_i}(A/2)\dot{n}_i\, p_{\dot{n}_i}(\dot{n}_i)\, d\dot{n}_i\, dt$. Finally, since an upward noise crossing corresponds to positive values of $\dot{n}_i$, we integrate over $0 < \dot{n}_i$ to obtain the probability $\overline{P}_n$ of an upward noise crossing in the time interval dt:

$$\overline{P}_n = p_{n_i}(A/2)\, dt \int_0^{\infty} \dot{n}_i p_{\dot{n}_i}(\dot{n}_i)\, d\dot{n}_i. \tag{7.10-11}$$

Equation (7.10-11) is easily evaluated, since $\dot{n}_i$ is Gaussian when $n_i(t)$ is a Gaussian random process. We obtain

$$\overline{P}_n = \frac{W_{\text{rms}}}{2\pi}\, dt \exp\left(-\frac{A^2}{8N_i}\right), \tag{7.10-12}$$

where W_{rms} is the rms bandwidth, as defined by (4.9-35), of the input noise which has average

*We assume a threshold voltage level of half the received pulse's peak voltage A. It is the optimum level for many systems.

†The procedure for proving this fact is to use (2.5-2), with $n = 1$, to represent $\dot{n}_i(t)$ and develop the cross-correlation function $R_{\dot{n}_i n_i}(t, t + \tau)$. It will be found to be zero when $\tau = 0$. Thus, n_i and $\dot{n}_i$ are uncorrelated and therefore, for Gaussian noise, are statistically independent.

ISBN 0-201-05758-1

power N_i. Next, we recognize that this probability also must equal

$$\overline{P}_n = dt/\overline{T}_n. \tag{7.10-13}$$

On equating (7.10-12) and (7.10-13), $\overline{T}_n$ is found to be $(2\pi/W_{\rm rms})\exp(A^2/8N_i)$. Receiver noise power now becomes

$$N_{rec} = \frac{W_{\rm rms}\Delta\tau^3}{2\pi k_{\rm PPM}{}^2}\left[\frac{\overline{f_n{}^2(t)}}{|f_n(t)|_{\max}{}^2} + \frac{1}{3}\right]\exp\left(-\frac{A^2}{8N_i}\right), \tag{7.10-14}$$

since

$$N_{rec} = \overline{\epsilon_t{}^2}/k_{\rm PPM}{}^2. \tag{7.10-15}$$

★ Performance near Threshold

The receiver noise expression of (7.10-14) holds for any noise power spectrum, any zero-mean message having a symmetric probability density function and any arbitrary (narrow) pulse shape. Its substitution into (7.10-8) will lead to an expression for $(S_o/N_o)_{\rm PPM}$ valid near and above threshold. A useful form follows from the assumptions: guard-time and pulse duration both much smaller than $\Delta\tau$, and sampling at the Nyquist rate. We than have

$$\left(\frac{S_o}{N_o}\right)_{\rm PPM} = \frac{\dfrac{1}{4N^2}\left(\dfrac{W_{ch}}{W_f}\right)^2 \dfrac{\overline{f_n{}^2(t)}}{|f_n(t)|_{\max}{}^2}\left(\dfrac{\hat{S}_i}{N_i}\right)_{\rm PPM}}{1 + \dfrac{1}{16N^3}\left(\dfrac{W_{\rm rms}}{W_f}\right)\left(\dfrac{W_{ch}}{W_f}\right)^2\left[\dfrac{\overline{f_n{}^2(t)}}{|f_n(t)|_{\max}{}^2} + \dfrac{1}{3}\right]\left(\dfrac{\hat{S}_i}{N_i}\right)_{\rm PPM}\exp\left[-\dfrac{1}{8}\left(\dfrac{\hat{S}_i}{N_i}\right)_{\rm PPM}\right]},$$

where we define (7.10-16)

$$(\hat{S}_i/N_i)_{\rm PPM} = A^2/N_i \tag{7.10-17}$$

as the input *peak* signal-to-average noise power ratio. It can be shown to equal the output average signal-to-noise ratio of a baseband system having the same input average signal power and same input noise power density—that is,

$$(\hat{S}_i/N_i)_{\rm PPM} = (S_o/N_o)_{\rm B}. \tag{7.10-18}$$

The numerator of (7.10-16) is recognized as the signal-to-noise ratio above threshold. Figure 7.10-2 illustrates the behavior of (7.10-16) when a *single* message ($N = 1$) is transmitted for which $|f_n(t)|_{\max}{}^2 = 2\overline{f_n{}^2(t)}$, as in the case of sinusoidal modulation. Noise power is assumed uniformly distributed over the channel passband ($W_{\rm rms} = W_{ch}/\sqrt{3}$). Threshold effect is clearly present and occurs at $(\hat{S}_i/N_i)_{\rm PPM}$ from 17.4 dB to 20.4 dB for $5 \leqslant W_{ch}/W_f \leqslant 40$.

7.11. CROSSTALK IN PPM AND PDM SYSTEMS

As with PAM time-multiplexed systems, PPM and PDM systems may also have crosstalk. We briefly examine the crosstalk problem in both PPM and PDM, assuming message recovery is by reconstruction in each case.

ISBN 0-201-05758-1

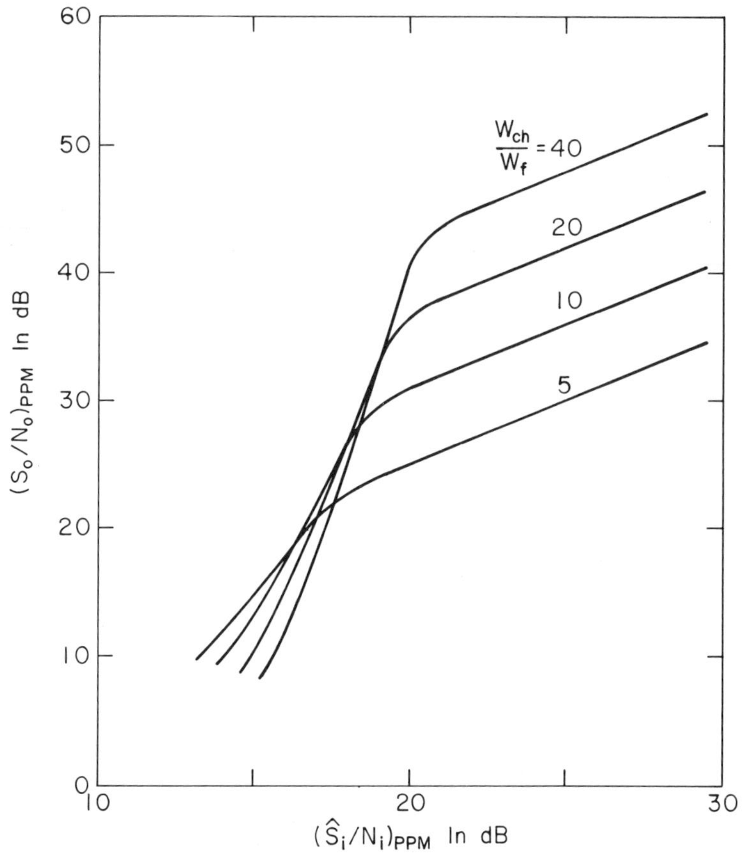

Fig. 7.10-2. Output signal-to-noise ratio in a PPM system transmitting one message for which $\overline{f^2(t)}/|f(t)|_{max}^2 = ½$.

Crosstalk in PPM

Consider typical sample pulses for two messages, n and $n + 1$, as depicted in Fig. 7.11-1. The sketches are grossly exaggerated for illustrative purposes. For simplicity, we shall assume the channel is a simple RC low-pass filter with 3-dB bandwidth W_{ch}. The rising and decaying portions of the pulses will have an exponential form with time constant $1/W_{ch}$. For convenience, the time origin has been selected at the unmodulated trailing-edge position of the ideal transmitted pulse corresponding to message $n + 1$.

Let the times of the PPM pulses be determined by position measurements made at the trailing edges. If we initially neglect the effect of the pulse due to message n, the trailing edge of the pulse due to message $n + 1$ is said to occur at time t_1, where the pulse falls below the threshold placed at half-amplitude. When the effect of message n is considered, the voltage level near t_1 is altered and a new crossing time t_2 occurs. The error in time-position caused by crosstalk is readily found to be

$$\epsilon_{ct}(t) = t_2 - t_1 = \frac{1}{W_{ch}} \ln \left\{ 1 + e^{-[(T_s/N) - \delta t_n + \delta t_{n+1}] W_{ch}} \right\}. \tag{7.11-1}$$

Now consider the maximum value of this function, which occurs when the deviations are extreme—that is, when $\delta t_n = \Delta\tau$ and $\delta t_{n+1} = -\Delta\tau$. If guard time τ_g is the smallest separation

ISBN 0-201-05758-1

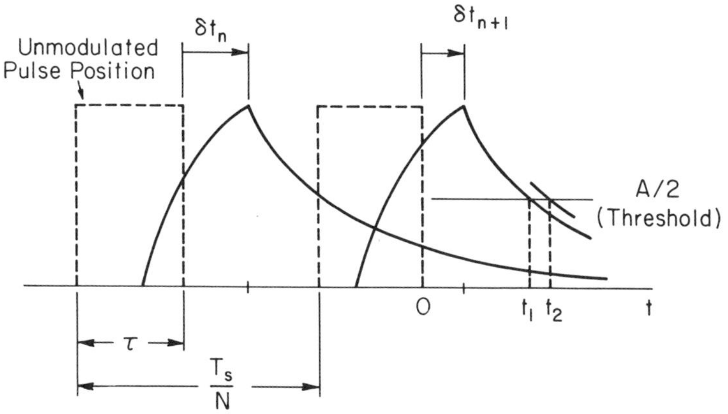

Fig. 7.11-1. Sample PPM pulses in two adjacent message time slots illustrating crosstalk.

between the trailing and leading edges of the two pulses (for rectangular pulses such as exist prior to channel distortion), the exponent of e equals $-W_{ch}(\tau + \tau_g)$. Since $W_{ch} \approx \pi/\tau$, then $-W_{ch}(\tau + \tau_g) \approx -\pi[1 + (\tau_g/\tau)]$, which means the exponential term is small relative to unity. The approximation $\ln(1+x) \approx x$ for $|x| \ll 1$ may be used to obtain

$$\epsilon_{ct} \leqslant \frac{\tau}{\pi} e^{-\pi[1+(\tau_g/\tau)]} = 0.014\tau e^{-\pi\tau_g/\tau}. \tag{7.11-2}$$

Crosstalk factor K_{ct} may be defined as the ratio of the peak time-position error to the peak time-position deviation $\Delta\tau$, namely,

$$K_{ct} = 0.014 \frac{\tau}{\Delta\tau} \exp(-\pi\tau_g/\tau). \tag{7.11-3}$$

This factor is small even for small guard times and relatively small deviations.

Crosstalk in PDM

For a trailing-edge modulated PDM system, (7.11-2) and (7.11-3) above apply to PDM if τ is replaced by τ_{min}. Of course, the deviations δt_n and δt_{n+1} now represent the trailing-edge modulations.

7.12. PULSE CODE MODULATION FUNDAMENTALS

One of the most valuable modern modulation types based on transmitting samples of a message is *pulse code modulation* (PCM). PCM is a widely different form of modulation compared with the analog pulse types such as PAM, PDM, and PPM. These types require that variation of some parameter of the pulse train be a linear function of the message sample. In PCM no such well-defined relationship exists. Instead, the presence or absence (or, alternatively, the sign) of pulses in a group of pulses is made to depend, in a somewhat arbitrary manner, on message samples.* The characteristics of the pulse group are related to the message sample through operations called *quantization* and *coding*.

*We do not imply a haphazard relationship here—only that there is arbitrariness allowed in relating pulse-group characteristics to the sample. Once chosen, the relationship becomes quite specific and may be the result of sound logic.

ISBN 0-201-05758-1

Quantization and Coding

Figure 7.12-1(a) is useful in understanding quantization. Assume some message is known to always fall in the range 0 to 7 volts as shown. As usual, samples are taken T_s seconds apart; four exact sample values are 2.21, 4.37, 5.22, and 2.83 volts in the figure. Quantization amounts to selecting a set of discrete voltage levels (say eight in the example, 1 volt apart) and rounding off the exact samples to the nearest discrete or *quantum level.* The quantum levels are separated by the *quantum step size* (1 volt in the example). The four *quantized samples* are then 2, 4, 5, and 3 volts.

A simple form of coding is to assign a *binary code* to represent the quantized sample value. A binary code is a finite sequence of zeros and ones. A number N having a three-digit code $k_2 k_1 k_0$ may be represented by $N = k_2(2)^2 + k_1(2)^1 + k_0(2)^0$. The k's are either zero or one. Numbers between 0 and 7 may be represented by a three-digit code. An example of a three-digit code is shown in Fig. 7.12-1(b). Each digit in the code is called a *bit* (after *bi*nary digi*t*). All bits of a code taken together form a *code word.* The coding process is completed by assigning a pulse to a code 1 and no pulse to a code 0. This results in *unipolar* PCM (*on-off* signaling). Alternatively, a positive pulse can be assigned to a code 1 and a negative pulse to a code 0 to give *polar* PCM.* Figure 7.12-1(b) illustrates unipolar and polar PCM pulses. If the

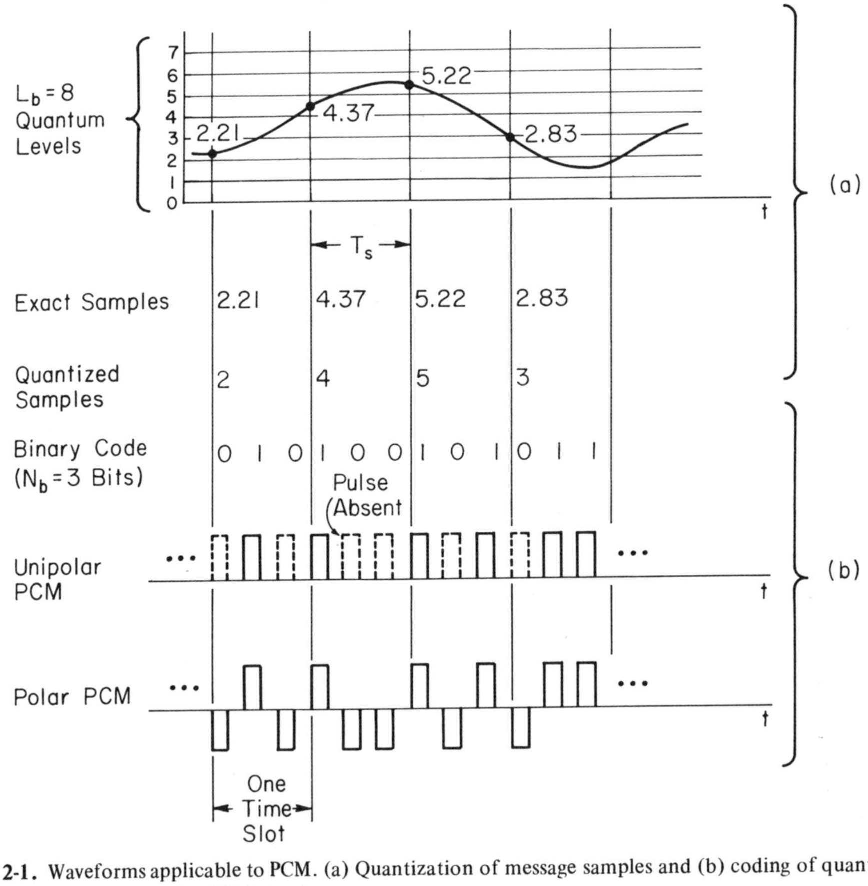

Fig. 7.12-1. Waveforms applicable to PCM. (a) Quantization of message samples and (b) coding of quantized samples into unipolar or polar PCM signals.

*The term bipolar PCM is tempting here. However, the name *polar* is customary to avoid confusion with a form of coding carrying the former name. It transmits a zero as no-pulse and transmits a one in the form of alternating positive and negative pulses [8].

ISBN 0-201-05758-1

various pulses are run together such that they have no gaps, the waveform is said to be *synchronous* [8]. A synchronous PCM signal may be either unipolar or polar. All these coding processes are referred to as *encoding* of the message. The circuit that generates the coded pulse train from quantized message samples is called an *encoder*.

We observe that a binary code has a finite number of bits. The number of levels that may be coded is likewise fixed. If N_b bits are involved, the number of levels L_b is

$$L_b = 2^{N_b}. \tag{7.12-1}$$

Table 7.12-1 gives L_b for a few values of N_b. A specific number $0 \leqslant N < L_b$ becomes

$$N = k_{N_b-1}(2)^{N_b-1} + \cdots + k_2(2)^2 + k_1(2)^1 + k_0(2)^0. \tag{7.12-2}$$

TABLE 7.12-1. Binary Levels Available for Various Code Lengths

Code Length N_b (bits)	Number L_b of Binary Levels
1	2
2	4
3	8
4	16
5	32
6	64
7	128
8	256
9	512
10	1024
11	2048
12	4096
13	8192
14	16384
15	32768

More generally, if a signal is to be quantized into L levels, where L may not be one of the discrete values of L_b in Table 7.12-1, we must use an N_b-bit code where

$$L \leqslant 2^{N_b} = L_b. \tag{7.12-3}$$

We take an example to illustrate the choice of L, L_b, and N_b.

Example 7.12-1

A message has zero mean value and a maximum magnitude $|f(t)|_{\max} = 10$ volts. It is to be quantized using a quantum step size of 0.1 volt with one level equal to zero volts. We find the necessary values of L, L_b, and N_b.

Since ths signal always exists between -10 and $+10$ volts, there are $L = (20/0.1) + 1 = 201$ levels required. The closest L_b value is 256 from (7.12-3) requiring an N_b = 8-bit code

ISBN 0-201-05758-1

from Table 7.12-1. The actual range of signal which could be handled is $(256 - 1)0.1 = 25.5$ volts, corresponding to, say, -12.5 to 13.0 volts. There is then an amplitude margin of safety of 25%.

The N_b pulses in a binary PCM signal only have two possible levels. If more pulse levels are allowed, say μ levels for each pulse, we have μ-ary PCM. The number L_b of quantum levels possible is now determined by the number L_b of *combinations* of N_b pulses with μ possible amplitudes each

$$L_b = \mu^{N_b}. \tag{7.12-4}$$

A choice $\mu = 3$ gives *trinary PCM.*

Binary is the most commonly used form of PCM. The most important reasons for this choice are: (1) binary PCM is the most noise-immune [13] form and (2) binary PCM can be processed quite easily using modern digital circuits. Except for a brief look at μ-ary PCM in Section 7.15 we shall hereafter limit discussions mainly to binary PCM.

Quantization Error and Performance Limitation

The process of quantization leads to unavoidable error. Clearly, information is lost in the process of rounding off exact sample values to the nearest quantum level. Perfect recovery of the message by the receiver is therefore impossible in PCM, even if noise is absent. Thus, a *quantization error* exists in PCM and, as will be seen, is the basic limitation in performance.

Quantization and sampling produce the same result as sampling and quantizing. Taking the former view gives a good insight into the behavior of the quantization error. Figure 7.12-2 illustrates appropriate waveforms. The quantized message $f_q(t)$ is the sum of the message $f(t)$ plus an error $\epsilon_q(t)$:

$$f_q(t) = f(t) + \epsilon_q(t). \tag{7.12-5}$$

In the receiver, only $f_q(t)$ can be reconstructed from the PCM signal, so that $\epsilon_q(t)$ is fundamental to overall performance. To see this fact, we observe that the receiver error $\epsilon_q(t)$ may be treated as a "noise." The maximum possible output signal-to-noise ratio is then the ratio of $\overline{f^2(t)}$ to quantization error noise power $\overline{\epsilon_q^{\,2}(t)}$:

$$(S_o/N_q)_{\mathrm{PCM}} = \overline{f^2(t)}/\overline{\epsilon_q^{\,2}(t)}. \tag{7.12-6}$$

The evaluation of $\overline{\epsilon_q^{\,2}(t)}$ involves averaging $\epsilon_q^{\,2}(t)$ over all time. If we assume that δv is small (a large number of levels spans the signal peak-to-peak variation) and if the error intervals at the signal extreme points are neglected, the error in each time interval will be approximately linear, extending over a range δv. Direct application of (4.8-35) with $\tau = 0$ will yield the mean-squared error*

*This result follows more correctly from a probability approach. The error may be considered random and, for small δv, assumed to be equally likely to fall anywhere in the interval $-\delta v/2$ to $\delta v/2$. The probability density function of error is then $1/\delta v$, and the mean-squared error is

$$\overline{\epsilon_q^{\,2}} = \int_{-\delta v/2}^{\delta v/2} \epsilon_q^{\,2}\, d\epsilon_q/\delta v = (\delta v)^2/12.$$

ISBN 0-201-05758-1

$$N_q = \overline{\epsilon_q^{\,2}(t)} = (\delta v)^2/12. \tag{7.12-7}$$

When (7.12-7) is used with (7.12-6) we have the maximum possible output signal-to-noise ratio, since receiver noise effects are absent.

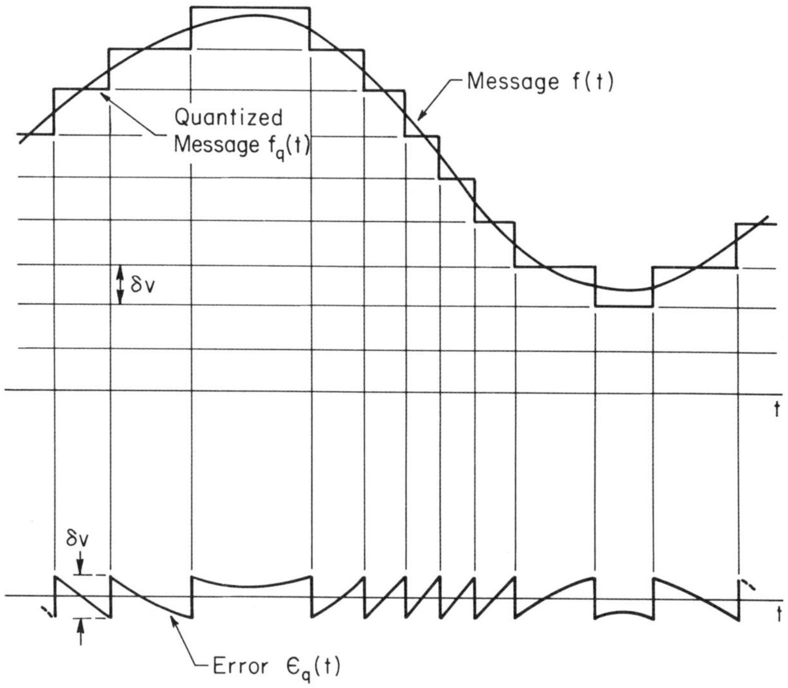

Fig. 7.12-2. A message, its quantized version, and the quantization error.

Companding

Let us define *crest factor* K_{cr} for a message as the ratio

$$K_{cr} = \left(\frac{|f(t)|_{\max}^{\,2}}{\overline{f^2(t)}} \right)^{1/2}. \tag{7.12-8}$$

For equal peak magnitudes, signals with large crest factors give poorer performance than those with small K_{cr}. To see this fact, assume a zero-mean message and note that δv must be determined from the peak magnitude by approximately

$$\delta v = 2|f(t)|_{\max}/L, \tag{7.12-9}$$

assuming a large number L of uniformly separated quantum levels spans the signal's variation. From (7.12-6) through (7.12-9) we obtain

$$(S_o/N_q)_{\text{PCM}} = 3L^2\overline{f^2(t)}/|f(t)|_{\max}^{\,2} = 3L^2/K_{cr}^{\,2}. \tag{7.12-10}$$

Thus, signals with large crest factor give poor performance.

ISBN 0-201-05758-1

A process called *companding* can be used to lower the crest factor of a waveform and produce better performance. The procedure is to use a nonlinear operation, known as a *compressor,* in the transmitter which tends to compress the message peaks. In the receiver the inverse operation is implemented by an *expandor* to restore the original message. Taken together a *com*pressor and ex*pandor* form a *compandor.*

The action of the compressor is to increase the rms signal value for a given peak magnitude. It can be implemented several ways. Perhaps the easiest to visualize is a nonlinear network acting on the message prior to quantization by a quantizer with uniformly separated quantum levels. A typical set of input-output characteristics for this form of compandor is illustrated in Fig. 7.12-3. However, in practice the equivalent effect is often achieved by direct quantization with a quantizer having *nonuniformly separated quantum levels.* In still other cases the equivalent effect is achieved by discarding certain levels of a uniform quantizer.

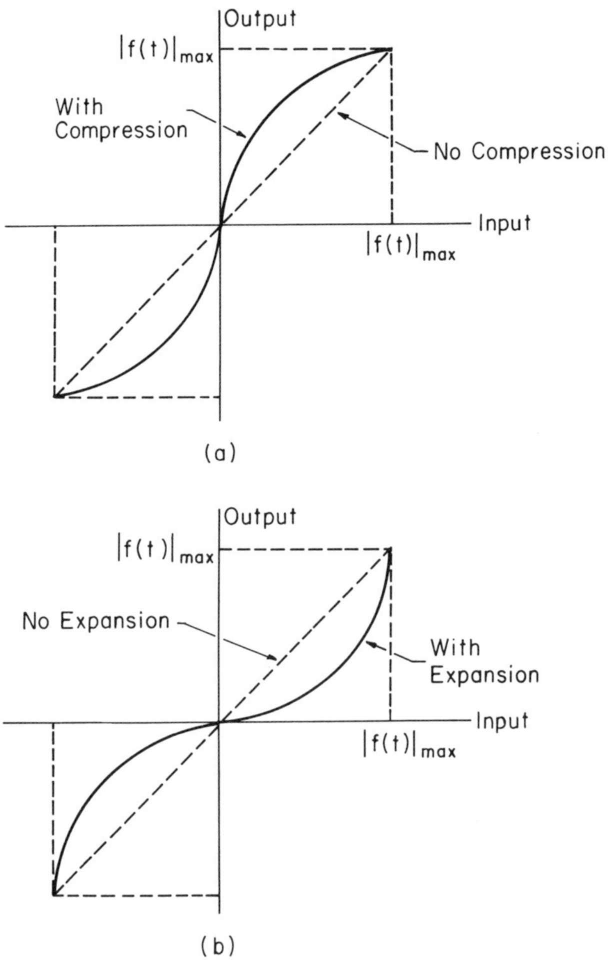

Fig. 7.12-3. Compandor characteristics. (a) Compressor and (b) expandor.

Time Multiplexing of Binary PCM

The PCM signals of Fig. 7.12-1(*b*) assume only one message. A unipolar multiplexed PCM signal assuming *N* similar messages and three-bit samples is illustrated in Fig. 7.12-4. More gen-

ISBN 0-201-05758-1

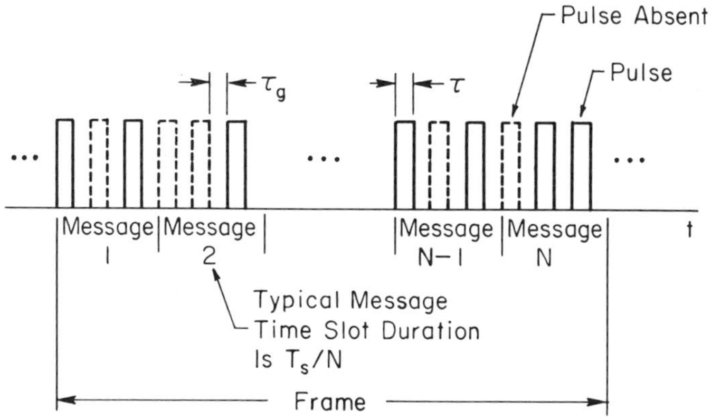

Fig. 7.12-4. N messages time-multiplexed using three-bit binary coding and unipolar PCM.

erally, for N_b bits per sample and messages bandlimited to W_f rad/s, the allowable number N of multiplexed messages follows from equating message time slot duration T_s/N with the required time per sample $N_b(\tau + \tau_g)$, where τ_g is the separation (if any) between pulses:

$$N = \frac{T_s}{N_b(\tau + \tau_g)} \leqslant \frac{\pi}{N_b W_f(\tau + \tau_g)}. \tag{7.12-11}$$

When used, a reasonable value for τ_g is to set it equal to τ. The equality holds if message sampling is at the Nyquist rate.

Channel Bandwidth

Equation (7.12-11) applies to a binary PCM signal. We may obtain from it an expression for the minimum channel bandwidth. For reasonable pulse fidelity at the channel output, the minimum pulse duration must satisfy*

$$\tau \geqslant \pi/W_{ch}. \tag{7.12-12}$$

Solving (7.12-11) for τ and substituting (7.12-12) we arrive at

$$W_{ch} \geqslant \frac{\pi}{(\pi/NN_b W_f) - \tau_g} \geqslant NN_b W_f. \tag{7.12-13}$$

Minimum bandwidth of NN_bW_f is approached by letting $\tau_g = 0$ and sampling at the Nyquist rate. For only one message the minimum PCM signal bandwidth is N_bW_f, which is usually much larger than the message bandwidth W_f.

Binary PCM Signal Generation

Many ways exist for implementing a PCM generation system. Figure 7.12-5 illustrates one method assuming N similar messages. Each message is low-pass filtered to assure bandlimiting and then sampled to generate N PAM waveforms, each in its assigned time slot. The com-

*Refer to the comments concerning pulse duration and channel bandwidth in the paragraph prior to (7.7-7).

ISBN 0-201-05758-1

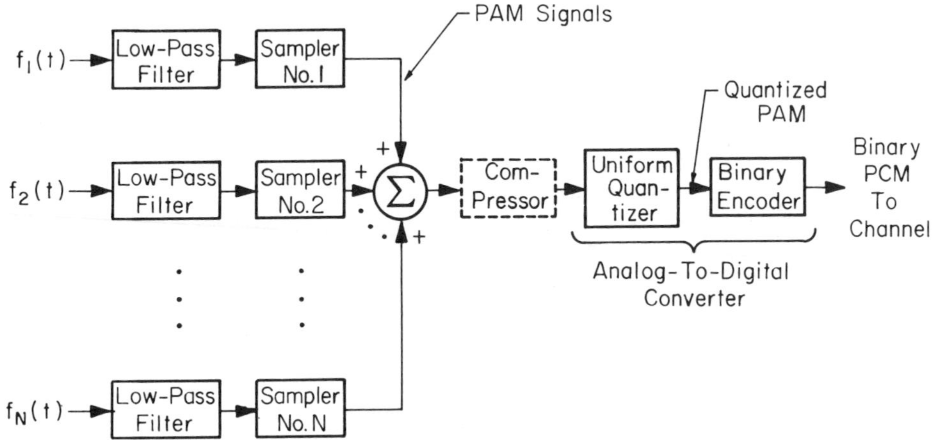

Fig. 7.12-5. Binary PCM signal generation system block diagram.

posite PAM signal is compressed, if companding is used. Next comes quantization,* and finally encoding of samples into the appropriate set of pulses for each time slot; these steps result in analog-to-digital (A-to-D) signal conversion and use an *A-to-D converter.* The N samplers perform the equivalent action of a commutator switch.

Signal Recovery

The receiver necessary to recover messages from the multiplexed PCM signal basically performs the inverse operations of the transmitter. Figure 7.12-6 illustrates a receiver based on the generator of Fig. 7.12-5. The first operation involves reconstructing the originally transmitted PCM signal as nearly as possible from the noise-contaminated received PCM waveform. Remaining operations are straightforward, and are just the inverses of those performed in the transmitter. A *digital-to-analog* (D-to-A) *converter* converts the reconstructed PCM signal to a quantized PAM waveform. If companding is used, the PAM signal is compressed and must

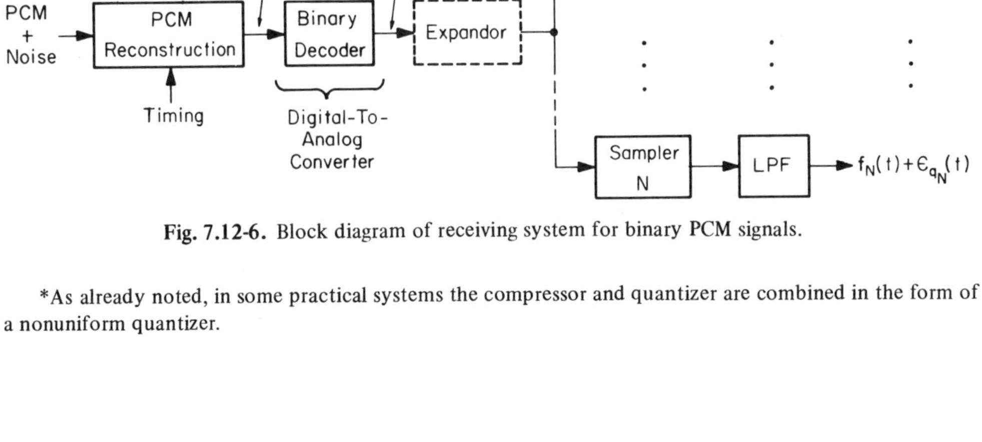

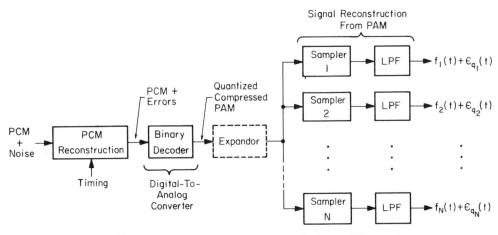

Fig. 7.12-6. Block diagram of receiving system for binary PCM signals.

*As already noted, in some practical systems the compressor and quantizer are combined in the form of a nonuniform quantizer.

ISBN 0-201-05758-1

next be expanded in the expandor. Finally, samplers, which behave collectively as a decommutator, serve to separate the multiplexed PAM waveforms. Each message is then reconstructed from its PAM waveform using any appropriate demodulation method as discussed in Section 7.4.

The reader will no doubt have already concluded that the most critical receiver operation is PCM reconstruction. This conclusion is correct, since it is at this point that the effect of noise is minimized through careful choice of circuit implementation. Some thought will reveal that the only knowledge required of the receiver to reconstruct the original PCM signal is whether the various transmitted bits are zeros or ones.* This requirement reduces to a determination of which of two voltage levels was transmitted for any given pulse of a code word. For example, with a unipolar stream of pulses of amplitude A, the two levels are A volts and zero volts. For polar pulses of amplitudes $\pm A$, the two levels are A and $-A$.

For each transmitted pulse, level determination reduces to a voltage measurement at some time instant. The measured voltage will equal the level caused by the pulse plus a noise voltage part. The ability of the receiver to make the correct level determination is optimized if the ratio of the pulse-caused voltage to the noise rms voltage is the largest possible at the measurement time. Recall that this condition is exactly that satisfied by the *matched filter* discussed in Section 4.12. Thus, the optimum implementation of PCM reconstruction consists of a matched filter, a sampler (level measurement), a decision device (to decide which of two levels is present), and a circuit to regenerate the PCM level decided upon. All these functions are assumed present in the PCM reconstruction block of Fig. 7.12-6.

The decision device consists in comparing the measured voltage to a preset *threshold voltage*. If the measured voltage is larger than the threshold, a one was assumed transmitted; if below, we decide a zero was sent. The optimum threshold is equal to the arithmetic average of the true voltages that correspond to transmissions of a zero and a one, if zeros and ones are equally likely and noise is Gaussian.

To illustrate the above conclusions, suppose the PCM signal consists of polar pulses of duration τ and amplitudes A or $-A$. The matched filter was determined for pulses of this form in example 4.12-1. For example, assuming a typical pulse as illustrated in Fig. 7.12-7(*a*), the matched filter impulse response required is shown in (*b*), where K and t_0 are arbitrary constants except that $t_0 \geqslant \tau$ is required (why?). The output $s_o(t)$ of the filter is sketched in (*c*). Since t_0 is the optimum time of measurement, the output level is $V_1 = KA\tau$ for a positive pulse. For a negative pulse the level would be $V_0 = -KA\tau$. The threshold voltage would be $(V_1 + V_0)/2 = 0$. For unipolar PCM the optimum threshold would be $KA\tau/2$.

Two facts are worth noting about the matched filter's signal response. First, the output signal level at the optimum measurement time is proportional to the *area* $A\tau$ of the input pulse. Second, the smallest value of t_0 is τ. The latter fact is useful in obtaining approximate realizations of the matched filter for a rectangular pulse.

Matched Filter Realizations

One realization for the case $t_0 = \tau$ was shown in Fig. 4.12-2. If the integrator "gain" is K, the pulse response is that of Fig. 7.12-7(*c*) with $t_0 = \tau$. The effect of the signal path containing the delay line in this realization occurs only for $\tau = t_0 < t$. This path contributes to causing the output to linearly return to zero following the end of the input pulse. Other realizations are possible where the delay line path is replaced by a "dump" control of the integrator which forces the output to zero immediately following the optimum sample time. Two such realiza-

*Strictly, the receiver must also know the *timing* of the incoming stream of pulses. We shall assume such timing exists such that the receiver is *synchronized* with the transmitter.

ISBN 0-201-05758-1

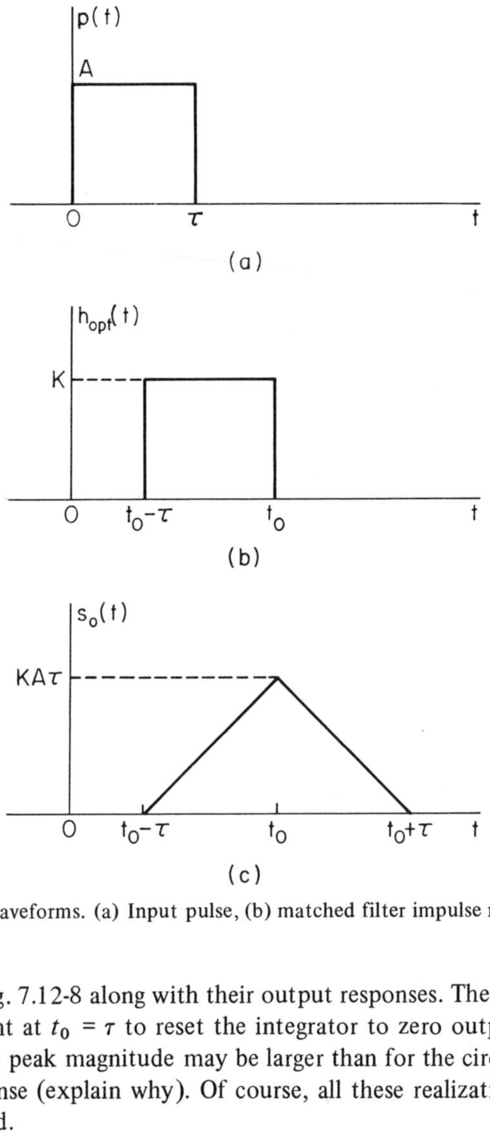

Fig. 7.12-7. Matched filter waveforms. (a) Input pulse, (b) matched filter impulse response, and (c) output waveform.

tions are illustrated in Fig. 7.12-8 along with their output responses. The dump switches close just after the measurement at $t_0 = \tau$ to reset the integrator to zero output. If $G >> 1$ in the method of (b), its output peak magnitude may be larger than for the circuit of (a) for a given nonlinearity in the response (explain why). Of course, all these realizations are only for the rectangular pulses assumed.

Bit Error Probability

The matched filter implementation of the PCM reconstruction stage of Fig. 7.12-6 leads to the optimum receiver. Such a receiver minimizes the number of errors made in deciding which pulse level was transmitted. Unfortunately, it cannot *eliminate* errors, and a finite probability P_e exists that an incorrect level decision will be made. A wrong decision is called a *bit error,* and the corresponding probability is the *bit error probability.*

To determine P_e, assume Gaussian noise at the receiver input. The filter output noise will also be Gaussian. Let the output average noise power be σ^2, and let polar PCM be assumed. If, at the optimum measurement times, the output signal voltages produced by the PCM pulses are V or $-V$, Fig. 7.12-9 is helpful in visualizing error probabilities. When a positive input pulse occurs, the conditional probability density function of output voltage is as shown in (a).

ISBN 0-201-05758-1

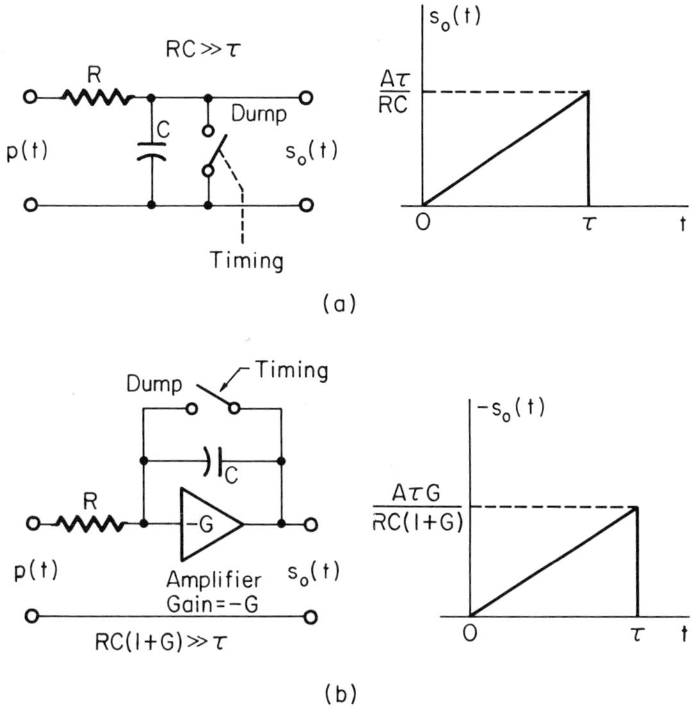

Fig. 7.12-8. Integrate-and-dump forms of matched filter. (a) Simple low-pass filter form and (b) high-gain amplifier form.

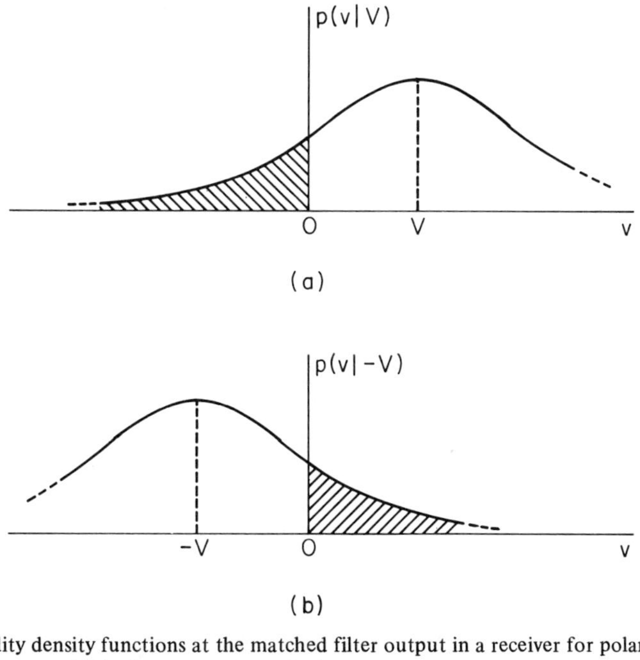

Fig. 7.12-9. Probability density functions at the matched filter output in a receiver for polar PCM. (a) When a positive pulse is present and (b) when a negative pulse is present.

ISBN 0-201-05758-1

If V and $-V$ are equally likely levels (the usual case), the optimum threshold for polar PCM is zero volts. Thus, the conditional probability of an error, when V is the true level, is equal to the area for $v \leqslant 0$. The product of this probability and the probability that level V occurs is the overall probability P_e^+ of error in correctly identifying level V:

$$P_e^+ = \frac{1}{2} \int_{-\infty}^{0} \frac{e^{-(v-V)^2/2\sigma^2}}{\sqrt{2\pi\sigma^2}} \, dv. \tag{7.12-14}$$

An equal probability $P_e^- = P_e^+$ occurs when a negative pulse is transmitted as shown in (*b*). Total probability of error P_e is $P_e^+ + P_e^- = 2P_e^+$. Hence, a simplification of (7.12-14) yields

$$P_e = \frac{1}{2} \operatorname{erfc} (V/\sqrt{2}\sigma), \qquad \text{polar PCM,} \tag{7.12-15}$$

where erfc $(\cdot)$ is the complementary error function defined by (6.12-8).

For unipolar PCM using pulses of amplitude A, (7.12-15) applies with V replaced by $V/2$ because the threshold is now set at $V/2$:

$$P_e = \frac{1}{2} \operatorname{erfc} (V/2\sqrt{2}\sigma), \qquad \text{unipolar PCM.} \tag{7.12-16}$$

To relate P_e to the matched filter responses, we note that the signal level is

$$V = KA\tau, \tag{7.12-17}$$

from Fig. 7.12-7(*c*). The filter's power transfer function is readily shown to be

$$|H_{\text{opt}}(\omega)|^2 = K^2\tau^2 \operatorname{Sa}^2 (\omega\tau/2), \tag{7.12-18}$$

so that output average noise power becomes [see (B-26) of Appendix B]

$$\sigma^2 = \frac{1}{2\pi} \int_{-\infty}^{\infty} \frac{\mathcal{N}_0}{2} |H_{\text{opt}}(\omega)|^2 \, d\omega = \frac{\mathcal{N}_0 K^2 \tau}{2}. \tag{7.12-19}$$

Here the input noise is assumed white with power density $\mathcal{N}_0/2$. Substitution of V and σ into (7.12-15) and (7.12-16) now gives

$$P_e = \frac{1}{2} \operatorname{erfc} [\sqrt{\epsilon}], \qquad \text{polar PCM,} \tag{7.12-20}$$

$$P_e = \frac{1}{2} \operatorname{erfc} [\sqrt{\epsilon/2}], \qquad \text{unipolar PCM,} \tag{7.12-21}$$

where ϵ is equal to the ratio of the average energy in an input pulse divided by twice the input noise power density. Assuming equally likely levels, the average energy in unipolar pulses is half that of bipolar pulses, since they are zero half the time. Hence,

$$\epsilon = A^2\tau/\mathcal{N}_0, \qquad \text{polar PCM,} \tag{7.12-22}$$

ISBN 0-201-05758-1

$$\epsilon = A^2\tau/2\mathcal{N}_0, \qquad \text{unipolar PCM.} \qquad (7.12\text{-}23)$$

Plots of (7.12-20) and (7.12-21) are given in Fig. 8.9-1 of Chapter 8.

Intersymbol Inteference

In PAM, PDM, and PPM only one pulse is used per sample of a message. We have already seen how a restricted channel bandwidth resulted in the pulse in one message time slot interfering with the pulse in an adjacent message time slot. Such interference was called crosstalk because two messages were involved. In PCM there can also be interference between adjacent pulses due to limited channel bandwidth. However, because adjacent pulses are more like symbols in the code representation of a single message sample, the term *intersymbol interference* is more appropriate.

As discussed earlier, the matched filter output voltage at the time of the measurement for a given pulse interval is proportional to the area of the input pulse. Ideally, for a rectangular pulse of duration τ and amplitude A, the output V is given by (7.12-17). With restricted channel bandwidth intersymbol interference causes a reduction in area. If a simple low-pass channel, defined by (7.5-7) and (7.5-8), is assumed, the transition between two levels for polar PCM is exponential as illustrated in Fig. 7.12-10 [7]. The time origin has been chosen for convenience.

A simple area calculation and a division by $A\tau$ gives the response of the matched filter relative to its maximum:

$$\text{relative area} = 1 - \frac{2}{W_{ch}\tau}\left(1 - e^{-W_{ch}\tau}\right) \approx 1 - \frac{2}{W_{ch}\tau}. \qquad (7.12\text{-}24)$$

The approximation holds for $W_{ch}\tau >> 1$. If this level reduction factor is to be maintained at 0.8 or better, W_{ch} must not be less than $10/\tau$. Thus, large channel bandwidth is required to maintain low intersymbol interference for a simple low-pass channel. In the next subsection it is shown that the transmitted pulses may purposely be shaped to aid in reducing intersymbol interference.

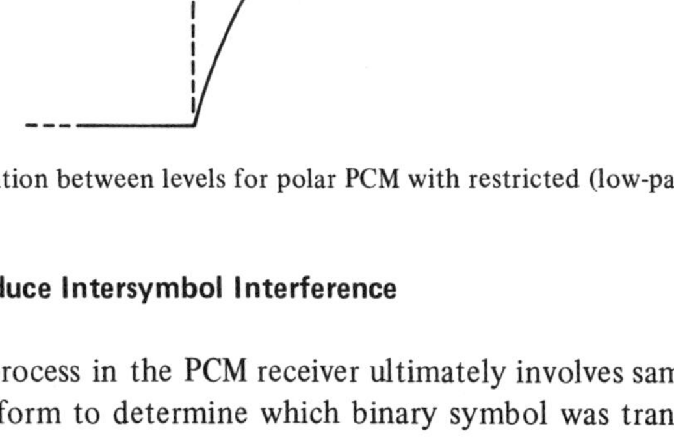

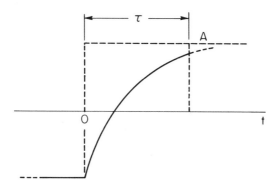

Fig. 7.12-10. Transition between levels for polar PCM with restricted (low-pass) channel bandwidth.

ISBN 0-201-05758-1

Pulse Shaping To Reduce Intersymbol Interference

The reconstruction process in the PCM receiver ultimately involves sampling a filtered version of the received waveform to determine which binary symbol was transmitted. If, at the time of any one sample, the voltage is entirely due to the symbol corresponding to the sample, there is no intersymbol interference. This situation corresponds to waveforms from *all* other transmitted symbols passing through a null at the sample time. However, since symbols are transmitted every T_s/NN_b seconds, we must also require that the waveforms for all symbols pass through *periodic* nulls to prevent interference regardless of which sample is taken.

One example of a theoretically suitable waveshape, having the form sin $(x)/x$, was encountered in PAM. It corresponded to the response of an ideal low-pass channel when excited by very narrow (impulsive) transmitted pulses. To employ this waveform in the more general problem, the *overall* product of the spectrum of the transmitted waveform, the channel transfer function, and the receiver matched filter would have to equal an ideal low-pass-type rectangular function of frequency. Such a result is not realizable and is difficult to approximate in practice. Even if it could be realized, there are still practical considerations in maintaining adequate synchronization of sample point timing that render the sin $(x)/x$ waveform unsuitable [13].

An alternative approach was developed in 1928 when H. Nyquist (*Transactions of the AIEE,* Vol. 47, February, 1928, pp. 617–644) demonstrated how a pulsed signal having the desired distribution of nulls could be achieved, while at the same time its spectrum magnitude approximated a realizable filter transfer function. Nyquist's result has been called the *vestigial-symmetry theorem* [8] owing to the form of the spectrum. It may have any shape, as illustrated in Fig. 7.12-11(a), so long as it is real and odd symmetry exists about the points $\omega = \pm W$. (Recall that this form of symmetry was assumed in vestigial-sideband modulation in order to obtain distortion-free message demodulation.) Clearly, the form of the spectrum is one that may be approximated by a realistic system having a gradual roll-off characteristic. It remains to show that the location of nulls is as claimed.

Proofs of Nyquist's vestigial-symmetry theorem usually follow the assumption $W_1 \leqslant W$, the case of usual practical interest. Making this assumption, we may decompose the spectrum $S_o(\omega)$ into the sum of a rectangular component and components with odd symmetry about $\pm W$, as illustrated in Fig. 7.12-11(b) and (c). Inverse transforms of the components are easily shown to be

$$h(t) = \frac{W}{\pi}\,\frac{\sin(Wt)}{Wt} \tag{7.12-25}$$

$$h_1(t) = \frac{-2}{\pi}\sin(Wt)\int_0^{W_1} H_1(\omega + W)\sin(\omega t)\,d\omega. \tag{7.12-26}$$

The time function $s_o(t)$ corresponding to the spectrum $S_o(\omega)$ becomes

$$s_o(t) = \frac{W}{\pi}\,\frac{\sin(Wt)}{Wt}\left[1 - 2t\int_0^{W_1} H_1(\omega + W)\sin(\omega t)\,d\omega\right]. \tag{7.12-27}$$

Now regardless of the precise value of the term within the brackets, the factor sin $(Wt)/Wt$ guarantees the existence of periodic nulls. Hence, for almost arbitrary choice of $H_1(\omega)$, intersymbol interference is zero if symbols in the PCM pulse train are conveyed in pulses having the shape of $s_o(t)$.

As an example, consider the *raised-cosine* spectrum defined by

$$S_o(\omega) = 1, \qquad |\omega| < W - W_1$$

ISBN 0-201-05758-1

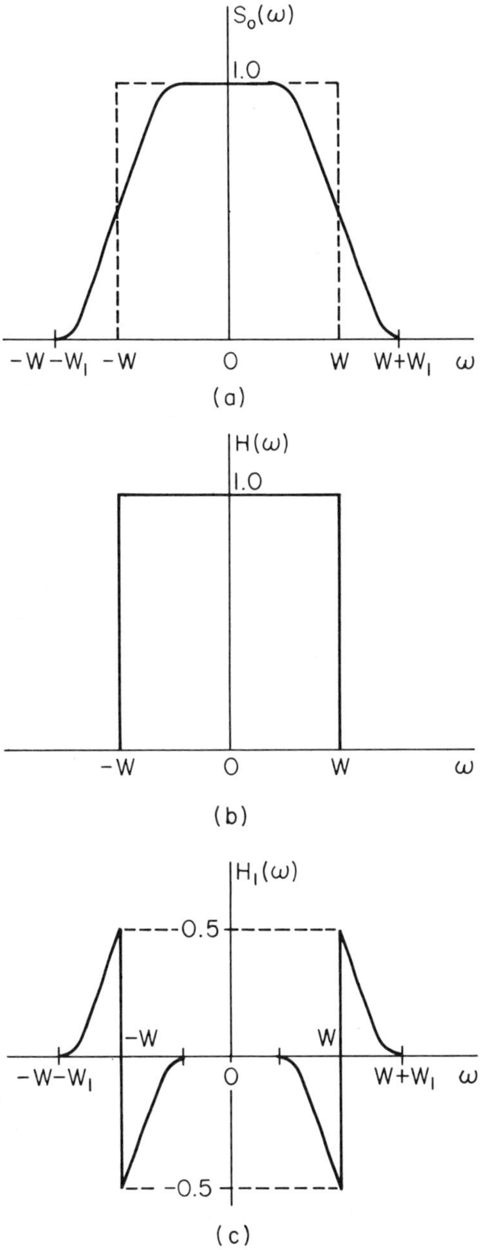

Fig. 7.12-11. Spectra associated with Nyquist's vestigial-symmetry theorem. (a) A spectrum with the required symmetry. (b) Its rectangular component, and (c) its components with odd symmetry about ±W.

$$= \frac{1}{2} + \frac{1}{2}\cos\left[\frac{\pi}{2W_1}(|\omega| - W + W_1)\right], \qquad W - W_1 < |\omega| < W + W_1$$

$$= 0, \qquad W + W_1 < |\omega|. \tag{7.12-28}$$

ISBN 0-201-05758-1

Straightforward calculation of (7.12-27) produces

$$s_o(t) = \frac{W}{\pi}\frac{\sin(Wt)}{Wt}\left[\frac{\cos(W_1 t)}{1-(2W_1 t/\pi)^2}\right]. \tag{7.12-29}$$

The above two functions are illustrated in Fig. 7.12-12 for $W_1/W = 0$, 0.5, and 1.0. Several important observations may be made about the pulses of (*b*). When $W_1/W = 1.0$, the sidelobes are very small (31.5 dB or more below the peak) indicating that intersymbol interference can be made small even in the presence of timing errors. Furthermore, additional nulls occur at $t = \pm(2n+1)\pi/2W$, $n = 1, 2, \ldots$, which also tend to lower interference. Finally, the half-amplitude pulse duration for $W_1/W = 1.0$ is exactly π/W, the time duration of one bit. Some reflection on the reader's part will verify that a polar signal constructed from this basic pulse will have possible nulls precisely π/W seconds apart.* This is an advantage, since *synchronization* signals may be extracted from a received digital waveform having this property.

With pulse shaping we may still transmit code symbols at one symbol per π/W seconds. In other words, our *signaling rate* is W/π. However, observe that the *total* band required with

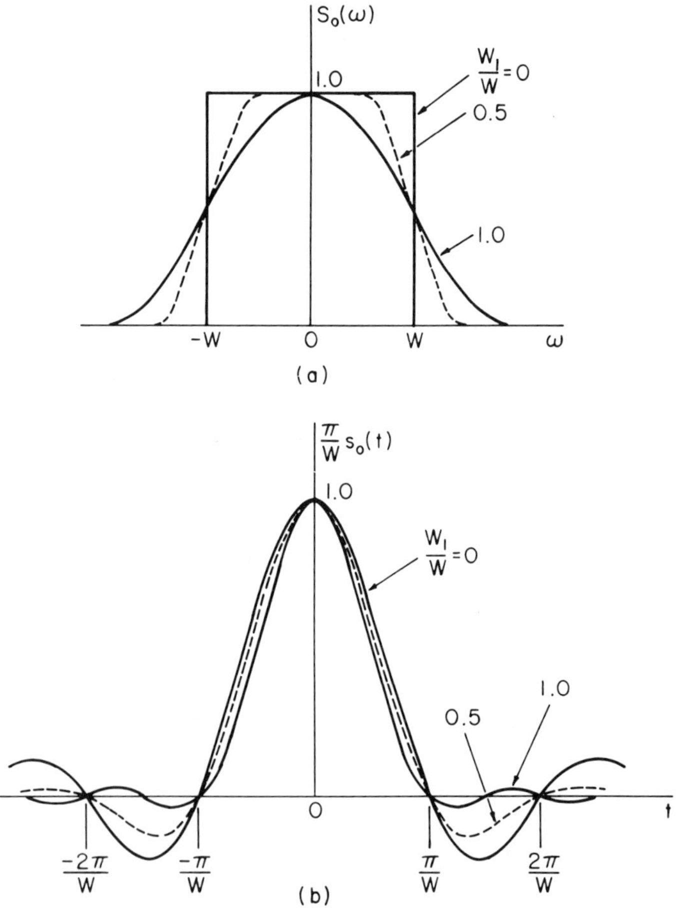

Fig. 7.12-12. Pulse shaping via Nyquist's vestigial-symmetry theorem. (a) Spectra and (b) shaped pulses.

*Actual null locations will correspond to the transition points in the binary sequence of ones and zeros.

ISBN 0-201-05758-1

pulse shaping becomes $W + W_1$ rather than W. Thus, compared with an ideal signaling rate of $(W + W_1)/\pi$ symbols per second, we suffer a reduction in rate by the factor $W/(W + W_1)$.

7.13. NOISE PERFORMANCE OF BINARY PCM

The measure of performance used thus far in this book for all modulation systems has been the ratio of output average signal to average noise powers. In the following paragraphs we find this ratio for a binary PCM system.

Performance above Threshold

The desired expression has already been found in (7.12-10) when noise is negligible and errors in recovering the message $f(t)$ are all due to quantization. The result is

$$\left(\frac{S_o}{N_q}\right)_{\text{PCM}} = 3L^2 \frac{\overline{f^2(t)}}{|f(t)|_{\max}^2} \tag{7.13-1}$$

for L uniformly separated quantum levels spanning the signal's variations.

Example 7.13-1

We calculate (7.13-1) for sinusoidal modulation and the parameters of example 7.12-1—that is, for $L = 201$. We have $\overline{f^2(t)}/|f(t)|_{\max}^2 = 1/2$ for sinusoidal signals, so $(S_o/N_q)_{\text{PCM}} = 3(201)^2/2 = 6.06(10^4)$, or 47.8 dB.

Effect of Receiver Noise

When receiver noise is negligible, the message is recovered, as given by (7.12-5), with only quantization error present. When receiver noise is not negligible, we may write the recovered signal as $\hat{f}_q(t)$:

$$\hat{f}_q(t) = f_q(t) + \epsilon_n(t) = f(t) + \epsilon_q(t) + \epsilon_n(t). \tag{7.13-2}$$

In other words, there is now an error $\epsilon_n(t)$ in the reconstructed message $\hat{f}_q(t)$ due to noise. Because the quantization and receiver errors arise from different mechanisms, they may be taken as statistically independent and the total output waveform power is the sum of individual powers.

From (7.12-7) the quantization noise power is $N_q = \overline{\epsilon_q^{\,2}(t)}$. Defining receiver error noise power by

$$N_{rec} = \overline{\epsilon_n^{\,2}(t)} \tag{7.13-3}$$

allows us to write

$$\left(\frac{S_o}{N_o}\right)_{\text{PCM}} = \frac{\overline{f^2(t)}}{N_q + N_{rec}}. \tag{7.13-4}$$

The presence of N_{rec} gives rise to a *threshold effect* in PCM.

ISBN 0-201-05758-1

Our basic problem is to find N_{rec}. To this end, we must first relate input noise to how it causes errors $\epsilon_n(t)$ to occur in the output. This relationship is found from a close look at the behavior of the PCM reconstruction and decoder blocks of Fig. 7.12-6. The former determines, on a pulse-to-pulse basis, whether pulses in a code word are present or not in unipolar PCM, or alternatively, whether pulses are positive or negative in polar PCM. At the end of each word, the decoder then generates an exact quantum level corresponding to the recovered code. The recovered code, and therefore $\hat{f}_q(t)$, will contain error *only* if an error is made in individual pulse decisions. The effect of receiver noise is to cause occasional false decisions. Such false decisions are called *bit errors,* and the average frequency of these errors is the *bit error rate.*

Thus, $\hat{f}_q(t)$ will possess a noise error if a code word is in error, and a code word is in error if any one or more bit errors are made. Most PCM systems operate such that it is highly unlikely that more than one bit is in error in any one word. Assuming this to be the case, let us examine the effect of bit errors on a bit-by-bit basis for an N_b-bit word.

If an error occurs in the least significant bit, an error in $\hat{f}_q(t)$ of δv occurs. If such an error occurs in m words out of M possible words the average (mean)-squared error is $(\delta v)^2 m/M$. For a large number of words m/M is interpreted as the probability P_e of the least significant bit's being in error. Average-squared error in $\hat{f}_q(t)$ is then $(\delta v)^2 P_e$.

For the next least significant bit, a bit error causes an error in $\hat{f}_q(t)$ of $2(\delta v)$. The mean-squared error becomes $(2\delta v)^2 P_e$. Continuing the logic to the ith next least significant bit, the mean-squared error in $\hat{f}_q(t)$ is $(2^i\ \delta v)^2 P_e$. Now, recognizing that the total mean-squared error $\epsilon_n{}^2(t)$ in $\hat{f}_q(t)$ is the sum of the contributions from each bit in the code word, we have

$$N_{rec} = \overline{\epsilon_n{}^2(t)} = (\delta v)^2 P_e \sum_{i=0}^{N_b-1} (2^2)^i = (\delta v)^2 P_e \frac{2^{2N_b} - 1}{3}. \tag{7.13-5}$$

Finally, since $2^{2N_b} >> 1$, we substitute (7.12-9) and obtain approximately

$$N_{rec} = \frac{4|f(t)|_{\max}{}^2\, 2^{2N_b} P_e}{3L^2}. \tag{7.13-6}$$

To complete the calculation of N_{rec}, we may substitute the expression for P_e given by (7.12-20) or (7.12-21). For polar PCM

$$N_{rec} = \frac{2|f(t)|_{\max}{}^2\, 2^{2N_b}}{3L^2} \operatorname{erfc}\left[\sqrt{A^2\tau/\mathcal{N}_0}\right]. \tag{7.13-7}$$

A similar result is achieved for unipolar PCM.

Performance near Threshold

Assume polar pulses arrive at the receiver input (matched filter input). The signal power per message in the PCM waveform is

$$S_i = A^2 N_b \tau / T_s \tag{7.13-8}$$

with T_s the sampling period and N_b the number of bits in a word. Noise power is easily found, for an ideal channel, to be

ISBN 0-201-05758-1

$$N_i = \mathscr{N}_0 W_{ch}/2\pi, \tag{7.13-9}$$

assuming white input noise with power density $\mathscr{N}_0/2$. Taking the ratio we have

$$\left(\frac{S_i}{N_i}\right)_{\text{PCM}} = \frac{2\pi A^2 N_b \tau}{\mathscr{N}_0 W_{ch} T_s}. \tag{7.13-10}$$

For the output signal-to-noise power ratio we substitute (7.12-20), (7.13-5), and (7.12-7) into (7.13-4) to arrive at

$$\left(\frac{S_o}{N_o}\right)_{\text{PCM}} = \frac{(S_o/N_q)_{\text{PCM}}}{1 + 2^{2N_b+1}\ \text{erfc}\ [\sqrt{A^2\tau/\mathscr{N}_0}]}, \quad \text{polar PCM}, \tag{7.13-11}$$

where, if $L = L_b$ is assumed,* (7.12-10) gives the largest value of $(S_o/N_q)_{\text{PCM}}$:

$$(S_o/N_q)_{\text{PCM}} = (3)2^{2N_b}/K_{cr}^2. \tag{7.13-12}$$

The basic result (7.13-11) may be put into other useful forms. By solving (7.13-10) for $A^2\tau/\mathscr{N}_0$, and observing that $T_s/N_b = N\tau$ when there is no time separation between pulses, we obtain

$$\frac{A^2\tau}{\mathscr{N}_0} = \left(\frac{S_i}{N_i}\right)_{\text{PCM}} \frac{NW_{ch}\tau}{2\pi}. \tag{7.13-13}$$

Now allowing $W_{ch} = \pi/\tau$, which is about the most pessimistic value we could choose,

$$\frac{A^2\tau}{\mathscr{N}_0} = \frac{N}{2}\left(\frac{S_i}{N_i}\right)_{\text{PCM}}. \tag{7.13-14}$$

Alternatively, assuming sampling is at the Nyquist rate,

$$\frac{A^2\tau}{\mathscr{N}_0} = \frac{1}{2N_b}\left(\frac{S_o}{N_o}\right)_{\text{B}} = \frac{NW_f}{2W_{ch}}\left(\frac{S_o}{N_o}\right)_{\text{B}}. \tag{7.13-15}$$

Here the output signal-to-noise ratio of a baseband system having the same average input signal power and noise power density as the PCM system is

$$\left(\frac{S_o}{N_o}\right)_{\text{B}} = \frac{2\pi A^2 N_b \tau}{\mathscr{N}_0 W_f T_s}. \tag{7.13-16}$$

Substitution of these expressions in (7.13-11) allows us to write specifically

*For some signal power the number of quantum levels spanning the signal will equal L_b, the maximum possible. For larger signal levels the quantizer amplitude range is exceeded and the PCM system becomes *amplitude-overloaded.*

ISBN 0-201-05758-1

$$\left(\frac{S_o}{N_o}\right)_{\text{PCM}} = \frac{\left(\frac{S_o}{N_q}\right)_{\text{PCM}}}{1 + 2^{(2W_{ch}/NW_f)+1}\ \text{erfc}\left[\sqrt{\frac{N}{2}\left(\frac{S_i}{N_i}\right)_{\text{PCM}}}\right]}, \tag{7.13-17}$$

$$\left(\frac{S_o}{N_o}\right)_{\text{PCM}} = \frac{\left(\frac{S_o}{N_q}\right)_{\text{PCM}}}{1 + 2^{(2W_{ch}/NW_f)+1}\ \text{erfc}\left[\sqrt{\frac{NW_f}{2W_{ch}}\left(\frac{S_o}{N_o}\right)_{\text{B}}}\right]}, \tag{7.13-18}$$

for polar PCM. Figure 7.13-1 gives a few plots of (7.13-17) for a single transmitted message ($N = 1$). Figure 7.13-2 is a replot of the data using (7.13-18). A sinusoidal signal has a crest

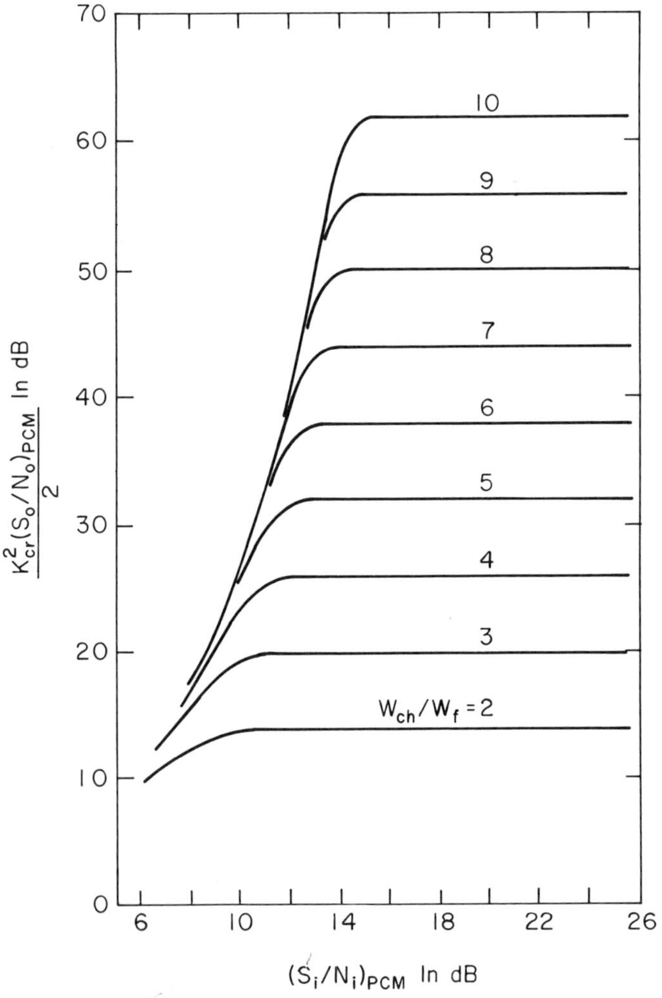

Fig. 7.13-1. Performance curves for a polar, binary PCM system when a single message is transmitted.

ISBN 0-201-05758-1

factor of $\sqrt{2}$ so that these plots correspond to the output-signal-to-noise ratio directly for such messages.

Threshold Calculation

The point of threshold will arbitrarily be defined as the point where $(S_o/N_o)_{PCM}$ drops 1 dB below the value of $(S_o/N_q)_{PCM}$. From (7.13-17) this occurs when the denominator becomes 1.26. Solving for the threshold signal-to-noise ratio $(S_i/N_i)_{PCM,th}$ gives

$$\left(\frac{S_i}{N_i}\right)_{PCM,th} = \frac{2}{N}\left[\mathrm{erfc}^{-1}\left(0.13/2^{2W_{ch}/NW_f}\right)\right]^2. \tag{7.13-19}$$

Figure 7.13-3 sketches (7.13-19) for $N = 1$.

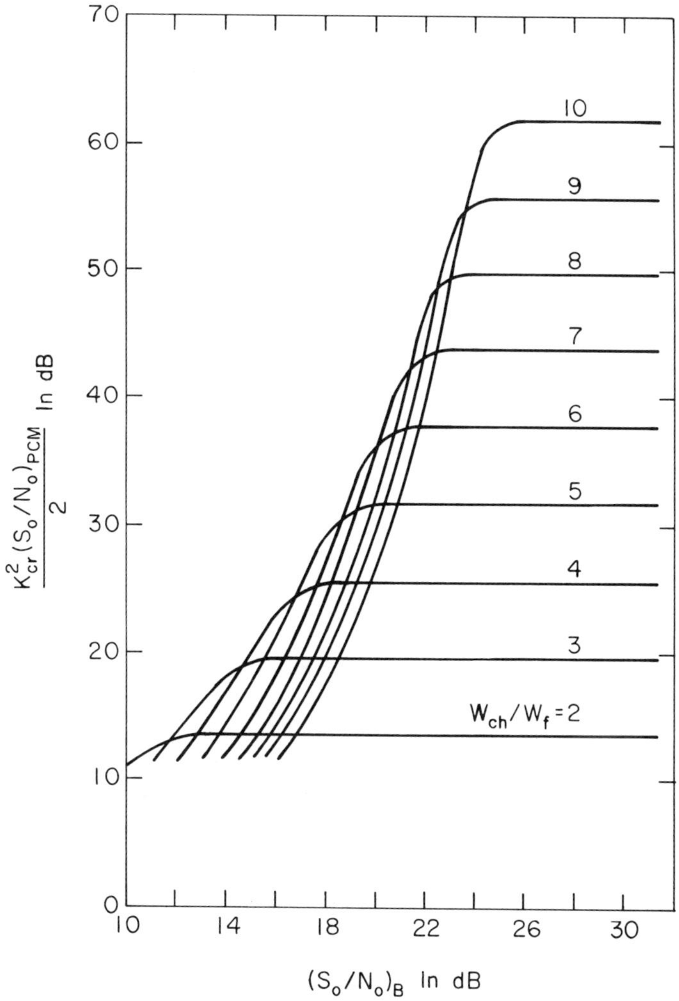

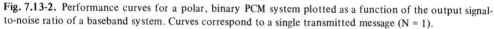

Fig. 7.13-2. Performance curves for a polar, binary PCM system plotted as a function of the output signal-to-noise ratio of a baseband system. Curves correspond to a single transmitted message (N = 1).

ISBN 0-201-05758-1

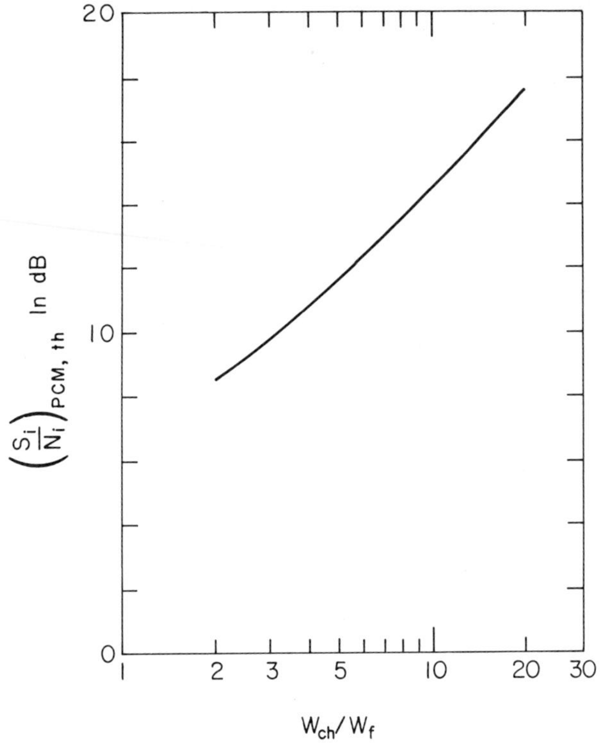

Fig. 7.13-3. Threshold signal-to-noise ratio for a polar, binary PCM system transmitting a single message.

7.14. PERFORMANCE COMPARISON OF BASEBAND, FM, PDM, PPM, AND BINARY PCM SYSTEMS

Let us now compare noise performance of some of the various modulation systems when operating above threshold. For convenience we bring together the various equations for single-message transmission.* They are (7.13-12) for PCM, (7.10-6) for PPM, (7.8-21) for PDM, and (7.8-22) for wideband FM:

$$\left(\frac{S_o}{N_o}\right)_{\text{PCM}} = 3(2^{2W_{ch}/W_f})\,\frac{\overline{f^2(t)}}{|f(t)|_{\max}{}^2}, \tag{7.14-1}$$

$$\left(\frac{S_o}{N_o}\right)_{\text{PPM}} = \frac{1}{4}\left(\frac{W_{ch}}{W_f}\right)^2 \frac{\overline{f^2(t)}}{|f(t)|_{\max}{}^2}\left(\frac{S_o}{N_o}\right)_{\text{B}}, \tag{7.14-2}$$

$$\left(\frac{S_o}{N_o}\right)_{\text{PDM}} = \frac{1}{2}\left(\frac{W_{ch}}{W_f}\right) \frac{\overline{f^2(t)}}{|f(t)|_{\max}{}^2}\left(\frac{S_o}{N_o}\right)_{\text{B}}, \tag{7.14-3}$$

*We omit PAM and all AM systems discussed in Chapter 5, since they give performance not better than that of a baseband system.

ISBN 0-201-05758-1

$$\left(\frac{S_o}{N_o}\right)_{FM} = \frac{3}{4}\left(\frac{W_{ch}}{W_f}\right)^2 \frac{\overline{f^2(t)}}{|f(t)|_{max}^{\;2}} \left(\frac{S_o}{N_o}\right)_B . \tag{7.14-4}$$

Figure 7.14-1 plots these results for $W_{ch}/W_f = 7$. Several important points may be made based on these example curves. First, all analog pulse modulation systems are poorer than FM. Next, coded modulation (PCM) near the PCM threshold is better than all analog pulse modulation systems and FM. Finally, far above threshold, PCM is poorer than all other systems owing to the quantization error limit. Thus, the payoff in PCM occurs for operation near threshold. Some practical systems may wish to operate near threshold for economic reasons. An example would be a deep space communication system where each additional decibel of power may be quite expensive.

Another point in favor of PCM is that it interfaces ideally with binary (computer-type) logic and is very flexible. Because of this fact, such a system designed for multiplexing analog information signals might also be used to convey digital data.

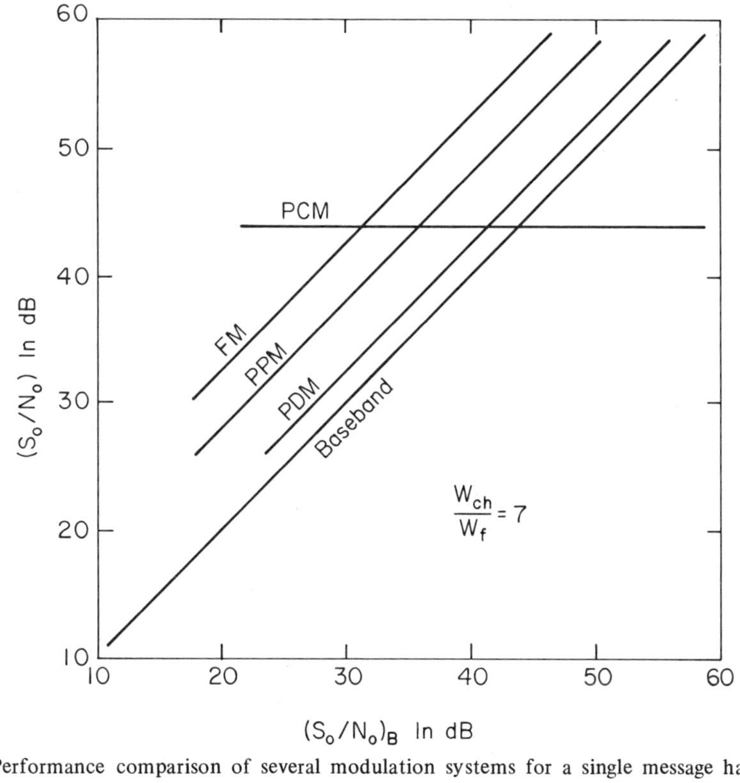

Fig. 7.14-1. Performance comparison of several modulation systems for a single message having a crest factor of $\sqrt{2}$.

★ 7.15. μ-ARY PCM

The more general form of pulse code modulation, where each pulse, in a sample group of pulses, may have μ amplitude levels, was briefly touched on in Section 7.12. We now complete the discussion of μ-ary PCM by summarizing some of the more elementary aspects involved.

ISBN 0-201-05758-1

Quantization Levels and Average Signal Power

Let the separation between pulse amplitude levels be Δv, and assume polar transmissions. The pulse amplitudes assume values A_j given by

$$A_j = \pm j\,\Delta v, \qquad \mu \text{ odd}, \tag{7.15-1}$$

$$A_j = \pm\left(\frac{2j+1}{2}\right)\Delta v, \qquad \mu \text{ even}, \tag{7.15-2}$$

$j = 0, 1, \ldots, (\mu - 1)/2$ or $(\mu - 2)/2$, respectively. Since a typical pulse may have any of these levels, average power per pulse must be found by averaging over all levels. If the levels are all equally likely and we take the case of μ odd,

$$\overline{A^2} = \sum_{j=-(\mu-1)/2}^{(\mu-1)/2} \frac{(j\,\Delta v)^2}{\mu} = \frac{2(\Delta v)^2}{\mu} \sum_{j=1}^{(\mu-1)/2} j^2. \tag{7.15-3}$$

The sum is known to be $\mu(\mu^2 - 1)/24$ (Appendix B), giving an average power per pulse of

$$\overline{A^2} = (\mu^2 - 1)(\Delta v)^2/12. \tag{7.15-4}$$

If all pulses of duration τ in a sample group of N_b pulses are equally likely, the average power per message is $N_b\tau\overline{A^2}/T_s$. This power is the average input power per message to the receiver:

$$S_i = \frac{N_b\tau(\mu^2 - 1)(\Delta v)^2}{12T_s}. \tag{7.15-5}$$

The reader may verify that this result also holds when μ is even.

Channel Bandwidth

Recall that the number L_b of message quantum levels is related to the number N_b of pulses in a sample group by $L_b = \mu^{N_b}$. Now consider two extremes for μ. When $\mu = L_b$, its largest value corresponding to $N_b = 1$, the PCM sample becomes simply a PAM sample of a quantized message. In this case, the minimum channel bandwidth is that of a PAM signal, which is NW_f (for N similar multiplexed messages).

In the other extreme, when μ has its smallest value 2, we revert to binary PCM for which the minimum channel bandwidth is N_bNW_f. Thus, in going from a one-pulse sample to an N_b-pulse sample, the bandwidth increased by a factor equal to the number of pulses. For in-between values of μ, we may then take the minimum channel bandwidth as [14, p. 230],

$$W_{ch} \geqslant N_bNW_f = NW_f \log_\mu (L_b). \tag{7.15-6}$$

Our comments are not intended as a proof of (7.15-6). Indeed, the main value of the foregoing discussion is to attach a sort of intuitive picture to the bandwidth problem.

Performance above Threshold

If we presume that input noise to the receiver is white noise of spectral density $\mathscr{N}_0/2$ passed over an ideal (rectangular passband) channel of bandwidth W_{ch}, the input average noise power

ISBN 0-201-05758-1

is

$$N_i = \mathcal{N}_0 W_{ch}/2\pi. \tag{7.15-7}$$

Assuming polar transmission, we use (7.15-5) to obtain

$$\left(\frac{S_i}{N_i}\right)_{\mu\text{-PCM}} = \frac{N_b \tau (\mu^2 - 1)(\Delta\nu)^2 \pi}{6 T_s \mathcal{N}_0 W_{ch}}. \tag{7.15-8}$$

For a sufficiently large input signal-to-noise ratio the only error made in the recovery of the message is due to quantization noise, just as in binary PCM. Thus, (7.12-10) also applies to μ-ary PCM where the number of quantum levels is μ^{N_b}.

$$\left(\frac{S_o}{N_q}\right)_{\mu\text{-PCM}} = 3\mu^{2N_b} \frac{\overline{f^2(t)}}{|f(t)|_{\max}^2}. \tag{7.15-9}$$

Another form is

$$\left(\frac{S_o}{N_q}\right)_{\mu\text{-PCM}} = 3\mu^{2W_{ch}/NW_f} \frac{\overline{f^2(t)}}{|f(t)|_{\max}^2}, \tag{7.15-10}$$

assuming minimum channel bandwidth and sampling at the Nyquist rate. Equation (7.15-10) is especially instructive because it illustrates that an *exponential* trade-off may be made between channel bandwidth and signal-to-noise ratio. An example illustrates this point.

Example 7.15-1

A system having $\mu = 3$ and $W_{ch}/W_f = 3$ is used to transmit one message which has been quantized into twenty-seven levels. The necessary number of pulses is $N_b = 3$. If all else remains constant, but the channel bandwidth is increased by a factor of 2 to allow using $N_b = 6$ pulses (and therefore 729 quantum levels), (7.15-10) is used to find that the signal-to-noise ratio increases by a factor $3^6 = 729$, or 28.6 dB.

The bandwidth-performance exchange is somewhat analogous to that noticed in wideband FM. There performance increased only as W_{ch}^2, while here it is considerably larger owing to the exponential relationship in (7.15-10). The exchange cannot continue indefinitely, however. Inspection of (7.15-8) reveals that as W_{ch} increases $(S_i/N_i)_{\mu\text{-PCM}}$ decreases. A point, the *threshold,* will ultimately be reached where input noise is not negligible, making (7.15-10) invalid.

Threshold Signal-to-Noise Ratio

Threshold occurs when noise causes false quantum-level decisions to happen with significant probability. Taking this probability to be about 10^{-5}, it can be shown that operation above threshold corresponds to about

$$\Delta\nu \geqslant 8.8\sqrt{N_i}. \tag{7.15-11}$$

ISBN 0-201-05758-1

Squaring and substituting for $(\Delta\nu)^2$ from (7.15-5), we get a rough rule-of-thumb for threshold signal-to-noise ratio:

$$(S_i/N_i)_{\mu\text{-PCM},th} = 6.5(\mu^2 - 1)/N. \tag{7.15-12}$$

Example 7.15-2

As a check on (7.15-12) let $\mu = 2$ and $N = 1$. We have $(S_i/N_i)_{2\text{-PCM},th} = 19.5$ or 12.9 dB. Comparing to Fig. 7.13-3, we find that (7.15-12) appears pessimistic for $W_{ch}/W_f < 7$ and optimistic for larger values.

7.16. DELTA MODULATION

Delta modulation (DM), a European invention, was apparently first discussed in the English-language literature by de Jager [15]. It may be described as a type of PCM where only one-bit encoding is used. One-bit coding is made possible by a feedback loop which is an integral part of the encoding process. In the ensuing development of DM principles, the simplest form of implementation is assumed. It has an ideal integrator in the loop feedback path and will be called a *basic* delta modulator. At the end of this section we consider some alternative configurations based on practical integrators.

System Block Diagram

Figure 7.16-1 illustrates the components of a DM system applicable to a single message. If time multiplexing of several messages is to be used, the separate pulse trains for the various messages may be added at point *A*. They are separated in the receiver by a suitable decommutator at point *B*. We shall assume in this section that N similar messages are multiplexed, each being bandlimited with maximum spectral extent W_f.

At the transmitter end of the channel, the pulse modulator accepts an input train of constant-amplitude clock pulses occurring at a rate $\omega_s/2\pi$ pulses per second. It converts these pulses to an output train of polar pulses with polarity determined by the "loop error" signal $-\epsilon_q(t)$. If $-\epsilon_q(t)$ is positive at the time of any given input pulse, the output pulse is positive. If $-\epsilon_q(t)$ is negative, a negative output pulse is generated. The error $-\epsilon_q(t)$ is the difference $f(t) - f_q(t)$ between the input message $f(t)$ and $f_q(t)$, which is the integrated version of the

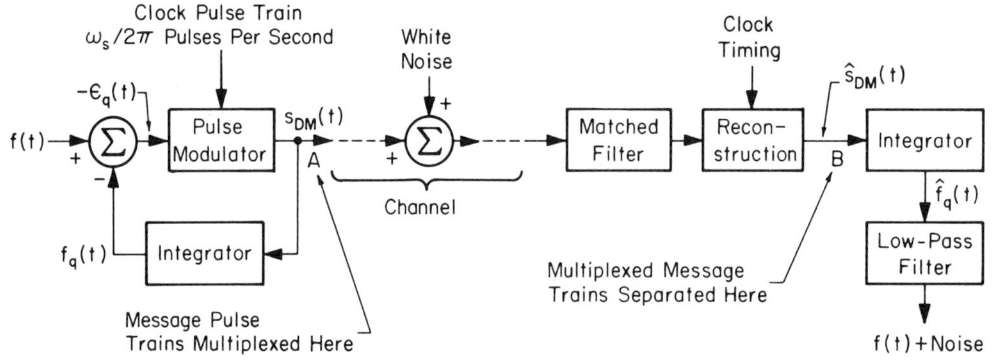

Fig. 7.16-1. Block diagram of a delta modulation system.

ISBN 0-201-05758-1

generated pulse train $s_{DM}(t)$. We assume that these generated pulses have a duration τ which is short relative to the time $T_s = 2\pi/\omega_s$ between "sample" pulses. Thus, the (ideal) integrator output $f_q(t)$ will be a "stair-step" type of waveform with an up-going step for each positive pulse and a down-going step for a negative pulse. Each transmitted pulse is clearly a one-bit encoded version of the signal $-\epsilon_q(t)$. Now since the action of the loop will be to force $f_q(t)$ to be an approximation to $f(t)$, a fact made more obvious in the following paragraph, the overall action of the modulator is to transmit one-bit information about the *changes* in $f(t)$ with time.

To better understand operation, assume $f(t)$ is applied, as shown in Fig. 7.16-2(*a*), at $t = 0$. Let the input clock pulse train timing be such that a pulse occurs at $t = 0$. At $t = 0$, and for a short time thereafter, $f(t) > f_q(t)$ and output pulses are all positive, as illustrated in (*b*). During this initial or *transient period* the signal $f_q(t)$ increases in steps of δv approaching the message until it finally exceeds it in voltage. After the transient period, $f_q(t)$ will become approximately equal to $f(t)$ and both positive and negative output pulses are generated with about equal regularity. The behavior of the error $\epsilon_q(t)$ is illustrated in (*c*). Except during the transient period the error never exceeds $\pm\delta v$, where δv is called the *step-size.*

At the receiver end of the channel a matched filter is shown in Fig. 7.16-1 to provide maximum noise immunity (see Chapter 4). The next step involves a reconstruction of the transmitted pulse train to remove noise. The operation is similar to reconstruction in PCM where, at the times of the maxima of the pulses emerging from the matched filter, samples of the received signal are compared to a voltage threshold level (zero volts here) to

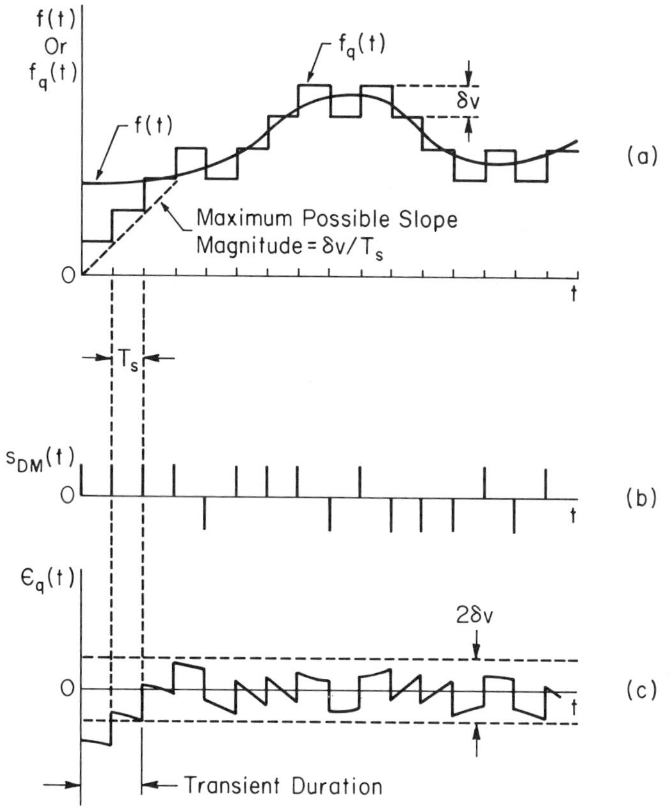

Fig. 7.16-2. Waveforms applicable to delta modulation. (a) f(t) and $f_q(t)$. (b) The DM signal, and (c) the error signal.

ISBN 0-201-05758-1

determine which polarity of pulse was received. Based on these pulse-to-pulse decisions, a version $\hat{s}_{DM}(t)$ of the original signal $s_{DM}(t)$ is reconstructed. In a practical receiver the reconstruction stage might be omitted (especially if only one message is involved). We include it here because it aids in both the understanding and the mathematical analysis of DM. Now if multiplexing is involved, the pulse trains are separated at point *B*, and each is applied to an integrator. For a typical message, the integrator output $\hat{f}_q(t)$ is approximately equal to the signal $f_q(t)$ in the transmitter (except for noise errors in reconstruction it would be exactly equal to $f_q(t)$ in principle). Thus, $\hat{f}_q(t) \approx f_q(t)$. Finally, a low-pass filter, having a bandwidth W_f, is used to smooth out the sharp step-behavior of $\hat{f}_q(t)$. The final output becomes $f(t)$ plus errors which we may collectively consider to be noise.

Errors and Slope Overload

Even if receiver random noise is negligible, the reconstructed message $\hat{f}_q(t)$ in Fig. 7.16-1 can at best equal $f_q(t)$. It can never equal the original message $f(t)$ because of the one-bit quantization involved. However, since

$$f_q(t) = f(t) + \epsilon_q(t), \tag{7.16-1}$$

the receiver may be thought of as recovering $f(t)$ with an error $\epsilon_q(t)$. This concept for handling errors was already introduced in the analysis of PCM, where $\epsilon_q(t)$ was called the quantization error. It was the only nonnoise type of error in PCM. On the other hand, in DM there are *two* types of nonnoise errors. One, called *granular error* $\epsilon_g(t)$, corresponds to the error due to quantization in the one-bit coding process. The second is called *slope overload error* $\epsilon_{so}(t)$. The composite error due to both granular and slope overload errors is called *quantization error* in DM. Thus, in terms of the component errors making up $\epsilon_q(t)$ we may write (7.16-1) as

$$f_q(t) = f(t) + \epsilon_g(t) + \epsilon_{so}(t). \tag{7.16-2}$$

Granular error occurs when $f_q(t) \approx f(t) + \epsilon_g(t)$–that is, during periods of time when there are no transient errors occurring. Slope overload errors occur when the input message at the delta modulator has a slope exceeding the slope capability of the feedback signal $f_q(t)$. From Fig. 7.16-2(*a*) the maximum generated slope capability is $\delta v/T_s$. Thus, to prevent slope overload

$$|df(t)/dt|_{max} \leqslant \delta v/T_s \tag{7.16-3}$$

is required. If, for short periods of time, (7.16-3) is not satisfied, slope overload errors occur. Slope overload may be corrected for a given message, by increasing the step-size δv, or by decreasing T_s (sample at a higher rate). Both these solutions are later found to be undesirable, and slope overload is one of the fundamental limitations of DM.

To better appreciate the effects of slope overload consider the sinusoidal message

$$f(t) = A_f \cos(\omega_f t). \tag{7.16-4}$$

From (7.16-3) the region of satisfactory performance corresponds to signal levels

$$\frac{A_f}{\delta v} \leqslant \frac{1}{2\pi}\left(\frac{\omega_s}{\omega_f}\right). \tag{7.16-5}$$

ISBN 0-201-05758-1

The right side of this expression may be viewed as a kind of spectral characteristic for a basic delta modulator, since it is a function of frequency ω_f. Taken in this light, we expect that DM would be especially applicable to waveforms that have a power density spectrum that decreases as the reciprocal of the square of frequency. For such signals the various spectral components could be made to "match" or operate in close proximity to the overload level, thereby optimizing performance. Indeed, this is the case, and basic DM has been found to perform well with speech or monochrome television video messages both of which possess average power spectra approximating the type described.

In order not to convey excessive granular noise $A_f/\delta v >> 1$ is necessary in (7.16-5). This condition determines the sample rate as the following simple example points out.

Example 7.16-1

For an 800-Hz sinusoidal message we find the required sample rate to prevent overload when $A_f/\delta v = 20$ is to be realized. From (7.16-5)

$$f_s = \omega_s/2\pi \geqslant 20\omega_f = 40\pi(800) = 100.5 \text{ kHz}.$$

Note that this sample rate is 62.8 times as large as the Nyquist sample rate of 1.6 kHz.

Channel Bandwidth

Assuming N similar messages are multiplexed, the time per message sample is T_s/N. Within this time slot the sample pulse duration is $\tau \leqslant T_s/N$ seconds. Since the minimum channel bandwidth W_{ch} required to pass pulses of duration τ is π/W_{ch}, we have

$$W_{ch} \geqslant \pi/\tau \geqslant \pi N/T_s = N\omega_s/2. \tag{7.16-6}$$

In practice, one would probably design the pulse duration as large as possible (T_s/N) to convey maximum energy per sample, especially if receiver (thermal) noise was significant. Thus, W_{ch} would not have to be radically larger than the minimum value $N\omega_s/2$.

Other Configurations

There are many variations of the basic delta modulator which uses an ideal integrator. In one case the integrator may be replaced by a time-varying network which adjusts step-size with time to reduce slope overload. Such a modulator is called an *adaptive delta modulator,* and it is discussed in Section 7.19. In another variation, the message is integrated prior to modulation by the basic delta modulator. This variation, called *delta-sigma modulation,* is discussed in Section 7.18. It is a means of approximately matching the overload characteristics of DM to messages having relatively flat power spectra (owing to integration the encoded message power spectrum decreases with frequency and therefore approximately matches the overload characteristic of the delta modulator).

Some variations of the basic delta modulator involve alteration of the feedback path to include more general linear networks that do not change with time. Clearly, from a practical standpoint, one cannot realize a perfect integrator. A simple low-pass one-section RC filter is often substituted instead. Johnson [16] has analyzed such a DM system and found that noise performance is unchanged over the basic modulator which has an ideal integrator. There is a change in the slope overload characteristic, however. If the filter time constant is $T_c = RC$ and its 3-dB frequency is $\omega_c = 1/T_c$, the slope overload characteristic for a sinusoidal message

ISBN 0-201-05758-1

of peak amplitude A_f and frequency ω_f becomes

$$A_f/\delta v = 1/\sqrt{1 + (\omega_f/\omega_c)^2}. \tag{7.16-7}$$

For frequencies above ω_c the behavior is similar to the basic delta modulator, since a $1/\omega_f$ characteristic is evident. The region $\omega_f > \omega_c$ again corresponds to slope overload. For frequencies $\omega_f < \omega_c$ the characteristic becomes flat, corresponding to *amplitude overload,* a condition not present in basic DM. For speech, this behavior is not serious, since the speech spectrum also tends to reach a maximum at lower frequencies.

A delta modulator may also contain two integrators in the feedback path. In this case message slope differences are encoded as opposed to amplitude differences. Instabilities may result, however, for two perfect integrators, and some practical modifications are required. The usual practical circuit is a two-section *RC* low-pass filter having an extra (stability-controlling) resistor placed in series with the output capacitor [15, 17]. Such modulators are called *delta modulators with double integration.* They may give improved noise performance compared to single-integration DM for speech. The article by Schindler [17] is a good summary of the various DM configurations. In a later paper [18] he includes a fairly extensive reference list which will serve those readers interested in additional depth on DM systems.*

All the delta modulators discussed above use regularly spaced samples. It is also possible to transmit in the form of irregularly spaced samples. Such a system, called *asynchronous delta modulation,* is described by Hawkes and Simonpieri [19].

7.17. NOISE PERFORMANCE OF DELTA MODULATION

In this section we consider mainly the noise performance of a single (ideal) integrator form of DM system as depicted in Fig. 7.16-1.

Performance with Granular Noise Only

Let us first consider receiver input (channel) noise to be negligible. From Fig. 7.16-1 and (7.16-2) we have

$$\hat{f}_q(t) = f_q(t) = f(t) + \epsilon_g(t) + \epsilon_{so}(t). \tag{7.17-1}$$

Here the recovered message $f(t)$ is corrupted by granular and slope overload errors $\epsilon_g(t)$ and $\epsilon_{so}(t)$. We shall treat the errors as "noises" and seek the amount of noise power due to each which reaches the final output through the low-pass filter. So far as signal power is concerned, the (ideal) filter bandwidth W_f is chosen just large enough to pass the signal undistorted. Output signal power is then

$$S_o = \overline{f^2(t)}. \tag{7.17-2}$$

Granular noise is essentially uncorrelated with the signal [20] and is approximately uniformly distributed [20, 21], over the interval $2\delta v$. We may gain some intuitive agreement with this fact from the behavior of the error in Fig. 7.16-2(*c*) during the nontransient time interval. The granular noise power prior to low-pass filtering becomes

$$\overline{\epsilon_g^{\,2}(t)} = (\delta v)^2/3. \tag{7.17-3}$$

*Since completion of the manuscript for this text, a book has become available that covers most aspects of DM systems [40].

ISBN 0-201-05758-1

For the filtered output granular noise power N_g, we adopt an heuristic argument given by Taub and Schilling [7]. By observing the form of $\epsilon_g(t)$ from Fig. 7.16-2(*c*) we conclude that there is considerable high-frequency content in its spectrum. To make an estimate of its extent, imagine $\epsilon_g(t)$ passed through a filter with a cutoff frequency ω_c chosen to pass $\epsilon_g(t)$ with only reasonable distortion. The cutoff frequency is then a good estimate of the spectral extent of $\epsilon_g(t)$. If we select the point of "reasonable" distortion as that where the filter rise time π/ω_c is equal to half the sample interval T_s, the spectrum of $\epsilon_g(t)$ is found to extend to a cutoff frequency $\omega_c = \omega_s = 2\pi/T_s$. Now since the power spectrum of $\epsilon_q(t)$ is more or less flat [20, p. 2120] at lower frequencies, we presume the total power to be uniformly distributed over $-\omega_c < \omega < \omega_c$. Filtered output power now easily calculates to

$$N_g = \frac{(\delta v)^2}{3}\left(\frac{W_f}{\omega_s}\right). \tag{7.17-4}$$

Thus, in general, the signal-to-granular noise power ratio is

$$\left(\frac{S_o}{N_g}\right)_{\text{DM}} = \frac{3\overline{f^2(t)}}{(\delta v)^2}\left(\frac{\omega_s}{W_f}\right). \tag{7.17-5}$$

In order to prevent slope overload it is necessary to select δv so that

$$\delta v \geqslant \left|\frac{df(t)}{dt}\right|_{\max} T_s \tag{7.17-6}$$

from (7.16-3). By substituting the minimum value of δv in (7.17-4), we get

$$N_g = \frac{4\pi^2}{3}\left|\frac{df(t)}{dt}\right|^2_{\max}\frac{W_f}{\omega_s^{\,3}}. \tag{7.17-7}$$

The maximum possible output signal-to-noise ratio becomes the ratio of (7.17-2) to (7.17-7):

$$\left(\frac{S_o}{N_g}\right)_{\text{DM}} = \frac{3\overline{f^2(t)}\,\omega_s^{\,3}}{4\pi^2\,|df(t)/dt|_{\max}^{\,2}\,W_f}. \tag{7.17-8}$$

This equation may be put in another useful form by defining a *slope crest factor* $\dot{K}_{cr}$, analogous to the amplitude crest factor of (7.12-8), by*

$$\dot{K}_{cr}^{\,2} = \frac{|df(t)/dt|_{\max}^{\,2}}{\overline{[df(t)/dt]^2}} \tag{7.17-9}$$

and observing that the rms bandwidth W_{rms} of the signal power density spectrum $\mathscr{S}_f(\omega)$ is

*Note that the dot does not imply the time derivative of K_{cr}. It is simply to denote a crest factor for a time-differentiated waveform.

ISBN 0-201-05758-1

$$W_{\mathrm{rms}}^2 = \frac{\overline{[df(t)/dt]^2}}{\overline{f^2(t)}} = \frac{\int_{-\infty}^{\infty} \omega^2 \mathcal{S}_f(\omega)\, d\omega}{\int_{-\infty}^{\infty} \mathcal{S}_f(\omega)\, d\omega}. \tag{7.17-10}$$

The result is

$$\left(\frac{S_o}{N_g}\right)_{\mathrm{DM}} = \frac{3}{4\pi^2 \dot{K}_{cr}^2} \left(\frac{W_f}{W_{\mathrm{rms}}}\right)^2 \left(\frac{\omega_s}{W_f}\right)^3. \tag{7.17-11}$$

Equation (7.17-11) gives the largest possible output signal-to-noise ratio for a single-integration DM system. The only system parameter appearing explicitly in the expression is the sample rate ω_s. By use of (7.16-6) we may put it in terms of channel bandwidth:

$$\left(\frac{S_o}{N_g}\right)_{\mathrm{DM}} = \frac{6}{\pi^2 \dot{K}_{cr}^2} \left(\frac{W_f}{W_{\mathrm{rms}}}\right)^2 \frac{1}{N^3} \left(\frac{W_{ch}}{W_f}\right)^3. \tag{7.17-12}$$

Thus, performance in DM increases as the third power of channel bandwidth.

Response of a DM system to a sinusoidal message has special application to speech transmission. An example is next used to specialize (7.17-11) for a single message ($N = 1$).

Example 7.17-1

We obtain an expression for $(S_o/N_g)_{\mathrm{DM}}$ for a single sinusoidal message $f(t) = A_f \cos(\omega_f t)$. Using the result of problem 7-33, we obtain $W_{\mathrm{rms}} = \omega_f$. For $\dot{K}_{cr}^2$ we have $|df(t)/dt|_{\max}^2 = (A_f\omega_f)^2$ and $\overline{[df(t)/dt]^2} = (A_f\omega_f)^2/2$, so $\dot{K}_{cr}^2 = 2$. Finally, (7.17-11) becomes

$$\left(\frac{S_o}{N_g}\right)_{\mathrm{DM}} = \frac{3}{8\pi^2} \left(\frac{W_f}{\omega_f}\right)^2 \left(\frac{\omega_s}{W_f}\right)^3, \qquad \text{sinusoidal modulation.}$$

For speech signals it is experimentally found [15] that noticeable overload does not occur as long as signal amplitude does not exceed the amplitude of an 800-Hz sine wave set just at the slope overload value. If $W_f/2\pi = 3.5$ kHz is assumed and the typical spectral distribution of speech is considered [22], 800 Hz turns out to be $W_{\mathrm{rms}}/2\pi$, the rms speech bandwidth. It is also known [15] that good speech transmission requires a sample rate of 40 kHz, while very good results are achieved for a 100-kHz sample rate. From example 7.17-1 these values show that we must have $(S_o/N_g)_{\mathrm{DM}} = 1086$ or 30.4 dB for good results or 42.3 dB for very good results.

For a DM system with double integration and operating with a sine wave, $(S_o/N_g)_{\mathrm{DM}}$ may be put in the form [15]

$$\left(\frac{S_o}{N_g}\right)_{\mathrm{DM}} = \frac{0.0178(3)}{8\pi^2} \left(\frac{W_f}{\omega_f}\right)^2 \left(\frac{\omega_s}{W_f}\right)^5. \tag{7.17-13}$$

ISBN 0-201-05758-1

Forming the ratio of (7.17-13) to the expression in example 7.17-1 shows that the double-integration system signal-to-granular noise power ratio is better by a factor of 0.0178 $(\omega_s/W_f)^2$. For $W_f/2\pi = 3.5$ kHz and $\omega_s/2\pi = 40$ kHz (good speech reproduction) or 100 kHz (very good speech) the factor evaluates to 3.7 dB and 11.6 dB, respectively. These improvements make the added complexity worth while for speech systems.

Performance with Slope Overload Noise Added

In general the slope overload error $\epsilon_{so}(t)$ in (7.17-1) is correlated with the message. A number of analyses to find the slope overload noise power N_{so} at the receiver low-pass filter output have been conducted for a single-integrator DM system. These analyses [20, 23–25] are complicated and beyond the scope of this book. Hence, we only state and use appropriate results. Greenstein [25] has given apparently the most general solution. It is difficult to apply, however, and in the important region of heavy overload Greenstein's work suggests (see his Fig. 7) that Abate's [24] empirical solution is reasonably accurate. Because Abate's results are easy to use, we present them here as taken from Greenstein's paper. In present notation, the slope overload noise power can be shown to be

$$N_{so} = \left[\frac{(\delta v)^2}{3}\,\frac{W_f}{\omega_s}\right]\frac{2}{9}\left(\frac{\omega_s}{W_f}\right)^3\left(\frac{1+3Z}{Z^2}\right)e^{-3Z}, \tag{7.17-14}$$

where Z, as given by

$$Z^2 = \frac{(\delta v)^2}{T_s^2\,\overline{[df(t)/dt]^2}}, \tag{7.17-15}$$

is known as the *slope overload factor.* Equation (7.17-14) was derived for Gaussian random or noise-like messages. The bracketed factor is recognized as the granular noise power from (7.17-4).

The output signal-to-quantization noise power ratio is

$$\left(\frac{S_o}{N_q}\right)_{DM} = \frac{\overline{f^2(t)}}{N_g + N_{so}}. \tag{7.17-16}$$

To put this equation in a useful form we must observe that a Gaussianly distributed message has no well-defined maximum amplitude or maximum slope. This means that granular noise power cannot be expressed in the exact form of (7.17-7). However, for some reference power level $\overline{f_r^2(t)}$ of the message, let us set the step-size according to

$$\delta v = \dot{K}\sqrt{\overline{\dot{f}_r^{\,2}(t)}}\,T_s, \tag{7.17-17}$$

where we define

$$\dot{f}_r(t) = df_r(t)/dt. \tag{7.17-18}$$

Thus, at the reference signal power level, $\delta v/T_s$ is proportional to the square root of the power

ISBN 0-201-05758-1

in the derivative of the reference signal. The proportionality constant is $\dot{K}$. It is *analogous* to the slope crest factor $\dot{K}_{cr}$ for messages having a definite maximum slope but not equal to it.*

Now by substituting (7.17-17) into the expressions for N_g and N_{so}, as given by (7.17-4) and (7.17-14), we may find $(S_o/N_q)_{\text{DM}}$ for any signal power level different from the reference level:

$$\left(\frac{S_o}{N_q}\right)_{\text{DM}} = \frac{x^2\dot{K}^2(S_o/N_g)_{\text{DM}}}{1 + \dfrac{2}{9}\left(\dfrac{\omega_s}{W_f}\right)^3 (3 + x)x \exp\left(-\dfrac{3}{x}\right)}, \tag{7.17-19}$$

where

$$x = \frac{1}{\dot{K}}\sqrt{\frac{\overline{f^2(t)}}{\overline{f_r^{\,2}(t)}}}, \tag{7.17-20}$$

$$\left(\frac{S_o}{N_g}\right)_{\text{DM}} = \frac{3}{4\pi^2}\frac{1}{\dot{K}^2}\left(\frac{W_f}{W_{\text{rms}}}\right)^2\left(\frac{\omega_s}{W_f}\right)^3, \tag{7.17-21}$$

and W_{rms} is the rms bandwidth of the power density spectrum of the signal $f(t)$. By substitution of minimum channel bandwidth we have

$$\left(\frac{S_o}{N_q}\right)_{\text{DM}} = \frac{x^2\dot{K}^2(S_o/N_g)_{\text{DM}}}{1 + \dfrac{16}{9N^3}\left(\dfrac{W_{ch}}{W_f}\right)^3 (3 + x)x \exp\left(-\dfrac{3}{x}\right)}. \tag{7.17-22}$$

The quantity $(S_o/N_g)_{\text{DM}}$ is the signal-to-granular noise ratio when the message is at its reference power level. The remaining factor represents the change in this ratio caused by the presence of slope overload noise and the fact that the message power may be different from the reference level.

Figure 7.17-1 provides several plots† of (7.17-22) for $N = 1$. To illustrate the practical application of the curves we take an example.

Example 7.17-2

In an assumed DM system $W_{ch}/W_f = 5$ and $\delta v/T_s$ is arbitrarily selected to be four times the rms signal derivative. Thus, $\dot{K} = 4$ from (7.17-17). If the system transmits only one message, $\omega_s = 2W_{ch}$ and $\omega_s/W_f = 10$. Now suppose the message power density spectrum is uniform over $-W_f < \omega < W_f$. Then $W_{\text{rms}}^{\,2} = W_f^{\,2}/3$. From (7.17-21)

$$\left(\frac{S_o}{N_g}\right)_{\text{DM}} = \frac{3(3)(10)^3}{4\pi^2(4)^2} = 14.25, \text{ or } 11.54 \text{ dB},$$

*A more exact interpretation follows a comparison of (7.17-17) with (7.17-15). $\dot{K}$ may be considered as a reference value of the slope overload factor.

†Problem 7-36 develops optimum values of x in terms of W_{ch}/W_f such that operation is at the curve peaks.

ISBN 0-201-05758-1

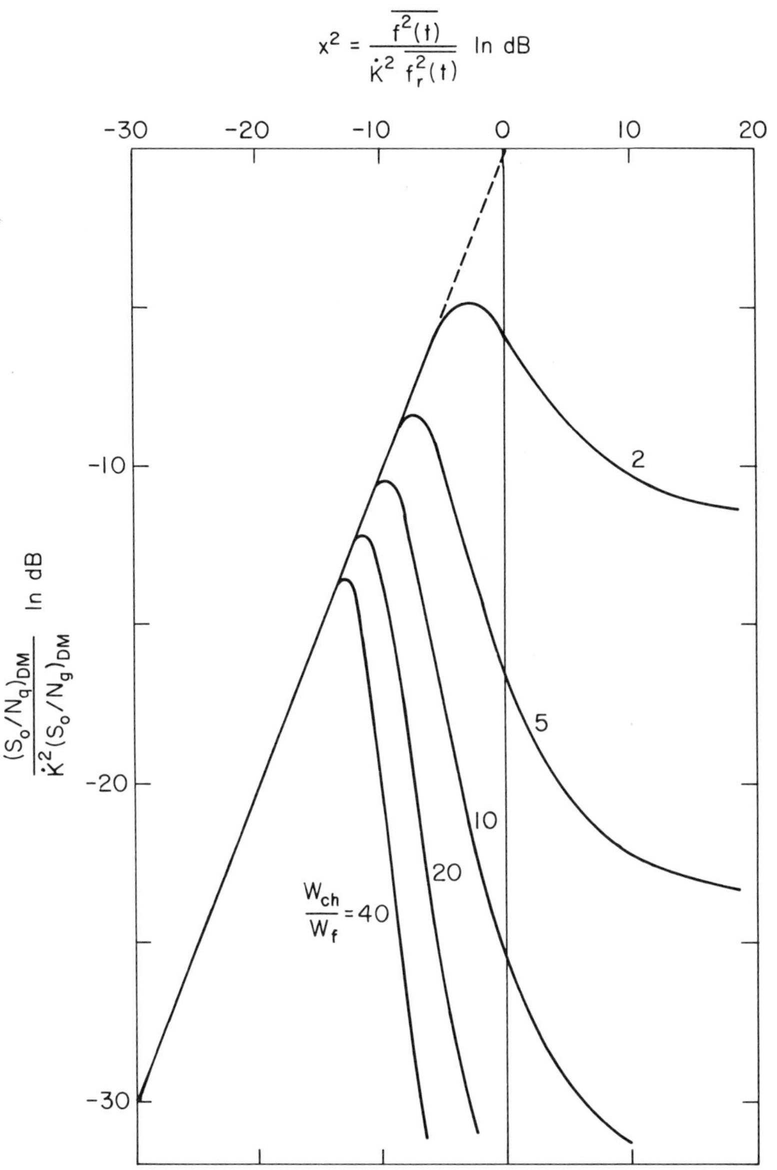

Fig. 7.17-1. Loss in DM system signal-to-noise ratio due to message level variations and presence of slope overload noise.

so $\dot{K}^2(S_o/N_g)_{\mathrm{DM}}$ = 228.0, or 23.58 dB. Thus $(S_o/N_q)_{\mathrm{DM}}$ will be less than 23.58 dB by the amount shown in Fig. 7.17-1 which depends on signal level. To see where the reference level falls on the $W_{ch}/W_f = 5$ curve, let $\overline{f^2(t)} = \overline{f_r^{\,2}(t)}$. Thus $x^2 = 1/\dot{K}^2 = 1/16$. The curve value is -12.05 dB at $x^2 = 1/16$, or -12.04 dB. Hence $(S_o/N_q)_{\mathrm{DM}}$ becomes $23.58 - 12.05 = 11.53$ dB.

Note in example 7.17-2 that the operating point is not near the peak of the $W_{ch}/W_f = 5$ curve. The following example optimizes the system.

ISBN 0-201-05758-1

Example 7.17-3

Let us alter the choice of $\delta v/T_s$ in the last example to operate at the peak of the curve when $\overline{f^2(t)} = \overline{f_r^2(t)}$. We require x^2 to be -7.5 dB. Thus, $x = 0.422$ or $\dot{K} = 2.37$. We now have

$$\left(\frac{S_o}{N_g}\right)_{\text{DM}} = 14.25\left(\frac{4}{2.37}\right)^2 = 40.59, \text{or } 16.08 \text{ dB},$$

and $\dot{K}^2(S_o/N_g)_{\text{DM}} = 228.0$, or 23.58 dB, the same as before. Now the loss, however, is only 8.51 dB. We obtain $(S_o/N_q)_{\text{DM}}$ equal to $23.58 - 8.51 = 15.07$ dB, an improvement of 3.54 dB over the previous example.

Effect of Receiver Noise on Performance

It only remains to determine the effect of receiver (channel) noise on the performance of a DM system with a single integrator. Presence of noise means that the reconstruction stage in Fig. 7.16-1 will occasionally make the wrong decision as to which polarity pulse is received. If a positive pulse is received and an error is made, the output generated is a negative pulse. We may consider the output as actually containing the true positive pulse plus a negative error pulse of *twice* the magnitude of the signal pulse. In a similar manner, a negative received pulse erroneously reconstructed results in a negative signal pulse plus a positive double-amplitude error pulse. The overall effect at the input to the receiver integrator is an uncorrupted stream of signal pulses plus a stream of occasional double-amplitude noise pulses.

Since the pulses are short duration and the response of the integrators to any given noise pulse will be a step of amplitude $2\delta v$ or $-2\delta v$, we may consider the noise pulses as impulses with amplitudes $\pm 2\delta v$. The energy density spectrum for a single impulse is a constant level $4(\delta v)^2$. Now if the average time between noise pulses in any one message channel is $\overline{T}_{n1}$, then the power density spectrum of the noise pulse train may be taken approximately as $4(\delta v)^2/\overline{T}_{n1}$ at the integrator input. The effect of the integrator is to divide this power density by ω^2. Finally, we may integrate to obtain output power.

The integral depends on the low-pass filter characteristic. Because DM with ideal integrators does not transmit a dc signal well (the reader should verify this point for himself), there is no need for the filter to have very low-frequency response. If W_L is its low-frequency cutoff and $W_L \ll W_f$, the output noise power is

$$N_{rec} = \frac{4(\delta v)^2}{\pi \overline{T}_{n1} W_L}. \tag{7.17-23}$$

$\overline{T}_{n1}$ may be related to the probability P_e of *any* given received pulse's being erroneously reconstructed by approximately $P_e = N\tau/\overline{T}_{n1}$, where τ is the received pulse duration and N is the number of multiplexed messages. To determine P_e we may use the same methods as were applied to PCM in Section 7.12. For received pulses of amplitudes A or $-A$ and white input noise of spectral density $\mathcal{N}_0/2$ we obtain

$$P_e = \frac{1}{2} \operatorname{erfc}\left(\sqrt{A^2\tau/\mathcal{N}_0}\right), \tag{7.17-24}$$

where erfc $(\cdot)$ is as defined in (6.12-8). Thus, the receiver noise power becomes

ISBN 0-201-05758-1

$$N_{rec} = N_g \frac{6\omega_s}{\pi W_f W_L N\tau} \operatorname{erfc}\left(\sqrt{A^2\tau/\mathcal{N}_0}\right) \tag{7.17-25}$$

on use of (7.17-4).

The worst-case or largest value of N_{rec} occurs for the smallest value of τ ($\tau \geqslant \pi/W_{ch}$) and the largest value of ω_s ($\omega_s \leqslant 2W_{ch}/N$). It is

$$N_{rec} \leqslant N_g \frac{12}{\pi^2 N^2}\left(\frac{W_{ch}}{W_f}\right)^2\left(\frac{W_f}{W_L}\right) \operatorname{erfc}\left[\sqrt{\frac{1}{2}\left(\frac{\hat{S}_i}{N_i}\right)_{\text{DM}}}\right], \tag{7.17-26}$$

where we define *peak* input signal-to-noise ratio by

$$\left(\frac{\hat{S}_i}{N_i}\right)_{\text{DM}} = \frac{A^2}{N_i} = \frac{A^2 2\pi}{\mathcal{N}_0 W_{ch}}. \tag{7.17-27}$$

Finally, we neglect slope overload noise for convenience and write

$$\left(\frac{S_o}{N_o}\right)_{\text{DM}} = \frac{\overline{f^2(t)}}{N_g + N_{rec}}. \tag{7.17-28}$$

By substitution of (7.17-26) we have

$$\left(\frac{S_o}{N_o}\right)_{\text{DM}} = \frac{(S_o/N_g)_{\text{DM}}}{1 + \dfrac{12}{\pi^2 N^2}\left(\dfrac{W_{ch}}{W_f}\right)^2\left(\dfrac{W_f}{W_L}\right)\operatorname{erfc}\left[\sqrt{\dfrac{1}{2}\left(\dfrac{\hat{S}_i}{N_i}\right)_{\text{DM}}}\right]} \tag{7.17-29}$$

for worst-case conditions of receiver noise. The numerator of (7.17-29) is

$$\left(\frac{S_o}{N_g}\right)_{\text{DM}} = \frac{3\overline{f^2(t)}}{(\delta v)^2}\left(\frac{\omega_s}{W_f}\right), \tag{7.17-30}$$

which will reduce to either (7.17-11) or (7.17-21), depending on how step-size δv is chosen. Typical evaluation of (7.17-29) ($N = 1$ and $W_f/W_L = 20$) shows that the 1-dB threshold signal-to-noise ratio $(\hat{S}_i/N_i)_{\text{DM},th}$ is not a sensitive function of W_{ch}/W_f. It varies from about 10 dB to about 14 dB for W_{ch}/W_f from 3 to 140, respectively.

Comparison of DM and PCM

When receiver noise is negligible and equal channel bandwidths are considered, PCM is always superior to DM for large values of W_{ch}/W_f. For small values of W_{ch}/W_f it is possible for DM to become superior to PCM. The point of changeover depends on signal conditions. The following example illustrates that a PCM system output signal-to-noise ratio is 3.3 dB larger than that of a DM system when $W_{ch}/W_f = 5$ and the message is sinusoidal. For $W_{ch}/W_f < 4$ it can be found that the DM system becomes slightly superior.

Example 7.17-4

We compare output signal-to-noise ratios of DM and PCM systems for a single sinusoidal message when $W_{ch}/W_f = 5$. For this message $K_{cr}^2 = 2$ and $\dot{K}_{cr}^2 = 2$. Assume the message fre-

ISBN 0-201-05758-1

quency is $\omega_f/2\pi$ = 800 Hz and $W_f/2\pi$ = 3.5 kHz, so that parameters approximate those of speech transmission in the DM system. Since $\omega_s/2W_f = W_{ch}/W_f$ for DM, the equation of example 7.17-1 gives

$$\left(\frac{S_o}{N_g}\right)_{\text{DM}} = \frac{3}{8\pi^2}\left(\frac{3.5}{0.8}\right)^2 (10)^3 = 727.3 \text{ (or 28.6 dB)}.$$

For PCM, assume $N_b = W_{ch}/W_f$. Equation (7.13-12) gives

$$\left(\frac{S_o}{N_q}\right)_{\text{PCM}} = \frac{3(2^{10})}{2} = 1536 \text{ (or 31.9 dB)}.$$

Thus, PCM is superior in performance to DM by 3.3 dB. Recalculation for $W_{ch}/W_f < 4$ shows DM slightly superior.

7.18. DELTA-SIGMA MODULATION

It was noted in Section 7.16 that a system using delta modulation performs most favorably with messages having a power density spectrum that decreases with increasing frequency (as $1/\omega^2$). It is not as well suited to messages having an approximately flat power spectrum. Furthermore, another difficulty arises if the message possesses a dc component, since it cannot easily be conveyed with DM having ideal integrators. Inose and Yasuda [26] have described a method whereby DM may be modified to perform well with messages having both a flat power spectrum and a dc component. The system is called *delta-sigma modulation* (D-SM).

A D-SM system is obtained, in concept, by adding one integrator and one differentiator to a basic DM system. With reference to Fig. 7.16-1, the integrator is added at the input so that an integrated version of the message $f(t)$ is applied to the summing point. The differentiator is provided in the receiver following the low-pass filter.*

Overload Characteristic

The D-SM system may be overloaded, just as the DM system, if the slope of the signal emerging from the input integrator is too large. This fact should be obvious, since from the summing point to the transmitter output we still have a basic DM system. Thus, (7.16-3) applies. However, the maximum slope is now that of the *integral* of the message so that, in order to prevent overload, we require that

$$|f(t)|_{\max} \leqslant \delta v/T_s \tag{7.18-1}$$

be satisfied.

For a sinusoidal message, which may be used to infer the overload characteristic, (7.18-1) becomes

*Our description will presume these two components are actually added, since this approach facilitates understanding of the operation of D-SM. In practice, no additional equipment is needed as compared to DM. On the contrary, D-SM can actually require one *fewer* components than DM. A practical implementation is discussed in reference [26].

ISBN 0-201-05758-1

$$A_f/\delta v \leqslant \omega_s/2\pi. \tag{7.18-2}$$

Here A_f is the message peak value, δv is the step-size, as before, and $T_s = 2\pi/\omega_s$ is the time between transmitted pulses. For transmitted pulses of short duration τ and amplitude A, step-size is given by

$$\delta v = A\tau. \tag{7.18-3}$$

Equation (7.18-2) is independent of the frequency of the sinusoidal message. Thus, although there is an upper bound to the amplitude that should be employed as a function of frequency, the bound is constant with frequency. We conclude, then, that D-SM will perform well with messages having relatively flat spectra.

Channel Bandwidth

The output pulse stream in D-SM has the same form as that of basic DM. Thus, minimum channel bandwidth is again given by (7.16-6).

Granular Noise Performance Limitation

As in DM, the final output of the D-SM receiver will be the recovered message $f(t)$ plus noise. The noise will again possess components of receiver (thermal) noise, granular noise, and overload noise. We treat only the first two and assume that overload noise is small. Even if receiver noise is small, granular noise is ever-present and ultimately limits the achievable output signal-to-noise ratio.

To determine the granular noise power reaching the output it is easiest to retrace some of the steps leading to (7.17-4) for a DM system. As noted earlier, the noise power spectrum at the low-pass filter input is approximately flat over $-W_f < \omega < W_f$, where W_f is the maximum frequency extent of the message. By working backward through the filter with the aid of (7.17-4), the power density is found to be $\pi(\delta v)^2/3\omega_s$. This power density also applies at the filter output, since it is presumed ideal. Adding a factor ω^2 to account for the effect of the differentiator the output granular noise power becomes

$$N_g = \frac{1}{2\pi} \int_{-W_f}^{W_f} \frac{\pi(\delta v)^2}{3\omega_s} \omega^2 \, d\omega \tag{7.18-4}$$

or

$$N_g = \frac{(\delta v)^2 W_f^3}{9\omega_s}. \tag{7.18-5}$$

Signal-to-noise ratio is obtained by dividing output signal power $\overline{f^2(t)}$ by (7.18-5). When we also substitute for the minimum value of δv, from (7.18-1), we have

$$\left(\frac{S_o}{N_g}\right)_{\text{D-SM}} = \frac{9}{4\pi^2} \frac{\overline{f^2(t)}}{|f(t)|_{\max}^2} \left(\frac{\omega_s}{W_f}\right)^3. \tag{7.18-6}$$

In terms of the crest factor defined by (7.12-8), we have

ISBN 0-201-05758-1

$$\left(\frac{S_o}{N_g}\right)_{\text{D-SM}} = \frac{9}{4\pi^2}\,\frac{1}{K_{cr}^{\,2}}\left(\frac{\omega_s}{W_f}\right)^3. \tag{7.18-7}$$

For a sinusoidal message $K_{cr}^{\,2} = 2$, and this expression reduces to

$$\left(\frac{S_o}{N_g}\right)_{\text{D-SM}} = \frac{9}{8\pi^2}\left(\frac{\omega_s}{W_f}\right)^3, \tag{7.18-8}$$

which is identical to that obtained by Johnson [16]. By comparing (7.18-7) with its delta modulation counterpart (7.17-11), we find that both depend on the cube of the sampling rate.

Johnson has also studied the case of practical integrators. That is, the integrators are simple, single-stage, *RC* low-pass filters having a 3-dB bandwidth ω_c. For a sinusoidal message, his result is

$$\left(\frac{S_o}{N_g}\right)_{\text{D-SM}} = \frac{9}{8\pi^2}\left(\frac{\omega_s}{W_f}\right)^3 \frac{1}{[1 + 3(\omega_c/W_f)^2]}. \tag{7.18-9}$$

Finally, we note that Inose and Yasuda [26] have studied both single and double practical *RC* integrator forms of D-SM.

Performance with Receiver Noise Added

Receiver noise will occasionally cause the reconstruction stage in Fig. 7.16-1 to produce a pulse of incorrect polarity. The probability of such an incorrect decision is denoted as P_e. Just as in DM, the output of the reconstruction stage may be viewed as a stream of correctly recovered message pulses plus a stream of error or noise pulses having magnitude twice that of the message stream. By applying the previous discussion of the error pulse stream for DM, we may justify that the noise power density spectrum at the integrator input is $4(\delta v)^2 P_e/N\tau$. Here N is the number of multiplexed messages and τ is the duration of transmitted pulses.

Since the integrator cancels the effect of the differentiator (and both can actually be eliminated in practice), the output noise may easily be found by integration:

$$N_{rec} = \frac{1}{2\pi}\int_{-W_f}^{W_f} \frac{4(\delta v)^2 P_e}{N\tau}\, d\omega. \tag{7.18-10}$$

Upon recognizing that P_e is again given by (7.17-24), N_{rec} evaluates to

$$N_{rec} = \frac{2(\delta v)^2 W_f}{\pi N\tau}\ \text{erfc}\left(\sqrt{A^2\tau/\mathcal{N}_0}\right). \tag{7.18-11}$$

Now the worst-case or largest value of N_{rec} occurs for the smallest value of τ, given by π/W_{ch}. Furthermore, if we use the largest value of ω_s, given by $2W_{ch}/N$, this choice tends to make N_{rec} large while minimizing granular noise from (7.18-5). Thus, we may arrive at

ISBN 0-201-05758-1

$$\left(\frac{S_o}{N_o}\right)_{\text{D-SM}} = \frac{(S_o/N_g)_{\text{D-SM}}}{1 + \frac{36}{\pi^2 N^2}\left(\frac{W_{ch}}{W_f}\right)^2 \operatorname{erfc}\left[\sqrt{\frac{1}{2}\left(\frac{\hat{S}_i}{N_i}\right)_{\text{D-SM}}}\right]}, \tag{7.18-12}$$

where we have defined peak input signal-to-noise ratio by

$$\left(\frac{\hat{S}_i}{N_i}\right)_{\text{D-SM}} = \frac{A^2}{N_i} = \frac{A^2 2\pi}{\mathcal{N}_0 W_{ch}} \tag{7.18-13}$$

and $(S_o/N_g)_{\text{D-SM}}$ is given by (7.18-6). By comparing (7.18-12) with (7.17-29) for delta modulation it is found that the threshold input peak signal-to-noise ratio in D-SM is smaller than that of DM for equal channel bandwidths and the same messages.

7.19. ADAPTIVE DELTA MODULATION

In previous discussions we found that a delta modulation system became overloaded when the instantaneous (short-term) slope of the message was too large. Specifically, if the signal slope exceeded the ratio of step-size δv to the time T_s between samples, overload occurred. For a given sample rate $\omega_s = 2\pi/T_s$, we could reduce slope overload by increasing step-size or decreasing the message input power level (and thereby reducing signal slope). Both measures are undesirable, because they give decreased performance. Another disadvantage in DM is its sensitivity to signal level variations. Even if slope overload is absent, the signal-to-granular noise power ratio decreases when signal power level decreases [see (7.17-5)]. Thus, in DM there exists only a finite range of signal-to-noise ratios, which exceeds whatever minimum one might establish as acceptable performance (for example, a good telephone link requires a signal-to-noise ratio of about 30 dB minimum). This finite interval is called the *dynamic range.*

Adaptive delta modulation (ADM) is a variation of DM which offers relief from the above disadvantages by adapting the step-size to accommodate changing signal conditions. If the input signal's slope is large, step-size is caused to increase, thereby reducing slope overload effects. On the other hand, step-size is decreased if the message is changing slowly or decreases in power level, thereby reducing granular noise. This latter fact leads to one of the principal advantages of ADM, increased dynamic range. Clearly, from (7.17-5), if $(\delta v)^2$ decreases in proportion to decreases in message power level $\overline{f^2(t)}$, we may cause signal-to-noise ratio to become constant, or nearly so, over some range of signal power values. In practice, 40 dB to 50 dB of such *dynamic range extensions* may readily be achieved.

System Block Diagram

Step-size in ADM can be changed by placing a variable-gain circuit in the feedback path of the basic DM generator (Fig. 7.16-1). The circuit may be made to give discrete [24, 27] or continuous [28–30] gain variations. Regardless of the type of gain variation, the main problem in ADM is the manner in which gain is controlled. At least two basic methods are possible. In the first, called *forward control,* a control voltage for step-size adjustment is derived from the input message. In order for the receiver to have step-size knowledge, this voltage must be transmitted over the channel separately from the ADM output pulse stream. A typical system [31] encodes the voltage in the form of a second DM pulse stream (at a low sample rate) and

ISBN 0-201-05758-1

multiplexes the two streams in time. Disadvantages of the method are the need for two bit streams and the problems associated with separating the two streams at the receiver.

The second method of controlling gain, called *feedback control,* does not have the above disadvantages. The applicable block diagram is illustrated in Fig. 7.19-1(a). It is the same as in Fig. 7.16-1 for DM (up to point A) except for the addition of the variable gain circuit and the step-size controller. The purpose of the controller is to sense the slope condition of the message conveyed in the transmitted pulse train $s_{ADM}(t)$. If the slope is large, the controller output causes the variable gain circuit to have a large gain. Loop step-size is then large. If the slope is small, the controller output causes a small gain and thereby a small step-size. Because the loop will force $f_q(t)$ to approximately equal $f(t)$, all that is necessary in the receiver for message recovery is to reproduce the operations in the feedback path of the encoder (modulator). Thus, the receiver blocks shown in (b) are identical to their modulator counterparts. The receiver for an overall ADM system would be similar to that of Fig. 7.16-1 but with all the components to the right of point B replaced by the receiver of Fig. 7.19-1(b).

It is no doubt obvious that the central problem in ADM is the design of the step-size controller. There are two important considerations. One is the functional law relating gain changes to signal slope changes. The other is the form of the implementation. In the following two subsections we discuss implementations. The natural course of discussion will involve the gain law as well. However, choice of a gain law is not inherent to any given implementation. The two can be separately chosen as far as principle goes. Some comments regarding an optimum gain law are included in the later consideration of noise performance of ADM.

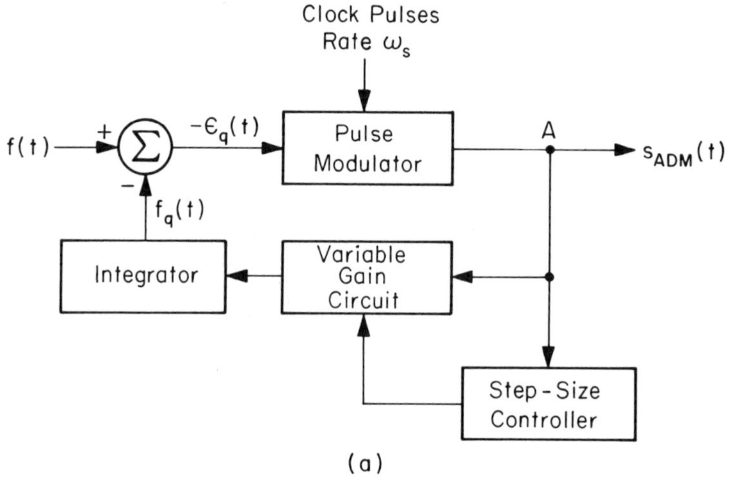

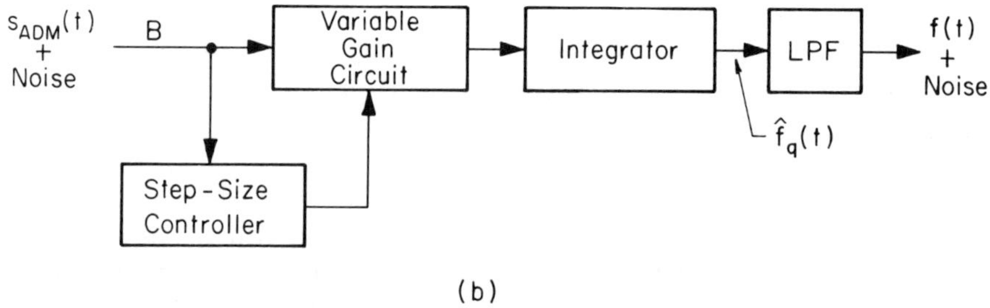

Fig. 7.19-1. (a) Adaptive delta modulator using feedback control and (b) the corresponding receiver decoder.

ISBN 0-201-05758-1

Instantaneous Step-Size Control

When the variable gain circuit is discrete, the usual method of implementing the step-size controller involves digital logic. Gain changes occur on a pulse-to-pulse basis and are said to be *instantaneous.* Several variations of design are possible [18, 24, 27, 32, 33]. In one approach [27] the controller has a one-bit memory. By considering the fact that slope overload leads to a sequence of output pulses of identical polarity, it compares the polarity of each transmitted pulse with the immediately preceding one. If they have the same polarity the controller indexes the gain circuit to a larger gain value (typically by a factor P larger than the previous gain). If the polarities are not the same, a sign reversal has taken place and gain is indexed to a lower value (typically by a factor Q lower). Jayant [27, 33] finds, from computer simulations with speech messages, that $PQ = 1$ and $P \approx 1.5$ represent optimum conditions to minimize quantization error power. Later simulations of Zetterberg and Uddenfeldt [32], however, indicate that $PQ = 1.05$ and $P = 1.1$ are more optimum, giving both minimum quantization error and maximum dynamic range. The difference in the two results apparently is due to the difference in the messages assumed in the simulations [32]. On the other hand, from reference [32], signal-to-noise ratio is not a sensitive function of P when optimum Q is used. It changes by about 3 dB as P changes* from 1.1 to 1.5. Optimum PQ remains in the vicinity of 1.1.

The above technique can be extended to controllers having more than a one-bit memory [18, 24, 32]. Data of reference [32] indicate that, for up to three bits of memory, final receiver output signal-to-noise ratio does not change greatly (less than 1.5 dB for $1.1 < P < 1.5$ and the optimum Q in each case of P when ω_s/W_f is near 13.7). The decision logic chosen was to increase gain by the factor P if, for k bits of memory, all $k + 1$ pulses were of the same sign and to decrease gain by the factor Q otherwise.

Schindler [18] has shown that, even with no slope overload, there may be cases where two consecutive pulses transmitted may have the same polarity. On this basis he justifies that two (or more) bits of memory are better suited to control step-size for nonstationary messages such as speech waveforms. A decision logic based on four total bits (three-bit memory) is given. It is most easily understood by assigning a four-bit binary word to the four most recent pulse transmissions. If the most recent pulse corresponds to the least-significant digit in the word, then gain is increased by 2 dB if the digital number is 0 or 15. Gain is increased by 1 dB if the number is 4, 7, 8, or 12. Gain is decreased by 0.125 dB for all other numbers. In a practical system, the decreases are accumulated until a total decrease of 1 dB is called for; the gain then is decreased by 1 dB.

Syllabic Step-Size Control

Many adaptive delta modulators do not change step-size on a pulse-to-pulse basis. Rather, changes are made much more slowly. Since most research has been directed toward use of DM and ADM for speech encoding, such slow control is referred to as *syllabic.* The usual implementation involves a continuously variable gain circuit controlled by an analog voltage from the step-size controller. The input to the controller is the ADM output pulse stream. This input may be processed either digitally [17, 29, 34, 35] or with analog circuitry [7, 28–31, 36] to obtain the analog control voltage.

The system of Greefkes and Riemens [35] is a good example of digital processing. Whenever the system output pulse stream contains four consecutive pulses of either state (on/off unipolar rather than polar pulses were used in reference [35]), a dc level is initiated. The level

ISBN 0-201-05758-1

*These conclusions assume $7 \leqslant \omega_s/W_f \leqslant 13.7$.

is maintained for a time equal to the total number of consecutive sampling intervals that the pulse state is preserved. The result is a sequence of "pulses" which are averaged by passing through a low-pass filter (bandwidth of $100/\pi$ Hz) to obtain the slowly varying control voltage. Clearly, the control voltage is proportional to the average fraction of time that the system output pulse stream contains four or more consecutive pulses of the same state. However, we may show that the voltage is also proportional to the fraction of time that the slope (derivative) of the message exceeds half the average slope capability of the ADM system. The proof consists in recognizing that, for four pulses, half the average slope capability is exceeded when the *average* number of pulses of any one state exceeds two. Only the case where all four pulses has the same state satisfies this condition. Greefkes and Riemens define the ratio of the signal derivative to the maximum permissible value of this derivative as *modulation index.* Thus, the analog voltage at the output of the step-size controller is proportional to the average fraction of time that the modulation index exceeds $1/2$.

A good example of an analog processor is the system of Tomozawa and Kaneko [28]. First, the system output pulse stream is passed through an integrator to obtain a signal representing the modulated signal component contained in the coded pulses. The integrator output is then envelope-detected. The detector feeds a low-pass filter which produces a slowly varying output voltage proportional to the modulating signal average level. The low-pass filter has a time constant of about 5 ms or a bandwidth (3 dB) of $100/\pi$ Hz. Thus, the controller output is a voltage proportional to the average message voltage magnitude over a 5-ms time interval. The voltage, plus a small fixed bias, controls the gain of the variable gain circuit. The bias helps prevent instabilities [28] and prevents the variable gain from becoming zero when the detector output is zero.

It should be observed that the analog control voltages in the above two example systems are different. One is related to message *slope,* the other is related to message *magnitude.* In the following subsection, we discuss noise performance assuming syllabic step-size control, and show, at least for stationary random signals, that either approach leads to the same performance.

Quantization Noise Performance of ADM

The performance of a DM system may easily be extended to give the performance of ADM. We model the message as a zero-mean stationary Gaussian random signal in order to make use of previous work. The model will form a reasonable first approximation for other messages. The assumption of stationarity allows a fixed relationship to exist between signal power and the power in the signal derivative $\dot{f}(t) = df(t)/dt$. The relationship is

$$\overline{\dot{f}^2(t)} = W_{\mathrm{rms}}{}^2\, \overline{f^2(t)}, \tag{7.19-1}$$

where W_{rms} is the rms bandwidth of the power density spectrum of $f(t)$, as given by (7.17-10). Thus, a certain average slope implies a certain signal power, and vice versa.

We treat ADM as DM with a variable step-size δv that is a function of message slope, or the equivalent, message level. Receiver output signal-to-quantization noise power ratio must again be given by either (7.17-19) or (7.17-22) if suitable account is made for a variable step-size. Suppose we begin by placing an upper bound or "reference" value δv_r on δv by choosing the ratio of δv_r to sampling interval T_s to be proportional to the rms slope of the message when it is at some reference level $f_r(t)$. In other words,

$$\delta v_r = \dot{K}\sqrt{\overline{\dot{f}_r^{\,2}(t)}}\, T_s, \tag{7.19-2}$$

ISBN 0-201-05758-1

where $\dot{K}$ is a constant of proportionality. Equation (7.19-2) is just an extension of (7.17-17) for fixed δv in delta modulation.

Now the signal-to-quantization noise power ratio that we seek is given by

$$\left(\frac{S_o}{N_q}\right)_{\text{ADM}} = \frac{\overline{f^2(t)}}{N_g + N_{so}}, \tag{7.19-3}$$

where the granular noise power N_g is given by (7.17-4) for any δv, and the slope overload noise power N_{so} is given by (7.17-14) for any δv and signal power. In order to find N_{so}, which is a function of Z as defined in (7.17-15), we first write Z^2 as

$$Z^2 = \frac{(\delta v)^2}{T_s^2\overline{\dot{f}^2(t)}} = \left(\frac{\delta v}{\delta v_r}\right)^2 \frac{\dot{K}^2\overline{f_r^2(t)}}{\overline{\dot{f}^2(t)}} = \left(\frac{\delta v}{\delta v_r}\right)^2 \frac{\dot{K}^2\overline{\dot{f}_r^2(t)}}{\overline{\dot{f}^2(t)}}. \tag{7.19-4}$$

In obtaining this expression, T_s from (7.19-2) has been substituted and (7.19-1) was used. In terms of x, previously defined as

$$x^2 = \overline{\dot{f}^2(t)}/\dot{K}^2\overline{\dot{f}_r^2(t)}, \tag{7.19-5}$$

we have

$$Z^2 = \left(\frac{\delta v}{\delta v_r}\right)^2 \frac{1}{x^2}. \tag{7.19-6}$$

Substitution into (7.17-14) gives the desired expression for N_{so}. Performing somewhat similar operations on the ratio $\overline{f^2(t)}/N_g$ will allow us to write (7.19-3) as

$$\left(\frac{S_o}{N_q}\right)_{\text{ADM}} = \frac{\dot{K}^2\left(\frac{S_o}{N_g}\right)_{\text{ADM},r}\left[\left(\frac{\delta v_r}{\delta v}\right)x\right]^2}{1 + \frac{2}{9}\left(\frac{\omega_s}{W_f}\right)^3\left[3 + \left(\frac{\delta v_r}{\delta v}\right)x\right]\left(\frac{\delta v_r}{\delta v}\right)x\exp\left[-3/\left(\frac{\delta v_r}{\delta v}\right)x\right]}. \tag{7.19-7}$$

Here

$$\left(\frac{S_o}{N_g}\right)_{\text{ADM},r} = \frac{3\overline{f_r^2(t)}\,\omega_s}{(\delta v_r)^2 W_f} = \frac{3}{4\pi^2\dot{K}^2}\left(\frac{W_f}{W_{\text{rms}}}\right)^2\left(\frac{\omega_s}{W_f}\right)^3 \tag{7.19-8}$$

is the "reference" signal-to-granular noise power ratio. In terms of minimum channel bandwidth $N\omega_s/2$, which is the same as in DM, we write

$$\left(\frac{S_o}{N_q}\right)_{\text{ADM}} = \frac{\dot{K}^2\left(\frac{S_o}{N_g}\right)_{\text{ADM},r}\left[\left(\frac{\delta v_r}{\delta v}\right)x\right]^2}{1 + \frac{16}{9N^3}\left(\frac{W_{ch}}{W_f}\right)^3\left[3 + \left(\frac{\delta v_r}{\delta v}\right)x\right]\left(\frac{\delta v_r}{\delta v}\right)x\exp\left[-3/\left(\frac{\delta v_r}{\delta v}\right)x\right]}. \tag{7.19-9}$$

ISBN 0-201-05758-1

We pause to examine our results. Observe that (7.19-7) is identical to (7.17-19) in form. In fact, x in the DM expression is replaced here by $x(\delta v_r/\delta v)$. Now since δv_r is the maximum value of δv, and it occurs whenever $\overline{f^2(t)} \geqslant \overline{f_r^2(t)}$, then, for this condition, $x^2(\delta v_r/\delta v)^2$ here is identical to x^2 for DM. Thus, whenever $x^2 \geqslant 1/\dot{K}^2$ we expect the normalized signal-to-noise ratio of ADM to be the same as in DM. When $x^2 < 1/\dot{K}^2$ the two are different. In DM, $(S_o/N_q)_{\text{DM}}$ decreases as x^2 decreases. In ADM, $(S_o/N_q)_{\text{ADM}}$ may be made nearly constant as x^2 decreases if $x^2(\delta v_r/\delta v)^2$ is nearly constant.

The condition required for $x^2(\delta v_r/\delta v)^2$ to be constant is readily found. When $\delta v = \delta v_r$, x^2 must equal $1/\dot{K}^2$. For $\delta v < \delta v_r$ we require

$$x^2(\delta v_r/\delta v)^2 = 1/\dot{K}^2, \tag{7.19-10}$$

or, on substitution of (7.19-5),

$$\delta v = \sqrt{\overline{f^2(t)}/\overline{f_r^2(t)}}\,\delta v_r. \tag{7.19-11}$$

In words, δv should be proportional to $[\overline{f^2(t)}]^{1/2}$, a condition which can be approximated in practice.

Unfortunately a normalized plot of (7.19-9) is not possible, as was done for its DM counterpart (7.17-22), because $\delta v_r/\delta v$ cannot readily be written in terms of x. We may, however, work toward developing some representative curves for a speech message to see the typical behavior of (7.19-9). First we assume the following approximation for (7.19-11):

$$\begin{aligned}\frac{\delta v}{\delta v_r} &= \frac{\delta v_0}{\delta v_r} + \left(1 - \frac{\delta v_0}{\delta v_r}\right)\sqrt{\frac{\overline{f^2(t)}}{\overline{f_r^2(t)}}}, \qquad & \overline{f^2(t)} < \overline{f_r^2(t)} \\ &= 1, & \overline{f^2(t)} \geqslant \overline{f_r^2(t)}.\end{aligned} \tag{7.19-12}$$

Here δv_0 is the smallest value that δv is allowed to have. The ratio $\delta v_r/\delta v_0$ when expressed in decibels may be called the *dynamic range factor* F_{dy}. That is,

$$F_{dy} = 20\log_{10}(\delta v_r/\delta v_0). \tag{7.19-13}$$

F_{dy} is the maximum amount that system dynamic range is increased over that of a basic DM system.

Next, we model the speech signal as a Gaussian random message having a simple low-pass power density spectrum. Below the 3-dB frequency $\omega = W_3$ it is constant. Above W_3 it decreases at -6 dB per octave frequency increase out to a point W_f where it abruptly becomes zero for $\omega > W_f$. For this model it is readily shown that

$$\left(\frac{W_{\text{rms}}}{W_f}\right)^2 = \left[\frac{(W_3/W_f)}{\tan^{-1}(W_f/W_3)} - \left(\frac{W_3}{W_f}\right)^2\right]. \tag{7.19-14}$$

For speech signals W_f/W_3 is approximately 4.35, corresponding to $W_{\text{rms}} = 0.344W_f$ [24].

Figure 7.19-2 has been constructed from (7.19-9) for a single message using (7.19-12), $W_{\text{rms}} = 0.344W_f$, $W_{ch} = 10W_f$, and $\dot{K} = 2.985$, which is the value that corresponds to the maximum point in Fig. 7.17-1. The effects of dynamic range factors of 10, 20, 30, and 40 dB are shown. For comparison, the response of a basic delta modulation system is shown dashed.

ISBN 0-201-05758-1

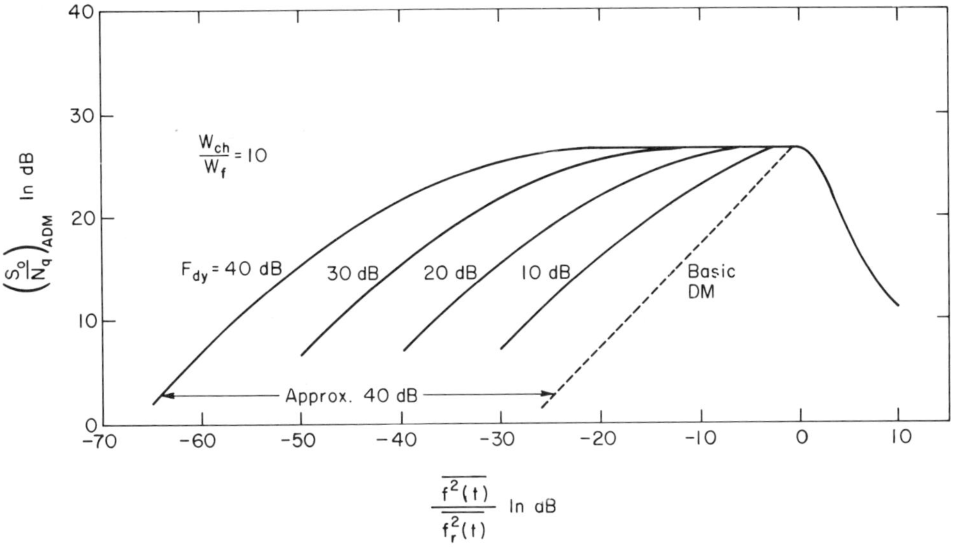

Fig. 7.19-2. Receiver output signal-to-noise ratio in adaptive delta modulation. A speech message is assumed for which $W_{rms} = 0.344W_f$.

Example 7.19-1

For the system of Fig. 7.19-2 we calculate the increase in dynamic range of the ADM system relative to the DM system when the minimum required signal-to-noise ratio is 15 dB. $F_{dy} =$ 40 dB is assumed.

In the slope overload region, (S_o/N_q) of 15 dB occurs for both systems at a relative signal strength of +6.8 dB. For smaller signal levels the DM and ADM systems reach 15 dB at relative signal strengths of − 12.5 dB and − 50.5 dB, respectively. Thus, the dynamic range of DM is 6.8 + 12.5 = 19.3 dB. That of ADM is 6.8 + 50.5 = 57.3 dB. Dynamic range extension becomes 57.3 − 19.3 = 38.0 dB, or slightly below the maximum of 40 dB.

7.20. DIFFERENTIAL PCM

From our earlier study of a delta modulator, we found that each transmitted pulse corresponded to a one-bit encoding of the difference between the message $f(t)$ and an approximation $f_q(t)$ to it. The overall effect of the pulse modulator (Fig. 7.16-1) was to quantize the difference signal into two levels, sample the quantized difference, and generate a one-bit coded pulse stream having one pulse per sample. If, instead, the difference signal is quantized into multiple levels, sampled, and each sample coded into an N_b-bit binary code, the result is called *differential pulse code modulation* (DPCM). The code is transmitted over the channel as binary pulses with N_b pulses per sample. DPCM is sometimes known as *delta PCM.*

DPCM System Block Diagram

DPCM systems may take on many forms. We shall discuss only the most elementary version and reserve comments on the variations for the end of this section. A simple DPCM modulator is illustrated in Fig. 7.20-1(*a*). The demodulator is shown in (*b*). When these components re-

ISBN 0-201-05758-1

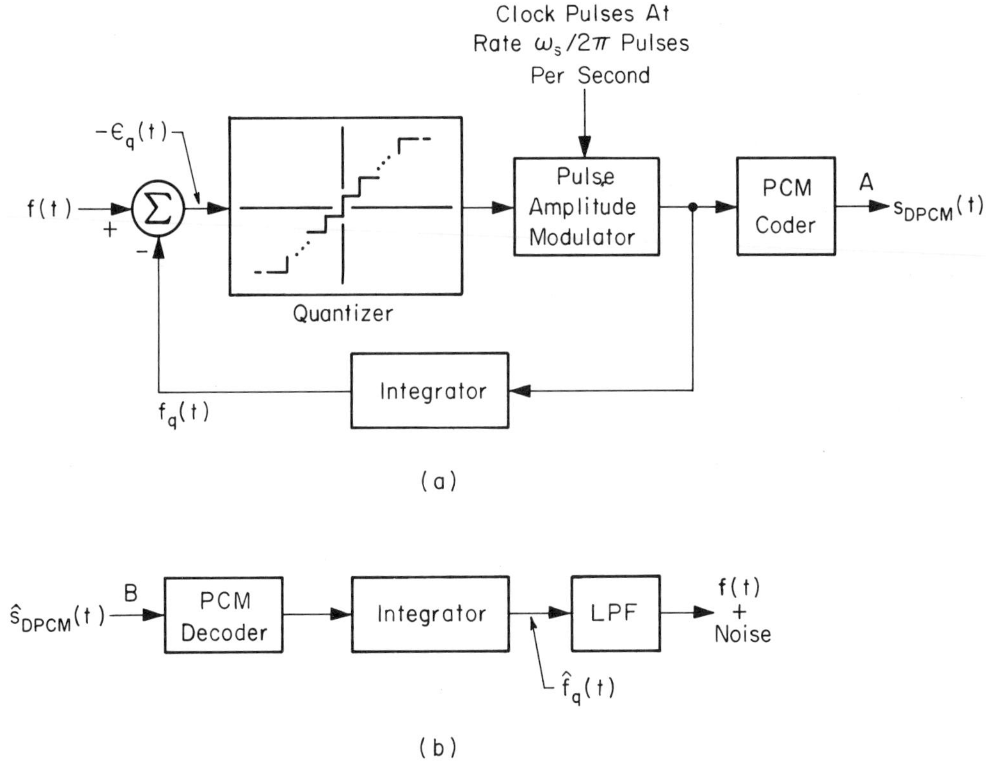

Fig. 7.20-1. Block diagrams applicable to a differential PCM system. (a) Modulator and (b) demodulator. The components fit into Fig. 7.16-1 to the left of point A and to the right of point B, respectively, to form a DPCM system.

place their counterparts in Fig. 7.16-1, a DPCM system is formed. Consider first the modulator. The error $-\epsilon_q(t)$ is rounded off to the nearest of 2^{N_b} equally spaced levels where N_b is a positive integer. The quantized version of $-\epsilon_q(t)$ acts as a modulating signal to the pulse amplitude modulator. Assuming the clock pulses have a short duration and constant amplitude, the modulator output is a train of pulse amplitude-modulated pulses. Each pulse becomes a quantized sample of the error $-\epsilon_q(t)$. Samples occur at the sample rate ω_s (every $T_s = 2\pi/\omega_s$ seconds). The PCM coder converts each sample pulse into an N_b-bit binary pulse sequence which is transmitted over the channel. The sample pulses are also applied to the integrator which closes the loop by producing $f_q(t)$ at its output.

Since the sample pulses are narrow, the integrator generates a staircase waveform as in a basic delta modulation system except there are now 2^{N_b} possible step-sizes available (including sign). The steps are [37]

$$\pm\,\delta v_0,\ \pm\,3\delta v_0,\ \ldots,\ \pm\,(2^{N_b}-3)\delta v_0,\ \pm\,(2^{N_b}-1)\delta v_0, \tag{7.20-1}$$

where δv_0 is the magnitude of the smallest step. Steps and pulse polarities are related through a simple binary code. Figure 7.20-2 sketches some typical waveforms of interest in generating a DPCM signal. A two-bit code is assumed; that is, $N_b = 2$. Quantization error magnitude $|\epsilon_q(t)|$ may be as large as $3\delta v_0$, as shown in (*c*). The pulse stream of (*b*) assumes that steps of $3\delta v_0$, δv_0, $-\delta v_0$, and $-3\delta v_0$ are coded with pulses having polarities ++, +−, −+, and −−, respectively.

ISBN 0-201-05758-1

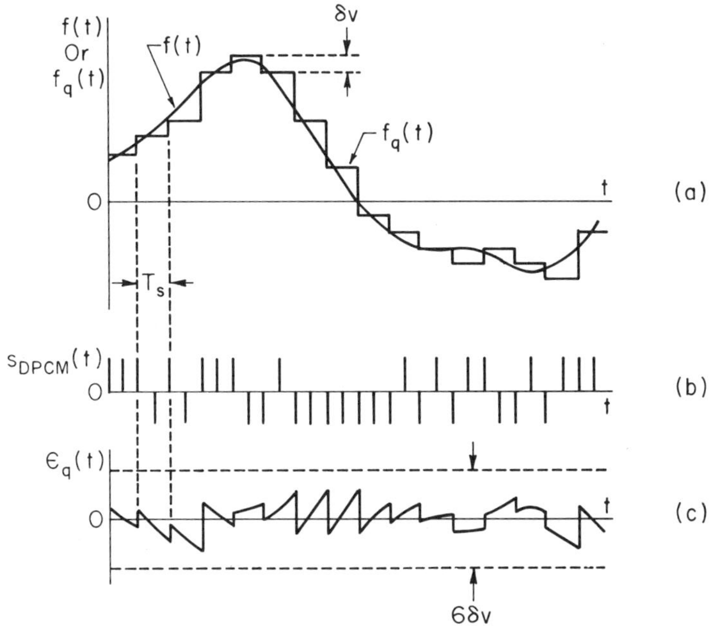

Fig. 7.20-2. Waveforms applicable to DPCM. (a) f(t) and $f_q(t)$. (b) DPCM pulse train, and (c) the error signal.

More generally, quantization error magnitude can be as large as $(2^{N_b} - 1)\delta v_0$, and $f_q(t)$ may still remain a good approximation to $f(t)$. Note, however, that periods of large error correspond to rapid changes occurring in the message. Thus, if the signal's slope exceeds the maximum slope capability of the system, we may have *slope overload* just as in DM. To prevent slope overload

$$|\dot{f}(t)|_{max} = |df(t)/dt|_{max} \leqslant (2^{N_b} - 1)\delta v_0/T_s \tag{7.20-2}$$

is required if the message has a maximum slope. If not, we may select a constant $\dot{K}$ such that

$$\dot{K}\sqrt{\overline{\dot{f}^2(t)}} \leqslant (2^{N_b} - 1)\delta v_0/T_s. \tag{7.20-3}$$

These two conditions are analogous to (7.16-3) and (7.17-17) for DM.

At the receiver end of the system, a PCM decoder recovers the amplitude-modulated sample pulses which, when integrated, become $\hat{f}_q(t)$, which is approximately equal to $f_q(t)$ back at the modulator. Except for receiver noise errors it would be the same. The low-pass filter smooths out the sharp steps in $\hat{f}_q(t)$, and the response is $f(t)$ pluse noise. As in DM the noise has granular, slope overload, and receiver noise components.

Channel Bandwidth

The channel bandwidth required in DPCM is that of an N_b-bit PCM system. Assuming N similar messages are time-multiplexed, we have

$$W_{ch} \geqslant \pi/\tau \geqslant \pi N N_b/T_s = N N_b \omega_s/2. \tag{7.20-4}$$

Here the PCM pulses have duration τ.

ISBN 0-201-05758-1

Performance with Granular Noise Only

Several analyses may be cited which deal with the various noise components in the final receiver output signal [20, 21, 23, 37]. If only granular noise is present, which corresponds to times when the message is slowly varying such that the system easily follows the changes, van de Weg [37] has determined the output signal-to-granular noise power ratio. By using a random message model he obtains

$$\left(\frac{S_o}{N_g}\right)_{\text{DPCM}} = \frac{3}{2k}\left(\frac{\omega_s}{W_f}\right), \qquad N_b \geqslant 2, \tag{7.20-5}$$

where W_f is the maximum spectral extent of the signal and

$$k = (\delta v_0)^2/\overline{f^2(t)}. \tag{7.20-6}$$

Equation (7.20-5) is valid as a first approximation independent of the power spectrum of the message [37, p. 383], and van de Weg even justifies its application to a nonrandom sinusoidal signal [37, p. 384]. Thus, we may apply (7.20-5) in a broad sense.

For signals having a definite maximum magnitude it is convenient to set the maximum step-size $(2^{N_b} - 1)\delta v_0$ to prevent slope overload according to the maximum or reference slope value $\dot{f}_r(t)$ of $\dot{f}(t)$:

$$|\dot{f}_r(t)|_{\max}^2 = [(2^{N_b} - 1)\delta v_0]^2/T_s^2. \tag{7.20-7}$$

For messages having no well-defined maximum magnitude, such as Gaussian random signals, we may choose the maximum step-size such that system slope capability is larger than the mean-squared signal slope by some factor $\dot{K}^2 > 1$:

$$\dot{K}^2\, \overline{\dot{f}_r^{\,2}(t)} = [(2^{N_b} - 1)\delta v_0]^2/T_s^2. \tag{7.20-8}$$

For this latter case k is easily evaluated. Substitution into (7.20-5) gives

$$\left(\frac{S_o}{N_g}\right)_{\text{DPCM}} = \frac{3(2^{N_b} - 1)^2}{8\pi^2\dot{K}^2}\left(\frac{W_f}{W_{\text{rms}}}\right)^2\left(\frac{\omega_s}{W_f}\right)^3, \qquad N_b \geqslant 2. \tag{7.20-9}$$

For the former case, we only replace $\dot{K}^2$ by $\dot{K}_{cr}^{\,2}$, the slope crest factor defined in (7.17-9). W_{rms} is defined in (7.17-10).

We note that $(S_o/N_g)_{\text{DPCM}}$ increases as $\omega_s^{\,3}$, which was also found for DM, D-SM, and ADM systems. If $N_b = 1$ we might expect that (7.20-9) would give results near those for DM. However, the value of (7.20-9) is actually 3 dB smaller than obtained in DM. This occurs because the derivation of (7.20-9) neglects correlation between successive steps when $N_b = 1$. For $N_b \geqslant 2$ the correlation becomes small [37]. Other important observations are that performance depends on the signal's power spectrum through the term $(W_f/W_{\text{rms}})^2$. The only system parameters* appearing explicitly in (7.20-9) are ω_s and N_b. Increasing either leads to better performance.

We now give three examples to illustrate use of (7.20-9) for three different information signals.

*$\dot{K}$ is a parameter related to both the system and the message.

ISBN 0-201-05758-1

Example 7.20-1

Let us assume a random noise-like information signal having a uniform (constant) power density over $-W_f < \omega < W_f$ and a slope crest factor $\dot{K}_{cr} = \dot{K} = 4$. Using (7.17-10) we find $W_{\mathrm{rms}}{}^2 = W_f^2/3$. From (7.20-9)

$$\left(\frac{S_o}{N_g}\right)_{\mathrm{DPCM}} = \frac{9(2^{N_b} - 1)^2}{128\pi^2}\left(\frac{\omega_s}{W_f}\right)^3, \qquad N_b \geqslant 2.$$

If we now let $N_b = 2$ and $\omega_s/W_f = 10$, we have $(S_o/N_g)_{\mathrm{DPCM}} = 64.1$, or 18.1 dB.

Example 7.20-2

Suppose $f(t)$ is a sinusoid of frequency $\omega_f \leqslant W_f$. From (7.17-9) we easily find that $\dot{K}_{cr}{}^2 = 2$. From problem 7-33 $W_{\mathrm{rms}} = \omega_f$. From (7.20-9)

$$\left(\frac{S_o}{N_g}\right)_{\mathrm{DPCM}} = \frac{3(2^{N_b} - 1)^2}{16\pi^2}\left(\frac{W_f}{\omega_f}\right)^2\left(\frac{\omega_s}{W_f}\right)^3, \qquad N_b \geqslant 2.$$

Example 7.20-3

As we have previously stated, results for delta modulation of speech signals may be found from the sinusoidal signal equation if $\omega_f/2\pi = 800$ Hz and $W_f/2\pi = 3.5$ kHz. Using these values in the equation of example 7.20-2 we obtain

$$\left(\frac{S_o}{N_g}\right)_{\mathrm{DPCM}} = \frac{3.59(2^{N_b} - 1)^2}{\pi^2}\left(\frac{\omega_s}{W_f}\right)^3, \qquad N_b \geqslant 2.$$

Again letting $N_b = 2$ and $\omega_s/W_f = 10$, we calculate $(S_o/N_g)_{\mathrm{DPCM}} = 3274$, or 35.2 dB. Comparing with example 7.20-1 we see that considerable difference in results can occur for different signal characteristics.

Performance Comparison with DM and PCM

DPCM is readily compared with DM. On the basis of equal channel bandwidth, the same values of $\dot{K}$, and equal message characteristics, we obtain

$$\left(\frac{S_o}{N_g}\right)_{\mathrm{DPCM}} = \frac{(2^{N_b} - 1)^2}{2N_b{}^3}\left(\frac{S_o}{N_g}\right)_{\mathrm{DM}}, \qquad (7.20\text{-}10)$$

which shows that DPCM is superior to DM if $N_b \geqslant 4$.

Van de Weg [37] has compared DPCM performance with that of PCM for a random message having a flat power spectrum and a slope crest factor of four. Assuming equal values of ω_s and the same number of bits per sample interval, it is found that DPCM performance is equal to or better than that of PCM when $\omega_s/W_f \geqslant 4$ and $N_b \geqslant 4$. On the other hand, for speech signals DPCM is superior to PCM for all values of ω_s/W_f and N_b when the comparison again assumes equal values of ω_s/W_f and N_b.

ISBN 0-201-05758-1

Performance with Slope Overload Noise Added

When receiver noise is negligible the total quantization noise power at the receiver output is the sum of granular and slope overload noise powers. O'Neal and Rice [23] have determined slope overload noise for a delta modulation system ($N_b = 1$). According to O'Neal [23], slope overload noise is determined from the delta modulation result by replacing the DM system step-size by the maximum value $(2^{N_b} - 1)\delta v_0$ in the multilevel quantization (DPCM) case. Thus, if we apply the same procedure to Abate's slope overload noise power which is given by (7.17-14), algebraic manipulation assuming random messages reveals

$$\left(\frac{S_o}{N_q}\right)_{\text{DPCM}} = \frac{x^2 \dot{K}^2 (S_o/N_g)_{\text{DPCM}}}{1 + \dfrac{(2^{N_b} - 1)^2}{9}\left(\dfrac{\omega_s}{W_f}\right)^3 (3 + x)xe^{-3/x}} . \tag{7.20-11}$$

Here $\dot{K}$ is determined by (7.20-8), x is given by

$$x^2 = \frac{\overline{f^2(t)}}{\dot{K}^2 \overline{f_r^{\,2}(t)}} , \tag{7.20-12}$$

$\overline{f^2(t)}$ is the power in the message, which may be different from the reference level $\overline{f_r^{\,2}(t)}$, and $(S_o/N_g)_{\text{DPCM}}$ is given by (7.20-9). A plot of (7.20-11) for a fixed value of N_b shows a behavior versus x similar to that for DM given by (7.17-19).

Although we shall omit a discussion of receiver (channel) noise effects, it is known [38] that performance for a well-designed DPCM system is considerably better than that of a well-designed PCM system operating on the same digital channel, even if the channel is noisy.

Other System Configurations

As with most basic modulation techniques, additional complexity may lead to better performance. Such is the case with DPCM. Good discussions of the more advanced DPCM system configurations which result are given by Jayant [33] and Bayless *et al.* [39]. Our comments shall refer to data in reference [33]. For original sources and a good list of other references the reader is referred to Jayant.

We found that, by using a more complicated linear feedback network in a DM system, performance could be improved. The network could be either stationary (double versus single integration) or time-varying (adaptive DM). The analogous configurations also exist in DPCM. If the basic DPCM integrator is viewed as a prediction circuit, a more general predictor can be implemented that is either stationary or adaptive. Typical stationary or fixed predictors involve feedback networks which are specified by n constant parameters. The optimum values of these parameters are related to the correlation (and therefore spectral) properties of the message. The case $n = 1$ corresponds to the basic DPCM system. In practice $n \leqslant 3$ would be implemented because little performance improvement follows $n > 3$. For example, with $n = 3$ and speech messages, DPCM system performance (receiver output signal-to-noise ratio) is about 8.4 dB better than in a PCM system [33]. Only a 0.2-dB additional gain is obtained when n is increased to 5.

If the correlation properties of the message change with time, such as with speech waveforms, the optimum predictor must change with time. Thus, the n parameters of the predictor may be made adaptive by performing periodic (typically on the order of every 4 ms for

ISBN 0-201-05758-1

speech) analysis of short segments of the message to derive new optimum values. Such an *adaptive DPCM* system for $n = 3$ and speech signals is capable of about 1.6 dB improvement over a basic DPCM system with $n = 3$. For $n = 5$ and 10 the gains increase to 2.9 dB and 4.0 dB, respectively.

Finally, we note that configurations are possible which alter the quantizer characteristics. Because a DPCM system can slope overload similar to DM, the quantizer may be made adaptive such that a variable maximum step-size (quantization level) occurs to accommodate signals which vary greatly in level (and therefore slope).

PROBLEMS

7-1. Show by application of the low-pass sampling theorem that the bandlimited signal

$$f(t) = \frac{\pi \cos (W_f t)}{(\pi/2)^2 - (W_f t)^2}$$

is the sum of two sampling functions separated in time by the sampling time $T_s = \pi/W_f$. (Hint: Choose $T_s/2$ and $-T_s/2$ as the sample times.)

★ 7-2. A signal $f(t)$ has the following bandlimited spectrum:

$$F(\omega) = 1 + \sum_{n=1}^{N} a_n \cos\left(\frac{n\pi\omega}{W_f} + \theta\right), \qquad |\omega| < W_f$$

$$= 0, \qquad |\omega| > W_f.$$

(*a*) If $\theta = 0$ and the coefficients a_n are real, find $f(t)$. What effect does each cosine term have on $f(t)$? (*b*) Repeat (*a*), except let $\theta = -\pi/2$.

★ 7-3. Prove that the sampling theorem approximation (7.1-15) for a low-pass wide-sense stationary random process satisfies (7.1-14)—that is, has zero mean-squared error.

7-4. Show that

$$f(t) \sum_{n=-\infty}^{\infty} \delta(t - nT_s) = \sum_{n=-\infty}^{\infty} f(nT_s)\delta(t - nT_s),$$

where T_s is a constant.

7-5. Start with (7.2-14) and show that the spectrum of $s_p(t)$ may be written as

$$S_p(\omega) = \omega_s \sum_{n=-\infty}^{\infty} P(n\omega_s)\delta(\omega - n\omega_s),$$

where $P(\omega)$ is the Fourier transform of $p(t)$.

7-6. Show that C_0 of (7.2-17) equals $K\tau/T_s$ with τ given by (7.2-18).

7-7. A TDM-PAM system uses flat-top sampling pulses 0.7 μs wide. If a guard time of 0.34 μs is allowed and $N = 120$ telephone messages are multiplexed: (*a*) What is the minimum sampling rate, and (*b*) what channel bandwidth is required if crosstalk factor between adjacent time slots is required to be 10^{-3}? (*c*) Compare your calculation of W_{ch} with NW_f for typical telephone signals if $W_f/2\pi = 3.3(10^3)$.

ISBN 0-201-05758-1

★ 7-8. A multiplexed PAM signal uses pulses which may be represented as impulses. The channel transfer function is

$$H_{ch}(\omega) = \left[\frac{1}{1 + j(\omega/W_{ch})}\right]^2.$$

(*a*) Find the response of the channel to the PAM signal. (*b*) Determine an appropriate time for the receiver to resample the channel response in reconstructing messages. (*c*) Determine crosstalk factor using results of (*b*).

7-9. (*a*) If finite duration rectangular pulses are transmitted over the channel of problem 7-8, determine and sketch the response of the channel to a single pulse. (*b*) Discuss crosstalk.

7-10. A time-multiplexed PAM system is designed to produce a receiver input power signal-to-noise ratio of 20 dB when transmitting ideal samples at the Nyquist rate over an ideal channel. (*a*) If all messages satisfy $\overline{f_n^2(t)}/A^2 = 0.16$ and the output signal-to-noise ratio is 30.15 dB, how many messages are multiplexed? (*b*) If channel bandwidth is doubled but all other system parameters are the same, what are the new input and output signal-to-noise ratios?

7-11. A message having a crest factor of 5 produces an output signal-to-noise ratio of 40 dB when transmitted over a baseband link. If the same total power is transmitted over a unipolar TDM-PAM link having the same noise power density at the receiver input as in the baseband system, what is the output signal-to-noise ratio?

7-12. Use (7.7-4) to show that the bandwidth of spectral components of $s_{\text{PDM}}(t)$, as given by (7.7-3), which cluster around frequencies $k\omega_s$ is approximately $2W_f$ so long as $k \ll N/3$.

7-13. For transmission of a single message sampled at the Nyquist rate, show that the performance of a TDM-PDM system using a low-pass filter for signal recovery is related to performance of a similar system using reconstruction approximately by

$$\left(\frac{S_o}{N_o}\right)_{\text{PDM (reconstruction)}} = \frac{W_{ch}}{W_f}\left(\frac{S_o}{N_o}\right)_{\text{PDM (filter)}}.$$

7-14. Show that, if message sampling is done at the Nyquist rate and minimum channel bandwidth is used, (7.8-17) and (7.8-18) may be written, respectively, as

$$\left(\frac{S_o}{N_o}\right)_{\text{PDM}} = \left(\frac{\Delta\tau_{\text{rms}}}{\tau_{\text{min}}}\right)^2 \frac{T_s}{\tau_0}\left(\frac{S_i}{N_i}\right)_{\text{PDM}}$$

$$\left(\frac{S_o}{N_o}\right)_{\text{PDM}} = \left(\frac{\Delta\tau_{\text{rms}}}{\tau_{\text{min}}}\right)^2 \frac{\tau_{\text{min}}}{\tau_0}\left(\frac{S_o}{N_o}\right)_{\text{B}},$$

where τ_{min} is the minimum duration of sample pulses in the PDM signal.

7-15. A time-multiplexed PDM system transmits 100 similar messages sampled at a rate of 400 samples per second (Nyquist rate) over a channel having bandwidth $W_{ch} = \pi(10^6)$ and noise power density $\mathcal{N}_0/2 = 0.44(10^{-15})$. Pulses in the train have amplitude $A = 10^{-3}$ and have rms duration changes of $\Delta\tau_{\text{rms}} = 2(10^{-6})$ when the messages have a crest factor of 5. (*a*) Find the minimum, average, and maximum pulses durations τ_{min}, τ_0, and τ_{max} of pulses in the train, respectively. (*b*) What is the output signal-to-noise

ISBN 0-201-05758-1

ratio? (*c*) What is the input signal-to-noise ratio? (*d*) What is the guard time used? Assume in all cases that $W_{ch}\tau_{min} = \pi$.

7-16. Determine how the maximum possible output signal-to-noise ratio of a PDM system for which $W_f = 400\pi$, $W_{ch} = \pi(10^6)$, and $|f_n(t)|_{max}^2/\overline{f_n^2(t)} = 25$ compares to that of a baseband system having the same input signal power. Assume only one message is transmitted and sampling is at the Nyquist rate in both cases.

7-17. In a PPM system transmitting a single message it is necessary to include a synchronizing signal. Similar PDM or PAM systems do not require this. Explain why.

7-18. Six bandlimited signals are to be time-multiplexed. Four have spectral extent W_f, one $4W_f$, and one $8W_f$. Develop a commutator arrangement which will interlace samples of each signal occurring at the various Nyquist rates.

7-19. (*a*) Obtain an initial design of a TDM–PDM system to find the largest number N of similar messages which may be multiplexed assuming: $W_{ch} = 4\pi(10^6)$, $W_f = 16\pi(10^3)$, guard time per time slot $\tau_g = 2.5\tau_{min}$, $\Delta\tau_{rms} = 4\tau_{min}$, sampling at the Nyquist rate, pulse duration matched to the channel rise time, and a message crest factor of 3. (*b*) Using N of part (*a*), find the maximum new allowable value of $\Delta\tau_{rms}$; this amounts to a final design. (*c*) Determine the output signal-to-noise ratio possible per message if the input ratio is 12 dB.

7-20. Work problem 7-19 for a TDM–PPM system.

★7-21. A single message having a crest factor of $\sqrt{2}$ is transmitted by PPM over a channel for which $W_{ch}/W_f = 5$ and noise is uniformly distributed over the channel passband. (*a*) If the input peak signal-to-average noise power ratio is 56, what is the output signal-to-noise ratio? Is the system operating above threshold? (*b*) If channel bandwidth is doubled to obtain better performance, what is the new output signal-to-noise ratio? Explain your result.

7-22. A signal $f(t) = 9 + A_f \cos(\omega_f t)$, with $A_f \leqslant 10$, is to be quantized into exactly 41 binary levels with one level set at the smallest value of $f(t)$. (*a*) Find the necessary values of N_b and L_b. (*b*) What are the values of the extreme quantum levels if centered as nearly as possible on the signal variations?

7-23. Work problem 7-22 for trinary quantization ($\mu = 3$).

7-24. Find the signal-to-quantization noise power ratio for problem 7-22 if $A_f = 10$ volts.

7-25. A certain message is known to produce a signal-to-quantization noise power ratio of 50 dB when used with a certain PCM quantizer having uniformly separated quantum levels. If a compressor is used and the message crest factor is lowered by a factor of 2, what signal-to-quantization noise power ratio is possible?

7-26. Show why polar (binary) PCM uses less power than unipolar PCM for the same pulse peak-to-peak amplitude separation. Assume equally likely binary levels.

7-27. Use arguments based on probability density functions to show that polar PCM as described in problem 7-26 gives the same bit error probability as unipolar PCM with less average power. Assume both systems have the same noise power.

7-28. A TDM–PCM system (binary) multiplexes similar messages bandlimited to $W_f = (\pi/180)$ 10^6 rad/s and having a crest factor of 6. Channel bandwidth is $W_{ch} = \pi(10^6)$ rad/s. Pulse spacing equal to the bit pulse duration is used, and pulse duration equals channel rise time. (*a*) How many bits per sample are required to give a signal-to-quantization noise power of at least 40 dB? (*b*) How many messages may be sent over the channel if sampling is at the Nyquist rate?

7-29. A single message is transmitted by polar binary PCM. Assume no pulse spacing is used, the message has a crest factor of $\sqrt{2}$ and is sampled every 50 μs (Nyquist rate). (*a*) What

ISBN 0-201-05758-1

is the largest value of the minimum channel bandwidth that will allow performance at or above threshold for an input signal-to-noise ratio of 12 dB? (*b*) What will the output signal-to-noise ratio be, and how many bits are used in the binary code?

7-30. Carry out the steps indicated in the text to prove (7.12-7) by use of (4.8-35).

7-31. Show that (7.15-4) applies also when μ is an even number.

7.32. Show that $\overline{[df(t)/dt]^2}/\overline{f^2(t)} = W_{rms}{}^2$ if $W_{rms}{}^2$ is given by the right side of (7.17-10). (Hint: Consider the power into and out of a differentiating network.)

7-33. Using (7.17-10) show that the rms bandwidth for a sinusoidal signal of frequency ω_f is ω_f.

7-34. Assuming operation above threshold in both PCM and DM systems, show that

$$\frac{(S_o/N_q)_{\text{PCM}}}{(S_o/N_g)_{\text{DM}}} = \frac{\pi^2 2^{2N_b}}{2N_b{}^3}\left(\frac{\dot{K}_{cr}}{K_{cr}}\right)^2\left(\frac{W_{\text{rms}}}{W_f}\right)^2,$$

where N_b is the number of pulses in a PCM system sample. Assume equal channel bandwidths and Nyquist rate sampling in the PCM case.

7.35. Assume sinusoidal modulation and determine how much poorer a DM system is than an eight-bit PCM system using the result of problem 7-34.

7-36. By use of (7.17-22) show that the (optimum) values of x which define the maximum points in Fig. 7.17-1 are given by

$$\frac{1}{x_{\text{opt}}} = \dot{K} = \ln\left[\frac{2}{N}\left(\frac{W_{ch}}{W_f}\right)\right]$$

for multiplexing of N similar signals using delta modulation.

7-37. A delta modulation system is designed to transmit three similar messages by time multiplexing. What is the optimum value of $\dot{K}$ to be used if $W_{ch}/W_f = 81.9$? For messages characterized by $W_f/W_{\text{rms}} = 2$, what optimum output signal-to-noise ratio can be achieved?

7-38. Work problem 7-37 for $W_{ch}/W_f = 30.13$.

7-39. Using (7.17-29) show that the threshold value (1-dB definition) of input peak signal-to-noise ratio in DM is

$$\left(\frac{\hat{S}_i}{N_i}\right)_{\text{DM},th} = 2\left\{\text{erfc}^{-1}\left[0.214N^2\left(\frac{W_f}{W_{ch}}\right)^2\left(\frac{W_L}{W_f}\right)\right]\right\}^2.$$

7-40. A delta modulation system uses a nearly ideal single integrator that is replaced by a double integrator. Determine the improvement in output signal-to-granular noise power ratio that can be expected for sampling rates of $\omega_s = 5W_f$, $10W_f$, $20W_f$, and $40W_f$ when a sinusoidal message is transmitted.

7-41. A single-integrator delta modulation system is to be optimally designed (problem 7-36) for transmitting a single message. If $W_{ch}/W_f = 27$ for the system, what must W_f/W_{rms} be for the message in order to achieve a 30-dB output signal-to-quantization noise power ratio? Assume the minimum channel bandwidth.

ISBN 0-201-05758-1

7-42. For the system of problem 7-41, at what input peak signal-to-noise power ratio does threshold occur if $W_L/W_f = 0.05$?

7-43. For a sinusoidal message of frequency ω_f show that, for equal sampling rates, a delta-sigma modulation system's performance can be no larger than three times that of a delta modulation system, assuming both systems use a single integrator. That is, show that

$$\left(\frac{S_o}{N_g}\right)_{\text{D-SM}} = 3\left(\frac{\omega_f}{W_f}\right)^2 \left(\frac{S_o}{N_g}\right)_{\text{DM}} \leqslant 3\left(\frac{S_o}{N_g}\right)_{\text{DM}}.$$

7-44. Show that the threshold input peak signal-to-noise ratio (1-dB definition) in a delta-sigma modulation system is given by

$$\left(\frac{\hat{S}_i}{N_i}\right)_{\text{D-SM},th} = 2\left\{\text{erfc}^{-1}\left[0.0713N^2\left(\frac{W_f}{W_{ch}}\right)^2\right]\right\}^2.$$

7-45. What maximum dynamic range improvement can be achieved in an adaptive delta modulation system if the step-size can be reduced from its maximum by a factor of 20? What is the improvement for factors of 40, 80, and 160?

7-46. Using (7.20-11) show that the value of x which optimizes $(S_o/N_q)_{\text{DPCM}}$ for a given ratio W_{ch}/W_f is given by

$$\frac{1}{x_{\text{opt}}} = \ln\left[\left(\frac{2^{N_b} - 1}{\sqrt{2}}\right)^{2/3}\left(\frac{\omega_s}{W_f}\right)\right].$$

REFERENCES

[1] Linden, D. A., A Discussion of Sampling Theorems, *Proceedings of the IRE,* July, 1959, pp. 1219–1226.

[2] Black, H. S., *Modulation Theory,* McGraw-Hill Book Co., New York, 1953.

[3] Sakrison, D. J., *Communication Theory: Transmission of Waveforms and Digital Information,* John Wiley & Sons, New York, 1968.

[4] Kohlenberg, A., Exact Interpolation of Band-limited Functions, *Journal of Applied Physics,* December, 1953, pp. 1432–1436.

[5] Petersen, D. P., and Middleton, D., Sampling and Reconstruction of Wave-Number-Limited Functions in N-Dimensional Euclidean Spaces, *Information and Control,* Vol. 5, 1962, pp. 279–323.

[6] Ragazzini, J. R., and Franklin, G. F., *Sampled-Data Control Systems,* McGraw-Hill Book Co., New York, 1958.

[7] Taub, H., and Schilling, D. L., *Principles of Communication Systems,* McGraw-Hill Book Co., New York, 1971.

[8] Carlson, A. B., *Communication Systems: An Introduction to Signals and Noise in Electrical Communication,* McGraw-Hill Book Co., New York, 1968. See also the Second Edition, 1975.

[9] Rowe, H. E., *Signals and Noise in Communication Systems,* D. Van Nostrand Co., Princeton, New Jersey, 1965.

[10] Friedman, B., *Principles and Techniques of Applied Mathematics,* John Wiley & Sons, New York, 1956.

[11] Wozencraft, J. M., and Jacobs, I. M., *Principles of Communication Engineering,* John Wiley & Sons, New York, 1965.

ISBN 0-201-05758-1

[12] Blachman, N., The SNR Threshold in PPM Reception, *IEEE Transactions on Communications,* Vol. COM-22, No. 8, August, 1974, pp. 1094–1098.

[13] Schwartz, M., *Information Transmission, Modulation, and Noise,* Second Edition, McGraw-Hill Book Co., New York, 1970.

[14] Bennett, W. R., *Introduction to Signal Transmission,* McGraw-Hill Book Co., New York, 1970.

[15] de Jager, F., Deltamodulation, A Method of P.C.M. Transmission Using the 1-Unit Code, *Phillips Research Reports,* Vol. 7, 1952, pp. 442–466.

[16] Johnson, F. B., Calculating Delta Modulator Performance, *IEEE Transactions on Audio and Electroacoustics,* Vol. AU-16, No. 1, March, 1968, pp. 121–129.

[17] Schindler, H. R., Delta Modulation, *IEEE Spectrum,* Vol. 7, No. 10, October, 1970, pp. 69–78.

[18] Schindler, H. R., Linear, Nonlinear, and Adaptive Delta Modulation, *IEEE Transactions on Communications,* Vol. COM-22, No. 11, November, 1974, pp. 1807–1823.

[19] Hawkes, T. A., and Simonpieri, P. A., Signal Coding Using Asynchronous Delta Modulation, *IEEE Transactions on Communications,* Vol. COM-22, No. 3, March, 1974, pp. 346–348.

[20] Protonotarios, E. N., Slope Overload Noise in Differential Pulse Code Modulation Systems, *Bell System Technical Journal,* Vol. 46, No. 9, November 1967, pp. 2119–2161.

[21] Goodman, D. J., Delta Modulation Granular Quantizing Noise, *Bell System Technical Journal,* Vol. 48, No. 5, May–June, 1969, pp. 1197–1218.

[22] French, N. R., and Steinberg, J. C., Factors Governing the Intelligibility of Speech Sounds, *Journal of the Acoustical Society of America,* Vol. 19, January, 1947, pp. 90–119.

[23] O'Neal, J. B., Jr., Delta Modulation Quantizing Noise Analytical and Computer Simulation Results for Gaussian and Television Input Signals, *Bell System Technical Journal,* Vol. 45, No. 1, January, 1966, pp. 117–141.

[24] Abate, J. E., Linear and Adaptive Delta Modulation, *Proceedings of the IEEE,* Vol. 55, No. 3, March, 1967, pp. 298–308.

[25] Greenstein, L. J., Slope Overload Noise in Linear Delta Modulators with Gaussian Inputs, *Bell System Technical Journal,* Vol. 52, No. 3, March, 1973, pp. 387–421.

[26] Inose, H., and Yasuda, Y., A Unity Bit Encoding Method by Negative Feedback, *Proceedings of the IEEE,* Vol. 51, November, 1963, pp. 1524–1535.

[27] Jayant, N. S., Adaptive Delta Modulation with a One-Bit Memory, *Bell System Technical Journal,* Vol. 49, No. 3, March, 1970, pp. 321–342.

[28] Tomozawa, A., and Kaneko, H., Companded Delta Modulation for Telephone Transmission, *IEEE Transactions on Communication Technology,* Vol. COM-16, No. 1, February, 1968, pp. 149–157.

[29] Chakravarthy, C. V., and Faruqui, M. N., Two Loop Adaptive Delta Modulation Systems, *IEEE Transactions on Communications,* Vol. COM-22, No. 10, October, 1974, pp. 1710–1713.

[30] Cartmale, A. A., and Steele, R., Calculating the Performance of Syllabically Companded Delta-Sigma Modulators, *Proceedings of the IEE* (London), Vol. 117, No. 10, October, 1970, pp. 1915–1921.

[31] Brolin, S. J., and Brown, J. M., Companded Delta Modulation for Telephony, *IEEE Transactions on Communication Technology,* Vol. COM-16, No. 1, February, 1968, pp. 157–162.

[32] Zetterberg, L. H., and Uddenfeldt, J., Adaptive Delta Modulation with Delayed Decision, *IEEE Transactions on Communications,* Vol. COM-22, No. 9, September, 1974, pp. 1195–1198.

[33] Jayant, N. S., Digital Coding of Speech Waveforms: PCM, DPCM, and DM Quantizers, *Proceedings of the IEEE,* Vol. 62, No. 5, May, 1974, pp. 611–632.

[34] Greefkes, J. A., A Digitally Controlled Delta Codec for Speech Transmission, *Conference Record, 1970 IEEE International Conference on Communications,* pp. 7-33 to 7-48.

[35] Greefkes, J. A., and Riemens, K., Code Modulation with Digitally Controlled Companding for Speech Transmission, *Philips Technical Review,* Vol. 31, No. 11/12, 1970, pp. 335–353.

[36] Greefkes, J. A., and de Jager, F., Continuous Delta Modulation, *Phillips Research Reports,* R664, Vol. 23, 1968, pp. 233–246.

[37] van de Weg, H., Quantizing Noise of a Single Integration Delta Modulation System with an N-Digit Code, *Phillips Research Reports,* Vol. 8, 1953, pp. 367–385.

[38] Chang, K-Y, and Donaldson, R. W., Analysis, Optimization, and Sensitivity Study of Differential PCM Systems Operating on Noisy Communication Channels, *IEEE Transactions on Communications,* Vol. COM-20, No. 3, June, 1972, pp. 338–350.

[39] Bayless, J. W., Campanella, S. J., and Goldberg, A. J., Voice Signals: Bit-by-Bit, *IEEE Spectrum,* Vol. 10, No. 10, October, 1973, pp. 28–34.

[40] Steele, R., *Delta Modulation Systems,* John Wiley & Sons, New York, 1975.

ISBN 0-201-05758-1

Chapter 8

Carrier Modulation by Digital Signals

8.0. INTRODUCTION

Digital signals, such as the PCM, DM, ADM, D-SM, and DPCM waveforms of the preceding chapter, must sometimes be transmitted over a channel requiring carrier modulation. For example, it would be impractical to attempt direct transmission of a time-multiplexed PCM signal between earth and some distant planet, satellite, or space station. The main reason, of course, is that antennas of practical size do not radiate efficiently in the baseband spectral region occupied by such signals. The solution is to translate the information signal spectrum to a higher frequency region by carrier modulation so that the antenna becomes efficient and its size is reasonable.

This chapter is dedicated to the study of carrier modulation by digital signals. We shall restrict our attention exclusively to binary waveforms. For additional information on multilevel (or μ-ary) systems, the reader is referred to some of the literature [1-5] as a starting point.

Three fundamental methods exist for carrier modulation. They are *amplitude, phase,* and *frequency modulation.* As one might expect from our past experience, there are many variations possible in implementing systems. Some of the most important of these are subsequently studied. First, however, we look at what is required to *optimally* recover the modulating waveform from the modulated carrier regardless of the type of modulation. The work is useful in understanding various system implementations.

8.1. OPTIMUM SIGNAL RECEPTION BY MATCHED FILTERS

The *matched filter* was introduced in Chapter 4 (Section 4.12). Recall that it was the filter that produced the largest possible ratio of output signal peak power, at some selectable instant in time, to total output noise power. This optimum signal-to-noise ratio could be achieved for any Fourier transformable input signal* and, in principle, any input noise power density spectrum. Such filters are especially attractive for processing digital signals because we are only interested in determining at the receiver whether a code zero or code one was sent by the transmitter. In a practical sense, this determination is made by sampling the receiver output voltage at one instant in time for each digital symbol transmitted and deciding which of the two possible levels was received. With a matched filter the sample time corresponds to the time of maximum output signal-to-noise ratio. Therefore, an optimum decision can be made.

Let us define waveforms $s_0(t)$ and $s_1(t)$ as those transmitted when a given symbol is a zero or one, respectively. Thus, in any given bit interval either $s_0(t)$ or $s_1(t)$ is impressed upon the

References start on p. 437.

*We refer here to a theoretical solution. A practical implementation requires a realizable solution.

ISBN 0-201-05758-1

receiver input. These *symbol waveforms* will be different for different methods of carrier modulation and may also depend on whether signal shaping is used for intersymbol interference reduction (Section 7.12). For present purposes they may remain arbitrary, although later in considering specific systems they are specialized.

Matched Filters for Digital Modulation

Suppose $s_0(t)$ and $s_1(t)$ invoke responses $s_{r0}(t)$ and $s_{r1}(t)$, respectively, when passed through a receiver filter having a transfer function $H(\omega)$. We shall find the optimum transfer function for processing carrier waveforms modulated by binary digital signals. Furthermore, we determine the optimum relationship that should exist between the two symbol waveforms $s_0(t)$ and $s_1(t)$. Toward these goals, we *define* an optimum *filter* as that which maximizes the difference between the two possible signal outputs according to

$$\left(\frac{S}{N}\right) = \frac{|s_{r1}(t_0) - s_{r0}(t_0)|^2}{N_o} = \text{maximum}, \tag{8.1-1}$$

where t_0 is the time at which the maximum occurs and N_o is the total average noise power emerging from the filter.

The maximization of (8.1-1) follows the same procedures used in Chapter 4 where a signal $f(t)$ caused a filter response $f_o(t)$ and the ratio

$$\left(\frac{S_o}{N_o}\right) = \frac{|f_o(t_0)|^2}{N_o} \tag{8.1-2}$$

was maximized. The optimum filter was found to be

$$H_{\text{opt}}(\omega) = K\frac{F^*(\omega)}{\mathscr{S}_N(\omega)}e^{-j\omega t_0}, \tag{8.1-3}$$

where $F(\omega)$ was the Fourier transform of $f(t)$, K was an arbitrary real constant, and $\mathscr{S}_N(\omega)$ was the power density spectrum of the input noise. The maximum value of (S_o/N_o) was determined to be

$$\left(\frac{S_o}{N_o}\right)_{\text{opt}} = \frac{1}{2\pi}\int_{-\infty}^{\infty}\frac{|F(\omega)|^2}{\mathscr{S}_N(\omega)}\,d\omega. \tag{8.1-4}$$

If we identify $f(t)$ with $s_1(t) - s_0(t)$, then (8.1-1) is identical to (8.1-2) in form. The optimum filter is easily seen by analogy to be

$$H_{\text{opt}}(\omega) = K\frac{[S_1^*(\omega) - S_0^*(\omega)]}{\mathscr{S}_N(\omega)}e^{-j\omega t_0} \tag{8.1-5}$$

from (8.1-3) after substitution of the Fourier transforms $S_0(\omega)$ and $S_1(\omega)$ of $s_0(t)$ and $s_1(t)$. Thus, the optimum *filter* consists of the difference between the optimum filters for the separate symbol waveforms as illustrated in Fig. 8.1-1(*a*). The optimum *receiver* is *defined* as that which samples the filter output at time $t = t_0$ and compares the sample with an optimum threshold voltage (chosen to minimize bit error probability). If larger than the threshold, one symbol was presumed sent (a code one); if lower, the other was assumed transmitted.

ISBN 0-201-05758-1

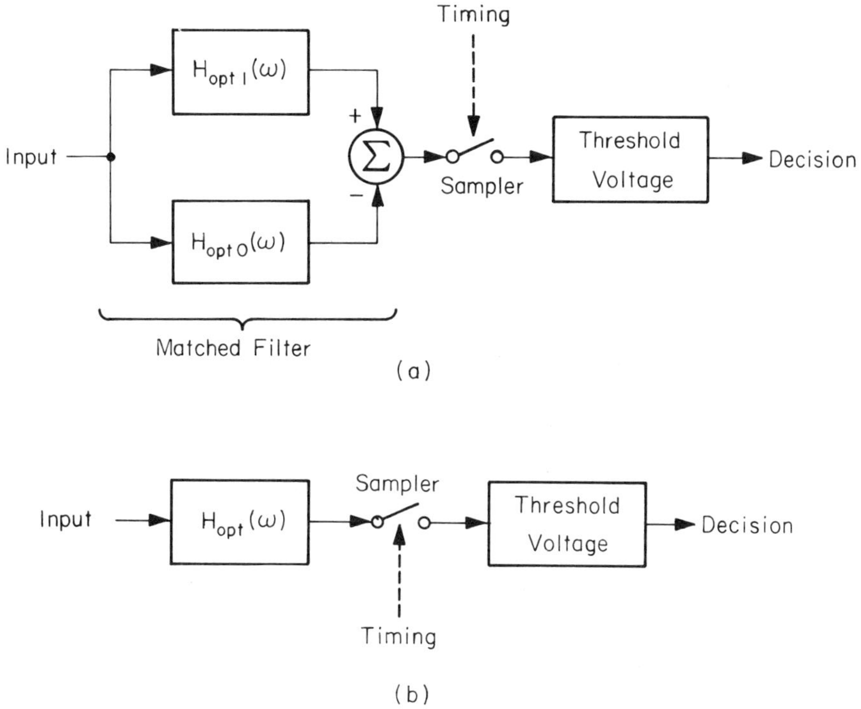

Fig. 8.1-1. Optimum matched filter receivers. (a) For arbitrary symbol waveforms $s_1(t)$ and $s_0(t)$. (b) For the case where $s_0(t)$ is proportional to $-s_1(t)$.

The optimum system of Fig. 8.1-1(*a*) applies for arbitrary $s_0(t)$ and $s_1(t)$. We next show that, for arbitrary $s_1(t)$, we may select $s_0(t)$ to provide the largest possible value of the ratio (S/N) as given by (8.1-1). The ratio is given by (8.1-4) on substitution of $F(\omega) = S_1(\omega) - S_0(\omega)$. It may be written as

$$\left(\frac{S}{N}\right)_{\text{opt}} = \frac{1}{2\pi} \int_{-\infty}^{\infty} \frac{|S_1(\omega)|^2 + |S_0(\omega)|^2}{\mathscr{S}_N(\omega)}\, d\omega$$

$$+ 2\text{Re}\left[\frac{-1}{2\pi} \int_{-\infty}^{\infty} \frac{S_1(\omega)S_0^*(\omega)}{\mathscr{S}_N(\omega)}\, d\omega\right]. \tag{8.1-6}$$

The first integral represents the sum of two energies from Parseval's theorem. One is related to the energy in $s_1(t)$, and one is related to that in $s_0(t)$. The second term, where Re $[\cdot]$ represents real part, may be maximized under a fixed energy constraint by proper choice of $S_0(\omega)$ to make $(S/N)_{\text{opt}}$ attain its largest allowable value. By applying the Schwarz inequality it may readily be shown that the maximum occurs if

$$S_0(\omega) = -K_0 S_1(\omega), \tag{8.1-7}$$

where K_0 is an arbitrary real constant. The optimum filter now becomes

$$H_{\text{opt}}(\omega) = K_1 \frac{S_1^*(\omega)}{\mathscr{S}_N(\omega)} e^{-j\omega t_0} \tag{8.1-8}$$

ISBN 0-201-05758-1

from (8.1-5), where $K_1 = K(1 + K_0)$ is an arbitrary real constant. The optimum receiver now reduces to that shown in Fig. 8.1-1(*b*).

For the important case of a flat noise power spectrum at the receiver input (white noise), (8.1-5) becomes

$$H_{\text{opt}}(\omega) = K[S_1^*(\omega) - S_0^*(\omega)]e^{-j\omega t_0} \tag{8.1-9}$$

when $s_1(t)$ and $s_0(t)$ are arbitrary. Similarly, (8.1-8) will reduce to

$$H_{\text{opt}}(\omega) = K_1 S_1^*(\omega) e^{-j\omega t_0} \tag{8.1-10}$$

for white input noise. These two expressions are useful in next demonstrating that the equivalent of a matched filter receiver for white channel noise is a correlation system.

Correlation Receiver Equivalent of Matched Filter

By inverse transforming (8.1-9) we obtain the optimum filter impulse response for flat (white) noise:

$$h_{\text{opt}}(t) = K[s_1^*(t_0 - t) - s_0^*(t_0 - t)]. \tag{8.1-11}$$

Now the symbol waveforms of principal interest are those that exist only in one bit interval having a duration that we call T_b. If we select the time origin as the start of a bit interval, then some thought will reveal that, since $s_1(t)$ and $s_0(t)$ are nonzero only for $0 < t < T_b$, then $h_{\text{opt}}(t) \neq 0$ only for $t_0 - T_b < t < t_0$, where $T_b \leqslant t_0$ is necessary for realizability. This last condition requires that we wait at least until the complete symbol waveform has entered the receiver before we may expect the output to achieve its maximum. By using the convolution integral and (8.1-11), we may determine the response $s_r(t_0)$ at time t_0 to either of the possible symbol waveforms:

$$s_r(t_0) = K\int_0^{T_b} [s_1^*(u) - s_0^*(u)]\, s_i(u)\, du, \qquad i = 0 \text{ or } 1. \tag{8.1-12}$$

This response is implemented by the correlation receiver of Fig. 8.1-2(*a*) if the arbitrary constant K is set to unity. The necessary integration is easily achieved with an integrate-and-dump circuit as described in Section 7.12.

By retracing the preceding steps it can be shown that, when $s_0(t)$ is proportional to $-s_1(t)$, the optimum filter of (8.1-10) may be implemented by the correlation receiver of Fig. 8.1-2(*b*).

Optimum Threshold Voltage

The above matched filter and correlation receiver discussions were not intended to be rigorous proofs for the forms of optimum receivers. Indeed, a rigorous analysis would also include developments leading to the optimum threshold voltages. These developments require the introduction of hypothesis testing, and multisample statistical decision theory, all considered beyond our scope. Therefore, we only state some important conclusions and refer the reader to the literature for more detail. Simpson and Houts [4] give a simple introduction to the more detailed problem.

ISBN 0-201-05758-1

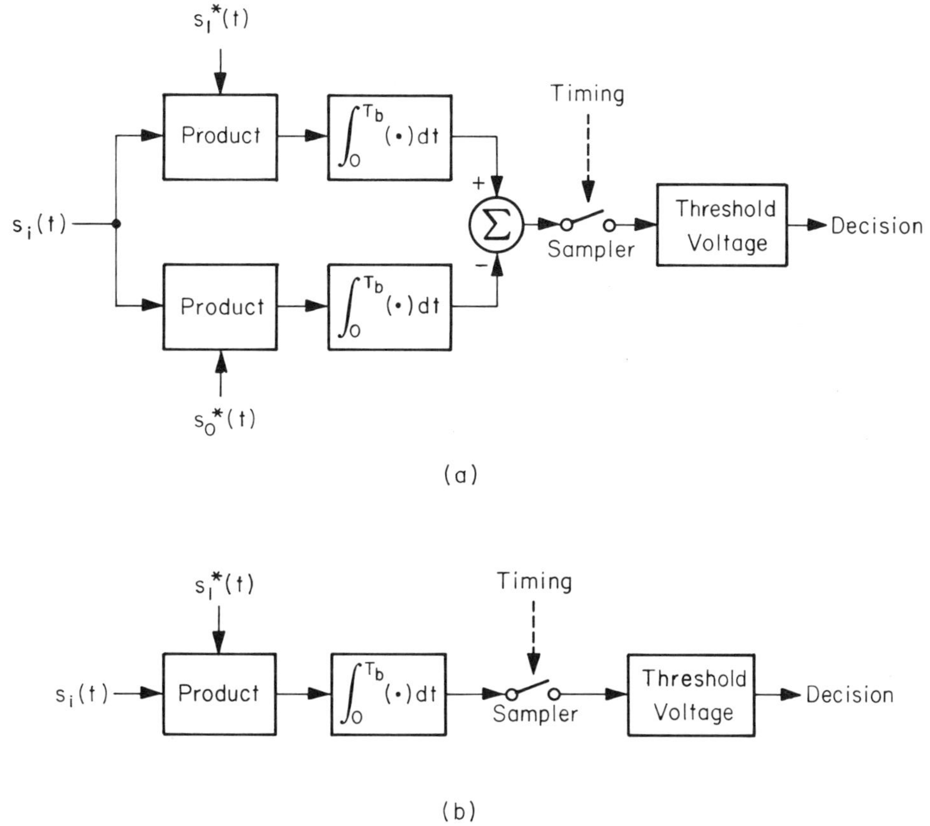

Fig. 8.1-2. Optimum correlation receivers. (a) For arbitrary symbol waveforms $s_1(t)$ and $s_0(t)$. (b) For the case where $s_0(t)$ is proportional to $-s_1(t)$.

For the important situation where input noise is Gaussian (zero mean) with a flat power density spectrum, the optimum threshold voltages for the optimum receivers of Figs. 8.1-1 and 8.1-2 are zero when the two symbol waveforms have equal energies and are equally likely to occur [4, p. 337]. This situation is encountered in many practical systems.

In the more general situation optimum threshold depends on symbol waveform and noise characteristics as well as the criterion adopted as "optimum" in the decision process.

8.2. AMPLITUDE SHIFT KEYING

In binary *amplitude shift keying* (ASK) the ampltude of a carrier is switched between two levels, usually the extremes of full *on* and full *off.* As a result, ASK is sometimes referred to as *on-off keying* (OOK). The on condition might typically correspond to a code one, while off corresponds to a code zero. In this and the following section we examine the OOK version of ASK assuming a rectangular transmitted pulse envelope and no spaces between pulses. The receiving system input noise is assumed to be Gaussian and white.

The optimum receivers of Figs. 8.1-1(*a*) and 8.1-2(*a*) may or may not apply directly to ASK depending on the form of the modulator. If, for *every* code one transmitted, the same waveform is used, then the optimum systems apply. For example, let the time origin be placed at the start of a bit interval and let the interval contain the transmission $A \cos(\omega_0 t + \theta_0)$, $0 < t < T_b$. Then for a code one in bit interval n, $n = 0, \pm 1, \ldots$, the required transmission is

ISBN 0-201-05758-1

$A \cos[\omega_0(t - nT_b) + \theta_0]$. In other words, the symbol waveform $s_1(t)$ must be exactly the same, except for a delay, regardless of what bit interval the signal is used in. These assertions are proved in problem 8-4. They are satisfied when ω_0 is an exact multiple of $2\pi/T_b$. Unfortunately, the ASK modulator is most easily implemented in a second form for which the optimum receivers do not apply without change.

The most simplified form of ASK modulator switches a running carrier on or off. For this type of system it can be shown (see problem 8-5) that samples of the matched filter output, taken at the proper sample times, are not necessarily maximum but depend on carrier phase. In fact, a maximum output may be realized only if the carrier frequency is an exact multiple of the bit rate (which reduces to the case described in the last paragraph). The problem may be corrected by inserting an envelope detector between the matched filter and the sampler to remove the carrier phase variation. The detector may be either coherent or noncoherent, the latter being more common.

System Block Diagrams

A coherent ASK system is illustrated in Fig. 8.2-1. The ASK signal $s_{\text{ASK}}(t)$ plus channel noise is applied to a filter matched to a single-bit carrier pulse. The magnitude of its output component due to signal is as large as possible at the sample time. The product device and wideband low-pass filter* form a synchronous detector. Its signal response is the magnitude (envelope) of the matched filter output due to signal. Its noise response is the in-phase component of the narrowband Gaussian noise emerging from the matched filter. The low-pass filter output is sampled, compared with a voltage threshold, and, if the sample is larger than the threshold, a code one was presumed sent. If it is smaller, a code zero is assumed. The final operation is to reconstruct the original digital message as $\hat{s}_{\text{DIG}}(t)$ which equals the true signal $s_{\text{DIG}}(t)$ plus errors due to noise. Optimum threshold voltage equals half the signal component voltage at the low-pass filter output, taken at the sample time, if equally likely symbol waveforms are assumed.

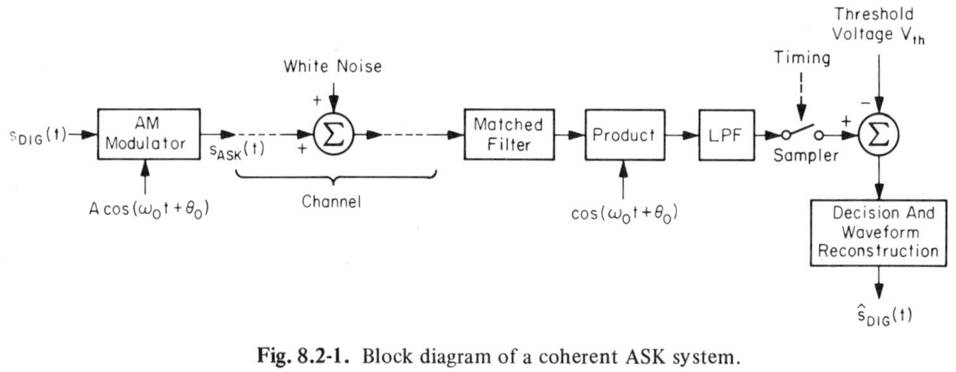

Fig. 8.2-1. Block diagram of a coherent ASK system.

Naturally other configurations of the coherent system are possible. For example, the matched filter can be replaced by a simple band-pass filter (so as to not admit excessive noise to the product device) and the low-pass filter can be made a matched filter (matched to a one-bit video pulse). In still another situation, the sampling operation can follow threshold voltage comparison.

*Bandwidth (radians per second) should be large relative to $2\pi/T_b$ and small relative to ω_0, conditions usually possible in practice.

ISBN 0-201-05758-1

A serious disadvantage of the coherent ASK system is the need to generate a locally coherent carrier. A noncoherent system may be formed, which does not require the local carrier, by replacing the product device and low-pass filter of Fig. 8.2-1 with a simple envelope detector. For a small detector input noise power level the remainder of the system is unchanged. For a large input noise level the optimum threshold voltage changes, as is shown later when noise performance is considered.

The noncoherent system is to be preferred to the coherent one on the basis of equipment simplicity. As we shall subsequently see (Section 8.9), there is only a difference of approximately 1 dB in signal strength required to give similar noise performance in the two systems for the usual noise power levels of interest. The coherent system gives the better performance and requires the least signal power.

Local Carrier Generation for Coherent ASK

The coherent system requires a local carrier having the same phase as the incoming carrier wave. The need for generating this local signal is a main disadvantage of the coherent system. Conceptually, all we require to generate the local signal is a very narrowband band-pass filter centered at frequency ω_0 followed by an amplifier-limiter as shown in Fig. 8.2-2(*a*). Because

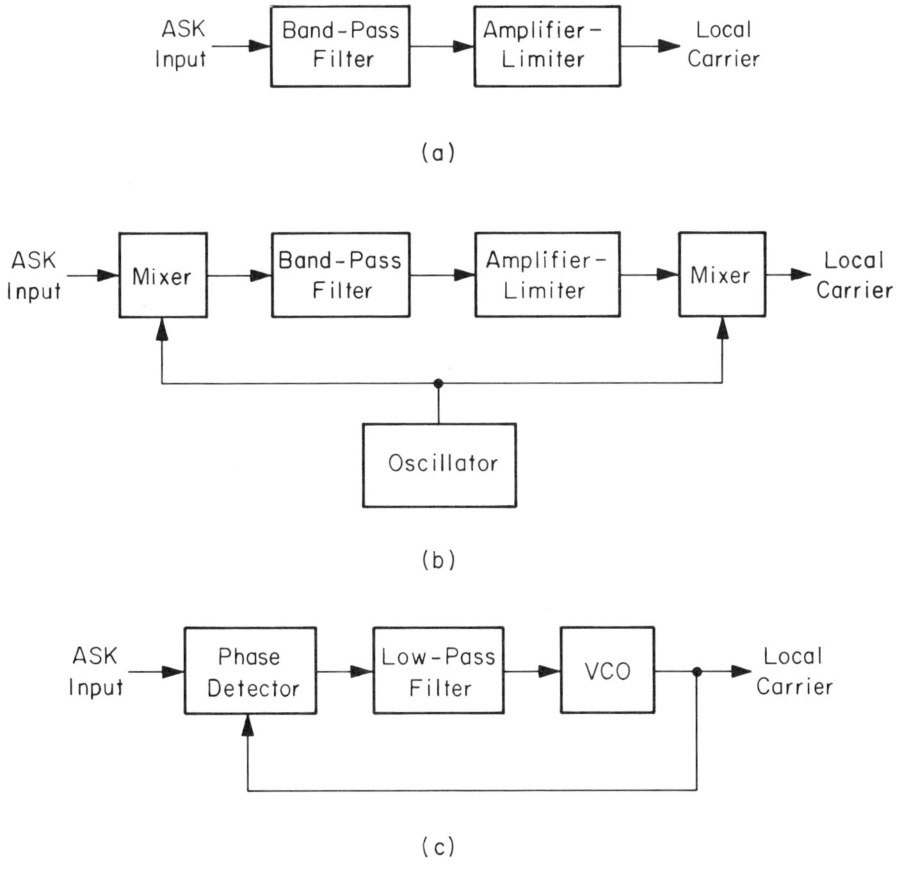

Fig. 8.2-2. Local carrier generation methods for ASK. (a) Using narrowband filter, (b) narrowband filter with mixers for center frequency shifting, and (c) the phase-locked loop.

ISBN 0-201-05758-1

the ASK waveform will contain a large amount of carrier, a filter can select out the carrier and reject information sidebands if its bandwidth is small enough. An amplifier–limiter is then used only to provide a fixed-level output. Implementation of the method may be difficult in practice owing to the need for the very narrow bandwidth filter at large center frequencies. A more practical method would be to first mix down to a lower frequency, use a filter–amplifier–limiter, and then mix back up to the carrier frequency as illustrated in (*b*). Still another scheme uses the phase-locked loop of (*c*); it behaves as a narrow filter at center frequency ω_0.

Signal Spectrum and Channel Bandwidth

Required channel bandwidth is determined by the power density spectrum of the ASK signal. We may treat the signal as standard AM where the carrier modulating signal is a dc voltage plus a random digital message that combine to give 100% modulation. Thus, if A is the peak signal amplitude when transmitting a pulse, then

$$s_{\text{ASK}}(t) = \frac{A}{2}\,[1 + s_{\text{DIG}}(t)]\cos(\omega_0 t + \theta_0), \tag{8.2-1}$$

where $s_{\text{DIG}}(t)$ is a polar waveform having levels of +1 and −1. If the levels are equally likely, it can be shown [6] that the power density spectrum of $s_{\text{DIG}}(t)$ is

$$\mathcal{S}_{\text{DIG}}(\omega) = T_b\,\frac{\sin^2(\omega T_b/2)}{(\omega T_b/2)^2}. \tag{8.2-2}$$

By adding a component $2\pi\delta(\omega)$ to $\mathcal{S}_{\text{DIG}}(\omega)$ to account for the dc component in the bracketed term of (8.2-1), and treating the waveform as a product, it can be shown that the power density spectrum of $s_{\text{ASK}}(t)$ is

$$\mathcal{S}_{\text{ASK}}(\omega) = \frac{\pi A^2}{8}\left\{\delta(\omega - \omega_0) + \delta(\omega + \omega_0) + \frac{T_b}{2\pi}\,\frac{\sin^2[(\omega - \omega_0)T_b/2]}{[(\omega - \omega_0)T_b/2]^2} + \frac{T_b}{2\pi}\,\frac{\sin^2[(\omega + \omega_0)T_b/2]}{[(\omega + \omega_0)T_b/2]^2}\right\}. \tag{8.2-3}$$

From (8.2-3) we find that channel bandwidth, as taken between first nulls in the $\sin^2(x)/x^2$ function centered on ω_0, is $2\omega_b$, where $\omega_b = 2\pi/T_b$. Taken between half-power (−3-dB) points the bandwidth would be $0.886\omega_b$. Between the −3.92-dB points it is ω_b, the bit rate. Thus, minimum channel bandwidth is on the order of one to two times the bit rate ω_b for an ASK system.

8.3. NOISE PERFORMANCE OF ASK

In all the analog modulation systems described in earlier work, the output signal-to-noise power ratio was taken as the measure of performance. In carrier systems modulated by digital signals it is the average probability of making an error in deciding what symbol was transmitted that is most important. Indeed, even for digital baseband signals, such as PCM, this *bit error probability* P_e may be used as a performance measure. In this section we determine P_e

ISBN 0-201-05758-1

for both coherent and noncoherent ASK systems when transmitted pulses are assumed to have a rectangular envelope and input noise is Gaussian and white.

Error Probability of Coherent ASK

The important quantities required to calculate P_e are: (1) the voltage level due to signal at the input to the sampler after a carrier pulse has arrived, and (2) the distributions of noise at the sampler for bit intervals with, and without, a carrier pulse. We only need to know these quantities at the sample time. The applicable system is illustrated in Fig. 8.2-1. Because the detector is coherent, its responses to signal and noise may be found separately.

For carrier pulses of peak amplitude A at the matched filter input, it is readily found that the voltage V at the sampler due to a carrier pulse (at the sampling time) is

$$V = A/2, \tag{8.3-1}$$

if the matched filter impulse response is assumed to be

$$\begin{aligned} h_{\text{opt}}(t) &= \frac{2}{T_b}\cos(\omega_0 t), \qquad 0 < t < T_b \\ &= 0, \qquad \text{elsewhere.} \end{aligned} \tag{8.3-2}$$

For noise, we observe that at the product device input the noise $n_i(t)$ is narrowband Gaussian and may be written as

$$n_i(t) = n_c(t)\cos(\omega_0 t + \theta_0) - n_s(t)\sin(\omega_0 t + \theta_0) \tag{8.3-3}$$

from (5.9-4). Here $n_c(t)$ and $n_s(t)$ are in-phase and quadrature-phase low-pass noises having powers equal to the power N_i of $n_i(t)$. That is, $\overline{n_c^2(t)} = \overline{n_s^2(t)} = \overline{n_i^2(t)} = N_i$. The noise power N_i is given by

$$N_i = \mathcal{N}_0/T_b. \tag{8.3-4}$$

We prove this relationship by means of an example.

Example 8.3-1

To obtain (8.3-4) we use pairs 3, 4, and 17 of Table 2.10-1 to obtain the Fourier transform of (8.3-2) as follows:

$$\begin{aligned} H_{\text{opt}}(\omega) &= e^{-j(\omega-\omega_0)T_b/2}\,\frac{\sin[(\omega-\omega_0)T_b/2]}{(\omega-\omega_0)T_b/2} \\ &\quad + e^{-j(\omega+\omega_0)T_b/2}\,\frac{\sin[(\omega+\omega_0)T_b/2]}{(\omega+\omega_0)T_b/2}. \end{aligned}$$

Forming the magnitude squared function and neglecting the cross terms, since $\omega_0 \gg 2\pi/T_b$ usually, we have

ISBN 0-201-05758-1

$$N_i = \frac{\mathcal{N}_0}{4\pi} \int_{-\infty}^{\infty} |H_{\text{opt}}(\omega)|^2 \, d\omega = \frac{\mathcal{N}_0}{4\pi} \int_{-\infty}^{\infty} \frac{\sin^2 [(\omega - \omega_0)T_b/2]}{[(\omega - \omega_0)T_b/2]^2} \, d\omega$$

$$+ \frac{\mathcal{N}_0}{4\pi} \int_{-\infty}^{\infty} \frac{\sin^2 [(\omega + \omega_0)T_b/2]}{[(\omega + \omega_0)T_b/2]^2} \, d\omega.$$

Here $\mathcal{N}_0/2$ is the power spectral density of the white noise at the input to the matched filter. From tables (Appendix B), each of the integrals evaluates to $2\pi/T_b$ giving the desired result (8.3-4).

Next, we recall from Section 4.11 of Chapter 4 that the response of a coherent detector to the noise $n_i(t)$ is $n_c(t)/2$. Thus, the noise power, denoted as σ^2, at the sampler is one-fourth of that given in (8.3-4):

$$\sigma^2 = \mathcal{N}_0/4T_b. \tag{8.3-5}$$

All quantities necessary to define the distributions of voltages at the input to the sampler at any one sample time are now available. When no pulse is received, only Gaussian noise with variance σ^2 is present as illustrated by curve A in Fig. 8.3-1(a). When a pulse is received, the

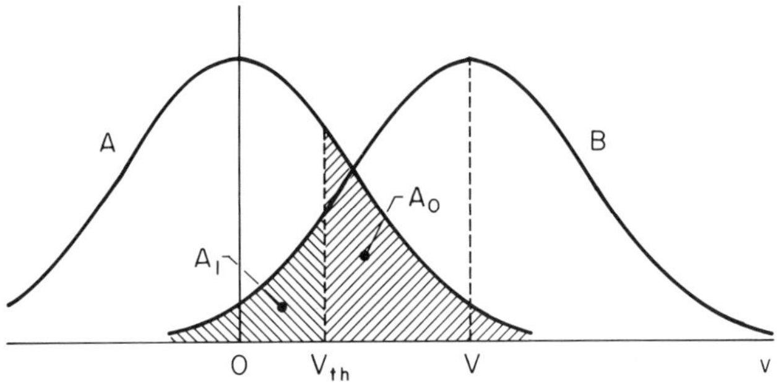

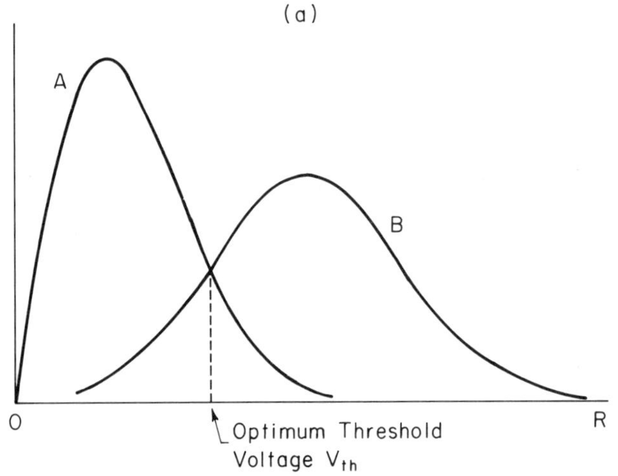

Fig. 8.3-1. Probability density functions applicable to ASK. (a) for coherent system and (b) for non-coherent system.

ISBN 0-201-05758-1

same distribution is shifted to a new mean value V as shown in curve B. For equally likely binary levels, the average error probability is

$$P_e = \frac{1}{2} A_0 + \frac{1}{2} A_1, \tag{8.3-6}$$

where A_0 and A_1 are the areas defined in the figure for the threshold voltage V_{th} shown. For the Gaussian curves we find, after some minor manipulation, that

$$P_e = \frac{1}{2} \int_{V_{th}}^{\infty} \frac{e^{-v^2/2\sigma^2}}{\sqrt{2\pi\sigma^2}}\, dv + \frac{1}{2} \int_{V-V_{th}}^{\infty} \frac{e^{-v^2/2\sigma^2}}{\sqrt{2\pi\sigma^2}}\, dv. \tag{8.3-7}$$

By differentiating these integrals with respect to V_{th}, by means of Leibnitz's rule, and setting the result to zero, we find that the optimum threshold voltage is

$$V_{th} = V/2. \tag{8.3-8}$$

This value corresponds to the intersection point of curves A and B. Notice that V_{th} is independent of σ^2. This fact makes adjustment to maintain optimality a simple matter when the signal level varies owing to, say, a fluctuating channel attenuation. One simple approach is to use an *automatic gain control* (AGC) loop to keep the signal level at the sampler constant in the presence of varying channel attenuation.

For the optimum threshold (8.3-7) reduces to

$$P_e = \frac{1}{2} \operatorname{erfc} (\sqrt{\epsilon/2}\,), \tag{8.3-9}$$

where (8.3-1) and (8.3-5) have been used and

$$\epsilon = A^2 T_b / 4\mathcal{N}_0 \tag{8.3-10}$$

is the ratio of the average energy per bit at the receiver input to twice the noise power density at the input. Equation (8.3-9) is plotted in Fig. 8.9-1.

Error Probability of Noncoherent ASK

P_e in a noncoherent system is more difficult to determine than in the coherent system. Two probability density functions are again required, this time at the matched filter output which is also the input point to the envelope detector. Both may be found by determining the probability density function applicable to the envelope of a sinusoid added to narrowband, bandpass, Gaussian noise. We consider the general problem first and then specialize the result to our ASK system.

Let a sinusoid $V_{mf} \cos (\omega_0 t + \theta_0)$ be added to the narrowband Gaussian noise of (8.3-3) to form the signal $s_{mf}(t)$:

$$s_{mf}(t) = [n_c(t) + V_{mf}] \cos (\omega_0 t + \theta_0) - n_s(t) \sin (\omega_0 t + \theta_0), \tag{8.3-11}$$

which may also be written as

ISBN 0-201-05758-1

$$s_{mf}(t) = x(t)\cos(\omega_0 t + \theta_0) - y(t)\sin(\omega_0 t + \theta_0). \tag{8.3-12}$$

Now since n_c and n_s are statistically independent, zero-mean, Gaussian random variables at time t (see Section 4.11), so are x and y except x has a mean value V_{mf}. The joint density of x and y is therefore the product of individual densities:

$$p(x, y) = \frac{1}{2\pi N_i} \exp\left[-\frac{(x - V_{mf})^2 + y^2}{2N_i}\right], \tag{8.3-13}$$

where N_i is the variance of both n_c and n_s as well as equaling the noise power in the band-pass noise waveform. That is, $N_i = \overline{n_c^2} = \overline{n_s^2}$. Our results, (8.3-12) and (8.3-13), are in terms of rectangular coordinates. We prefer to know the joint density of envelope $R(t)$ and phase component $\psi(t)$ of $s_{mf}(t)$. We then write (8.3-12) in terms of polar variables R and ψ as

$$s_{mf}(t) = R(t)\cos[\omega_0 t + \theta_0 + \psi(t)], \tag{8.3-14}$$

where x and y are related to R and ψ by

$$x(t) = R(t)\cos[\psi(t)], \tag{8.3-15}$$

$$y(t) = R(t)\sin[\psi(t)]. \tag{8.3-16}$$

To transform (8.3-13) to polar variables we equate elemental probabilities $p(x, y)\,dx\,dy = p(R, \psi)\,dR\,d\psi$ and observe that differential areas satisfy $dx\,dy = R\,dR\,d\psi$. Algebraic substitution of (8.3-15) and (8.3-16) into (8.3-13) gives

$$p(R, \psi) = \frac{R}{2\pi N_i} \exp\left[-\frac{R^2 + V_{mf}^2 - 2RV_{mf}\cos(\psi)}{2N_i}\right], \tag{8.3-17}$$

for $0 \leqslant R$ and $0 \leqslant \psi \leqslant 2\pi$. Integrating to obtain the marginal density function of envelope R alone, which we call $p_1(R)$, we obtain

$$p_1(R) = \frac{R}{N_i} e^{-(R^2 + V_{mf}^2)/2N_i} \frac{1}{2\pi} \int_0^{2\pi} e^{(RV_{mf}/N_i)\cos(\psi)}\,d\psi. \tag{8.3-18}$$

The integral is recognized as $2\pi I_0(RV_{mf}/N_i)$, where $I_0(\cdot)$ is the modified Bessel function of the first kind of order zero. Our principal result becomes

$$p_1(R) = \frac{R}{N_i} e^{-(R^2 + V_{mf}^2)/2N_i} I_0\left(\frac{RV_{mf}}{N_i}\right), \qquad 0 \leqslant R, \tag{8.3-19}$$

which is recognized as the Rice distribution (4.5-28). When $V_{mf} = 0$, corresponding to no sinusoid, $I_0(0) = 1$ and the Rice distribution reduces to the Rayleigh density, which we label $p_0(R)$:

$$p_0(R) = \frac{R}{N_i} e^{-R^2/2N_i}, \qquad 0 \leqslant R. \tag{8.3-20}$$

ISBN 0-201-05758-1

Returning to the ASK problem, (8.3-20) is the envelope distribution at the detector when no pulse is received. It is sketched in Fig. 8.3-1(*b*), curve *A*. When a pulse is received, (8.3-19) describes the envelope as sketched in curve *B*. The noise parameter N_i is given by (8.3-4). The signal magnitude V_{mf} *at the time of the sample* may be shown to be

$$V_{mf} = A. \tag{8.3-21}$$

Average error probability becomes

$$P_e = \frac{1}{2}\int_{V_{th}}^{\infty} p_0(R)\,dR + \frac{1}{2}\int_0^{V_{th}} p_1(R)\,dR. \tag{8.3-22}$$

We may differentiate these integrals with respect to V_{th} according to Leibnitz's rule to find V_{th} that gives minimum P_e. We find V_{th} must satisfy

$$\frac{1}{2}\,[-p_0(V_{th}) + p_1(V_{th})] = 0, \tag{8.3-23}$$

which corresponds to the *intersection point* of the two density functions. Reduction of (8.3-23) produces the transcendental equation

$$I_0\left(\frac{V_{th}V_{mf}}{N_i}\right) = \exp\left(\frac{V_{mf}^2}{2N_i}\right), \tag{8.3-24}$$

that the optimum V_{th} must satisfy. An excellent approximation for the solution [7, Chapter 7 by Stein] is

$$V_{th} = \frac{V_{mf}}{2}\left(1 + \frac{8N_i}{V_{mf}^2}\right)^{1/2}. \tag{8.3-25}$$

Notice that optimum threshold depends on noise as well as on signal level. Thus, if signal fluctuates owing to a varying channel attenuation, it is no longer a simple matter to adjust and maintain optimality as in coherent ASK. This is one price that must be paid for the otherwise simplicity of the noncoherent system.

Finally, P_e may be found approximately. We assume a large signal-to-noise ratio such that $V_{th} \approx V_{mf}/2$ and the following approximations may be used:

$$I_0(x) \approx e^x/\sqrt{2\pi x}, \qquad 1 \ll x, \tag{8.3-26}$$

$$\operatorname{erfc}(x) \approx e^{-x^2}/\sqrt{\pi}\,x, \qquad 1 \ll x. \tag{8.3-27}$$

The integrals of (8.3-22) may be evaluated to give

$$P_e \approx \frac{1}{2}\,(1 + \sqrt{1/2\pi\epsilon})e^{-\epsilon/2}, \tag{8.3-28}$$

where

ISBN 0-201-05758-1

$$\epsilon = A^2 T_b / 4\mathcal{N}_0 \tag{8.3-29}$$

is the average energy per bit at the receiver input divided by twice the input noise power density. Fig. 8.9-1 illustrates the behavior of (8.3-28).

8.4. FREQUENCY SHIFT KEYING

Generation of a carrier waveform by binary *frequency shift keying* (FSK) consists in switching between two constant frequency sources with the source selected being dependent on the binary code. Thus, the transmission of a code one may correspond to a pulse of frequency $\omega_0 + \Delta\omega$, while a code zero selects a frequency $\omega_0 - \Delta\omega$. We call $\Delta\omega$ the *peak frequency deviation.* It is analogous to $\Delta\omega$ in analog FM.

System Block Diagrams

The FSK waveform may be viewed as two interlaced ASK signals at different frequencies. Because of this fact, all the problems associated with the way in which the transmitted waveform is generated are the same as in ASK. We shall assume simple switching between two continuous sources $A \cos(\omega_0 t + \Delta\omega t + \theta_1)$ and $A \cos(\omega_0 t - \Delta\omega t + \theta_0)$, where θ_0 and θ_1 are arbitrary phase angles. A typical FSK receiver, which may again be coherent or noncoherent, becomes two ASK receivers in parallel. The noncoherent system block diagram is shown in Fig. 8.4-1.

Allow noise to first be negligible, and assume a pulse of frequency $\omega_0 + \Delta\omega$ is received. The detector output R_1 will provide a nonzero voltage to the sampler while $R_0 = 0$. A pulse of frequency $\omega_0 - \Delta\omega$ will produce the reverse situation except the (same magnitude) voltage is negative at the sampler. After sampling, the decision device simply determines if the sample is positive or negative and generates a digital signal $\hat{s}_{DIG}(t)$. Without noise $\hat{s}_{DIG}(t)$ is an exact replica of the original digital message $s_{DIG}(t)$. For nonzero noise the procedures are the same. The decision device still senses whether samples are positive or negative (for equally likely digital symbols). Now, however, noise may occasionally cause incorrect decisions.

The coherent system is identical to Fig. 8.4-1 except the envelope detectors are replaced by synchronous detectors. Two coherent local oscillators are required as opposed to one in ASK. Local carrier recovery follows the same methods as described for ASK. As we shall subsequently find (see Section 8.9), the coherent system enjoys only about a 1-dB signal power advantage over the noncoherent system for similar noise performance. For this reason the noncoherent system is often preferred for its simplicity.

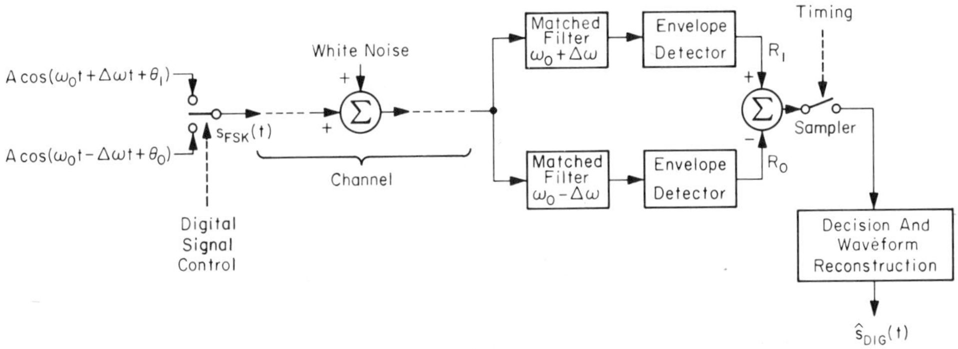

Fig. 8.4-1. Block diagram of a noncoherent FSK system.

ISBN 0-201-05758-1

We note, in concluding this subsection, that FSK could also be implemented using FM at the transmitter and a discriminator at the receiver. This technique does not fare well because of problems in controlling intersymbol interference [1, p. 398]. The approach becomes more feasible for multilevel (μ-ary) systems.

Signal Spectrum and Channel Bandwidth

Because the FSK signal is the same as two interlaced ASK waveforms, we may use (8.2-1) to write

$$s_{FSK}(t) = \frac{A}{2}\,[1 + s_{DIG}(t)]\cos[(\omega_0 + \Delta\omega)t + \theta_1] + \frac{A}{2}\,[1 - s_{DIG}(t)]\cos[(\omega_0 - \Delta\omega)t + \theta_0], \qquad (8.4\text{-}1)$$

where $s_{DIG}(t)$ is the digital message assumed to have equally likely levels +1 and −1. The power density spectrum $\mathscr{S}_{FSK}(\omega)$ of this signal may be found by direct Fourier transformation of its autocorrelation function. Calculations are relatively straightforward if θ_1 and θ_0 are assumed to be uniformly distributed random variables (over 2π), statistically independent of $s_{DIG}(t)$. This is a fair assumption for independent oscillators switched on at arbitrary times. Our result is

$$\mathscr{S}_{FSK}(\omega) = \frac{A^2\pi}{8}\Big[\delta(\omega - \omega_0 - \Delta\omega) + \delta(\omega - \omega_0 + \Delta\omega) + \delta(\omega + \omega_0 - \Delta\omega) + \delta(\omega + \omega_0 + \Delta\omega)\Big] + \frac{A^2 T_b}{16}\Big\{\mathrm{Sa}^2\Big[(\omega - \omega_0 - \Delta\omega)\frac{T_b}{2}\Big] + \mathrm{Sa}^2\Big[(\omega - \omega_0 + \Delta\omega)\frac{T_b}{2}\Big] + \mathrm{Sa}^2\Big[(\omega + \omega_0 - \Delta\omega)\frac{T_b}{2}\Big] + \mathrm{Sa}^2\Big[(\omega + \omega_0 + \Delta\omega)\frac{T_b}{2}\Big]\Big\}. \qquad (8.4\text{-}2)$$

This expression indicates that the power spectrum is just the sum of two ASK spectra symmetrically displaced by amounts $\pm\Delta\omega$ from ω_0.

Minimum signal, and therefore channel, bandwidth is determined by how small $\Delta\omega$ may be made. For the rectangular pulses occupying the entire bit interval, as assumed, it can be shown that $\Delta\omega$ should be a multiple of π/T_b. That is,

$$\Delta\omega = m\pi/T_b = m\omega_b/2, \qquad m = 1, 2, \ldots, \qquad (8.4\text{-}3)$$

where $\omega_b/2\pi$ is the bit rate. The criterion used to arrive at (8.4-3) is zero response at one matched filter output (at the sample time) when a pulse is being received in the other filter.

Figure 8.4-2(a) illustrates the positive frequency portion of (8.4-2) when $\Delta\omega$ has its minimum value. Bandwidths between −3-dB and −6-dB points are $1.94\omega_b$ and $2.27\omega_b$, respectively, while the separation between first nulls is $3\omega_b$. Thus, depending on interpretation, required channel bandwidth is from $1.94\omega_b$ to about $3\omega_b$. Sometimes in coherent FSK

ISBN 0-201-05758-1

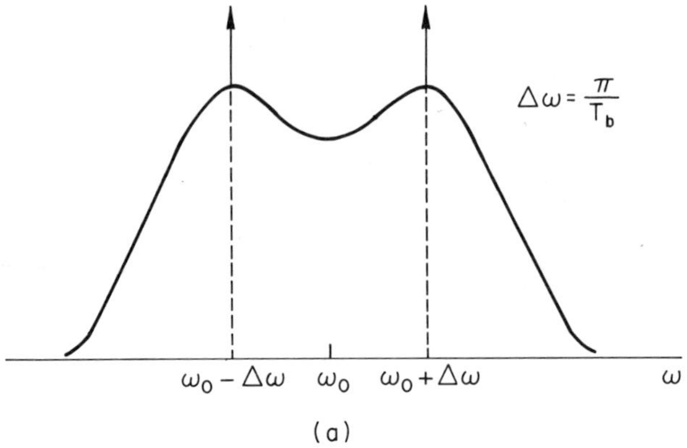

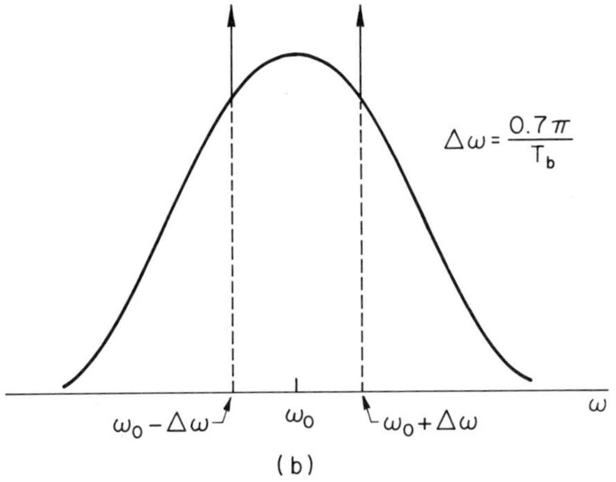

Fig. 8.4-2. Power density spectra for FSK. (a) When $\Delta\omega = \pi/T_b$, and (b) when $\Delta\omega = 0.7\pi/T_b$.

a small amount of coupling between channels can be used to reduce error probability. The optimum for this case is $\Delta\omega = 0.7\pi/T_b$ [7, p. 82], for which the corresponding power spectrum is illustrated in (*b*). Approximate bandwidths for this function at −3-dB, −6-dB, and −10-dB points are $1.41\omega_b$, $1.86\omega_b$, and $2.22\omega_b$, respectively.

8.5. NOISE PERFORMANCE OF FSK

In this section we determine the bit error probability for both coherent and noncoherent binary FSK. In each case we assume rectangular pulses of peak amplitude A arriving at the matched filters and equally likely digital signal levels. Noises emerging from the two filters are assumed to be uncorrelated and Gaussian with zero mean values.

Error Probability of Coherent FSK

The coherent FSK analysis is quite similar to that of ASK, and the expression for P_e will ultimately be the same. For discussion assume a code one is received. The signal response of the corresponding matched filter and coherent detector as seen by the summer prior to sampling, but *at the sampling time,* is

ISBN 0-201-05758-1

$$V = A/2. \tag{8.5-1}$$

The noise response of the same circuitry is a Gaussian, zero-mean noise having average power

$$\sigma^2 = \mathcal{N}_0/4T_b. \tag{8.5-2}$$

Call this noise n_1. The response of the filter and synchronous detector that does not contain the signal pulse passes only noise to the summing point. Call this noise n_0. Its power is also σ^2 given by (8.5-2).

Now the voltage out of the sum, at the sample time, is $V + n_1 - n_0$. An error is made by the decision device if this voltage is less than zero. The probability of such an error is the probability of $V + n_1 - n_0 < 0$. Since n_1 and n_0 are assumed to be uncorrelated Gaussian noises, $n_0 - n_1$ is a Gaussian noise having average power $2\sigma^2$. Define $n = n_0 - n_1$. Error probability becomes the probability that $V < n$. For the alternative case of a code zero bit received, the same condition results. Thus, average bit error probability P_e is the probability of $V < n$. We easily find

$$P_e = \frac{1}{\sqrt{4\pi\sigma^2}} \int_V^\infty e^{-n^2/4\sigma^2}\, dn = \frac{1}{2} \operatorname{erfc}\left(\frac{V}{2\sigma}\right). \tag{8.5-3}$$

On substitution of (8.5-1) and (8.5-2),

$$P_e = \frac{1}{2} \operatorname{erfc}\left(\sqrt{\epsilon/2}\right), \tag{8.5-4}$$

where

$$\epsilon = A^2 T_b/2\mathcal{N}_0 \tag{8.5-5}$$

is the average energy per bit of the input signal divided by twice the input noise power density. P_e is plotted in Fig. 8.9-1.

Example 8.5-1

A coherent FSK system involves a transmitter signal having a peak voltage of 2 volts, a channel with a loss of 105 dB, $\mathcal{N}_0$ at the receiver input of $4.0(10^{-18})$ W/Hz. The bit interval is 1 μs. We determine P_e for this system. The value of A is 105 dB below 2 volts. Thus $A = 2/1.78(10^5) = 1.124(10^{-5})$ volts. From (8.5-5), $\epsilon = 15.8$ or 12.0 dB. Using accurate approximations for erfc (x) given in reference [8] we obtain $P_e = 3.5(10^{-5})$. The approximation (8.3-27) gives $3.7(10^{-5})$ for comparison.

Error Probability of Noncoherent FSK

For noncoherent detection the block diagram of Fig. 8.4-1 applies. Let us assume a pulse is received with frequency $\omega_0 + \Delta\omega$. The distribution of envelope R_1 at the time of the sample will again be described by (8.3-19) with V_{mf} and N_i given by (8.3-21) and (8.3-4), respectively. The response R_0 of the other detector will be described by (8.3-20). An error will

ISBN 0-201-05758-1

result if $R_1 < R_0$. Because the noises at the matched filter outputs are assumed statistically independent, this probability becomes

$$P_e = \int_0^\infty p_1(R_1)\left[\int_{R_0=R_1}^\infty p_0(R_0)\,dR_0\right]dR_1. \tag{8.5-6}$$

These integrals evaluate to [7, p. 298],

$$P_e = \frac{1}{2}\exp[-\epsilon/2], \tag{8.5-7}$$

where ϵ is given by (8.5-5). This result is plotted in Fig. 8.9-1.

Example 8.5-2

We rework example 8.5-1 for the noncoherent case for comparison. Here $P_e = 0.5\exp(-15.8/2) = 1.85(10^{-4})$. The noncoherent system will have a considerably poorer error rate than the coherent system. In fact, it is 5.29 times as large.

If the systems of the above two examples are compared on the basis of equal values of P_e, we find from Fig. 8.9-1 that the coherent system has less than a 1-dB power advantage. This remains true for all values of P_e of practical interest.

8.6. PHASE SHIFT KEYING

In binary *phase shift keying* (PSK) the phase of a carrier is switched between two values according to the two levels of the digital signal. The two values are usually separated by π radians corresponding to the two signal levels. This is the only choice we shall consider. It is sometimes called *phase reversal keying* (PRK).

The effect of phase reversal modulation is to produce a double sideband suppressed-carrier amplitude-modulated waveform where the information signal is digital. Thus, we may write the PSK signal as

$$s_{\text{PSK}}(t) = As_{\text{DIG}}(t)\cos(\omega_0 t + \theta_0), \tag{8.6-1}$$

where $s_{\text{DIG}}(t)$ is +1 or −1 according to the digital code during any bit interval having a duration* T_b, and A is the peak carrier amplitude. If the carrier frequency ω_0 is an exact multiple of the bit rate $\omega_b = 2\pi/T_b$, PSK is an optimum system of the type described in Section 8.1. The waveforms $s_1(t)$ and $s_0(t)$ can be written as

$$\begin{aligned} s_1(t) &= A\sin(\omega_0 t + \theta_0), && 0 < t < T_b \\ &= 0, && \text{elsewhere}, \end{aligned} \tag{8.6-2}$$

$$s_0(t) = -s_1(t), \tag{8.6-3}$$

with $\omega_0 T_b/2\pi$ equal to a positive integer. In this case the receiving system could be implemented by either the matched filter of Fig. 8.1-1(b) or the correlator of Fig. 8.1-2(b) for

*We consider only the case of no spaces.

ISBN 0-201-05758-1

white noise. In either case the optimum threshold voltage is zero for white noise and equally likely symbol waveforms.

System Block Diagram

More generally, the carrier may not be harmonically related to the bit rate, and coherent detectors must be used to remove carrier phase variations at the sample times. The block diagram of an appropriate system is given in Fig. 8.6-1. Up to the sampler, this system is the same as a DSB amplitude modulation system, except we use a matched filter in place of a band-pass filter. As a consequence of this similarity, we conclude that local carrier recovery may be accomplished by the methods described in Section 5.7. The system is also essentially the same as for coherent ASK (Fig. 8.2-1). However, because the information is carried in the phase of the PSK signal, the system must necessarily be coherent. No noncoherent variety is possible in the sense that envelope detectors are used. There is, however, a semicoherent implementation that is described in Section 8.8.

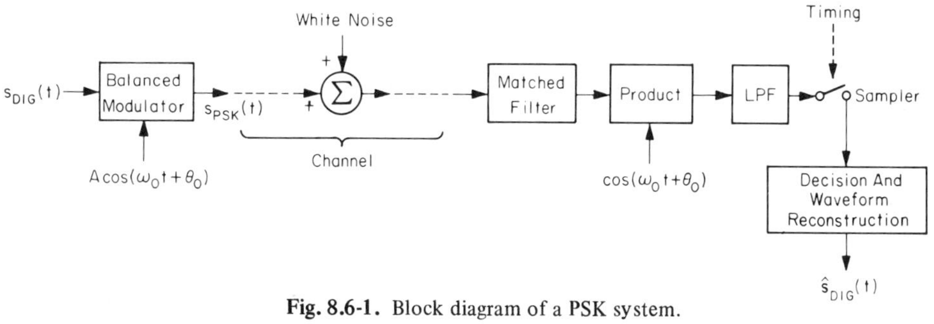

Fig. 8.6-1. Block diagram of a PSK system.

Spectrum and Channel Bandwidth

By treating $s_{\text{DIG}}(t)$ as a random signal with equally likely levels of +1 and −1, we treat (8.6-1) as a product and recall some earlier work of Section 4.11. There the power density spectrum of such a product was found in (4.11-6).* Making use of (8.2-2) we obtain the PSK signal power spectrum

$$\mathscr{S}_{\text{PSK}}(\omega) = \frac{A^2 T_b}{4}\left\{\text{Sa}^2\left[(\omega - \omega_0)\frac{T_b}{2}\right] + \text{Sa}^2\left[(\omega + \omega_0)\frac{T_b}{2}\right]\right\}. \quad (8.6\text{-}4)$$

Taken at −3-dB points the bandwidth of this function is $0.886\omega_b = 5.57/T_b$. Between first nulls it is $2\omega_b$. These values place minimum constraints on channel bandwidth.

8.7. NOISE PERFORMANCE OF PSK

One of the main advantages of PSK is its superior noise performance. As will be seen (see Fig. 8.9-1), for the same average probability of bit error PSK requires about 3 to 4 dB less power than any of the ASK or FSK systems that have been discussed. The somewhat universal rule of having to pay for what you get applies to PSK, however, in the form of the complexity of coherent detection.

*Equation (4.11-6), as derived, applies only to a stationary random input signal. However, it can be shown to apply to the present case if θ_0 in (8.6-1) is uniformly distributed over 2π radians and is independent of the nonstationary message $s_{\text{DIG}}(t)$.

ISBN 0-201-05758-1

Error Probability

Calculation of average probability P_e of a bit error follows the steps used for the analysis of coherent ASK almost exactly. When a code one is transmitted, the voltage at the sampler input (at the sample time) is the sum of a signal component V and a Gaussian noise, as described by curve B in Fig. 8.3-1(a). The noise power σ^2 is given by (8.3-5), while V is obtained from (8.3-1). A difference in PSK arises when a code zero arrives. The sampler voltage is again described by a curve like B except it is centered at $-V$. The PSK threshold voltage is zero volts if we assume equally likely symbols. A little reflection will show that P_e is given by

$$P_e = \int_0^{\infty} \frac{1}{\sqrt{2\pi\sigma^2}} \exp\left[-\frac{(v + V)^2}{2\sigma^2}\right] dv, \qquad (8.7\text{-}1)$$

which reduces to

$$P_e = \frac{1}{2} \operatorname{erfc}\,(\sqrt{\epsilon}). \qquad (8.7\text{-}2)$$

Here ϵ is given by (8.5-5). To gain some feeling for numbers we take an example. More general behavior of P_e can be seen from the PSK curve plotted in Fig. 8.9-1.

Example 8.7-1

We calculate P_e for the system parameters described in example 8.5-1 for coherent FSK:

$$P_e = \frac{1}{2} \operatorname{erfc}\,(\sqrt{15.8}) = 9.75(10^{-9}).$$

This value is smaller than that which the similar FSK system produces by a factor of more than 3500.

Effects of Phase Errors

Since PSK is a coherent system, a local carrier must be reconstructed in the receiver. If errors exist in the local carrier phase, a degradation in error probability occurs. Effects of local phase errors in coherent detectors were studied in some detail in Chapter 5 [see (5.4-6)]. Specifically, it was found that signal level at the detector output decreases from its maximum by a factor cos (ϕ_e), where ϕ_e is the phase error. Inserting this factor into (8.7-2) gives

$$P_e = \frac{1}{2} \operatorname{erfc}\left(\sqrt{\epsilon \cos^2(\phi_e)}\right). \qquad (8.7\text{-}3)$$

Because erfc (x) is a monotonically decreasing function of increasing x, P_e is larger when phase error occurs. For example, a fixed error of 27 degrees or less is required to keep ϵ within 1 dB of its maximum.

If phase error is a random quantity with a probability density function $p(\phi_e)$, the average error probability, denoted as $\overline{P}_e$, is found by averaging P_e as given by (8.7-3):

ISBN 0-201-05758-1

$$\overline{P_e} = \int_{-\pi}^{\pi} P_e p(\phi_e)\, d\phi_e. \tag{8.7-4}$$

Numerical methods may be required to evaluate $\overline{P}_e$ in many cases.

8.8. DIFFERENTIAL PHASE SHIFT KEYING

An ingenious technique called *differential phase shift keying* (DPSK) employs special preprocessing of the digital signal prior to PSK modulation of the carrier such that no local carrier is required in the receiver. The entire demodulation process may take place using the received signal alone.

System Block Diagram

Figure 8.8-1 is a block diagram of one form of DPSK system. The preprocessing amounts to converting the binary digital signal $s_{\rm DIG}(t)$ to the bit stream $b(t)$. For purposes of discussion, assume $s_{\rm DIG}(t)$ is a polar waveform as illustrated in Fig. 8.8-2(a). A system is also possible for unipolar signals. The digital message is multiplied by a one-bit-interval delayed version $b(t - T_b)$ of $b(t)$. Prior to the start of $s_{\rm DIG}(t)$ (at $t = 0$ in the figure) the value of $b(t - T_b)$ is arbitrary; +1 is assumed in the example waveforms. During any bit interval after $s_{\rm DIG}(t)$ is applied, $b(t)$ is determined by multiplying $s_{\rm DIG}(t)$ by $b(t)$ in the previous interval. The result is a bit stream $b(t)$ which assumes a level different from its previous level only when $s_{\rm DIG}(t)$ is at level −1. The bit stream $b(t)$ modulates the carrier as in ordinary PSK.

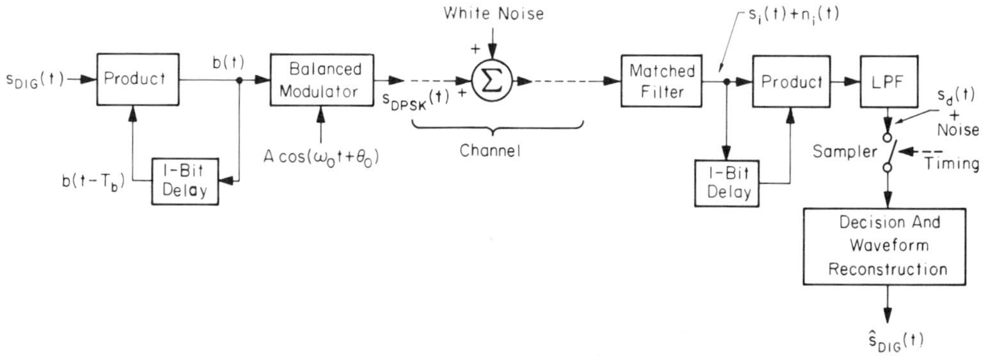

Fig. 8.8-1. Block diagram of a DPSK system.

In the receiver the key to operation is the synchronous detector (product and low-pass filter). *Both* inputs to the detector are derived from the output of a filter matched to the pulse of a single-bit interval. Let $s_i(t)$ be the signal component of the filter output, and let $s_d(t)$ represent the signal component at the detector output. The action of the detector is such that $s_d(t)$ is a positive voltage if the phases of both $s_i(t)$ and $s_i(t - T_b)$ are the same—that is, have like phases. On the other hand, if the phases of $s_i(t)$ and $s_i(t - T_b)$ differ by π radians, $s_d(t)$ will be a negative voltage. At the sample times these voltages will have maximum amplitude. Thus, the *sign* of the voltage at the sampler depends on the phase relationship between

ISBN 0-201-05758-1

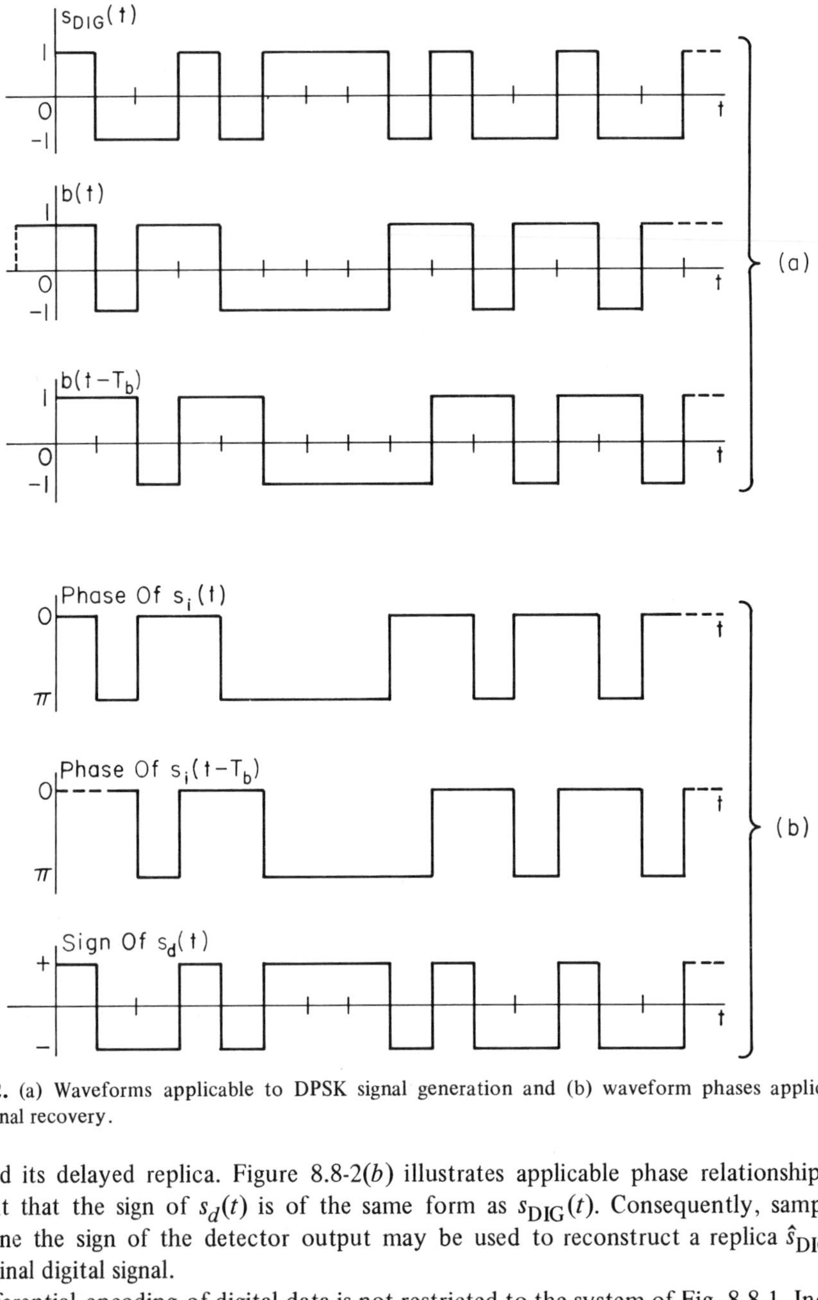

Fig. 8.8-2. (a) Waveforms applicable to DPSK signal generation and (b) waveform phases applicable to DPSK signal recovery.

$s_i(t)$ and its delayed replica. Figure 8.8-2(*b*) illustrates applicable phase relationships. It is apparent that the sign of $s_d(t)$ is of the same form as $s_{DIG}(t)$. Consequently, sampling to determine the sign of the detector output may be used to reconstruct a replica $\hat{s}_{DIG}(t)$ of the original digital signal.

Differential encoding of digital data is not restricted to the system of Fig. 8.8-1. Indeed, it may be used for purposes other than local carrier elimination. For example, in the ordinary PSK system of Fig. 8.6-1, $s_{DIG}(t)$ could be a differentially encoded waveform. The decoder would be placed at the receiver output to operate on $\hat{s}_{DIG}(t)$; it would appear as shown in Fig. 8.8-3. The careful reader will ask, Why add the complexity to the PSK system? The answer is that it relieves a principal problem in local carrier recovery. Various recovery schemes suffer a phase ambiguity of π radians. Even if carrier synchronism is initially established, various factors may occasionally cause local phase to slip by π radians, resulting in a sign inversion in the

ISBN 0-201-05758-1

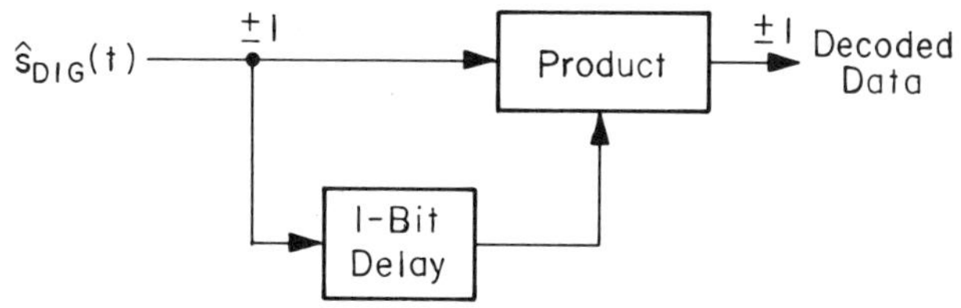

Fig. 8.8-3. Decoder for a differentially coded digital waveform.

recovered data. With differential encoding, the output of the decoder of Fig. 8.8-3 will be correct even in the presence of a sign change in $\hat{s}_{DIG}(t)$.

Noise Performance of DPSK

The analysis necessary to find the average probability P_e of bit error is somewhat messy. An excellent discussion of details is given by Stein [7, p. 306]. A somewhat simpler discussion, using Stein's approach, is also given by Carlson [1, p. 401]. Omitting details, we have

$$P_e = \frac{1}{2} \exp(-\epsilon), \tag{8.8-1}$$

where

$$\epsilon = A^2 T_b / 2\mathcal{N}_0. \tag{8.8-2}$$

Here A is the peak amplitude of the DPSK signal and $\mathcal{N}_0$ is twice the input noise power density. Equation (8.8-1) is plotted in Fig. 8.9-1.

Oberst and Schilling [9] have studied the performance of several self-synchronized PSK systems which use differential data encoding. These were DPSK, *decision feedback* PSK (a system combining signals received during the q previous bit intervals to derive an estimate of carrier phase), *phase-doubling* PSK [one form uses the circuit of Fig. 5.7-1(c) to develop the local carrier], and an optimum system among all systems that do not use past information. It was found that (8.8-1) is a tight upper bound on P_e for these systems. A tight lower bound also exists such that

$$\operatorname{erfc}(\sqrt{\epsilon})\left[1 - \frac{1}{2}\operatorname{erfc}(\sqrt{\epsilon})\right] \leqslant P_e \leqslant \frac{1}{2}\exp(-\epsilon). \tag{8.8-3}$$

8.9. COMPARATIVE PERFORMANCE OF DIGITAL MODULATION METHODS

In all systems of carriers keyed by digital signals, the affect of random noise is to cause occasional bits to be in error in the reconstructed digital signal. Indeed, this is the case even for direct transmission of PCM. Thus, the important measure to be used for comparison of noise performance of these systems is the probability P_e of a bit error.

The various expressions that have been developed have been in terms of ϵ, the average transmitted energy per bit divided by twice the input noise power density to the receiver. It forms a reasonable basis for comparing performance of the various systems. We bring together the applicable formulas for easy reference:

ISBN 0-201-05758-1

$$P_e = \frac{1}{2}\,\mathrm{erfc}\,(\sqrt{\epsilon}), \qquad \text{PCM (polar)}, \qquad (8.9\text{-}1)$$

$$P_e = \frac{1}{2}\,\mathrm{erfc}\,(\sqrt{\epsilon/2}), \qquad \text{PCM (unipolar)}, \qquad (8.9\text{-}2)$$

$$P_e = \frac{1}{2}\,\mathrm{erfc}\,(\sqrt{\epsilon/2}), \qquad \text{ASK (coherent)}, \qquad (8.9\text{-}3)$$

$$P_e \approx \frac{1}{2}\left[1 + \frac{1}{\sqrt{2\pi\epsilon}}\right]\exp(-\epsilon/2), \qquad \text{ASK (noncoherent)}, \qquad (8.9\text{-}4)$$

$$P_e = \frac{1}{2}\,\mathrm{erfc}\,(\sqrt{\epsilon/2}), \qquad \text{FSK (coherent)}, \qquad (8.9\text{-}5)$$

$$P_e = \frac{1}{2}\exp(-\epsilon/2), \qquad \text{FSK (noncoherent)}, \qquad (8.9\text{-}6)$$

$$P_e = \frac{1}{2}\,\mathrm{erfc}\,(\sqrt{\epsilon}), \qquad \text{PSK}, \qquad (8.9\text{-}7)$$

$$P_e = \frac{1}{2}\exp(-\epsilon), \qquad \text{DPSK}. \qquad (8.9\text{-}8)$$

Figure 8.9-1 shows plots of the above expressions suitable for comparing performances. In the region of usual interest, $P_e \leqslant 10^{-4}$, several facts become clear. For direct transmission (no carrier modulation) polar PCM is better than unipolar PCM by a factor of 2 (or 3 dB).* None of the carrier modulation systems gives better performance than direct polar PCM transmission, although performance is equaled by a PSK system. DPSK is the next best system, requiring about 1 dB or less of additional power for equal values of P_e. Thus, the PSK systems outperform those based on frequency or amplitude.

Of the FSK and ASK systems the coherent versions give equal performance. By the same token, the noncoherent versions are almost identical for $P_e \leqslant 10^{-4}$. The coherent systems provide at most a 1-dB power saving over the noncoherent versions for $P_e \leqslant 10^{-4}$. The best FSK or ASK system is only as good as direct unipolar PCM.

PROBLEMS

8-1. Apply Fig. 8.1-1(*a*) to the reception of a polar PCM signal (no carrier modulation) and find the filter transfer functions $H_{\text{opt}1}(\omega)$ and $H_{\text{opt}0}(\omega)$. Assume white noise, rectangular pulses, and no spaces.

8-2. Explain how Fig. 8.1-1(*a*) changes if the received signal is a unipolar PCM signal (no carrier modulation).

8-3. Explain why the product operations in Fig. 8.1-2 can be eliminated if the received signal is either a polar or a unipolar PCM signal (no carrier modulation) having perfectly rectangular symbol waveforms.

*Signal power is proportional to ϵ so that comparison of required ϵ is equivalent to comparing required power.

ISBN 0-201-05758-1

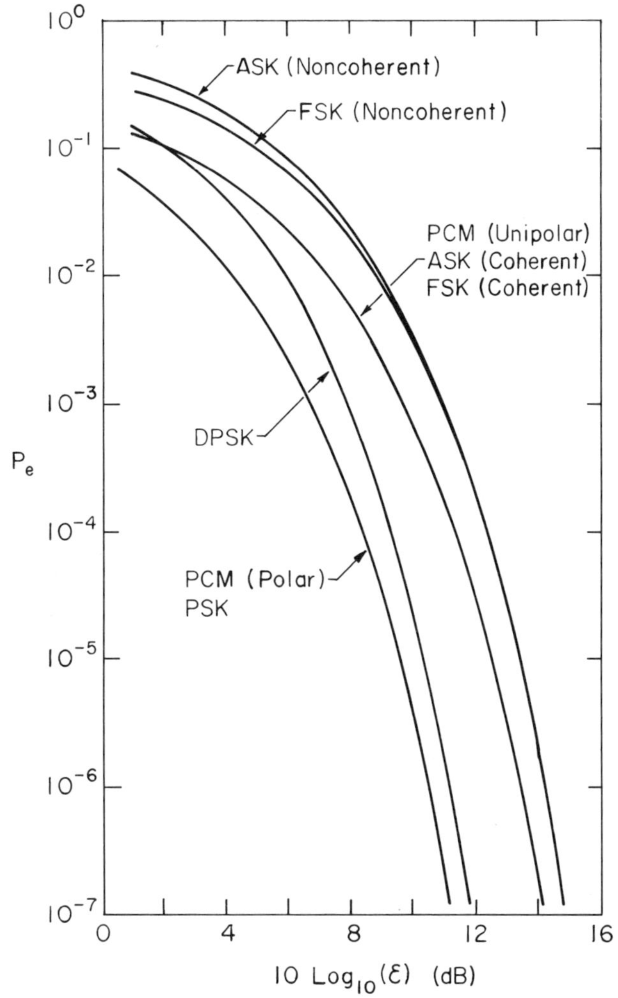

Fig. 8.9-1. Bit error probability P_e for various digital systems.

★ 8-4. Let $s_1(t) = A \cos [\omega_0(t - nT_b) + \theta_0]$, $nT_b < t < (n + 1)T_b$, be the symbol waveform in ASK when a code one occurs in bit interval n, $n = 0, \pm 1, \pm 2, \ldots$, and zero elsewhere. Show that Fig. 8.1-1(a) applies to ASK by proving that the samples taken at times $t = (n + 1)T_b$ are constants equal to the maximum possible value A. Assume white input noise, a mid-band gain of unity for the optimum filter, and $\omega_0 >> 2\pi/T_b$.

★ 8-5. Assume an ASK receiver of the form of Fig. 8.1-1 is designed for a symbol waveform $s_1(t) = A \cos (\omega_0 t + \theta_0)$, $0 < t < T_b$ (zero elsewhere). If the transmitter sends a waveform $s_i(t) = A \cos (\omega_0 t + \theta_0)$, $nT_b < t < (n+1)T_b$, in bit interval n, where $n = 0, \pm 1, \ldots$, show that the sample of the matched filter response $s_r(t)$ taken at time $t = (n + 1)T_b$ is $s_r[(n + 1)T_b] = A \cos (\omega_0 nT_b)$. Assume white input noise, a mid-band gain of unity for the optimum filter, and $\omega_0 >> 2\pi/T_b$.

★ 8-6. For the filter specified by (8.3-2), prove (8.3-1) for a coherent ASK system. Assume ω_0 and the low-pass filter's bandwidth are large with respect to $2\pi/T_b$.

8-7. In a coherent ASK system the probability of any one bit's being a pulse is p and the probability of no pulse is $(1 - p)$. What threshold voltage should the receiver use to minimize average bit error probability? Assume white, Gaussian channel noise.

ISBN 0-201-05758-1

★ 8-8. Use the approximations given in (8.3-26) and (8.3-27) to show that P_e in noncoherent ASK is given approximately by (8.3-28).

8-9. By using the approximation (8.3-27) for the complimentary error function, derive an expression for the average bit error probability of a noncoherent ASK system in terms of the error probability of a coherent ASK system if both systems have similar parameters (same ϵ). If $\epsilon = 23.8$ for both systems, what are the two error probabilities? Check these values with Fig. 8.9-1.

8-10. An ASK system does not go to full off when a code zero is transmitted. If the two peak carrier levels are A_1 and A_0, obtain an expression for the modulated signal power density spectrum.

8-11. If the ASK system in problem 8-10 is coherent what will the optimum threshold voltage be if the receiver contains a matched filter? Obtain an expression for P_e if channel noise is white and Gaussian.

8-12. If the ASK system of problem 8-10 is noncoherent, explain how the probability density functions of Fig. 8.3-1(b) change. Prove that the optimum threshold changes from that of an OOK system but is still at the intersection point of the two density functions if symbols are equally likely.

8-13. A coherent ASK system transmits with a peak voltage of 5 volts over a channel having an unknown loss. If $\mathcal{N}_0 = 6.0(10^{-18})$ W/Hz, bit intervals are 0.5 μs in duration and the system performs with $P_e = 10^{-4}$, what is the loss?

8-14. Work problem 8-13 for a noncoherent ASK system.

8-15. Draw a block diagram of the receiver portion of Fig. 8.2-1 to show how an integrate-and-dump matched filter could be used in coherent ASK.

8-16. Write an expression for P_e in coherent ASK when the local carrier phase contains an error ϕ_e.

8-17. Sketch the power density spectrum of an FSK signal when $\Delta\omega = \omega_b$. Compare your result with Fig. 8.4-2.

8-18. Work problem 8-16 for a coherent FSK system where errors on the two local carriers are ϕ_{e0} and ϕ_{e1}.

8-19. Draw a block digram of the receiver of a coherent FSK system where integrate-and-dump matched filtering is used.

8-20. Work problem 8-13 for a coherent FSK system.

8-21. Work problem 8-13 for a noncoherent FSK system.

8-22. In a certain noncoherent FSK system $\epsilon = 20.0$. If the channel suddenly increases 1 dB in attenuation, by what factor will P_e increase? Find the two values of P_e.

8-23. In a PSK system ω_0 is a multiple m of the bit rate $\omega_b = 2\pi/T_b$. If $s_1(t) = A \sin(\pi t/T_b) \sin(\omega_0 t)$, $0 \leqslant t \leqslant T_b$, and $s_0(t) = -s_1(t)$, sketch the block diagram of an integrate-and-dump optimum receiver for white input noise. Find the signal output voltage at the sampler at the sample time.

8-24. Find the transfer function of the matched filter form of receiver for the system of problem 8-23.

8-25. In a PSK system for which $P_e = 10^{-7}$ the signal $s_{\text{DIG}}(t)$ is a polar PCM waveform. Each PSK pulse has a peak amplitude of $A = 10$ volts and is transmitted over a channel where $\mathcal{N}_0 = 3.69(10^{-7})$ W/Hz. How many similar messages may be time multiplexed if message spectral extent is $W_f = \pi(10^4)$ rad/s, sampling is at the Nyquist rate, and five-bit PCM encoding is used?

ISBN 0-201-05758-1

8-26. Find P_e for a PSK system having $\epsilon = 10$. If local carrier phase error causes the signal at the sampler to decrease by 1.5 dB, to what value will P_e change?

8-27. For a DPSK system calculate and plot the upper and lower bounds on P_e, as given by (8.8-3), for $4.0 < \epsilon < 14.0$. Over this range what is the approximate maximum difference in ϵ for the same values of P_e. Use the approximation of (8.3-27) in the calculations.

REFERENCES

[1] Carlson, A. B., *Communication Systems, An Introduction to Signals and Noise in Electrical Communication,* Second Edition, McGraw-Hill Book Co., New York, 1975.

[2] Lucky, R. W., Salz, J., and Weldon, E. J., Jr., *Principles of Data Communications,* McGraw-Hill Book Co., New York, 1968.

[3] Stein, S., and Jones, J. J., *Modern Communication Principles with Application to Digital Signaling,* McGraw-Hill Book Co., New York, 1967.

[4] Simpson, R. S., and Houts, R. C., *Fundamentals of Analog and Digital Communication Systems,* Allyn and Bacon, Boston, Massachusetts, 1971.

[5] Wozencraft, J. M., and Jacobs, I. M., *Principles of Communication Engineering,* John Wiley & Sons, New York, 1965.

[6] Thomas, J. B., *An Introduction to Statistical Communication Theory,* John Wiley & Sons, New York, 1969.

[7] Schwartz, M., Bennett, W. R., and Stein, S., *Communication Systems and Techniques,* McGraw-Hill Book Co., New York, 1966.

[8] Abramowitz, M., and Stegun, I. A. (Editors), *Handbook of Mathematical Functions with Formulas, Graphs, and Mathematical Tables,* National Bureau of Standards Applied Mathematics Series 55, U.S. Government Printing Office, Washington, D.C., 1964.

[9] Oberst, J. F., and Schilling, D. L., Performance of Self-Synchronized Phase-Shift Keyed Systems, *IEEE Transactions on Communication Technology,* Vol. COM-17, No. 6, December, 1969.

ISBN 0-201-05758-1

Chapter 9

System Power Transfer and Sensitivity

9.0. INTRODUCTION

In every communication system it is necessary to transfer the modulated signal from the transmitter (or sender) to the receiver (or user). In some cases this is a relatively simple task involving perhaps only telephone wires, telegraph lines, or coaxial cables. However, many systems, such as microwave telephone and television links, space communication links, and even commercial broadcast systems, require that the transmitter and receiver be coupled by a medium containing air or vacuum (or both). Such systems must, as a consequence, use some kind of antenna as a transducer to couple the two ends of the medium to the transmitter and receiver parts of the system.

One of the main purposes of this chapter is to define the basic principles which govern the power transferred to the output of the receiver. We shall concentrate primarily on microwave systems—that is, systems having carrier frequencies above 300 MHz. In many respects our work is a logical extension of the discussions of Chapter 8, since those principles apply to a large class of present-day systems. For example, a designer would be hard put to create an effective satellite communication link without being able to calculate received signal power and system noise, the two things that we concentrate on here.

Attention is centered mainly on theoretical principles as opposed to circuit design. Thus, various system components become boxes in a suitable system block diagram.

9.1. COMMUNICATION SYSTEM ELEMENTS

Nearly all microwave systems may be broken down into the elements diagramed in Fig. 9.1-1. The purpose of the transmitter is to drive the transmitting antenna with the final carrier modulated signal at the required power level. It may consist of nothing more than a high-power linear radio-frequency (RF) amplifier in some systems. In others, it may actually form part of the modulation process. At any rate, it simply accepts whatever form of information-bearing signal is appropriate to the system and transforms it to the power level P_t. In some cases, such as ASK, we may wish to think of P_t as the *peak* transmitted power. In others, such as in analog modulation, P_t is the average power.

Some means must be provided for the power P_t to flow to the antenna. Typically, the path may contain various microwave components (such as isolators, hybrids, impedance matching structures, phase shifters, diplexers, and rotary joints) coupled together by *RF* transmission lines (such as waveguides, coaxial cables, and strip lines). These items all have attenuation. Whatever the total power loss from the transmitter to antenna terminal, it is called L_t. Power loss is defined as a number larger than unity. Thus, a loss L_t = 2.0 or 3 dB corresponds to an attenuation of 0.5. The similar receive-path loss is labeled L_r.

ISBN 0-201-05758-1

References start on p. 469.

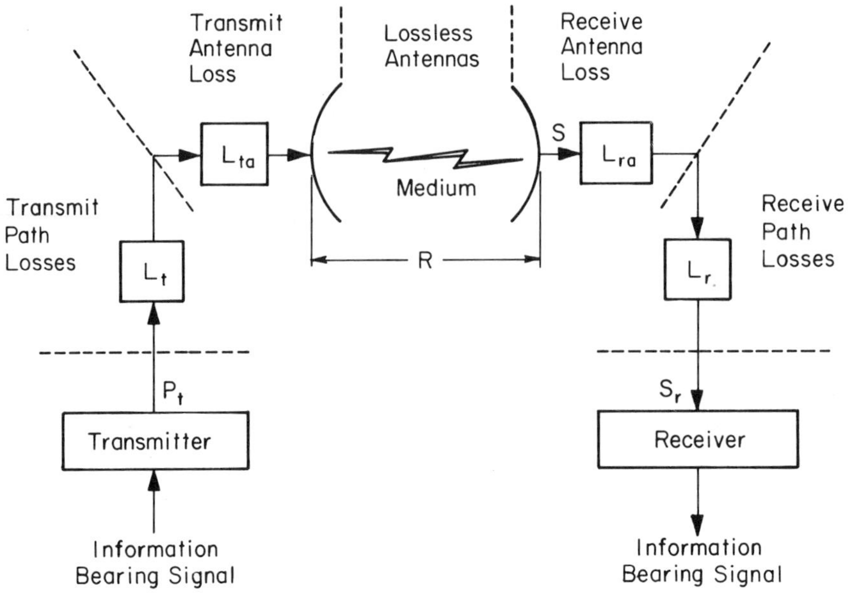

Fig. 9.1-1. Elements of a microwave communication system.

To couple energy into the medium, an antenna is necessary. Most antennas do not have great loss. There is always some, however, due to such things as impedance mismatch, ohmic losses, and other effects related to detailed antenna theory. Whatever the total loss, as compared to an antenna without such practical effects, it is labeled as L_{ta}. Thus, we shall model a practical lossy antenna as a loss in cascade with an ideal lossless antenna which otherwise has all the properties of the actual antenna. Similar comments apply to the receiving antenna where the loss is L_{ra}.

Finally, the antennas are separated by a distance R, and the signal power available at the lossless antenna output is S while that at the receiver input is S_r. There is also noise accompanying the signal, and one of our principal efforts will be its determination. First, however, we concentrate on determining S and S_r. Toward this end we must first establish certain basic antenna principles.

9.2. BASIC ANTENNA PRINCIPLES

It is not our purpose to develop in-depth details on antenna design. Rather, we shall be interested only in those principles that lead to an understanding of the roll of an antenna as a power-coupling element of a system. In effect we treat the antenna as a black box which functions as a transducer, coupling power from waveguide or coaxial lines to space, and vice versa.

The basic principles of antennas follow easily once we understand what is meant by the radiation intensity pattern.

Radiation Intensity Pattern

Antennas do not radiate poweı equally well in all directions in space. It is the ***radiation intensity pattern*** that describes the power intensity* of radiation in any spatial direction. Using spherical coordinates centered on the antenna A, as shown in Fig. 9.2-1(a), the radiation

*Power intensity is power per unit solid angle. The unit is watts per steradian.

ISBN 0-201-05758-1

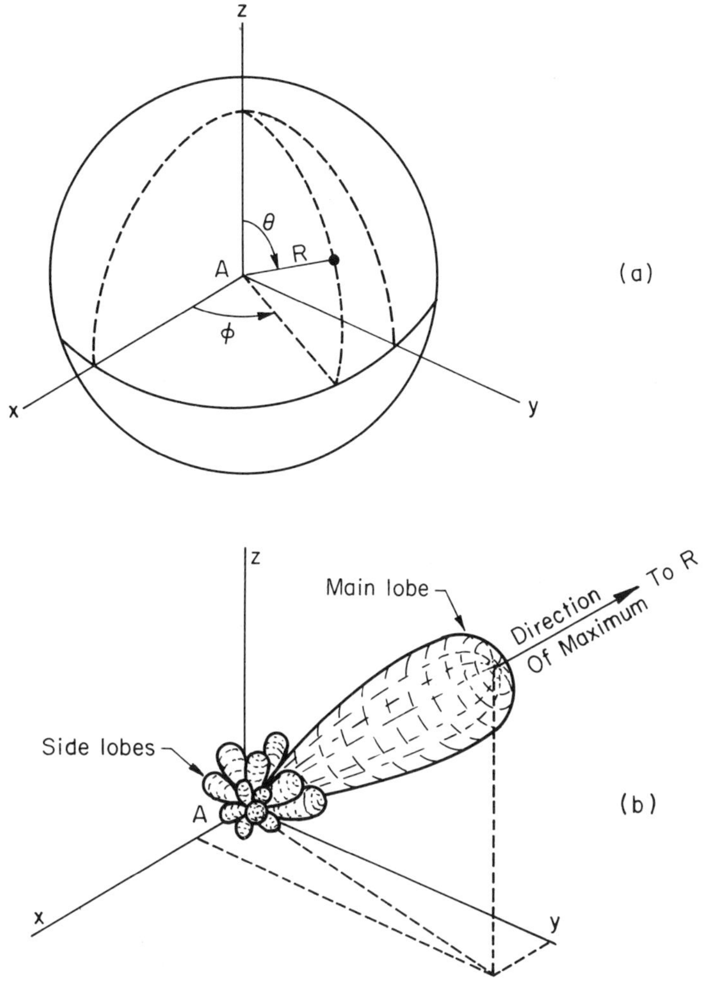

Fig. 9.2-1. (a) Spherical coordinates applicable to the radiation intensity pattern of (b).

intensity pattern might appear as shown in (*b*). The power intensity at any distant point *R* having angles θ and ϕ is plotted as a magnitude *P* from *A*. Clearly the radiation pattern is a power intensity function $P(\theta, \phi)$ of the angular directions θ and ϕ and appears as a surface as illustrated in (*b*), when all possible values of θ and ϕ are considered.

In many communication problems it is desirable that the transmitting antenna concentrate most of its radiated power in one direction, that of the receiver. The radiation intensity pattern will then contain one large *main lobe* in its surface. Other minor lobes or *sidelobes* typically occur as well. The receiver will receive the maximum possible power if the transmit (receive) antenna is physically oriented so that the peak of the main lobe is in the direction of the receiver (transmitter).

It is obvious that the sidelobes represent wasted power—that is, power radiated in unwanted directions. Hence, it is desirable to have these lobes as small as feasible. With reasonable care in design the sidelobe peaks can be maintained 20 to 30 dB below the main lobe peak. With very careful design this range might be extended from 30 to over 40 dB.

An antenna is a *reciprocal element.* That is, the power output when used as a receiving antenna for a wave from a distant radiator will be determined by the radiation intensity pattern. Thus, the power output will be maximum for sources in the direction of the main

ISBN 0-201-05758-1

lobe maximum and will vary according to the intensity pattern for other directions assuming a constant source distance.

If the radiation intensity pattern is scaled such that the maximum intensity is unity, the result is commonly called simply the *radiation pattern.* In many problems where the radiation pattern has one dominant main lobe it is not necessary to discuss the full three-dimensional pattern. Behavior may adequately be described by behavior in two orthogonal planes containing the maximum of the main lobe. Figure 9.2-2 illustrates a possible pattern along one of these planes for both polar (*a*) and linear (*b*) angle plots. The angular separation between points on the radiation pattern that are 3 dB down from the maximum is called the *beam width.* In the two orthogonal plane patterns, called *principal plane patterns,* we shall call these beam widths θ_B and ϕ_B. Their units are radians unless otherwise stated.

Directive Gain

An important antenna parameter is its gain. Gain is a number related to how intense the radiation is at the maximum of the main lobe of the radiation pattern. There are two types of gain worth considering here; one is *directive gain* and the other is *power gain.*

As the name implies, directive gain G_D is a measure of how directive the pattern is. It is defined by

$$G_D = \frac{\text{maximum radiation intensity}}{\text{average radiation intensity}}. \tag{9.2-1}$$

We note that this definition is based entirely on *radiated* powers and therefore does not include any losses that might occur in the antenna. Since

$$\begin{aligned}\text{average radiation intensity} &= \text{total radiated power}/(4\pi) \\ &= P_{\text{rad}}/(4\pi),\end{aligned} \tag{9.2-2}$$

then

$$G_D = 4\pi \frac{\text{maximum radiation intensity}}{P_{\text{rad}}}. \tag{9.2-3}$$

For many purposes it is more useful to employ a gain definition which includes the effects of antenna losses. Such a gain is the power gain.

Power Gain

We define an *isotropic antenna* as a lossless antenna which radiates power equally well in all directions.* For such an antenna, if P is the input power, the radiation intensity in any direction is a constant given by

$$\text{radiation intensity} = P/(4\pi), \qquad \text{isotropic antenna.} \tag{9.2-4}$$

*An isotropic antenna is not a device that can be realized; it can only be approximated in practice. However, it is convenient to imagine such an antenna and use it as a reference for comparison with actual antennas.

ISBN 0-201-05758-1

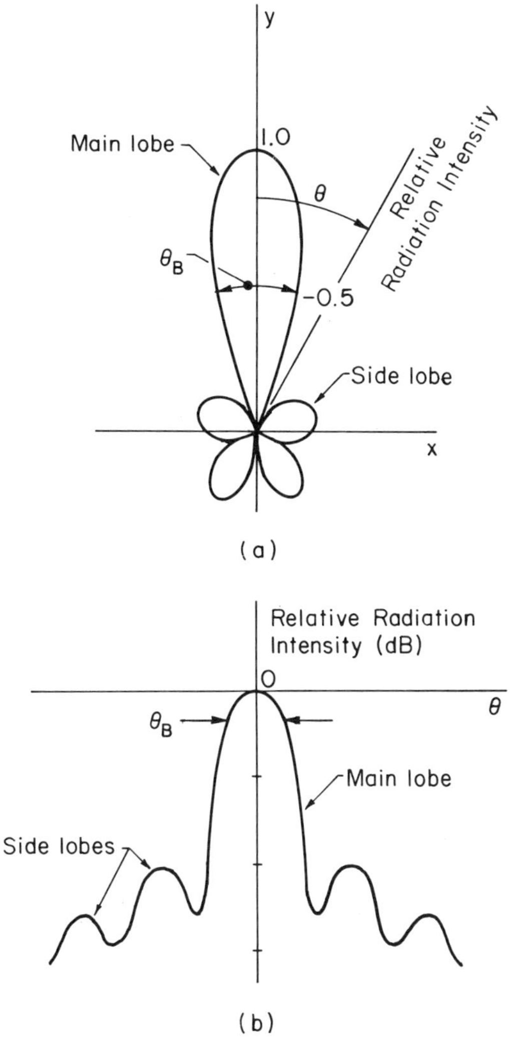

Fig. 9.2-2. Radiation patterns. (a) In polar and (b) in linear coordinates.

Power gain G is a measure of maximum radiation intensity of an antenna as compared to that which would result from an isotropic antenna *with the same power input.* Hence G is defined by the ratio

$$G = \frac{\text{maximum radiation intensity}}{\text{radiation intensity of isotropic source with same power input}}. \tag{9.2-5}$$

Using (9.2-4) this becomes

$$G = \frac{4\pi(\text{maximum radiation intensity})}{P}. \tag{9.2-6}$$

We observe that if the antenna is lossless all the input power P is radiated power P_{rad}, so that

ISBN 0-201-05758-1

directive and power gains are the same from (9.2-3):

$$G = G_D, \qquad \text{lossless antenna.} \tag{9.2-7}$$

When losses are present in the actual antenna, then $G < G_D$. The amount that G is less than G_D is taken into account by the *radiation efficiency factor* ρ_r:

$$G = \rho_r G_D, \tag{9.2-8}$$

where $0 < \rho_r \leqslant 1.0$.

Throughout this work we shall represent a real lossy antenna by a lossless one in cascade with an equivalent loss. The loss is equal to $1/\rho_r$. Thus, the antenna losses of Fig. 9.1-1 are given by

$$L_{ta} = 1/\rho_{rt}, \tag{9.2-9}$$

$$L_{ra} = 1/\rho_{rr}, \tag{9.2-10}$$

where ρ_{rt} and ρ_{rr} are the radiation efficiencies of the transmit and receive antennas, respectively. Because of handling actual losses in this manner, we shall hereafter refer simply to antenna gain and use the symbol G, meaning either power or directive gain, since they are the same.

A useful, although approximate, expression for antenna gain is

$$G = 4\pi A_e/\lambda^2, \tag{9.2-11}$$

where λ is the wavelength being transmitted or received and A_e is called the *effective aperture* or *effective area* of the antenna.

Another useful result relates gain to the beam widths θ_B and ϕ_B of the pattern main lobe. The result, which is only approximate [1], is

$$G = 4\pi/\theta_B\phi_B \tag{9.2-12}$$

if beam widths are in radians, or

$$G = 41.3(10^3)/\theta_B\phi_B \tag{9.2-13}$$

if units are degrees.

Example 9.2-1

An antenna has beam widths of 3 degrees and 10 degrees in orthogonal planes and has a radiation efficiency factor of 0.5. We find the maximum radiation intensity if 1000 watts is applied to the antenna. From (9.2-13) gain is

$$G = 41.3(10^3)/3(10) = 1.38(10^3),$$

or 31.4 dB. If all power were radiated, then maximum intensity, from (9.2-3), would be $1.38(10^3)10^3/4\pi = 109.8$ kW per steradian. However, only half is effectively transmitted according to (9.2-8), so the maximum intensity is 54.9 kW per steradian.

ISBN 0-201-05758-1

Aperture Efficiency

Effective aperture A_e is related to the true physical radiating area A of the antenna by

$$A_e = \rho_a A, \tag{9.2-14}$$

where ρ_a is called the *aperture efficiency* and $0 < \rho_a \leq 1.0$. One would expect intuitively that, if all possible portions of the true physical surface were radiating power with equal intensity, an aperture with maximum possible efficiency would result ($\rho_a = 1.0$). Such is the case, as pointed out by Silver* [2], and it is called *uniform illumination.*

If some portions of the antenna's radiating plane (aperture) radiate at a lower intensity level than the maximum, they may be considered as contributing to a less efficient antenna. This is referred to as *nonuniform illumination,* and $\rho_a < 1.0$. Thus, nonuniformly illuminated antennas are less efficient than uniformly illuminated ones, having less gain. The lower gain is easily seen by using (9.2-14) with (9.2-11):

$$G = 4\pi\rho_a A/\lambda^2 . \tag{9.2-15}$$

Although they have the disadvantage of less gain, nonuniformly illuminated apertures may give lower sidelobes than a uniformly illuminated one.

Signal Reception

If a distant source produces radiation with *power density* $\mathcal{S}$ (unit of power per unit area) at a receiving antenna, then the antenna physical area A will intercept a power $S = \mathcal{S}A$. Because of reciprocity, if the antenna is uniformly illuminated, the whole physical area contributes effectively toward power collection. In general, however, if the antenna has nonuniform illumination it is the effective area that produces the output power S, where

$$S = \mathcal{S}A_e = \mathcal{S}\rho_a A. \tag{9.2-16}$$

The power S is actually the maximum *available* power that the antenna is capable of producing. The actual power developed depends on the receiver impedance. As we shall see shortly, it is the available power which is important in system calculations.

Polarization

Most antennas transmit only one polarization of electromagnetic wave. That is, the electric field of the propagating wave is oriented relative to the antenna in only one direction. For example, the main lobe is usually† normal to the aperture as illustrated in Fig. 9.2-3 for a circular aperture. If the local vertical at the antenna is the y axis and the electric field lies in the vertical plane as shown in (a), the radiation is said to be *vertically polarized.* If the electric field lies in the horizontal plane as shown in (b), the radiation is said to be *horizontally polarized.* Both vertical and horizontal are special cases of *linear polarization* as shown in (c).

*Strictly speaking, our stated result applies only to what we shall call *constant-phase illumination functions.* However, many practical antennas are of this type, and all our work is based on the constant-phase assumption.

†In most antennas the main lobe is in a direction normal to the plane of the aperture. However, for the important class of *phased-array antennas* the main lobe may be electronically steered to other angles away from the "broadside."

ISBN 0-201-05758-1

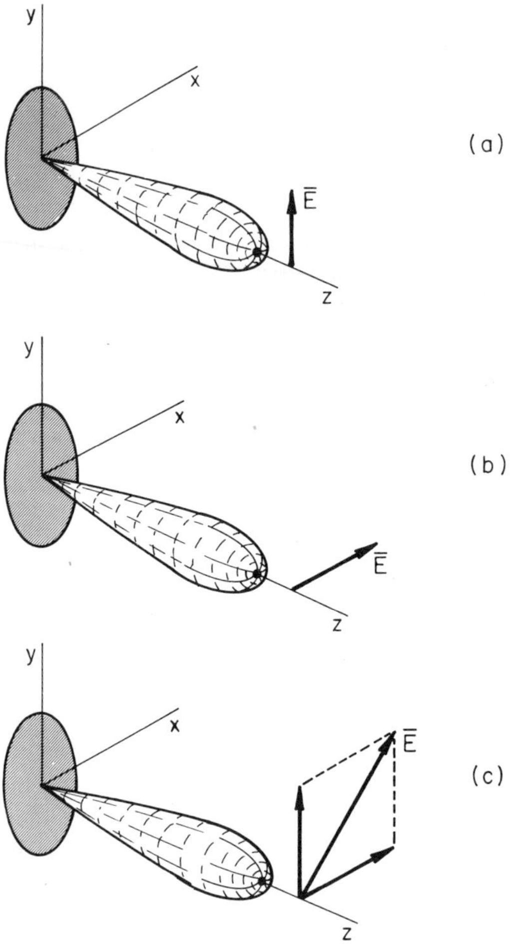

Fig. 9.2-3. Illustrations of electric field polarizations. (a) vertical, (b) horizontal, and (c) arbitrary linear.

Here the electric field is still in a plane but has horizontal and vertical components that are in time phase. If the antenna main lobe points other than horizontally, vertical and horizontal have no meaning and the above polarizations become simply linear polarizations.

Some systems are designed to transmit two linear orthogonal polarizations simultaneously that are not in time phase. *Circular polarization* is probably the most useful case where horizontal and vertical electric fields are 90 degrees out of time phase and have equal magnitude.* If the horizontal component of the radiation lags the vertical component by 90 degrees, the resultant field appears to rotate counterclockwise in the *xy* plane with time, as one located at the antenna views the wave leaving the antenna, giving rise to what is called *left-hand* circular polarization. If the horizontal component leads the vertical component by 90 degrees the resultant field rotates clockwise with time in the *xy* plane and is called *right-hand* circular polarization. Circular is a special case of the more general *elliptical polarization* where the two linear components have arbitrary relative amplitudes and arbitrary time phase. Elliptical polarization has seen little use in practice.

*Circular polarization gives distinct advantages when propagating through the ionosphere where Faraday rotation is a problem. It is also useful in reducing undesired echoes from rain and other forms of precipitation in radar. Finally it is useful in space applications where the receiving antenna orientation may change relative to that of the transmitter.

ISBN 0-201-05758-!

9.3. RECEIVED SIGNAL POWER

Figure 9.1-1 may now be used to determine the signal power S_r available to a receiver from its antenna when a power P_t/L_t is applied to the transmitting antenna. We start the discussion by considering a lossless isotropic transmitting antenna. The *radiation power density** (unit is watts per unit area) at a distance R from the transmitter is the total power divided by the surface of the sphere of radius R:

$$\text{power density} = \frac{P_t}{4\pi R^2 L_t}, \tag{9.3-1}$$

since the radiation intensity pattern is constant in all directions.

Now in reality the actual antenna will have directivity and therefore power gain. The actual power density will be larger than (9.3-1) by the gain factor for a lossless antenna.† An additional factor $1/L_{ta}$ accounts for the losses of the actual antenna. Power density for the actual antenna becomes

$$\text{power density} = \frac{P_t G_t}{4\pi R^2 L_t L_{ta}}, \tag{9.3-2}$$

where G_t denotes "transmit" antenna gain.

If the receive antenna is at distance R, the power density at its location would be given by (9.3-2) if there were no losses through the medium. Since there is loss over the medium, we account for this by a loss factor L_{at}. Actual power density $\mathscr{S}$ is then

$$\mathscr{S} = \text{power density} = \frac{P_t G_t}{4\pi R^2 L_t L_{ta} L_{at}}. \tag{9.3-3}$$

The available‡ power S from the lossless receiving antenna is the product of the wave power density and the capture antenna effective area A_{re}:

$$S = \frac{P_t G_t A_{re}}{4\pi R^2 L_t L_{ta} L_{at}}. \tag{9.3-4}$$

Finally, the receive-path loss $L_{ra}L_r$ is added and a total loss L is defined by

$$L = L_t L_{ta} L_{at} L_{ra} L_r, \tag{9.3-5}$$

so that available power S_r at the receiver becomes

$$S_r = \frac{P_t G_t A_{re}}{4\pi R^2 L}. \tag{9.3-6}$$

Other useful forms of (9.3-6) may be obtained. Using (9.2-11) we have

*Power density is radiation intensity divided by the square of the distance R.

†The transmitter (receiver) main lobe maximum is assumed to point in the direction of the receiver (transmitter).

‡All powers stated below are implied to be available power as discussed following (9.2-16).

ISBN 0-201-05758-1

$$S_r = \frac{P_t G_t G_r \lambda^2}{(4\pi)^2 R^2 L}, \tag{9.3-7}$$

where G_r represents the gain of the receiving antenna.

If we define various quantities expressed in decibels by

$$(\cdot)_{\text{dB}} = 10 \log_{10}(\cdot), \tag{9.3-8}$$

then (9.3-7) may be written as

$$(S_r)_{\text{dB}} = (P_t)_{\text{dB}} + (G_t)_{\text{dB}} + (G_r)_{\text{dB}} + 2(\lambda)_{\text{dB}}$$
$$- [22.0 + 2(R)_{\text{dB}} + (L)_{\text{dB}}]. \tag{9.3-9}$$

In applying (9.3-9), units must be consistent; that is, if $(S_r)_{\text{dB}}$ is power in decibels relative to a watt, then P_t is in watts, and λ and R are in the same units.

Let our principal result (9.3-7) be illustrated with an example.

Example 9.3-1

We calculate the available received power for two stations separated 48.28 km (30 miles), each having a circular aperture antenna with a diameter of 2 meters. Aperture efficiency is 0.5 at 4000 MHz. The total loss over the link is 7.94 (9 dB), and the transmitter generates 0.5 watt. For wavelength, λ equals 300 divided by the frequency in megahertz or 0.075 meter. From (9.2-15) $G = G_t = G_r = 4\pi(0.5)\pi/(0.075)^2 = 3509.2$. From (9.3-7)

$$S_r = \frac{0.5(3.509)^2\, 10^6\, (0.075)^2}{(4\pi)^2\, (48.28)^2\, 10^6\, (7.94)} = 1.18(10^{-8}) \text{ watt}.$$

All the factors required in (9.3-7) are more or less readily approximated except the atmospheric loss term L_{at}. This term is briefly considered in the next section.

9.4. ATMOSPHERIC LOSSES

When the transmitted wave passes through the atmosphere it may undergo losses due to various causes. Even if there are no effects due to weather, such as rain, snow, fog or clouds, there remains loss due to gas molecules in the clear atmosphere.

Clear Atmosphere

Attenuation of a wave in a clear atmosphere is a complicated function of altitude (gas molecules are more dense near the earth), temperature, humidity, etc. Below about 15 GHz attenuation is mainly due to oxygen absorption. Above this frequency, but below about 35 GHz, there is a water vapor loss which dominates, and severe attenuation occurs around 22.2 GHz where molecules resonate. From 35 to 80 GHz attenuation is again controlled mainly by oxygen absorption, and an extremely severe loss at 60 GHz occurs owing to resonance.

Below 10 GHz, where most communication systems operate, a fair approximation to published data [3] for loss in the clear atmosphere is

ISBN 0-201-05758-1

$$\text{loss (dB/km)} \approx 0.005[\log_{10}(10f_{\text{GHz}})]^2, \qquad 0.3 \leqslant f_{\text{GHz}} \leqslant 10. \qquad (9.4\text{-}1)$$

Here f_{GHz} is frequency in gigahertz. For problems where the wave path includes the entire troposphere at zero-degree elevation angle, typical data from Blake [4] may be approximated by

$$\text{loss (dB)} \approx 1.5 \log_{10}(10f_{\text{GHz}}), \qquad 0.2 \leqslant f_{\text{GHz}} \leqslant 10. \qquad (9.4\text{-}2)$$

For higher angles the attenuation decreases rapidly. This result shows that, for the worst case (low angle, whole atmosphere), the largest loss is about 3.0 dB for frequencies below 10 GHz.

Rain

Loss due to particles of rainfall increases with both increasing frequency or rainfall rate. It is most severe for frequencies above about 3 GHz, depending on path length. A computation [5] of attenuation based on modeling raindrops as spheres leads to the curves of Fig. 9.4-1. The definition of rainfall categories follows that used in references [5] and [1].

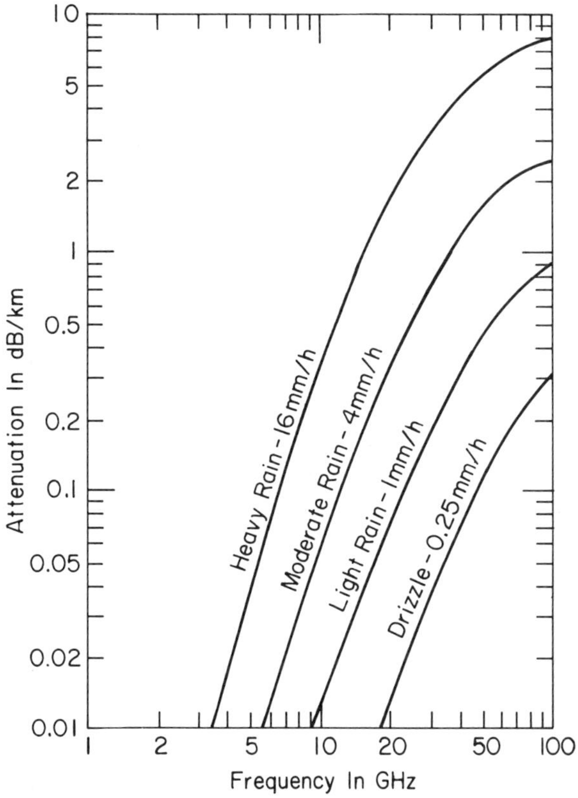

Fig. 9.4-1. Attenuation due to rainfall at 18°C. Curves are adapted from Fig. 8.12 of Kerr [5] with permission.

Fog or Clouds

From results given in reference [1] an expression for loss due to either clouds or fog may be put in the form

ISBN 0-201-05758-1

$$\text{loss (dB/km)} = 0.148 f_{\text{GHz}}{}^2 / V_m{}^{1.43}. \tag{9.4-3}$$

Here V_m is called *optical visibility* (meters). Equation (9.4-3) applies at a temperature of 18°C. For temperatures from 0°C to 40°C the loss varies by a factor of about 0.5 to 2 from its 18°C value. Some idea of numerical values may be obtained by choosing V_m according to the International Visibility Code [3, pp. 26–27], where for dense fog $V_m < 50$ meters, for thick fog $50 \leqslant V_m < 200$, and for moderate fog $200 \leqslant V_m < 500$. It is found that only dense or thick fog produces significant attenuation rates for frequencies below about 15 GHz. Actual attenuation, of course, depends on the extent of the fog or size of the cloud.

Snow

Loss through a portion of the atmosphere containing ice particles is generally less than that caused by rain of an equivalent water fall rate [1]. Gunn and East [6] give an expression for snow which may be put in the form

$$\text{loss (dB/km)} = 7.47(10^{-5}) f_{\text{GHz}} I [1 + 5.77(10^{-5}) f_{\text{GHz}}{}^3 I^{0.6}], \tag{9.4-4}$$

where I is snowfall intensity in millimeters of melted water content per hour. For most values of f_{GHz} and I the second term is negligible. In fact, for frequencies below 15 GHz only moderate (4 mm/h) or heavier snowfall produces significant loss in most cases.

We consider an example to observe the relative significance of atmospheric losses.

Example 9.4-1

For a link covering 16.09 km (10 miles) we determine the loss in a wave traveling a horizontal path near the earth when $f_{\text{GHz}} = 10$. From (9.4-1) the clear atmosphere produces $0.005[\log_{10}(100)]^2 16.09 = 0.32$ dB of loss. If the entire path is filled with rain falling at a 16-mm/h rate, the additional loss is $0.36(16.09) = 5.79$ dB. Total loss is 6.11 dB. If dense fog ($V_m = 50$ meters) had been present instead, (9.4-3) gives a loss of $[0.148(10)^2/(50)^{1.43}]16.09 = 0.886$ dB and a total loss of 1.21 dB. Finally, suppose heavy snow (16 mm/h) is falling. From (9.4-4) the added loss is

$$16.09(7.47)10^{-5}(10)16[1 + 5.77(10^{-5})10^3(16)^{0.6}] = 0.251 \text{ dB}.$$

Total loss becomes 0.57 dB.

The above example points out that rain may often be the most significant cause of atmospheric loss.

9.5. POWER FLOW IN NETWORKS

Having determined the received signal power, we next examine the noise power present in a communication system. This and the following three sections are devoted to the noise problem. We begin with some general topics governing power flow in networks.

Maximum Power Transfer Theorem

Consider a signal source and load as depicted in Fig. 9.5-1(a). The source has been replaced by its *Thévenin equivalent circuit,* which is a generator having a voltage e_s, equal to the source

ISBN 0-201-05758-1

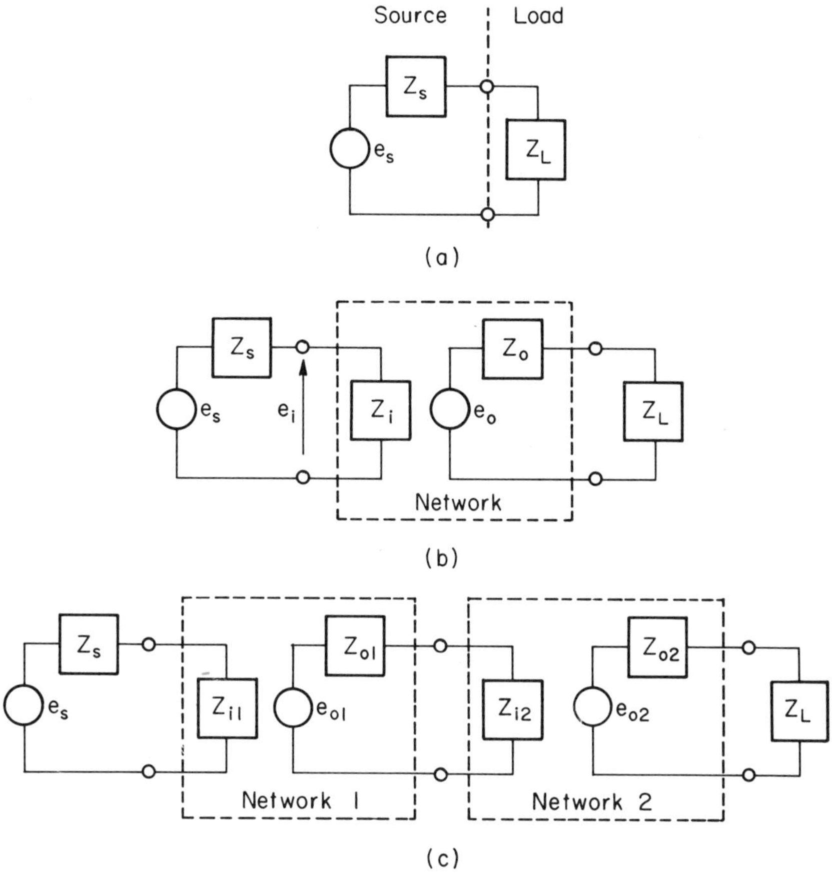

Fig. 9.5-1. Thévenin equivalent circuits. (a) Of a source and load. (b) Of a network representing an amplifier with load driven by a source, and (c) a cascade of two networks representing two amplifiers in cascade.

open-circuit voltage, in series with the source internal impedance. Suppose initially that e_s is a sinusoidal time function:

$$e_s = A \cos (\omega_0 t + \theta_0). \tag{9.5-1}$$

Now for a fixed source impedance Z_s and variable load impedance Z_L, the *maximum power transfer theorem* states that the largest possible power is delivered to the load when its impedance equals the complex conjugate of the source impedance. That is, $Z_L = Z_s^*$ for maximum load power. This power is called the *available power* P_{as}, since it is all that may be extracted from the source. It is given by

$$P_{as} = \frac{\overline{e_s^2(t)}}{4R_s} = \frac{A^2/2}{4R_s}. \tag{9.5-2}$$

Next, suppose $e_s(t) = e_n(t)$ is a random signal or noise having a power density spectrum $\mathcal{S}_n(\omega)$ that exists only in very narrow incremental bands $d\omega$ centered at frequencies $\pm\omega$. The *incremental available noise power* dN_{as} from this source is

$$dN_{as} = \frac{\overline{e_n^2(t)}}{4R_s}. \tag{9.5-3}$$

ISBN 0-201-05758-1

Now since $\overline{e_n^2(t)} = R_n(0)$, with $R_n(\tau)$ being the autocorrelation function of $e_n(t)$, we use (4.9-21) to obtain*

$$dN_{as} = \mathcal{S}_n(\omega)\, d\omega/4\pi R_s, \qquad 0 < \omega, \tag{9.5-4}$$

for the incremental available power in the band $d\omega$. More generally, the source power is spread over a larger band than the incremental band discussed here. In such a case, dN_{as} may be a function of frequency. Total available power must then be found by integration.

Available Power Gain

Next consider a source driving a linear amplifier as illustrated in Fig. 9.5-1(b). So far as the source is concerned, it simply sees an input impedance Z_i equal to the impedance between input terminals with the output terminals open-circuited. As seen from the output terminals, e_o and Z_o are the Thévenin equivalent voltage and internal impedance of the source-amplifier combination. Z_L terminates the network. The voltage e_o is, in general, related to e_i according to some function of frequency. However, for present purposes the exact relationship need not be specified. The *available power gain* G_a of the amplifier is defined as the ratio of incremental available power dP_{ao} at the output to the incremental available power dP_{as} from the source

$$G_a = \frac{dP_{ao}}{dP_{as}}. \tag{9.5-5}$$

Clearly, $dP_{ao} = \overline{e_o^2(t)}/4R_o$. Combining with (9.5-2), G_a becomes

$$G_a = \frac{R_s\overline{e_o^2(t)}}{R_o\overline{e_s^2(t)}} = \frac{R_s\mathcal{S}_o(\omega)}{R_o\mathcal{S}_s(\omega)}. \tag{9.5-6}$$

The last form results from application of (9.5-4), assuming a noise source.

Available power gain is important in the analysis of noise performance of networks in cascade. The extension to the case of two amplifiers is easily accomplished with the aid of Fig. 9.5-1(c). Overall available power gain becomes

$$G_a = \frac{dP_{ao2}}{dP_{as}} = \frac{dP_{ao1}}{dP_{as}}\frac{dP_{ao2}}{dP_{ao1}} = G_{a1}G_{a2}. \tag{9.5-7}$$

Here dP_{ao1} and dP_{ao2} are the incremental available powers from amplifiers 1 and 2 having available power gains G_{a1} and G_{a2}, respectively. Equation (9.5-7) shows that available power gain of two amplifiers in cascade equals the product of separate available power gains. More generally, for N linear networks in cascade

*The reader should exercise care in interpreting this result. $\mathcal{S}_n(\omega)$ is a *two-sided* power density. However, because $\mathcal{S}_n(\omega)$ is a real, even function of ω, a factor of 2 has been inserted to account for power due to negative frequencies. Thus, dN_{as} is total incremental power and $\mathcal{S}_n(\omega)$ is to be evaluated only for $\omega >$ 0.

ISBN 0-201-05758-1

$$G_a = G_{a1}G_{a2} \ldots G_{aN} = \prod_{n=1}^{N} G_{an}. \tag{9.5-8}$$

In each case the available power gain of a given network must be determined when all *preceding* networks are connected as in the overall cascade. Thus, G_a of a given stage depends on the impedance relationships of all previous stages but does not depend on the load impedance seen by the stage.

9.6. INCREMENTAL MODELING OF NOISE SOURCES

One of the most important types of noise in most communication sytems is thermal noise. It is generated in all conductors having finite temperature T. One source of this type of noise is a resistor where thermal agitation of electrons causes random motion within the material. The motion causes a random voltage to occur across the resistor. It has been well verified by both theory and measurement that the mean-squared voltage produced across the terminals is

$$\overline{e_n^2(t)} = 2kTR\; d\omega/\pi, \tag{9.6-1}$$

where $k = 1.38(10^{-23})$ joule per Kelvin is *Boltzmann's constant*, T is temperature in Kelvin, R is resistance in ohms, and $\overline{e_n^2}$ is due to energy in the incremental frequency band $d\omega$. An interesting property of thermal noise is that $\overline{e_n^2(t)}$ is a constant independent of where in frequency the incremental interval $d\omega$ is taken, at least for ω up to about $2\pi(10^{13})$ rad/s [7, p. 275]. Beyond this frequency $\overline{e_n^2(t)}$ decreases.

From an application of (4.9-21) the power density spectrum of the noise is

$$\mathscr{S}_n(\omega) = 2kTR \tag{9.6-2}$$

for $|\omega| \lesssim 2\pi(10^{13})$. Because this frequency range is so large (about 10,000 GHz), thermal noise behaves as white noise so far as almost all communication systems are concerned. In terms of power density $\mathscr{N}_0/2$, used so often in this book, $\mathscr{N}_0/2 = 2kTR$ for a resistive thermal source.

Resistor Noise Source

A noisy resistor may be modeled by a Thévenin equivalent circuit, where the resistor R is *noise-free*, as illustrated in Fig. 9.6-1(a). The Norton equivalent circuit of (b) is also valid. The noise power dN_L delivered to the load resistor in the band $d\omega$ is

$$dN_L = \frac{kT\,d\omega}{\pi}\,\frac{2R_L R}{(R_L + R)^2}. \tag{9.6-3}$$

This power is maximum when $R_L = R$. Let this incremental available noise power be called dN_{as}. It is

$$dN_{as} = kT\,d\omega/2\pi. \tag{9.6-4}$$

This expression shows that the maximum incremental noise power which can be extracted

ISBN 0-201-05758-1

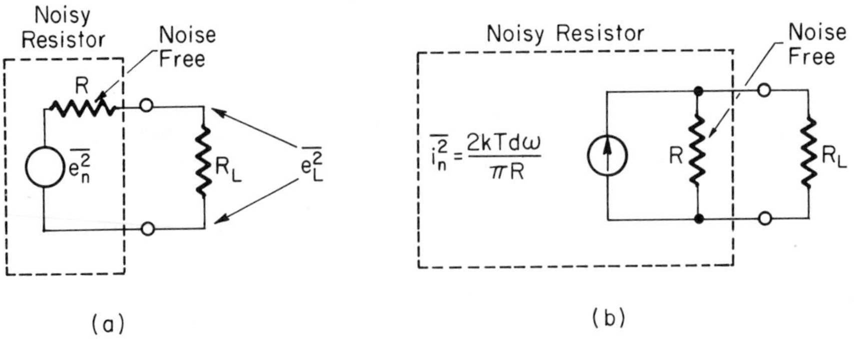

Fig. 9.6-1. Equivalent circuits of a noisy resistor. (a) Thévenin voltage model and (b) Norton current model.

from a resistor is *independent of its resistance*; it depends only on its physical temperature T. This somewhat remarkable result may be used as a basis for modeling an arbitrary source.

Arbitrary Noise Source, Effective Noise Temperature

We may model an arbitrary noise source by ascribing all its generated noise in an incremental band $d\omega$ to the real part of its internal impedance and *defining* an *effective noise temperature* T_{es} such that the available noise power is of the form of (9.6-4). That is,

$$T_{es} = \frac{2\pi \, dN_{as}}{k \, d\omega} . \tag{9.6-5}$$

If the available noise power dN_{as} varies with frequency, T_{es} becomes frequency dependent. For any network containing only *R-L-C* elements, all at the same physical temperature T, T_{es} will equal T [7, p. 303].

An example using two noisy resistors in parallel will be used to illustrate effective noise temperature.

Example 9.6-1

Two noisy resistors R_1 and R_2 are placed in parallel as illustrated in Fig. 9.6-2(a). R_1 is at temperature T_1, and R_2 is at T_2. Between terminals AB the two-resistor combination is equivalent to a single resistor $R = R_1 R_2/(R_1 + R_2)$ as shown. We now ask: What is the equivalent mean-squared noise voltage across R, and what is the effective noise temperature of R? To answer these questions we use the Norton equivalent circuits, as shown in (b), for analysis. Equating total dissipated power, and recognizing that powers add for independent noise sources, we have

$$\overline{i_1^2}\left(\frac{R_1 R_2}{R_1 + R_2}\right) + \overline{i_2^2}\left(\frac{R_1 R_2}{R_1 + R_2}\right) = (\overline{i_1^2} + \overline{i_2^2})R = \overline{i_n^2}R.$$

Hence,

$$\frac{\overline{e_n^2}}{R^2} = \overline{i_n^2} = \overline{i_1^2} + \overline{i_2^2} = 2k\left(\frac{T_1}{R_1} + \frac{T_2}{R_2}\right)\frac{d\omega}{\pi} .$$

ISBN 0-201-05758-1

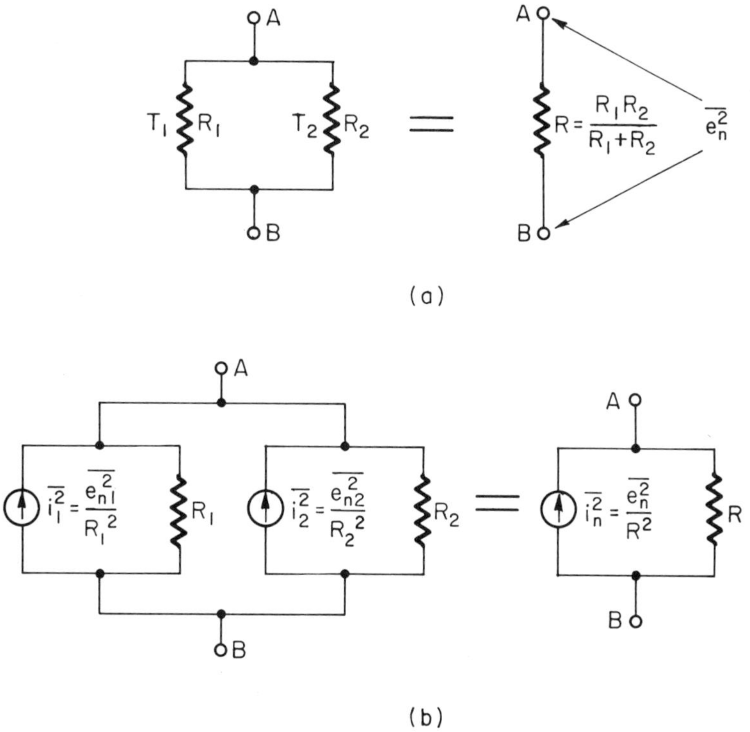

Fig. 9.6-2. Parallel resistors. (a) Two resistors at different temperatures, and (b) their Norton equivalent circuit.

Solving for $\overline{e_n^2}$ we get

$$\overline{e_n^2} = 2k\left(\frac{T_1}{R_1} + \frac{T_2}{R_2}\right)R^2\,\frac{d\omega}{\pi} = 2k\left(\frac{T_1R_2 + T_2R_1}{R_1 + R_2}\right)R\,\frac{d\omega}{\pi}.$$

Finally, because R is the internal impedance of the resistor combination as a source, the effective noise temperature must be

$$T_e = \frac{T_1R_2 + T_2R_1}{R_1 + R_2}$$

by analogy with (9.6-1)

In the above example it is to be noted that T_1 and T_2 were true physical temperatures while T_e *is not.* T_e is simply the temperature that a single resistance R *would* have to have to produce the equivalent effect of the actual circuit. If $T_1 = T_2 = T$, then $T_e = T$, as we would expect, since both resistors are at the same temperature and should be directly replaceable by a single resistor R at temperature T.

9.7. INCREMENTAL MODELING OF NOISY NETWORKS

All real networks generate noise internally. Thus, even if a network is not excited by external sources of noise, there will be noise present at the output point of interest due to the internal

ISBN 0-201-05758-1

sources. In this section we show how a noisy network can be modeled as a noise-free network plus a suitably defined external noise source. Also considered are methods of placing quality measures on the performance of noisy systems. All work is applicable to an incremental band $d\omega$.

Equivalent Noise-Free Network

Figure 9.7-1(a) depicts a noisy network excited externally by a source of noise with effective temperature T_s. The available output noise power dN_{aos} due to the source excitation is equal to the product of available gain G_a and available source noise power:

$$dN_{aos} = G_a k T_s \, d\omega/2\pi. \tag{9.7-1}$$

The available output *excess noise power* ΔN_{ao} is due to network internally generated noise.

The actual network may be replaced by an equivalent noise-free network as illustrated in (b) if we define *effective input noise temperature* T_e as the temperature *increase* that the external source would require to account for all the output available noise power, including ΔN_{ao}. From this definition it follows that

$$\Delta N_{ao} = G_a k T_e \, d\omega/2\pi. \tag{9.7-2}$$

In some following work it will occasionally be helpful to visualize the noise-free equivalent network as having available input powers from two sources as diagramed in Fig. 9.7-1(c). The second source, with effective temperature T_e, does not actually exist physically, but imagining its existence will materially aid in the calculation of system performance when a cascade of networks is involved.

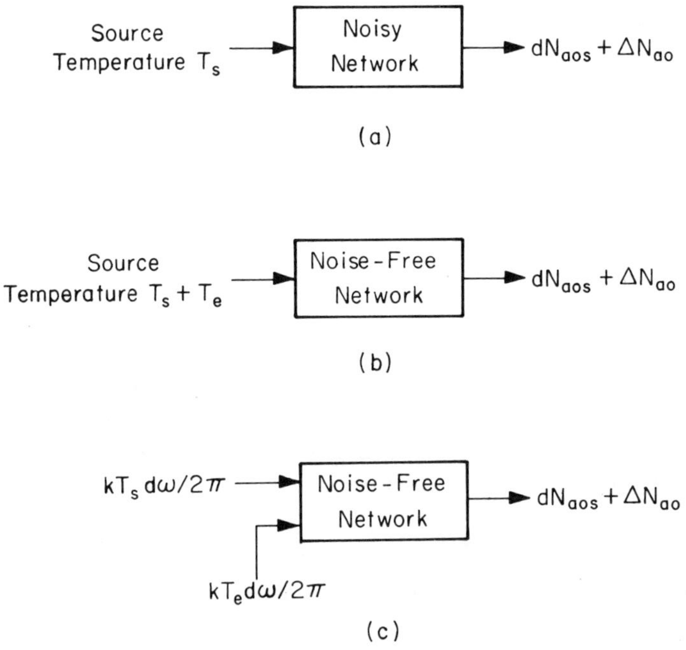

Fig. 9.7-1. A noisy network (a) and its noise-free equivalents (b) and (c).

ISBN 0-201-05758-1

Spot Noise Figure

Excess noise power ΔN_{ao} may be taken as a measure of noise quality of a network. In a poor network ΔN_{ao} will be large, while an ideal or perfect network has $\Delta N_{ao} = 0$. Another often-used quality measure is also based on the magnitude of ΔN_{ao}. It is the *incremental* or *spot noise figure* F defined as the total incremental available output noise power divided by the incremental available output noise power due to the external source alone:

$$F = \frac{dN_{ao}}{dN_{aos}} = \frac{dN_{aos} + \Delta N_{ao}}{dN_{aos}} = 1 + \frac{\Delta N_{ao}}{dN_{aos}}. \tag{9.7-3}$$

In an ideal network $F = 1$, since $\Delta N_{ao} = 0$; for any other network $F > 1$. From substitution of (9.7-1) and (9.7-2) we get the alternative form

$$F = 1 + \frac{T_e}{T_s}. \tag{9.7-4}$$

Operating and Standard Spot Noise Figures

Clearly, the spot noise figure depends on the source temperature. If the source is that for which the network is designed to operate, F is called *operating spot noise figure* and is given the symbol F_{op}. Hence

$$F_{op} = 1 + \frac{T_e}{T_s}. \tag{9.7-5}$$

It is sometimes convenient to specify the noisiness of a network in a manner not dependent on a specific operating source. For example, it would be desirable to know the quality of an amplifier without having to state whether its source was an antenna, mixer, other amplifier, etc. This convenience is possible by defining a *standard source* as having a temperature $T_s = T_0 = 290$ K. *Standard spot noise figure* F_0 is defined as

$$F_0 = 1 + \frac{T_e}{T_0}. \tag{9.7-6}$$

Operating and standard spot noise figures are related by

$$F_0 = 1 + \frac{T_s}{T_0}(F_{op} - 1) \tag{9.7-7}$$

or

$$F_{op} = 1 + \frac{T_0}{T_s}(F_0 - 1). \tag{9.7-8}$$

Example 9.7-1

An engineer purchases an amplifier quoted to have a mid-band spot (standard) noise figure

ISBN 0-201-05758-1

of 3.98 (6 dB). It is to be used with a source having an effective temperature $T_s = 100$ K. We find the operating spot noise figure. From (9.7-8)

$$F_{op} = 1 + \frac{290}{100}(3.98 - 1) = 9.64 \text{ (or 9.84 dB)}.$$

Cascaded Networks

The model of a noisy network diagramed in Fig. 9.7-1(c) can be used to analyze the noise characteristics of cascaded networks. A cascade of M networks is illustrated in Fig. 9.7-2(a). Available power gains and effective input noise temperatures are G_m and T_{em}, $m = 1, 2, \ldots,$ M. The overall cascade can be replaced by a single system, so far as noise is concerned, as depicted in (b) if overall effective input noise temperature T_e is chosen properly. To find T_e we may equate output available powers for the two representations. For the cascade

$$\begin{aligned} dN_{aos} + \Delta N_{ao} = k\{T_s G_1 G_2 \ldots G_M &+ T_{e1} G_1 G_2 \ldots G_M \\ &+ T_{e2} G_2 G_3 \ldots G_M \\ &\vdots \\ &+ T_{eM} G_M\} \, d\omega/2\pi. \end{aligned} \tag{9.7-9}$$

For the equivalent representation,

$$dN_{aos} + \Delta N_{ao} = k\{T_s G_1 G_2 \ldots G_M + T_e G_1 G_2 \ldots G_M\} \, d\omega/2\pi. \tag{9.7-10}$$

Thus, on equating these two expressions,

$$T_e = T_{e1} + \frac{T_{e2}}{G_1} + \frac{T_{e3}}{G_1 G_2} + \cdots + \frac{T_{eM}}{G_1 G_2 \ldots G_{M-1}}. \tag{9.7-11}$$

In many systems the cascade of Fig. 9.7-2(a) might represent the stages in an amplifier. Assuming each stage gives a gain, (9.7-11) shows that only the earlier stages are most im-

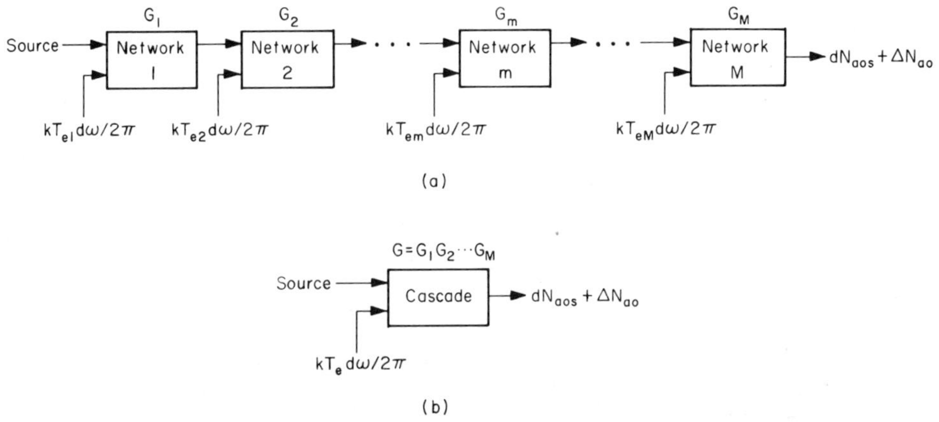

Fig. 9.7-2. A cascade of M networks (a), and (b) its equivalent network.

ISBN 0-201-05758-1

portant in determining amplifier noise properties. The contribution to T_e becomes small for later stages, since each stage's effective input noise temperature is decreased by the available gain of all prior stages. The following example demonstrates these facts.

Example 9.7-2

An amplifier consists of four stages. Each stage is similar, with T_{em} = 800 K and G_{am} = 10 for m = 1, 2, 3, and 4. The amplifier effective input noise temperature from (9.7-11) is

$$T_e = 800 + \frac{800}{10} + \frac{800}{10^2} + \frac{800}{10^3}$$

$$= 800 + 80 + 8 + 0.8 = 888.8 \text{ K}.$$

Contributions to T_e from stages after the second are approximately negligible, being less than 1% of the final value.

Procedures similar to the above lead to the overall operating noise figure F_{op} of a cascade. If F_{opm} is the operating noise figure of stage m *when driven by all previous stages* as a source, then application of (9.7-5) to (9.7-11) gives

$$F_{op} = F_{op1} + \frac{T_{s1}(F_{op2} - 1)}{T_s G_1} + \cdots + \frac{T_{s(M-1)}(F_{opM} - 1)}{T_s G_1 G_2 \ldots G_{M-1}}, \tag{9.7-12}$$

where $T_{s(m-1)}$ represents the temperature of all stages prior to stage m treated as a source.

Equation (9.7-12) is somewhat impractical to use, although valid. Another approach would be to first find the overall amplifier standard noise figure F_0 and then apply (9.7-8) to obtain F_{op}. If standard noise figures of the various stages are F_{0m}, m = 1, 2, . . . , M, then substitution of (9.7-6) into (9.7-11) gives

$$F_0 = F_{01} + \frac{(F_{02} - 1)}{G_1} + \cdots + \frac{(F_{0M} - 1)}{G_1 G_2 \ldots G_{M-1}}. \tag{9.7-13}$$

9.8. PRACTICAL MODELING OF NOISY NETWORKS

For an incremental frequency interval $d\omega$ quantities such as available power gain, effective input noise temperature, spot noise figures, and source effective noise temperature may be taken as constants. In a practical network, where the frequency band of interest is not incremental, these quantities become frequency-dependent. To account for the frequency dependence we define *average* noise figures and temperatures.

Average Operating Noise Figure

We extend the definition of spot operating noise figure by defining *average operating noise figure* $\overline{F}_{op}$ as the total output available noise power N_{ao} of a network divided by the total output available noise power N_{aos} due to the source alone. That is,

$$\overline{F}_{op} = N_{ao}/N_{aos}. \tag{9.8-1}$$

ISBN 0-201-05758-1

N_{aos} is obtained by summing all incremental noise power as given by (9.7-1):

$$N_{aos} = \int_0^\infty dN_{aos} = \frac{k}{2\pi} \int_0^\infty G_a T_s \, d\omega. \tag{9.8-2}$$

In a similar manner (9.7-3) produces

$$N_{ao} = \int_0^\infty F_{op} \, dN_{aos} = \frac{k}{2\pi} \int_0^\infty F_{op} G_a T_s \, d\omega. \tag{9.8-3}$$

Forming the ratio in (9.8-1) we obtain

$$\overline{F}_{op} = \frac{\displaystyle\int_0^\infty F_{op} G_a T_s \, d\omega}{\displaystyle\int_0^\infty G_a T_s \, d\omega}. \tag{9.8-4}$$

This equation is analogous to the expression of (4.4-4) for the mean value for a function of a random variable. Here ω and F_{op} are analogous to the random variable and function, respectively, while the normalized product $G_a T_s$ is analogous to the probability density function. Thus, the overbar represents an appropriate notation for average noise figure.

In many cases the source temperature may be constant, or at least approximately so. A good example is an antenna as a source feeding a low-noise amplifier over a relatively narrow band of interest. For these cases

$$\overline{F}_{op} = \frac{\displaystyle\int_0^\infty F_{op} G_a \, d\omega}{\displaystyle\int_0^\infty G_a \, d\omega}. \tag{9.8-5}$$

Average Standard Noise Figure

For a standard source T_0 is a constant. Letting $T_s = T_0$ in (9.8-4) gives *average standard noise figure* $\overline{F}_0$:

$$\overline{F}_0 = \frac{\displaystyle\int_0^\infty F_0 G_a \, d\omega}{\displaystyle\int_0^\infty G_a \, d\omega}. \tag{9.8-6}$$

ISBN 0-201-05758-1

Average Noise Temperatures

We define *average effective source temperature* $\overline{T}_s$ and *average effective input noise temperature* $\overline{T}_e$ as *constant* temperatures which produce the same output noise power as obtained by integration of the output incremental powers using incremental source and input effective temperatures. By the latter approach, total output noise power is developed by integrating the sum of dN_{aos} and ΔN_{ao} as given by (9.7-1) and (9.7-2). We have

$$\text{total noise power} = \frac{k}{2\pi} \int_0^\infty (T_s + T_e) G_a \, d\omega. \tag{9.8-7}$$

Assuming the equivalent source has temperature $\overline{T}_s + \overline{T}_e$, the total output noise power is

$$\text{total noise power} = \frac{k(\overline{T}_s + \overline{T}_e)}{2\pi} \int_0^\infty G_a \, d\omega. \tag{9.8-8}$$

Since the two total powers are required to be equal, we equate term by term to obtain

$$\overline{T}_s = \frac{\displaystyle\int_0^\infty T_s G_a \, d\omega}{\displaystyle\int_0^\infty G_a \, d\omega}, \tag{9.8-9}$$

and

$$\overline{T}_e = \frac{\displaystyle\int_0^\infty T_e G_a \, d\omega}{\displaystyle\int_0^\infty G_a \, d\omega}. \tag{9.8-10}$$

Noise Figure and Noise Temperature Interrelationships

Average noise figures and average noise temperatures are related as follows:

$$\overline{F}_{op} = 1 + \frac{\overline{T}_e}{\overline{T}_s}, \tag{9.8-11}$$

$$\overline{F}_0 = 1 + \frac{\overline{T}_e}{T_0}. \tag{9.8-12}$$

By equating $\overline{T}_e$ from these expressions we may also write

ISBN 0-201-05758-1

$$\bar{F}_0 = 1 + \frac{\bar{T}_s}{T_0}(\bar{F}_{op} - 1), \tag{9.8-13}$$

$$\bar{F}_{op} = 1 + \frac{T_0}{\bar{T}_s}(\bar{F}_0 - 1), \tag{9.8-14}$$

which are analogous to (9.7-7) and (9.7-8) for spot noise figures. The proof of (9.8-11) follows easily from substitution of (9.7-5) into (9.8-4). A similar proof of (9.8-12) follows use of (9.7-6) and (9.8-6).

Example 9.8-1

An amplifier with average standard noise figure of 8 (9 dB) is used with a source such that the average operating noise figure is 32 (15 dB). We determine its average effective input noise temperature and the source average temperature. For the former we use (9.8-12): $\bar{T}_e = 290(8 - 1) = 2030$ K. For the source we invoke (9.8-11): $\bar{T}_s = 2030/(32 - 1) = 65.5$ K.

Noise Bandwidth

By the concept of average effective temperatures we have been able to model a noisy network as a noise-free network excited by appropriately defined external noise sources. We now complete the modeling of frequency-dependent networks by introducing a quantity called *noise bandwidth.* Let the available power gain of the network be $G_a(\omega)$. Its value at center frequency is $G_a(\omega_0)$ ($\omega_0 = 0$ for a low-pass network). We define noise bandwidth W_n as the bandwidth of a network having an ideal (rectangular pass-band) transfer function with the same center-band available power gain as the actual network, and producing the same total available output power, when excited by a source of constant temperature.

If $\bar{T} = \bar{T}_e + \bar{T}_s$ is the constant source temperature, the available output power from the actual network is

$$\text{available output power} = \frac{k\bar{T}}{2\pi}\int_0^\infty G_a(\omega)\, d\omega. \tag{9.8-15}$$

That of the ideal network is

$$\text{available output power} = \frac{k\bar{T}}{2\pi} G_a(\omega_0) W_n. \tag{9.8-16}$$

Equating these two powers gives W_n:

$$W_n = \frac{\int_0^\infty G_a(\omega)\, d\omega}{G_a(\omega_0)}. \tag{9.8-17}$$

Losses

In many cases losses arising from antennas, waveguides, coaxial cables, various passive microwave components, etc., may be taken as constants independent of frequency because their

ISBN 0-201-05758-1

bandwidth is often much larger than the signal band of interest. Consider finding average effective input noise temperature of an impedance-matched network at physical temperature T_L with constant loss L at all frequencies. Thus, the network may be treated as resistive with available power gain $1/L$. If it is driven by a resistive (thermal) source at temperature T_L having available power $kT_L\ d\omega/2\pi$, then the available output power dN_{ao} from the network is also $kT_L\ d\omega/2\pi$ because its output appears to any load as just another resistive source at temperature T_L. Now since the component of dN_{ao} due to the source alone is $dN_{aos} = kT_L\ d\omega/2\pi L$, the excess output available power ΔN_{ao} due to the network alone is the difference $\Delta N_{ao} = dN_{ao} - dN_{aos}$:

$$\Delta N_{ao} = \frac{kT_L\ d\omega}{2\pi}\left(1 - \frac{1}{L}\right). \tag{9.8-18}$$

Input average effective noise temperature $\overline{T}_e$ is easily calculated by equating (9.7-2) with (9.8-18), solving for $G_a T_e$, and substituting into (9.8-10). We get

$$\overline{T}_e = T_L(L - 1). \tag{9.8-19}$$

If the loss is driven from a source with temperature different from its physical temperature, we may use (9.8-11), which applies for an arbitrary source of constant temperature $\overline{T}_s$, to obtain the average operating noise figure for the loss:

$$\overline{F}_{op} = 1 + \frac{T_L}{\overline{T}_s}(L - 1). \tag{9.8-20}$$

Note for no loss $L = 1$ and $\overline{F}_{op} = 1$ as we might expect. For a standard source

$$\overline{F}_0 = 1 + \frac{T_L}{T_0}(L - 1). \tag{9.8-21}$$

An interesting case corresponds to a source with temperature T_L, the same as the physical temperature of the loss:

$$\overline{F}_{op} = L. \tag{9.8-22}$$

Thus, when $\overline{T}_s = T_L$, the average noise figure of a loss is equal to the loss.

The Antenna as a Noise Source

Every antenna is capable of some available noise power. It can, as a consequence, be considered as a noise source. We may assign an effective noise temperature T_a, called the *antenna temperature,* to the real part of the antenna source impedance* to account for its available noise power. Thus, in an incremental band, this power is $kT_a\ d\omega/2\pi$. In most systems the frequency band occupied by the received information-bearing signal is small enough that T_a may be taken as constant.

Several sources contribute to antenna noise. Cosmic noise is one component resulting from radiation received from outer space. It decreases with increasing carrier frequency and

*Power flowing into an antenna is lost through radiation. The loss (as far as the power source goes) may be accounted for by a resistance, the *radiation resistance.* From reciprocity, if the antenna is a source, the real part of its internal impedance is the radiation resistance.

ISBN 0-201-05758-1

becomes negligible above 2 GHz even for the lowest noise systems (contribution to T_a is about 10 K or less). For $\omega/2\pi < 2$ GHz cosmic noise increases rapidly [8] and may become the dominant contributor to T_a for frequencies below about 300 MHz.

The atmosphere has attenuation as already discussed in Section 9.4. From the theory of blackbody radiation, any body which absorbs energy must radiate the same amount of energy or else the temperature of the body would change. The radiation from the attenuating atmosphere is seen by the antenna as noise. Even under wost-case conditions of an antenna pointed horizontally, the contribution to T_a from a clear atmosphere does not exceed about 200 K for frequencies below 15 GHz.* The contribution rapidly decreases for increasing elevation angle and becomes less than about 10 K for an antenna pointed vertically when $\omega/2\pi < 15$ GHz. As a function of decreasing frequency, the contribution to T_a decreases only slightly for 1 GHz $< \omega/2\pi \leqslant 15$ GHz. Below 1 GHz it decreases more rapidly.

Other sources of antenna noise are: radiation from the earth (blackbody effect), radiation from extraterrestrial bodies such as planets, the moon, and sun (an especially "hot" body), atmospherics such as lightning [1, p. 369], man-made sources such as ignition systems, razors, fluorescent lights, and attenuation due to precipitation.

Broadly speaking, for antennas with relatively narrow beam widths and a single polarization, T_a will range from a few tens of Kelvin to a few hundreds of Kelvin depending on elevation angle, frequency, and care in antenna design. Careful attention to sidelobe design is required to achieve low temperatures for earth-bound antennas (why?).

9.9. SYSTEM SENSITIVITY EQUATIONS

System performance is limited by the achievable input signal-to-noise ratio. Much of the work of this chapter has been directed toward developing methods whereby noise power in networks can easily be described. These methods, when combined with the signal power expressions found in Section 9.3, will allow the calculation of system input signal-to-noise ratio.

Overall System Model

Figure 9.9-1(*a*) illustrates a block diagram of a receiving system model adequate to describe overall system noise. Point A corresponds to the output of the lossless antenna in Fig. 9.1-1, and point B is the input to the receiver. The model combines all the previous noise-free network models that have been discussed. Thus, the whole receiver has an average input effective noise temperature of $\overline{T}_R$, available power gain $G(\omega_0)$, and noise bandwidth W_n. The total receive-path loss $L_{ra}L_r$ is assumed to be frequency-independent and have a physical temperature T_L. The antenna is modeled as a noise source with effective temperature T_a. Total available output noise power N_{ao} is readily found to be

$$N_{ao} = k[T_a + T_L(L_{ra}L_r - 1) + \overline{T}_R L_{ra}L_r] \frac{G(\omega_0)W_n}{L_{ra}L_r 2\pi}. \tag{9.9-1}$$

A convenient equivalent system noise model is diagramed in Fig. 9.9-1(*b*). Available output noise power becomes

$$N_{ao} = k\overline{T}_{sys} \frac{G(\omega_0)W_n}{L_r L_{ra} 2\pi}, \tag{9.9-2}$$

*This figure is for a single polarization. If the antenna responds to two orthogonal polarizations, such as circular, the number must be doubled.

ISBN 0-201-05758-1

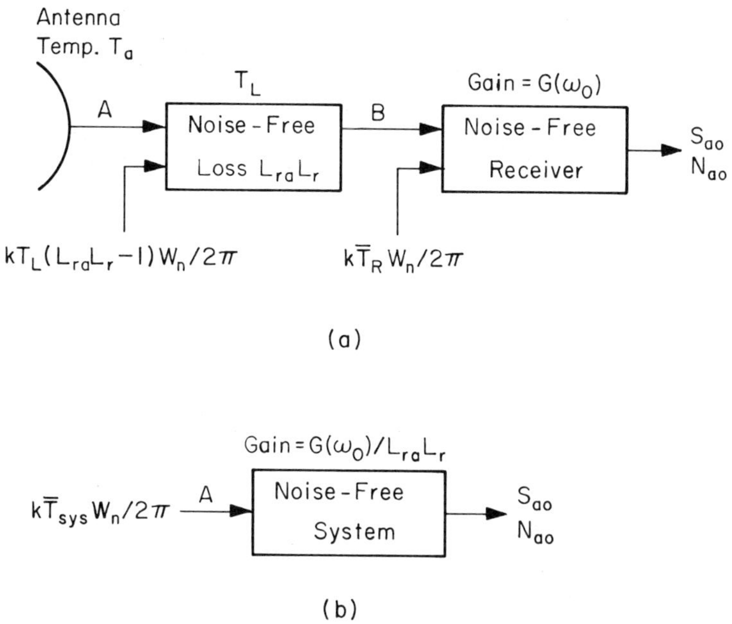

Fig. 9.9-1. A model of a receiving system for communications (a), and (b) its equivalent model.

if we define *system noise temperature* $\overline{T}_{sys}$ as that of an equivalent source which accounts for *all* available output power. Clearly,

$$\overline{T}_{sys} = T_a + T_L(L_{ra}L_r - 1) + \overline{T}_R L_{ra}L_r \tag{9.9-3}$$

by equating (9.9-2) and (9.9-1). The concept of system noise temperature allows us to imagine all noise in the system to enter at point A in Fig. 9.9-1(a) as though it came from the antenna. The available noise power N at point A is

$$N = k\overline{T}_{sys}W_n/2\pi. \tag{9.9-4}$$

Sensitivity Equations

The available signal power S at point A in our model has already been determined in (9.3-4). Thus, using (9.9-4),

$$\left(\frac{S}{N}\right) = \frac{P_t G_t A_{re}}{2R^2 L_t L_{ta} L_{at} k\overline{T}_{sys} W_n}, \tag{9.9-5}$$

or equivalently,

$$\left(\frac{S}{N}\right) = \frac{P_t G_t G_r \lambda^2}{8\pi R^2 L_t L_{ta} L_{at} k\overline{T}_{sys} W_n}. \tag{9.9-6}$$

This expression for available input signal-to-noise ratio is our principal result. It can be placed in another form related to noise figure.

Average noise figure has been defined as the ratio of total available noise power to that due to the source alone. Applying this definition to point A in Fig. 9.9-1(a) leads to *system*

ISBN 0-201-05758-1

noise figure $\overline{F}_{sys}$, where total available power is given by (9.9-4) and that due to the source is $kT_aW_n/2\pi$. Hence,

$$\overline{F}_{sys} = \overline{T}_{sys}/T_a \tag{9.9-7}$$

and (9.9-6) may be written as

$$\left(\frac{S}{N}\right) = \frac{P_tG_tG_r\lambda^2}{8\pi R^2 L_t L_{ta} L_{at} \overline{F}_{sys} k T_a W_n}. \tag{9.9-8}$$

PROBLEMS

9-1. A certain antenna has a "pencil beam" pattern—that is, one having the same shape in any plane containing the main lobe maximum. If the pattern is $P(\xi) = 4J_1{}^2(\xi)/\xi^2$, where $J_1(\xi)$ is the Bessel function of the first kind of order one and ξ is a normalized angle, sketch the pattern and graphically determine its beam width in the normalized coordinate ξ.

9-2. For the antenna pattern of problem 9-1 determine the level (in decibels) of the largest sidelobe (at its peak) relative to the main lobe maximum.

9-3. If an antenna radiates 1000 watts and has a directive gain of 10,000, what is its maximum radiation intensity? If it has the same beam width in two orthogonal planes containing the main lobe maximum, what is its beam width?

9-4. If the antenna of problem 9-3 has a radiation efficiency factor of 0.8, what is its power gain? If its frequency is 3 GHz, what is its effective area? If its aperture efficiency is 0.55, what is its true area? For a circular aperture compute its diameter. All else being equal, if the antenna is replaced by one with uniform illumination, how much will its power gain increase?

9-5. An antenna has a rectangular aperture with sides 3 meters and 2 meters. If its aperture efficiency is 0.75, what received radiation power density is necessary to produce 1 microwatt of available received power?

9-6. Two stations in a microwave link are separated by 32 km (20 miles). If they use a frequency of 10 GHz, antennas having effective areas of 1 m^2, and the total link loss is 10 dB, what received power is possible for a 1-watt transmitter? Use (9.3-7) to make the calculation.

9-7. Work problem 9-6 using (9.3-9) for the computation.

9-8. For the system of problem 9-6, how much of the link loss is due to the atmosphere on a clear day? If it rains over the whole link at a 4-mm/h rate, how much additional loss occurs?

9-9. In the system of problem 9-6, 5 dB of link loss was allowed for worst-case atmospheric loss. Worst case is defined as heavy rain (16 mm/h) over half the link *or* dense fog (30-meter visibility) over the whole link. Calculate the total atmospheric attenuation in each case. Is the link adequately designed?

9-10. For the system of problem 9-6, what link attenuation can be expected owing to snow's falling over the entire link at a rate $I = 4$ mm/h of melted water content?

★ 9-11. For the network of Fig. 9.5-1(*b*), let $H(\omega)$ be the transfer function relating e_o to e_i. Develop expressions for the available power gain and true power gain of the network if

ISBN 0-201-05758-1

true power gain is defined as the ratio of power delivered to Z_L to power delivered to Z_i. State conditions under which the two are the same.

★ 9-12. Work problem 9-11 where true power gain is defined as the ratio of load power when the network is present to load power when the network is removed. Such a power gain is called the *insertion power gain.*

9-13. For the network of Fig. 9.6-1(*a*), plot the ratio of available incremental power to actual incremental load power as a function of R_L/R when $\overline{e_n^2}$ is given by (9.6-1). How much can R_L differ from its optimum value if true power must be within 1 dB of the available power?

9-14. In Fig. 9.6-1(*a*) treat R and R_L as a network having $e_n(t)$ as its input. Let $e_L(t)$ be the output of the network. The network power transfer function is $R_L^{\,2}/(R+R_L)^2$. If the input noise power density spectrum (incremental) is given by (9.6-2) and if output incremental power is found from use of (4.10-11) over the incremental band, the result does not agree with power as given by (9.6-3). Explain.

9-15. A thermal noise source having the power density spectrum of (9.6-2) and resistive impedance R_s is applied to the network of Fig. P3-1. Find the mean-squared output noise voltage if the network resistor is noise-free and has a resistance different from that of the source.

9-16. Assume noise-free resistors in Fig. P3-10 with $C_1 = 0$, and work problem 9-15.

9-17. Consider the output of the network of problem 9-15 as a noise source, and find its effective noise temperature for any frequency. Assume R is a noisy resistor at the same temperature as R_s.

9-18. The two-port network of Fig. P9-18 is at temperature T_p and is driven at its input port by a thermal source at temperature T_s and resistance R_s. Find the effective source temperature of the system as seen from the output terminals.

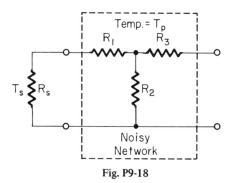

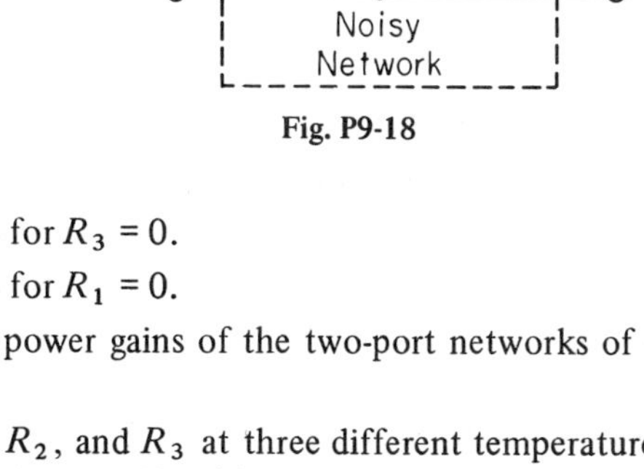

Fig. P9-18

9-19. Work problem 9-18 for $R_3 = 0$.

9-20. Work problem 9-18 for $R_1 = 0$.

9-21. Find the available power gains of the two-port networks of problems 9-18, 9-19, and 9-20.

9-22. Three resistors R_1, R_2, and R_3 at three different temperatures T_1, T_2, and T_3, respectively, are connected in parallel. (*a*) What is the effective noise temperature of the combination? (*b*) What is the mean-squared voltage across the parallel combination?

9-23. Work problem 9-22 when the resistors are in series.

9-24. An amplifier has a standard spot noise figure of 8 dB. If it is used with a source having a temperature of 100 K, what is the operating spot noise figure? What is its effective input noise temperature?

ISBN 0-201-05758-1

9-25. What amplifier operating spot noise figure is required if total available output noise power must not exceed forty times the source available noise power when its available power gain is 10 dB?

9-26. An amplifier has a spot noise figure of 10 dB when operating with a 200 K source. What is its standard spot noise figure?

9-27. Find the effective input noise temperature T_e of a cascade of three amplifier stages with effective noise temperatures of 1800 K (stage 1), 2700 K, and 4800 K (stage 3) and available power gains of 10 dB, 16 dB, and 20 dB, respectively.

9-28. For the amplifier of problem 9-27 what first stage gain would be necessary to reduce the contribution to T_e of all stages after the first to 10% or less of the first stage contribution?

9-29. A cascade has three amplifiers with incremental input effective noise temperatures of 100 K (stage 1), 500 K, and 3000 K (stage 3). The available power gains of stages 2 and 3 are 10 and 16, respectively. (*a*) What first-stage available power gain G_1 is required if the overall effective input noise temperature is to be 150 K? (*b*) What is the maximum percent change (stability) in G_1 that will maintain overall effective input noise temperature within 5% of the value in (*a*)?

9-30. A receiver sold by one manufacturer has a standard spot noise figure of 6 dB. Another sells a receiver having an operating spot noise figure of 9 dB when used with a source having an effective temperature of 150 K. (*a*) What are the input effective noise temperatures of the two amplifiers? (*b*) Which would be best to buy if all other parameters (gain, cost, etc.) are equal?

9-31. A two-stage amplifier cascade is driven by a source having a noise temperature of 200 K. Available power gains of the first and second stages are 6 dB and 9 dB, respectively. (*a*) If the first stage input effective noise temperature is 300 K and the second stage has a standard spot noise figure of 16 dB, what is the operating spot noise figure of the second stage if the first stage is considered its source? (*b*) What is the noise temperature of the first stage treated as a source?

9-32. A source with temperature $T_s = T_1/[1 + (\omega/\alpha)^2]$ drives the input of a network having an available power gain $G(\omega) = G_1/[1 + (\omega/\alpha)^2]^2$. Here T_1, G_1, and α are positive constants. Find an expression for the average source temperature.

9-33. Find the noise bandwidth of the network of problem 9-32.

9-34. An amplifier has an average standard noise figure of 6 dB. An attenuator is added to its input to control overload when large signals are present. If the attenuator is at standard temperature (290 K) what is the average standard noise figure of the attenuator–amplifier combination when the attenuator is set to 3 dB? What is it for 6 dB and 9 dB?

9-35. Work problem 9-34 for an attenuator cooled to 77 K.

9-36. Find the average operating noise figure for the network of problem 9-34 when connected to a lossless antenna with temperature 30 K. Assume the attenuator is set to 3 dB. What is the system noise temperature?

9-37. If the receive-path losses L_{ra} and L_r in Fig. 9.9-1(*a*) have temperatures T_{ra} and T_r, respectively, obtain a revised expression for system noise temperature analogous to (9.9-3).

9-38. A satellite is 184 km from a ground station. It transmits at 4 GHz using an antenna having an effective area of 1 m^2 and loss of 1 dB. The loss between transmitter and antenna is 2 dB. The ground station antenna effective area and loss are 10 m^2 and 1.5 dB, respectively. The loss between antenna and receiver is 3 dB and is at a temperature of

ISBN 0-201-05758-1

290 K. The receiver has a noise bandwidth of 10 MHz and average standard noise figure of 4 dB. If the antenna temperature is 140 K when the atmospheric attenuation is 3 dB, what satellite transmitter power is required to give a signal-to-noise power ratio of 50 dB? What is the system noise temperature? What is the system noise figure? How much would the system noise temperature improve if the 3-dB receive-path loss could be reduced to 1 dB?

REFERENCES

[1] Skolnik, M. I., *Introduction to Radar Systems,* McGraw-Hill Book Co., New York, 1962.

[2] Silver, S., *Microwave Antenna Theory and Design,* McGraw-Hill Book Co., New York, 1949.

[3] International Telephone and Telegraph Corp., *Reference Data for Radio Engineers,* Fifth Edition, Howard W. Sams. & Co., New York, 1970.

[4] Blake, L. V., A Guide to Basic Pulse-Radar Maximum Range Calculation (Part I), U.S. Naval Research Lab. Report 5868, December 28, 1962.

[5] Kerr, D. E. (Editor), *Propagation of Short Radio Waves,* Vol. 13, M.I.T. Radiation Laboratory Series, Boston Technical Publishers, Lexington, Massachusetts, 1964.

[6] Gunn, K. L. S., and East, T. W. R., The Microwave Properties of Precipitation Particles, *Quarterly Journal of the Royal Meteorological Society,* Vol. 80, October, 1954, pp. 522–545.

[7] Lathi, B. P., *Communication Systems,* John Wiley & Sons, New York, 1968.

[8] Hogg, D. C., and Mumford, W. W., The Effective Noise Temperature of the Sky, *Microwave Journal,* March, 1960, pp. 80–84.

ISBN 0-201-05758-1

Appendix A

Trigonometric Identities

$$\cos(x \pm y) = \cos(x)\cos(y) \mp \sin(x)\sin(y) \tag{A-1}$$

$$\sin(x \pm y) = \sin(x)\cos(y) \pm \cos(x)\sin(y) \tag{A-2}$$

$$\cos\left(x \pm \frac{\pi}{2}\right) = \mp \sin(x) \tag{A-3}$$

$$\sin\left(x \pm \frac{\pi}{2}\right) = \pm \cos(x) \tag{A-4}$$

$$\cos(2x) = \cos^2(x) - \sin^2(x) \tag{A-5}$$

$$\sin(2x) = 2\sin(x)\cos(x) \tag{A-6}$$

$$2\cos(x)\cos(y) = \cos(x - y) + \cos(x + y) \tag{A-7}$$

$$2\sin(x)\sin(y) = \cos(x - y) - \cos(x + y) \tag{A-8}$$

$$2\sin(x)\cos(y) = \sin(x - y) + \sin(x + y) \tag{A-9}$$

$$2\cos^2(x) = 1 + \cos(2x) \tag{A-10}$$

$$2\sin^2(x) = 1 - \cos(2x) \tag{A-11}$$

$$4\cos^3(x) = 3\cos(x) + \cos(3x) \tag{A-12}$$

$$4\sin^3(x) = 3\sin(x) - \sin(3x) \tag{A-13}$$

$$8\cos^4(x) = 3 + 4\cos(2x) + \cos(4x) \tag{A-14}$$

$$8\sin^4(x) = 3 - 4\cos(2x) + \cos(4x) \tag{A-15}$$

$$A\cos(x) - B\sin(x) = R\cos(x + \theta) \tag{A-16}$$

where

$$R = \sqrt{A^2 + B^2} \tag{A-17}$$

ISBN 0-201-05758-1

$$\theta = \tan^{-1}(B/A) \tag{A-18}$$

$$A = R\cos(\theta) \tag{A-19}$$

$$B = R\sin(\theta) \tag{A-20}$$

$$2\cos(x) = e^{jx} + e^{-jx} \tag{A-21}$$

$$2j\sin(x) = e^{jx} - e^{-jx} \tag{A-22}$$

ISBN 0-201-05758-1

Appendix B

Useful Integrals and Series

INDEFINITE INTEGRALS

Rational Algebraic Functions

$$\int x^n\,dx = \frac{x^{n+1}}{n+1}, \qquad 0 \leqslant n \tag{B-1}$$

$$\int \frac{dx}{x} = \ln|x| \tag{B-2}$$

$$\int \frac{dx}{x^n} = \frac{-1}{(n-1)x^{n-1}}, \qquad 1 < n \tag{B-3}$$

$$\int (a+bx)^n\,dx = \frac{(a+bx)^{n+1}}{b(n+1)}, \qquad 0 \leqslant n \tag{B-4}$$

$$\int \frac{dx}{a+bx} = \frac{1}{b}\ln|a+bx| \tag{B-5}$$

$$\int \frac{dx}{(a+bx)^n} = \frac{-1}{(n-1)b(a+bx)^{n-1}}, \qquad 1 < n \tag{B-6}$$

$$\int \frac{dx}{a^2+b^2x^2} = \frac{1}{ab}\tan^{-1}\left(\frac{bx}{a}\right) \tag{B-7}$$

$$\int \frac{dx}{(a^2+x^2)^2} = \frac{x}{2a^2(a^2+x^2)} + \frac{1}{2a^3}\tan^{-1}\left(\frac{x}{a}\right) \tag{B-8}$$

$$\int \frac{dx}{(a^2+x^2)^3} = \frac{x}{4a^2(a^2+x^2)^2} + \frac{3x}{8a^4(a^2+x^2)} + \frac{3}{8a^5}\tan^{-1}\left(\frac{x}{a}\right) \tag{B-9}$$

ISBN 0-201-05758-i

$$\int \frac{x\,dx}{a^2 + x^2} = \frac{1}{2} \ln(a^2 + x^2) \tag{B-10}$$

$$\int \frac{x^2\,dx}{a^2 + x^2} = x - a \tan^{-1}\left(\frac{x}{a}\right) \tag{B-11}$$

$$\int \frac{dx}{a^4 + x^4} = \frac{1}{4a^3\sqrt{2}} \ln\left(\frac{x^2 + ax\sqrt{2} + a^2}{x^2 - ax\sqrt{2} + a^2}\right) + \frac{1}{2a^3\sqrt{2}} \tan^{-1}\left(\frac{ax\sqrt{2}}{a^2 - x^2}\right) \tag{B-12}$$

$$\int \frac{x^2\,dx}{a^4 + x^4} = -\frac{1}{4a\sqrt{2}} \ln\left(\frac{x^2 + ax\sqrt{2} + a^2}{x^2 - ax\sqrt{2} + a^2}\right) + \frac{1}{2a\sqrt{2}} \tan^{-1}\left(\frac{ax\sqrt{2}}{a^2 - x^2}\right) \tag{B-13}$$

Trigonometric Functions

$$\int \cos(x)\,dx = \sin(x) \tag{B-14}$$

$$\int x\cos(x)\,dx = \cos(x) + x\sin(x) \tag{B-15}$$

$$\int x^2\cos(x)\,dx = 2x\cos(x) + (x^2 - 2)\sin(x) \tag{B-16}$$

$$\int \sin(x)\,dx = -\cos(x) \tag{B-17}$$

$$\int x\sin(x)\,dx = \sin(x) - x\cos(x) \tag{B-18}$$

$$\int x^2\sin(x)\,dx = 2x\sin(x) - (x^2 - 2)\cos(x) \tag{B-19}$$

ISBN 0-201-05758-1

Exponential Functions

$$\int e^{ax}\, dx = \frac{e^{ax}}{a} \tag{B-20}$$

$$\int xe^{ax}\, dx = e^{ax}\left[\frac{x}{a} - \frac{1}{a^2}\right] \tag{B-21}$$

$$\int x^2 e^{ax}\, dx = e^{ax}\left[\frac{x^2}{a} - \frac{2x}{a^2} + \frac{2}{a^3}\right] \tag{B-22}$$

DEFINITE INTEGRALS

$$\int_0^\infty e^{-a^2 x^2}\, dx = \sqrt{\pi}/2a, \qquad 0 < a \tag{B-23}$$

$$\int_0^\infty x^2 e^{-x^2}\, dx = \sqrt{\pi}/4 \tag{B-24}$$

$$\int_0^\infty \mathrm{Sa}\,(x)\, dx = \int_0^\infty \frac{\sin\,(x)}{x}\, dx = \frac{\pi}{2} \tag{B-25}$$

$$\int_0^\infty \mathrm{Sa}^2\,(x)\, dx = \pi/2 \tag{B-26}$$

FINITE SERIES

$$\sum_{n=1}^{N} n = \frac{N(N+1)}{2} \tag{B-27}$$

$$\sum_{n=1}^{N} n^2 = \frac{N(N+1)(2N+1)}{6} \tag{B-28}$$

$$\sum_{n=1}^{N} n^3 = \frac{N^2(N+1)^2}{4} \tag{B-29}$$

$$\sum_{n=0}^{N} x^n = \frac{x^{N+1} - 1}{x - 1} \tag{B-30}$$

ISBN 0-201-05758-1

Appendix C

Gaussian (Normal) Probability Density and Distribution Functions

For zero mean and unit variance:

$$p(x) = \frac{1}{\sqrt{2\pi}} e^{-x^2/2} \tag{C-1}$$

$$P(x) = \frac{1}{\sqrt{2\pi}} \int_{-\infty}^{x} e^{-u^2/2}\, du \equiv \Phi(x) \tag{C-2a}$$

$$\Phi(-x) = 1 - \Phi(x) \tag{C-2b}$$

TABLE C-1. The Gaussian or Normal Density and Distribution

x	$p(x)$	$\Phi(x)$	x	$p(x)$	$\Phi(x)$
0.00	0.3989	0.5000	1.40	0.1497	0.9192
0.05	0.3984	0.5199	1.50	0.1295	0.9332
0.10	0.3970	0.5398	1.60	0.1109	0.9452
0.15	0.3945	0.5596	1.70	0.0940	0.9554
0.20	0.3910	0.5793	1.80	0.0790	0.9641
0.25	0.3867	0.5987	1.90	0.0656	0.9713
0.30	0.3814	0.6179	2.00	0.0540	0.9772
0.35	0.3752	0.6368	2.10	0.0440	0.9821
0.40	0.3683	0.6554	2.20	0.0355	0.9861
0.45	0.3605	0.6736	2.30	0.0283	0.9893
0.50	0.3521	0.6915	2.40	0.0224	0.9918
0.55	0.3429	0.7088	2.50	0.0175	0.9938
0.60	0.3332	0.7257	2.60	0.0136	0.9953
0.65	0.3230	0.7422	2.70	0.0104	0.9965
0.70	0.3123	0.7580	2.80	0.0079	0.9974
0.75	0.3011	0.7734	2.90	0.0060	0.9981
0.80	0.2897	0.7881	3.00	0.0044	0.9987
0.85	0.2780	0.8023	3.20	0.0024	0.9993
0.90	0.2661	0.8159	3.40	0.0012	0.9997
0.95	0.2541	0.8289	3.60	0.0006	0.9998
1.00	0.2420	0.8413	3.80	0.0003	0.9999
1.10	0.2179	0.8643	4.00	0.0001	1.0000
1.20	0.1942	0.8849	4.50	0.0	1.0000
1.30	0.1714	0.9032			

ISBN 0-201-05758-1

For arbitrary mean m and variance σ^2:

$$p(x) = \frac{1}{\sqrt{2\pi\sigma^2}} e^{-(x-m)^2/2\sigma^2} \tag{C-3}$$

$$P(x) = \Phi\left(\frac{x - m}{\sigma}\right) \tag{C-4}$$

ISBN 0-201-05758-1

Appendix D

Table of Bessel Functions

$J_n(\beta)$

n \ β	0.5	1	2	3	4	6	8	10	12
0	0.9385	0.7652	0.2239	−0.2601	−0.3971	0.1506	0.1717	−0.2459	0.0477
1	0.2423	0.4401	0.5767	0.3391	−0.0660	−0.2767	0.2346	0.0435	−0.2234
2	0.0306	0.1149	0.3528	0.4861	0.3641	−0.2429	−0.1130	0.2546	−0.0849
3	0.0026	0.0196	0.1289	0.3091	0.4302	0.1148	−0.2911	0.0584	0.1951
4	0.0002	0.0025	0.0340	0.1320	0.2811	0.3576	−0.1054	−0.2196	0.1825
5	–	0.0002	0.0070	0.0430	0.1321	0.3621	0.1858	−0.2341	−0.0735
6		–	0.0012	0.0114	0.0491	0.2458	0.3376	−0.0145	−0.2437
7			0.0002	0.0025	0.0152	0.1296	0.3206	0.2167	−0.1703
8			–	0.0005	0.0040	0.0565	0.2235	0.3179	0.0451
9				0.0001	0.0009	0.0212	0.1263	0.2919	0.2304
10				–	0.0002	0.0070	0.0608	0.2075	0.3005
11					–	0.0020	0.0256	0.1231	0.2704
12						0.0005	0.0096	0.0634	0.1953
13						0.0001	0.0033	0.0290	0.1201
14						–	0.0010	0.0120	0.0650

ISBN 0-201-05758-1

Index